3 4028 08188 4786
HARRIS COUNTY PUBLIC LIBRARY

W9-AJR-813

Surfaces and Essences

Surfaces and Essences

Analogy as the Fuel and Fire of Thinking

Douglas Hofstadter & Emmanuel Sander

Basic Books
A Member of the Perseus Books Group
New York

Grateful acknowledgment is hereby made to the following individuals and organizations for permission to use material that they have provided or to quote from sources for which they hold the rights. Every effort has been made to locate the copyright owners of material reproduced in this book. Omissions that are brought to our attention will be corrected in subsequent editions.

Photograph of Mark Twain: © CORBIS
Photograph of Edvard Grieg: © Michael Nicholson/CORBIS
Photograph of Albert Einstein: © Philippe Halsman/Magnum Photos
Photograph of Albert Schweitzer: © Bettmann/CORBIS

We also most warmly thank Kellie Gutman and Tony Hoagland for their generous permission to publish their poems in this volume.

Published by Basic Books
A Member of the Perseus Books Group

Designed by Douglas Hofstadter
Cover by Nicole Caputo and Andrea Cardenas

Books published by Basic Books are available at special discounts for bulk purchases in the United States by corporations, institutions, and other organizations. For more information, please contact the Special Markets Department at the Perseus Books Group, 2300 Chestnut Street, Suite 200, Philadelphia, Pennsylvania 19103, or call (800) 810-4145, ext. 5000, or send an email to: special.markets@perseusbooks.com.

Library of Congress Control Number: 2013932688

ISBN: 978-0-465-01847-5 (hardcover)
ISBN: 978-0-465-02158-1 (e-book)

10 9 8 7 6 5 4 3 2 1

To Francesco Bianchini

and

To Michaël, Tom, and Talia

TABLE OF CONTENTS

Words of Thanks

❧ ❧ ❧

A Little Background

As we look back, fondly reliving the genesis of our book, we vividly recall the first key moment, which took place in mid-July, 1998, at an academic congress, in Sofia, Bulgaria. The occasion was the first international conference on the subject of analogy. Organized by Boicho Kokinov, Keith Holyoak, and Dedre Gentner, this memorable meeting assembled researchers from many countries, who, in an easy-going and lively atmosphere, exchanged ideas about their shared passion. Chance thus brought the two of us together for the first time in Sofia, and we found we had an instant personal rapport — a joyous bright spark that gradually developed into a long-term and very strong friendship.

In 2001–2002, Douglas Hofstadter spent a sabbatical year in Bologna, Italy, and during that period he was invited by Jean-Pierre Dupuy to give a set of lectures on cognition at the École Polytechnique in Paris. At that time, Emmanuel Sander had just published his first book — an in-depth study of analogy-making and categorization — and at one of the lectures he proudly presented a copy of it to his new friend, who, upon reading it, was delighted to discover how deeply similar was the vision that its author and he had of what cognition is really about. And then some time passed, with a couple of brief get-togethers in Paris and Toulouse, supplemented by email exchanges and phone calls, mixing intellectual content and friendly feelings.

In February of 2005, Doug invited Emmanuel to Bloomington for a gala birthday party he was throwing for his many friends, as he turned 60. One day during that event, he suggested to Emmanuel that he would be very happy to come to Paris for a few weeks to work with Emmanuel in translating his book into English. Emmanuel was very pleased, but shortly after Doug arrived in Paris in July, the original goal mutated into a considerably larger one — namely, that of co-authoring a book on the fundamental role of analogy in thought, approaching the topic in a non-technical manner and from many points of view, and using a large sampling of concrete examples to justify the theoretical positions. The book would hopefully be accessible to anyone interested in thinking, yet would also have the high ambition of reaching an academic audience and of putting forth a new and original stance towards cognition. This thus was the moment of our book's conception!

Over a three-week period in Paris, many ideas were tossed about, and the result was a forty-page document that featured snips of conversations between the two future co-authors, many notes on ideas for the book, and a very preliminary sketch of what its chapters might look like. And then, for the next four years — 2006 to 2009 — each of the authors made a month-long visit to the other one in his home town. Adding to that, Doug spent an eight-month sabbatical in Paris in 2010. During all this time, there was a constant exchange of ideas via email and via phone, allowing the book to evolve from a few cells into a viable complex organism.

As all this shows, the present book is the fruit of a long collaboration, and finally it has reached maturity. Its authors have invested in it the hope that it has a message with enduring value, even if it is clearly rooted in today's culture and style of life — in fact, it is rooted "in vibrant thought", as we fondly recall one of our friends putting it. But we hope that, despite the spatial and temporal specificity of its origins, its key ideas are universal enough that they will withstand the passage of time.

Ping-ponging between Languages and Cultures

We are quite proud of the fact that our joint book is the result of a very unusual creative process. Not just written by two people, it was written in two languages at the same time. To be more specific, this book has two originals — one in French and one in English. Each is a translation of the other, or perhaps neither of them is a translation. But however you choose to look at it, the two versions of this book have equal standing. They are two highly distinct concrete incarnations of one immaterial entity — namely, this book as it exists on the ethereal plane not of words but of ideas.

To be sure, the writing process involved countless acts of translation, but those acts took place at the very moment that the original text was being generated. Sometimes they carried ideas from English to French, and sometimes they went in the other direction, but what is key here is that these back-and-forth exchanges between the brains of the two authors were accompanied — and this is a rare thing — by back-and-forth exchanges between two languages, which led, in a convergent fashion, to many modifications of the original text, bringing it into closer alignment with its translation, and then the resulting text went once more around the bilingual, bicultural, bicerebral loop, until finally, after a good number of iterations back and forth, things reached a satisfactory equilibrium.

And thus the two versions — the English one that you are now looking at, and its counterpart in French — have gone back and forth many times through the filters of both languages. Often we would find that a high level of clarity emerged as a result of this special dynamic, as translation is nothing if not a merciless revealer of imprecision, vagueness, and lack of logical flow. Translation brings such defects out like a flashlight turned on in a dusty attic. A different metaphor is the sharpening of a knife, because our process of repeated exchanges became, for us, a constant act of sharpening of the ideas we were trying to express. And thus the fact that this book has two originals is not merely an amusing curiosity, but more importantly, it has been a guiding principle keeping us constantly focused on the goal of coherence and lucidity. At least we, the authors, see our book in this way, and we hope that our readers will see it as we do.

We encourage those of our English-speaking readers who are comfortable with French to try tackling a few passages in both versions, because each specific version takes advantage of ideas, images, and turns of phrase that are deeply rooted in the culture at which it is aimed. This fact made for a particularly enjoyable and stimulating exercise for both authors, in that we were constantly being challenged to come up with an apt analogue for, say, a given idiomatic phrase or a given situation, or perhaps a given speech error, and the quest for optimal examples really kept us on our toes. For anyone who loves languages, then, a parallel perusal of the two texts should provide, in addition to plenty of new ideas (which was of course our primary aim), a special experience of savoring ideas fleshed out in two contrasting ways — in short, a bit of delicious icing on the cake.

"Merci" to So Many!

Two authors, two languages, two lives. While this book was being written, many people were involved with us in many diverse ways, and life predictably followed its unpredictable course. We thus have many heartfelt feelings to express.

At the top of the list are our families, whom we cherish immeasurably. For Doug this means first of all Baofen, and for Emmanuel it means Cécile. They are our muses, amusing and amazing, loving and beloved. "À B., C. – D., E." says it all, using initials, in French. Next come our children. To Doug, his son Danny and daughter Monica mean everything. Both are rich in humor, verve, idealism, and artistic imagination, inherited largely from their loving mother Carol, who, alas, was torn from us so many years ago. On Emmanuel's side, there are Michaël, who is protective, intense, and impetuous, and Tom, who is sensitive, social, and solid, and Talia, who is impish, witty and creative; and there is Daniela, their devoted and loving mother. Into our families have recently come, on Doug's side, Baofen's son David, and on Emmanuel's side, Cécile's son Arthur, who grace our lives with their talents and their gentle natures.

Doug expresses many thanks to his sister Laura Hofstadter, her husband Len Shar, and their two sons, Nathaniel and Jeremy, both filled with intellectual brio. Over the years, their house has been the site of innumerable "jolly evenings" marked by crack croquet competitions, wild word-wizardry, and side-splitting semantic silliness, along with the yummiest of food and the chummiest of chatting. Somewhat symmetrically, Emmanuel warmly thanks David, trusted brother and insightful colleague, David's wife Véronique, and their daughters Hannah and Gabriela, his radiant nieces. Emmanuel also extends his deepest gratitude to his father and mother, Jean-Pierre and France Sander, for having fostered his growth, from his earliest youth, in the most generous fashion imaginable. And Doug likewise recalls with enormous thanks all the warmth and encouragement of his late parents, Robert and Nancy Hofstadter.

Over the course of these seven-plus years, each of us has experienced the grief of losing several people with whom the bonds ran very deep. Here we wish to honor the treasured memory of Raphaël Sander, Agnès Sander, Maurice Sander, Esther Sidi, Morgan Rogulski, and Lucie Cohen, on Emmanuel's side, and of Nancy Hofstadter, Helga Keller, Steve Larson, Valentino Braitenberg, and Paolo Bozzi, on Doug's.

It behooves us now to devote a paragraph or two to Paragraphe, the laboratory that, ever since this book was conceived, has been Emmanuel Sander's intellectual home at the University of Paris VIII. Its director Imad Saleh leads, with ebullience, generosity, and vigor, a laboratory where human and scientific values exist side by side. Within Paragraphe, the research group CRAC (a French acronym for "Understanding, Reasoning, and Knowledge Acquisition") is led jointly by Emmanuel Sander and Raphaële Miljkovitch. Emmanuel treasures his intellectual exchanges with Raphaële, and he is delighted to have used some of the plentiful harvest she has made of linguistic oddities issuing from the mouths of her two young sons. CRAC is a cooperative team whose members represent many diverse facets of developmental psychology and get along so well that many strong friendships have come to bloom within it.

And thus a big thank-you to Jean Baratgin (whose specialty is the study of reasoning), Christelle Bosc-Miné (problem-solving), Rémi Brissiaud (educational psychology), Sandra Bruno (conceptual development), Anne-Sophie Deborde (attachment), Corinne Demarcy (problem-solving), Sabine Guéraud (understanding), Caroline Guérini (theory of mind), Frank Jamet (naïve reasoning), Hélène Labat (learning to read), Annamaria Lammel (cultural psychology), Jean-Marc Meunier (knowledge representation), Sandra Nogry (conceptual development), and Carine Royer (learning to read). Emmanuel's doctoral students, current and former, have given much to him through their dedication and the freshness and openness of their thinking. They are a hard act to follow. In particular, we mention Valentine Chaillet, Laurence Dupuch, Sylvie Gamo, Khider Hakem, Bruno Martin, Évelyne Mengue, Lynda Taabane, and Emmanuel Trouche. We also keep in our hearts the memory of Justine Pélouard, who seemed to be headed for a wonderful scientific career, when all at once her life came to an end.

In other teams within Paragraphe, we would like to single out Anne Bationo, Ghislaine Azemard, Claude Baltz, Françoise Decortis, Hakim Hachour, Madjid Ihadjadene, Pierre Quettier, Alexandra Saemmer, Samuel Szoniecky, and Khaldoun Zreik for rich interactions with Emmanuel on numerous occasions. Over the years, some of them have become deeply appreciated friends. Emmanuel would also like to express his gratitude to a set of colleagues outside of Paragraphe, but still in the Psychology Department, for their lively ideas and their personal warmth. Above all, he thanks Marie-Carmen Castillo and Roxane Bordes, and then Aline Frey, Alain Blanchet, Samuel Demarchi, Sophie Frigout, Corinna Kohler, Michèle Montreuil, Tobie Nathan, Michael Pichat, Jean-Luc Picq, and Frédéric Rousseau.

The members of Doug's research group FARG ("Fluid Analogies Research Group"), early on in Ann Arbor but mostly in Bloomington have, over three decades, shed much light on the richness of that elusive mental phenomenon called "analogy-making". We are thinking of Marsha Meredith (who developed the computer model Seek-Whence),

Melanie Mitchell (Copycat), Robert French (Tabletop), Gary McGraw (Letter Spirit), John Rehling (Letter Spirit), James Marshall (Metacat), Harry Foundalis (Phaeaco), Francisco Lara-Dammer (George), Abhijit Mahabal (SeqSee), and Eric Nichols (Musicat). Standing on their shoulders and following in their footsteps are Matthew Hurley, Ben Kovitz, William York, and David Bender. Others who have brought ideas and insights to FARG over the years include Daniel Defays (Numbo), Alex Linhares (Capyblanca), David Moser (errors and humor), Donald Byrd, Gray Clossman, Steve Larson, Hamid Ekbia, David Chalmers, Wang Pei, Peter Suber, Yan Yong, Liu Haoming, Christoph Weidemann, Roy Leban, Liane Gabora, and Damien Sullivan.

Beyond FARG, Doug's life has been vitally enriched by so many good friends and sparkling colleagues in so many lands. Let's start with France, where, among the names that come to mind, are François Vannucci, Jacqueline Henry, Serge Haroche, Daniel Kiechle, Daniel Bougnoux, André Markowicz, Jacques Pitrat, Paul Bourgine, François Récanati, Gilles Cohen, Gilles Esposito-Farèse, Alain Zalmanski, Françoise Strobbe, Jean-Pierre Strobbe, Martine Lemonnier, Anne Bourguignon, Hubert Ceram, Karine Ceram, Liana Gourdjia, Marc Coppey, Geoff Staines, Silvia Busilacchi, Michelle Brûlé, and Denis Malbos. Turning to Italy, where he has always been so warmly received, he is reminded of Benedetto Scimemi, Luisa Scimemi, Giuseppe Trautteur, Pingo Longo, Giovanni Sambin, Alberto Parmeggiani, Francesco Bianchini, Maurizio Matteuzzi, Alex Passi, Sabrina Ardizzoni, Achille Varzi, Oliviero Stock, Enrico Predazzi, Cristina Peroni, Maurizio Codogno, Enrico Laeng, Paola Turina, Patrizio Frosini, Ozalp Babaoglu, Irene Enriques, Pietro Perconti, Andrea Padova, and *la famiglia* Genco.

But we shouldn't omit his friends and colleagues on the North American continent! Doug thus takes great pleasure in saluting (and in a fairly arbitrary order): Scott Buresh, Greg Huber, Karen Silverstein, Kellie Gutman, Richard Gutman, Caroline Strobbe, Grant Goodrich, Peter Rimbey, Scott Kim, Peter Jones, Steve Jones, Brian Jones, Iranee Zarb, Francis Zarb, David Policansky, Charles Brenner, Inga Karliner, Jon Thaler, Larry Tesler, Colleen Barton, Pentti Kanerva, Eric Hamburg, Michael Goldhaber, Rob Goldstone, Katy Börner, Rich Shiffrin, Jim Sherman, Colin Allen, John Kruschke, Mike Dunn, Breon Mitchell, Dan Friedman, George Springer, Mike Gasser, David Hertz, Willis Barnstone, Sumie Jones, Betsy Stirratt, Marc Hofstadter, Daniel Dennett, John Holland, Bob Axelrod, Dick Nisbett, Ken DeWoskin, Bill Cavnar, Gilles Fauconnier, Mark Turner, Lera Boroditsky, Mark Johnson, Bubal Wolf, Joe Becker, Donald Norman, Bernard Greenberg, Johnny Wink, Jay Curlin, Joseph Sevene, Anton Kuerti, Bill Frucht, Glen Worthey, Marilyn Stone, Jim Falen, Eve Falen, James Plath, Christopher Heinrich, Karen Bentley, Ann Trail, Sue Wunder, Julie Teague, Phoebe Wakhungu, Clark Kimberling, John Rigden, Leon Lederman, Jerry Fisher, Steve Chu, Peter Michelson, Bill Little, Paul Csonka, Sidney Nagel, Don Lichtenberg, Philip Taylor, Simone Brutlag, Doug Brutlag, Sandy Myers, Kristen Motz, and last but not least, Ollie (truly a *golden* retriever). Further afield, scattered hither and yon around the globe, are Francisco Claro, Peter Smith, Robert Boeninger, Cyril Erb, John Ellis, Alexander Rauh, Marina Eskina, Marek Karliner, Hakan Toker, and Michel Moutschen. To all of the above, Doug tips a deeply thankful hat.

Many friends and colleagues have likewise influenced Emmanuel's ideas about thought and have enriched his intellectual, professional, and personal life. Jean-François Richard occupies a special place of honor because of his constant presence, his gift of inspiring others, and his phenomenal creative drive. Emmanuel also wishes to cite the profound influence of colleagues with whom he has had long-time interactions and who have inspired him in many ways. These include Daniel Andler, Nicolas Balacheff, Jean-Marie Barbier, Claude Bastien, Luca Bonatti, Jean-François Bonnefon, Valérie Camos, Roberto Casati, Evelyne Clément, Jacques Crépault, Karine Duvignau, Michel Fayol, Jean-Paul Fischer, Bruno Gaume, Jean-Marc Labat, Jacques Lautrey, Ahn Nguyen Xuan, Jean-François Nicaud, Ira Noveck, Pierre Pastré, Sébastien Poitrenaud, Guy Politzer, Pierre Rabardel, Sandrine Rossi, Gérard Sensevy, Catherine Thevenot, Andrée Tiberghien, André Tricot, Jean-Baptiste Van der Henst, Gérard Vergnaud, Lieven Verschaffel, and Bruno Villette. On a more personal level, Emmanuel wishes to thank so many long-time friends for their loyalty and their indescribably valuable affection: Youri Beltchenko, Florence Deluca Boutrois, Patrick Grinspan, Michaël Jasmin, Audrey Norcia, Franck Lelong, Gaëlle Le Moigne, Philippe Pétiard, Caroline Sidi, Nadine Zahoui, and Marie-Hélène Zerah.

Our publishers on both sides of the Atlantic have been extremely cordial and helpful to us. On the American side, John Sherer was, for several years, a true believer in this book, and when he left Basic Books, T. J. Kelleher and Lara Heimert kept the enthusiastic fires warmly burning. We would also like to thank Nicole Caputo, Andrea Cardenas, Tisse Takagi, Michele Jacob, Cassie Nelson, and Sue Caulfield for their top-notch contributions to this work in its English-language incarnation. On the French side, Odile Jacob and Bernard Gotlieb welcomed us on board and made us know they would support us strongly. We owe special thanks to Jean-Luc Fidel for his meticulous reading of the manuscript and for his sensitive comments. We also wish to thank Jeanne Pérou, Cécile Andrier-Taverne, and Claudine Roth-Islert for their excellent work on the production and distribution of this book in France.

We expresss deeepest gratitude to David Bender, Steven Willliams, and Jane Stewart Adams for scouring *this* book for typos and non sequiturs, and *mutatis mutandis*, we thank Christellle Bosc-Miné and Karine Duvignau for scouring *that* book. And no, certainly not to be forgotten are Greg Huber, Tom Seeber, and D. Alvin Oyzeau, all of whom tinkered most helpfully with the figures (and sometimes with the facts as well!).

It goes without saying that many of Doug's friends have become friends of Emmanuel's, and vice versa, which of course blurs the borders of all these categories. Such mingling of two worlds has been one of the great side benefits of so many years of work together. Indeed, sometimes the process seemed so long it would never end, and yet here we are, putting the finishing touches on this book. We have learned a great deal about thinking, about writing, and about language from this process, and we hope our readers will take pleasure in, and hopefully inspiration from, our joint creation.

SURFACES AND ESSENCES

PROLOGUE

Analogy as the Core of Cognition

❧ ❧ ❧

Giving Analogy its Due

In this book about thinking, analogies and concepts will play the starring role, for without concepts there can be no thought, and without analogies there can be no concepts. This is the thesis that we will develop and support throughout the book.

What we mean by this thesis is that each concept in our mind owes its existence to a long succession of analogies made unconsciously over many years, initially giving birth to the concept and continuing to enrich it over the course of our lifetime. Furthermore, at every moment of our lives, our concepts are selectively triggered by analogies that our brain makes without letup, in an effort to make sense of the new and unknown in terms of the old and known. The main goal of this book, then, is simply to give analogy its due — which is to say, to show how the human ability to make analogies lies at the root of all our concepts, and how concepts are selectively evoked by analogies. In a word, we wish to show that analogy is the fuel and fire of thinking.

What Dictionaries Don't Say about Concepts

Before we can tackle this challenge, we need to paint a clear picture of the nature of concepts. It is easy — in fact, almost universal — to underestimate the subtlety and complexity of concepts, all the more so because the tendency to think of concepts in overly simple terms is reinforced by dictionaries, which claim to lay out the various different meanings of a given word by dividing the main entry into a number of subentries.

Take, for example, the noun "band". In any reasonably-sized dictionary, there will be, in the overall entry for this word, a subentry describing a band as a piece of cloth that can be wrapped around things, another subentry describing how a band can be a colored strip or stripe on a piece of cloth or other type of surface, another subentry describing a band as a smallish set of musicians who tend to play certain types of music or to use only certain types of instruments, another one for a group of people who work or play together, another one for a wedding ring, another one for a selection on a record or a compact disk, another one for a range of frequencies or energies or prices or ages (etc.), and perhaps a few others. The dictionary will clearly set out these various concepts, all fairly distinct from each other and all covered by the same word "band", and then it will stop, as if each of these narrow meanings had been made perfectly clear and were cleanly separable from all the others. All well and good, except that this gives the impression that each of these various narrower meanings of the word is, on its own, homogeneous and not in the least problematic, and as if there were no possible risk of confusion of any one of them with any of the others. But that's nowhere near the truth, because sub-meanings are often closely related (for instance, the colored stripe and the range of frequencies, or the wedding ring and the piece of cloth wrapped around something), and because each of these supposedly clear and separate senses of the word "band" constitutes on its own a bottomless chasm of complexity. Although dictionaries give the impression of analyzing words all the way down to their very atoms, all they do in fact is graze their surfaces.

One could spend many years compiling a huge anthology of photographs of highly diverse wedding bands, or, for that matter, an anthology of photos of headbands, or of jazz bands, or of bands of criminals — or then again, of photos of wildly different chairs or shoes or dogs or teapots or versions of the letter "A", and on and on — without ever coming close, in any such anthology, to exhausting the limitless possibilities implicitly inherent in the concept. Indeed, there are books of precisely this sort, such as *1000 Chairs*. If the concept *chair* were completely straightforward, it is hard to see what interest such a book could possibly have. To appreciate the beauty, the originality, the practicality, or the style of a particular chair requires a great deal of experience and expertise, of which dictionaries cannot convey even an iota.

One could of course make similar observations concerning the subtleties of various types of bands — thus, one could spend one's whole life studying jazz bands, or headbands, or criminal bands, and so forth. And even concepts that seem much simpler than these are actually endless swamps of complexity. Take the concept of the capital letter "A", for instance. One would need many pages of text in complex, quasi-legal language if one were trying to pin down just what it is that we recognize in common among the countless thousands of shapes that we effortlessly perceive as members of that category — something that goes way beyond the simple notion that most people have of the concept "A" — namely, that it consists of two oppositely leaning diagonal strokes connected by a horizontal crossbar.

Indeed, catalogues of typefaces are veritable gold mines for anyone interested in the richness of categories. In the facing figure, we have collected a sampler of capital "A"'s designed for use in advertising, and as is clear from a moment's observation, any *a priori* notion that one might have dreamt up of *A*-ness will be contradicted by one or more of these letters, and yet each of them is perfectly recognizable — if not effortlessly so when displayed all by itself, then certainly in the context of a word or sentence.

The everyday concepts *band*, *chair*, *teapot*, *mess*, and *letter 'A'* are very different from specialized notions such as *prime number* or *DNA*. The latter also have unimaginably many members, but what is shared by all their members is expressible precisely and unambiguously. By contrast, in the mental structure underpinning a word like "band", "chair", "mess", or "teapot" there lurks a boundless, blurry richness that is completely passed over by dictionaries, because spelling out such subtleties is not a dictionary's aim. And the fact is that ordinary words don't have just two or three but an *unlimited number* of meanings, which is quite a scary thought; however, the more positive side of this thought is that each concept has a limitless potential for variety. This is a rather pleasing thought, at least for people who are curious and who are stimulated by novelty.

Zeugmas: Amusing Revealers of Conceptual Subtlety

There is a linguistic notion called "zeugma" (also sometimes called "syllepsis") that, although it is fairly obscure, has a good deal of charm and brings out the hidden richness of words (and thus of concepts). The zeugma or syllepsis is one of the classical figures of speech, and is often — perhaps nearly always — used to humorous effect. It is characterized by the fact that more than one meaning of a word is exploited in a sentence, although the word itself appears only once. For example:

I'll meet you in five minutes and the garden.

This sentence exploits two different meanings of the preposition "in" — one temporal and the other spatial. When one imagines meeting someone *in* a garden, one sees in one's mind's eye two relatively small entities physically surrounded by a larger entity, whereas when one imagines a meeting taking place *in* five minutes, one thinks of the period of time that separates two specific moments from each other. Everyone understands with no trouble that these are two very different concepts associated with the same word, and the fact that the preposition "in" is used only once in the sentence despite the wide gap between the two meanings that it's conveying is what makes us smile when we read the sentence.

Here are a few other somewhat humorous examples of zeugmas:

Kurt was and spoke German.
The bartender gave me a wink and a drink.
She restored my painting and my faith in humanity.
I look forward to seeing you with Patrick and much joy.

In the first, the word "German" is forced to switch rapidly, in the reader's mind, from being an adjective denoting a nationality to being a noun denoting a language.

The second zeugma involves two different aspects of the notion of *transfer* between human beings. Does one person really *give* a wink to another person? Is a wink a material object like a drink, which one person can hand another?

In the third zeugma, the speaker's faith in humanity had disappeared and was made to come back, whereas the painting had not disappeared at all. Moreover, faith in humanity is far less palpable than a painting on one's wall. What gives this zeugma its flavor of oddness is that one of the meanings of the verb "restore" that it depends on is "to return something that has been lost", while the other meaning used is "to make something regain its former, more ideal state", and although these two senses of the same word are clearly related, they are just as clearly not synonymous.

Finally, the last zeugma in our quartet plays on two sharply contrasting senses of the preposition "with", one conjuring up the image of someone (Patrick) physically accompanying someone else (the speaker and the person being addressed), and the other communicating the emotional flavor (great pleasure) of a mental process (the anticipation of a reunion). As in the other cases, the zeugmatic use of "with" brings out the wide gap between two senses of one word, and to experience this distinction in such a crisp fashion is thought-provoking. We thus see that any well-designed zeugma will, by its very nature, automatically highlight certain semantic subtleties of the word (or phrase) around which it is built.

For example, what does the word "book" mean? One would at first tend to say that it designates an object made of printed sheets of paper bound together in some fashion, and having a cover (and so forth and so on). This is often correct, but the following zeugma brings out a different sense of the word:

> The book was clothbound but unfortunately out of print.

This sentence reminds us that the word "book" also denotes a more abstract concept — namely, the set of all copies available in stores or warehouses. Are we thus in the presence of *one* concept, or of *two*? And when someone says, "I'm translating this book into English", are they using a third sense of the word? How many subtly distinct concepts secretly coexist in the innocent word "book"? It would be an instructive exercise to try to construct more zeugmas based on yet other senses of the word "book", but we have other goals here, so we will leave that challenge to our readers.

Instead, let's look at a somewhat more complex zeugma:

> When they grew up, neither of those bullies ever had to pay
> for all the mean things that they did as, and to, younger kids.

Here the trickiness is in the strange, lightning-fast shifting of meaning of "younger kids" as a function of whether it is seen as part of the phrase "things that they did as younger kids" or as part of the phrase "things that they did to younger kids", since in the first case the *younger kids* are the ex-bullies themselves (or rather, the bullies that they once were), while in the latter case the *younger kids* are their victims.

Some Revealing Zeugmas

Although the zeugmas we've exhibited above are mostly quite amusing, it's not for entertainment but for enlightenment that we've brought up the topic. And so let's take a look at some cases that raise more serious issues.

> "You are always welcome in my home," he said in English and all sincerity.

This zeugma is clearly built around the word "in", and the natural question here is whether we are dealing with *one* sense or *two* senses of the word. In a respectable dictionary, these two meanings would probably have distinct subentries. However, what about the following sentence?

> "You are no longer welcome in my home," he said in anger and all sincerity.

Are the two meanings of "in" here exactly the same? Perhaps — after all, they both apply to the mental states of a single person; but then again perhaps not — after all, one could replace "in anger" by "in an outburst of anger" but certainly one could not say "in an outburst of sincerity". So it's rather tricky. As a matter of fact, it would be impossible to give a definitive judgment on this issue. Indeed, we chose this example precisely because it brings out certain subtle nuances of the concept *in.* How does one recognize those situations that match the English word "in"? To put it another way, how does one recognize *in*-situations? What do all *in*-situations have in common, and how do some of them differ from others, and why would it be next to impossible to make a precise and sharp classification of all the types of *in*-situation?

Let's shift our attention from a preposition to a verb. Does the following sentence strike you as innocuous and perfectly acceptable (*i.e.*, nonzeugmatic), or does it grate on your ears (thus it would be a zeugma)?

I'm going to brush my teeth and my hair.

Are the two types of brushing really just one thing deep down, or are they worlds apart? We might gain perspective on this question by looking at a similar example in another language. In Italian, one might easily and comfortably say:

Voglio lavarmi la faccia e i denti.

(In a fairly literal translation, this says, "I want to wash my face and my teeth.") The fact that Italian speakers say things this way sheds light on how they perceive the world — namely, it shows that they perceive the act of washing one's face and the act of brushing one's teeth as belonging to the same category (both are types of *washing*), and thus they are, in some sense, "the same act".

On the other hand, to speakers of English, brushing one's teeth is not a kind of washing (washing usually involves soap of some sort, and most people would hesitate to refer to toothpaste as "soap", though the two have much in common), so the sentence sounds zeugmatic (that is, its double application of the same word makes us smile). As for French, although occasionally one will hear "se laver les dents" ("to wash one's teeth"), it is more common to say (and hear) "se brosser les dents" ("to brush one's teeth"). The latter seems more natural to French speakers than the former. And thus we see that a phrase ("to wash one's teeth and one's face") can be very zeugmatic in one language (English), can have a faintly zeugmatic flavor in another language (French), and can be totally nonzeugmatic in a third language (Italian).

The preceding example shows how a zeugma can reveal a conceptual division that speakers of language A find blatantly obvious, while to speakers of language B it is difficult to spot. For instance, in English, we can say without any sense of oddness:

Sometimes I go to work by car, and other times on foot.

In German or Russian, however, these two forms of locomotion call for different verbs. When one takes a vehicle to arrive at one's destination, then the verb "fahren" is used in German, whereas when one goes somewhere on foot, then the verb "gehen" is used.

In Russian it's trickier yet, because not only is there a distinction between *going in a vehicle* and *going on foot*, but also the choice of verb depends on whether this kind of motion is undertaken frequently or just one time. Thus a completely innocuous-seeming verb in English breaks up into several different verbs in Russian. In other words, what to English speakers seems to be a monolithic concept splits into four distinct concepts to Russian speakers.

Let's take another very simple sentence in English:

The boy and the dog were eating bread.

This sentence is nonzeugmatic in English; that is, it simply *works*, sounding neither strange nor humorous to the English-speaking ear. On the other hand, it sounds wrong in German, because different verbs apply to animal and human ingestion — "fressen" for the beasts, and "essen" for humans. In other words, German speakers split up what to us anglophones is the monolithic concept of *eating*, breaking it into two varieties, according to the type of creature that is carrying out the act.

The "Natural" Conceptual Distinctions Provided by Each Language

These examples might inspire someone to imagine a language (and culture) that has no verb that applies both to men and to women. Thus it would have one verb that would apply to eating acts by *men* and a different one that would apply to eating acts by *women* — say, "to wolf down" for men and "to fox down" for women, as in "Petunia foxed down her sandwich with relish, gusto, and pickles". Speakers of this hypothetical language would find it jolting to learn that in English one can say, "My husband and I enjoy eating the same things" or "A girl and a boy were walking down the sidewalk." To them, such sentences would sound nonsensical. A language like this may strike you as ludicrous, but many languages do make just such gender-based lexical distinctions.

For instance, in French there is a clear-cut distinction between enjoyment partaken of by men and enjoyment partaken of by women, which shows up in, among other venues, the standard adjective meaning "happy": whereas a joyous man or boy will be "heureux", a joyous woman or girl will be "heureuse". And thus, a *curieux* French male might well wonder what it feels like to be *heureuse* — but he would do so in vain! A man simply cannot be *heureuse*! In like manner, a *curieuse* French woman might wonder what it feels like to be *heureux* — but her efforts, no matter how valiant, would be doomed to failure. A Venusian might as well try to imagine what it feels like to be Martian!

Does all this sound far-fetched to you? Well, consider that there is a famous Russian poem centered on what the poet, a man named Il'ya L'vovich Selvinsky, considered a very strange fact: namely, that every act of his lover — every single one of the mundane verbs that described her actions — was graced, when in the past tense, by a feminine ending (often the syllables or bisyllables "la", "ala", or "yala"). The poet describes various completely ordinary actions on her part (walking, eating, etc.), and then expresses wonderment at his own feeling of disorientation, since he, being a male, has never once performed a single one of these "uniquely feminine" acts, nor experienced a single one of these "uniquely feminine" sensations, and, alas, will never be able to do so. In making such observations, is Selvinsky expressing something deep, or is he merely playing with words?

One can easily enough imagine a language that, with a panoply of verbs, distinguishes between a vast number of different ways of eating — the eating of a famished boy, of a high-society lady, of a pig, a horse, a rabbit, a shark, a catfish, an eagle, a hummingbird, and so forth and so on. Such a fine-grained breakup of a concept that seems to *us* completely monolithic is perfectly imaginable, because we understand that there are genuine differences between these creatures' ways of ingesting food (indeed, if there weren't any, we would not have written "genuine differences"). Each language has the right and the responsibility to decide where it wishes to draw distinctions in the zone of semantic space that includes all of these distinct activities. After all, there are not, on earth (and never have been, and never will be) two creatures that eat in an exactly identical fashion, nor even two different moments in which a single creature eats in exactly the same manner, down to the tiniest detail.

Every act is unique, and yet there are resemblances between certain acts, and it is precisely these resemblances that give a language the opportunity to describe them all by the same label; and when a language chooses to do so, that fact creates "families" of actions. This is a subtle challenge to which every language reacts in its own fashion, but once this has been done, each group of people who share a common native language accepts as completely natural and self-evident the specific breakdown of concepts handed to them by their language. On the other hand, the conceptual distinctions that are part and parcel of *other* languages may strike them as artificial, pointlessly finicky, even incomprehensible or stupid, unless they find some interest in the subtleties of such distinctions, which may then make them see their own set of concepts in a fresh light.

Wordplay with the Word "Play"

The verb "to play" affords us a delightful sampler of zeugmas, or else, depending on a person's native language and on their own personal way of perceiving the actions involved, non-zeugmas. For example:

Edmond plays basketball and soccer.

This sentence, on first sight, might seem about as natural as they come, and very far from zeugmaticity, and yet the two activities involved, although they both belong to the category of *sports*, are different in numerous ways from each other. For instance, one involves a ball that is primarily in contact with the feet (and on occasion with the head), while the other involves a ball that is primarily in contact with the hands (and virtually never with the head). Certain speakers of English might therefore hear a trace of strangeness, albeit only very slight, in the application of the same verb to two rather disparate activities.

If *essen* (which is what people do when they eat food) and *fressen* (which is what, say, pigs and rabbits do with their food) are seen by German speakers as activities that belong to two different categories, then there is nothing to keep us from imagining a language in which one would say:

Edmondus snuoiqs basketballum pluss iggfruds soccerum.

The speakers of this hypothetical language would see the actions of basketball players — or rather, of basketball *snuoiqers* — as being just as different from the actions of soccer *iggfruders* as the sounds "snuoiq" and "iggfrud" are different from each other.

If this example's zeugmaticity seems too weak, then we can try another avenue of approach to the same issue:

Sylvia plays tennis, Monopoly, and violin.

This sentence involves a musical instrument and two types of game that are much more different from each other than are basketball and soccer. If one tried to measure the distances between these three concepts by asking people to estimate them, it's likely that most people would place *violin* quite a long ways from *tennis* and *Monopoly*, and those two games, though not extremely near each other, would be much closer than either of them is to *violin.* And finally, not too surprisingly, this matches the collective choice of Italian speakers, who would translate the above sentence as follows:

Sylvia *gioca* al tennis e a Monopoly, e *suona* il violino.

It would be unthinkable, in Italian, for anyone to *play* (in the sense of *giocare*) a musical instrument; the mere suggestion is enough to make an Italian smile. The kind of scene that such a phrase would conjure up is that of people playing catch with a Stradivarius, for instance. While it is natural for English and French speakers to see violin-playing as belonging to the same category as soccer-playing and basketball-playing, the idea would seem downright silly to Italian speakers.

In French, the verb *jouer* is used both for musical instruments and for sports, but it is followed by different prepositions in the two cases. Thus one plays *at* a sport but one plays *of* a musical instrument. Does this syntactic convention split the concept of *jouer* into two quite clear and distinct sub-meanings? In English, there is no similar syntactic convention that would create a mental division of the verb "to play" into two separate pieces; rather, it simply feels monolithic.

Playing Music and Sports in Chinese

The distinction made in Italian between "giocare" (for sports) and "suonare" (for musical instruments) might seem a bit precious. After all, not only English but plenty of other languages are happy to use exactly the same verb for both kinds of activities —

thus French uses "jouer", German uses "spielen", Russian uses "играть", and so on. What about Chinese?

It turns out that Mandarin speakers are considerably more finicky in this matter than Italian speakers: they linguistically perceive four broad types of musical instruments, each type meriting its own special verb. Thus for stringed instruments there is the verb "拉" (pronounced "lā"), meaning roughly "to pull", while for wind instruments one says "吹" ("chuī"), which means "to blow". Then for instruments such as the guitar, whose strings are plucked by the fingers, or the piano, whose keys are pushed by the fingers, the verb is "弹" ("tán") — and finally, for drums, which are banged, what one says is "打" ("dǎ").

Curiously enough, it's possible to apply the verb that means "to play" (as in "play with a toy") to any musical instrument (it is "玩", pronounced "wán"); unfortunately, however, the meaning is not what an English speaker might expect: it's essentially the idea of *fussing around* with the instrument in question, and moreover this usage of "玩" is extremely informal, indeed slangy.

One might naturally wonder how a Chinese speaker would ask a more generic question, such as "How many instruments does Baofen play?" But the best translations of this perfectly natural English sentence elegantly bypass the problem by making use of very broad verbs such as "学习" ("xuéxí") or "会" ("huì"), which mean, respectively, "to study" and "to be able; to know", and which have no particular connection with music. In short, there is no general verb in Mandarin that corresponds to the *musical* notion of playing, even though to us English speakers the concept seems totally logical, even inevitable; but the fact is that speakers of Chinese have no awareness of this lacuna in their lexicon, no matter how blatant it might seem to us.

Well, all right, then. But what about playing games and sports — surely there is just one verb in Chinese for this monolithic concept? To begin with, one does not, in Mandarin, play board games and sports with the same verb. For chess, one engages in the activity of "下" ("xià"), which one does not do with any kind of ball. And for a sport that uses a ball, it all depends on the kind of ball involved. For basketball, it's "打" ("dǎ"), the verb that applies to playing a drum (the connection may seem a bit strained to a non-Chinese), whereas for soccer it's "踢" ("tī"), which means "to kick". Thus one might say, "I prefer kicking soccer to beating basketball." Once again we see that in a domain that strikes an English speaker as monolithic — everything is *played*, and that's all there is to it! — distinctions are not just rife but necessary in Chinese.

For English speakers, despite our use of the single verb "to play", it's not terribly hard to see that this verb conflates two activities that are quite different — namely, making rhythmic noises and having fun — and that the conceptual union thus created is not inevitable, and might even be seen as being rather arbitrary. On the other hand, *within* each of these two domains, it's harder to see a lack of natural unity. If someone were to ask us if *playing dolls*, *playing chess*, and *playing soccer* are all really "the same activity", we could of course point out differences, but to focus on such fine distinctions would seem quite nitpicky. And when we learn that in Mandarin, *playing soccer* and *playing basketball* require different verbs, it is likely to strike us as really overdoing things, rather as if some exotic tongue insisted on using two different verbs to say "to drink", depending on whether it involved drinking *white* wine or *red* wine. But then again, this is an important distinction for wine-lovers, so it's conceivable that some of them would very much like the idea of having two such verbs.

Zeugmas and Concepts

Our brief excursion to Zeugmaland will come to a climax in the following bold prediction:

> You will enjoy this zeugma as much as a piece of chocolate or of music.

This sentence has a couple of zeugmatic aspects. Firstly, it plays on two senses of the noun "piece". In some readers recognition of this contrast will evoke a smile, even though there's no denying that both usages of the word are completely standard. Secondly, it plays on three senses of the verb "enjoy" — one involving a gustatory experience, another involving an auditory experience, and yet another involving the savoring of a linguistic subtlety. Each reader will of course have a personal feeling for how large the distinction between these three senses of the word is.

Aside from making us smile, zeugmas offer us the chance to reflect on the hidden structure behind the scenes of a word or phrase — that is, on the concept associated with the lexical item, or more precisely, on the *set of concepts* associated with it — and since most words could potentially be used to form a zeugma (including very simple-seeming words such as "go", as we saw above in the discussion of German and Russian), the phenomenon necessarily increases our sensitivity to the miracle of the human brain's ability to spontaneously assign just about anything it encounters to some previously known category. After all, despite the inevitable and undefinable blurriness of the "edges" of each one of our categories, and despite the enormous number of categories, our brains manage to carry out such assignments in a tiny fraction of a second and in a manner of which we are totally unaware.

The Nature of Categorization

The spontaneous categorizations that are continually made by and in our brains, and that are deeply influenced not just by the language we are speaking but also by our era, our culture, and our current frame of mind, are quite different from the standard image, according to which categorization is the placing of various entities surrounding us into preexistent and sharply-defined mental categories, somewhat as one sorts items of clothing into the different drawers of a chest of drawers. Just as one can easily stick one's shirts into a physical drawer labeled "shirts", so one would easily assign dogs to the mental drawer labeled "dog", cats to the nearby mental drawer labeled "cat", and so forth. Every entity in the world would fit intrinsically into one specific mental "box" or "category", and this would be the mental structure to which all the different entities of the same type would be assigned. Thus all bridges in the world would be unambiguously assigned to the box labeled "bridge", all situations involving motion would be assigned to the box labeled "move", and all situations involving things standing still would be assigned to the box labeled "stationary". This mechanism of "boxing" everything in the world would be both automatic and completely reliable, the *raison d'être* of mental categories being to assign entities objectively to their proper conceptual label in an objective, observer-independent fashion.

Such a vision of the nature of categorization is very far from what really goes on, and in the pages to come we will do our best to show why this is so. But hopefully, already from Chapter 1 onwards, readers will feel persuaded that mental categories are anything but drawers into which clear-cut items are automatically sorted, and this idea will be reinforced ever more strongly as the book proceeds.

What, then, do we mean in this book by "category" and "categorization"? For us, a category is a mental structure that is created over time and that evolves, sometimes slowly and sometimes quickly, and that contains information in an organized form, allowing access to it under suitable conditions. The act of categorization is the tentative and gradated, gray-shaded linking of an entity or a situation to a prior category in one's mind. (Incidentally, when we use the term "category", we always mean a category in someone's mind, as opposed to mechanical labels used in computer data bases or technical labels used in scientific taxonomies, such as lists of the names of biological species.)

The tentative and non-black-and-white nature of categorization is inevitable, and yet the act of categorization often feels perfectly definite and absolute to the categorizer, since many of our most familiar categories seem on first glance to have precise and sharp boundaries, and this naïve impression is encouraged by the fact that people's everyday, run-of-the mill use of words is seldom questioned; in fact, every culture constantly, although tacitly, reinforces the impression that words are simply automatic labels that come naturally to mind and that belong intrinsically to things and entities. If a category has fringe members, they are made to seem extremely quirky and unnatural, suggesting that nature is really cut precisely at the joints by the categories that we know. The resulting illusory sense of the near-perfect certainty and clarity of categories gives rise to much confusion about categories and the mental processes that underlie categorization. The idea that category membership always comes in shades of gray rather than in just black and white runs strongly against ancient cultural conventions and is therefore disorienting and even disturbing; accordingly, it gets swept under the rug most of the time. Since the nature of mental categories is much subtler than the naïve impression suggests, it is well worth examining carefully.

A category pulls together many phenomena in a manner that benefits the creature in whose mind it resides. It allows invisible aspects of objects, actions, and situations to be "seen". Categorization gives one the feeling of understanding a situation one is in by providing a clear perspective on it, allowing hidden items and qualities to be detected (by virtue of belonging to the category *person*, an entity is known to have a *stomach* and a *sense of humor*), future events to be anticipated (the glass that my dog's tail just knocked off the table is going to break) and the consequences of actions to be foreseen (if I press the "G" button, the elevator will go down to the ground floor). Categorization thus helps one to draw conclusions and to guess about how a situation is likely to evolve.

In short, nonstop categorization is every bit as indispensable to our survival in the world as is the nonstop beating of our hearts. Without the ceaseless pulsating heartbeat of our "categorization engine", we would understand nothing around us, could not reason in any form whatever, could not communicate with anyone else, and would have no basis on which to take any action.

Two Misleading Caricatures of Analogy-making

If categorization is central to thinking, then what mechanism carries it out? Analogy is the answer. But alas, analogy-making, like categorization, is also plagued by simplistic and misleading stereotypes. We therefore proceed straightaway to discuss those stereotypes, in the aim of quickly ridding ourselves of the contaminating and confusing visions that they give of the nature of the motor of cognition.

The first of these stereotypes of analogy-making takes the word "analogy" as the name of a certain very narrow class of sentences, seemingly mathematical in their precision, of the following sort:

West is to *east* as *left* is to *right.*

This can be made to look even more like a mathematical statement if it is written in a quasi-formal notation:

west : *east* :: *left* : *right*

Intelligence tests often employ puzzles expressed in this kind of notation. For example, they might pose problems of this sort: "*tomato* : *red* :: *broccoli* : *X*", or perhaps "*sphere* : *circle* :: *cube* : *X*", or "*foot* : *sock* :: *hand* : *X*", or "*Saturn* : *rings* :: *Jupiter* : *X*", or "*France* : *Paris* :: *United States* : *X*" — and so forth and so on. Statements of this form are said to constitute *proportional analogies*, a term that is itself based on an analogy between words and numbers — namely, the idea that an equation expressing the idea that one pair of numbers has the same ratio as another pair does ($A/B = C/D$) can be carried over directly to the world of words and concepts. And thus one could summarize this very analogy in its own terms:

proportionality : *quantities* :: *analogy* : *concepts*

There is no scarcity of people who believe that this, no more and no less, is what the phenomenon of analogy is — namely, a template always involving exactly four lexical items (in fact, usually four words), and which has the same rigorous, austere, and precise flavor as Aristotle's logical syllogisms (such as the classic "All men are mortal;

Socrates is a man; ergo, Socrates is mortal"). And indeed it was none other than Aristotle who first studied proportional analogies. For him, analogy, understood in this narrow fashion, was a type of formal reasoning belonging to the same family as deduction, induction, and abduction. The fact that many people today understand the word "analogy" in just this narrow way therefore has genuine and valid historical roots. Nonetheless, such a restrictive vision of the faculty of analogy-making leads almost ineluctably to the conclusion that it is such a precise, focused, and specialized type of mental activity that it will crop up only in very rare circumstances (particularly in intelligence tests!).

And yet analogy, as a natural form of human thought, is not by any means limited to this kind of case. Although each of the proportional analogies exhibited above was intended to have just one single correct response — the so-called *right answer* — the fact is that the world in which we live does anything but give us a long series of intelligence-test questions in the form of right-answer analogy puzzles. Thus in the case of the "Paris of the United States" puzzle given above, although we ourselves were thinking mostly of New York as "the right answer", we have collected, in informal conversations, quite a few other perfectly defensible answers, including Washington, Boston, Los Angeles, Las Vegas, Philadelphia, and — of course! — Paris, Texas.

Indeed, quite to the contrary, the world confronts us with a never-ending series of vague and ambiguous riddles, such as this one: "What disturbing experience in my life, or perhaps in the life of some friend of mine, is meaningfully similar to the sudden confiscation of my eight-year-old son's bicycle by the principal of his school?" It is by searching for strong, insight-providing analogues in our memory that we try to grasp the essences of the unfamiliar situations that we face all the time — the endless stream of curve balls that life throws at us. The quest for suitable analogues is a kind of art that certainly deserves the label "vital", and as in any other form of art, there seldom is a single right answer. For this reason, although proportional analogies may on occasion be gleaming jewels of precision and elegance, the image that they give of the nature of analogy-making is wildly misleading to anyone who would seek the crux of that mental phenomenon.

Another widely held view of analogy (and here we come to the second stereotype) is that when people make analogies, they call on sophisticated reasoning mechanisms that, through intricate machinations, somehow manage to link together far-flung domains of knowledge, sometimes in a conscious fashion; the conclusions reached thereby may be very subtle but will also be very tentative. This vision gives rise to the image of analogies as being the fruit of strokes of genius, or at least of deep and unusual insights. And there are indeed numerous famous cases of this sort that one can cite — great scientific discoveries resulting from sudden inspirations of people who found undreamt-of links between seemingly unrelated domains. Thus the mathematician Henri Poincaré wrote, "One day… the idea came to me very concisely, very suddenly, and with great certainty, that the transformations of indeterminate ternary quadratic forms were identical to those of non-Euclidean geometry." This flash of inspiration gave rise to much rich new mathematics. One can also admiringly recall various architects, painters, and designers who, thanks to some fresh analogy, were able to transport a concept from one domain to a distant one in such a fruitful way that people were amazed. From this perspective, the making of analogies is a cognitive activity that only a small number of extremely inventive spirits engage in; it happens only when a mind dares to explore highly unlikely connections between concepts, and it reveals relationships between things that no one had ever before thought of as being related.

This stereotype of analogy-making does not presume that such acts are limited to scientists, artists, and designers; much the same vision of sophisticated reasoning that connects distant domains and leads to daring but tentative conclusions applies to people in everyday life. For example, it is universally accepted that analogy plays a major role in teaching. Most everyone can recall analogies from their schooldays, such as between the atom and the solar system, between an electrical circuit and a circuit in which water flows, between the heart and a pump, or between the benzene molecule and a snake biting its own tail. All of these cases feature connections that link rather remote domains (or, to be more precise, domains that *seem* remote when only their surface is taken into account). One can also find examples of such analogies in everyday arguments, whether someone is supporting an idea or trying to knock it down. For instance, if everyone laughs in the face of a person who dares to reveal grand ambitions, a natural retort might be, "Laugh all you want; they all laughed at Christopher Columbus!" And in political debate, analogies between far-flung situations play a key role. Thus these days, likening the leader of a foreign country to Hitler as a way to ignite patriotic fervor has become a hackneyed stratagem (for example, the elder George Bush pulled the Hitler analogy out of his hat numerous times in order to justify the first Iraq war), whereas likening a war to the Vietnam war has played precisely the opposite role in the United States (the opponents of the second Iraq war called on the Vietnam analogy over and over again). One even finds such insightful analogies, full of freshness, in childish observations, such as when the daughter of one of this book's authors, rising to the full mental height of her seven summers, proudly declared, "School is like a staircase; each new grade is one step higher!" Such a joyous moment of enlightenment is, in its humble way, an insight comparable to Poincaré's joyous insight into abstruse mathematical phenomena.

To summarize, then, the first of these two stereotypes — proportional analogy — is so formally constrained that if that were all analogy-making amounted to, it would merely be the Delaware of cognition; by contrast, the second stereotype pinpoints a far more important mental phenomenon — namely, the selective exploitation of past experiences to shed light on new and unfamiliar things belonging to another domain. And thus we will spend very little time on proportional analogies in this book; however, it's quite another matter as far as rich interdomain analogies are concerned, and we will devote a great deal of attention to them. And yet, despite its clear relevance to our central topic, this second vision of analogy-making is still impoverished, since it vastly under-represents the wide range of mental phenomena to which analogy is connected. Indeed, it completely leaves out the idea that analogy-making is the machinery behind the pulsating heartbeat of thought: categorization.

Analogy-making and Categorization

Indeed, the central thesis of our book — a simple yet nonstandard idea — is that the spotting of analogies pervades every moment of our thought, thus constituting thought's core. To put it more explicitly, analogies do not happen in our minds just once a week or once a day or once an hour or even once a minute; no, analogies spring up inside our minds numerous times every second. We swim nonstop in an ocean of small, medium-sized, and large analogies, ranging from mundane trivialities to brilliant insights. In this book, we will show how the simplest and plainest of words and phrases that we come out with in conversations (or in writing) come from rapidly, unconsciously made analogies. This incessant mental sparkling, lying somewhere below the conscious threshold, gives rise to our most basic, humdrum, low-level acts of categorization, whose purpose is to allow us to understand the situations that we encounter (or at least their most primordial elements), and to let us communicate with others about them.

A substantial fraction of the myriads of analogies constantly being born and quickly dying in our heads are made in order to allow us to find the standard words that name mundane objects and activities, but by no means all of them are dedicated to that purpose. Many are created to try to make sense of situations that we face on a much larger scale. To pinpoint, in the form of a single previously known concept, the essence of a complex situation that has just cropped up for the first time involves a much more penetrating and global understanding of a situation than one gets from simply smacking labels on its many familiar constituents. And yet this far deeper process — the retrieval of a long-buried memory by an analogy — is so central and standard in our lives that we seldom think about it or notice it at all. It is an automatic process, and virtually no one wonders why it occurs, nor how, since it is so familiar. If asked "How come that particular memory popped to your mind right after I told you what happened to me?", a typical person might reply, with a bemused tone at being asked such a silly question, "Well, what I remembered *is very much like* what you told me. *That's* why I remembered it! How could it have been any other way?" It's as if they had been asked, "Why did you fall down?" and answered, "Because I tripped!" In other words, having X, which is in some sense very similar to Y, come to mind when Y occurs and seizes our attention seems as natural and inevitable as falling down when one is tripped — there is no mystery, hence there seems to be no need whatsoever for any explanation!

The triggering of memories by analogy lies so close to what seems to be the essence of being human that it is hard to imagine what mental life would be like without it. Asking why one idea triggers another similar one would be like asking why a stone falls if one lets go of it three feet above the ground. The phenomenon of gravity is so familiar and obvious to us, striking us as so normal and so inevitable, that no one, aside from a tiny minority of physicists obsessed with explaining what others take for granted, even sees that there is anything to ask about. For most non-physicists, it's hard to see why gravity needs an explanation — and the same holds for the triggering of memories by analogy. And yet, how many scientific discoveries can hold a candle to general relativity, Albert Einstein's wildly unexpected revelation of what gravity actually is?

Categorization and Analogy-making as the Roots of Thinking

The idea that we will here defend is that a certain mental phenomenon subsumes all the aforementioned stereotypes of categorization and analogy, but is much broader than any of them are, taken in isolation. To give a foretaste of this crucial idea, we turn once again to the theme of zeugmas, because these linguistic oddities have a great deal to do with categorization through analogy. Indeed, zeugmas provide a rich wellspring of examples running the full gamut from the most mundane to the most inspired of analogies; in their own small way, then, zeugmas perfectly reflect the ubiquity and uniformity of the mechanism of categorization by analogy.

Suppose you heard someone say, "The asparagus tips and the potato dumplings were delicious." Your ever-ready zeugma detector wouldn't register a thing, because it would seem self-evident that, in this context, *asparagus tips* and *potato dumplings* simply belong to one and the same very standard category (namely, that of *scrumptious edibles*). But it would feel rather different if someone were to say, "The asparagus tips and the after-dinner witticisms were delicious", because here one senses that the adjective "delicious" has been used in two quite different senses, and so the needle on one's zeugma detector would move a bit, and as a result you would feel that a slight analogical link had been suggested between the asparagus tips and the postprandial quips. Then again, were you to hear "The asparagus tips and the expression of surprise on Anna's face were delicious", your zeugma detector would register a yet higher reading, meaning that the semantic distance (or interconceptual stretch) was yet greater; this would lead you to see and feel an analogy between the asparagus tips and a certain friend's facial expression, rather than merely thinking that they both belong to the commonplace category of *delicious things*.

In brief, it is misleading to insist on a clear-cut distinction between analogy-making and categorization, since each of them simply makes a connection between two mental entities in order to interpret new situations that we run into by giving us potentially useful points of view on them. As we will show, these mental acts cover a spectrum running from the humblest recognition of an object to the grandest contributions of the human mind. Thus analogy-making, far from being merely an occasional mental sport, is the very lifeblood of cognition, permeating it at all levels, ranging all the way from mundane perceptions ("That is a table") to subtle artistic insights and abstract scientific discoveries (such as general relativity). Between these extremes lie the mental acts that we carry out all the time every day — interpreting situations, judging the quality of various things, making decisions, learning new things — and all these acts are carried out by the same fundamental mechanism.

All of these phenomena seem quite different, but underlying them all there is just one single mechanism of nonstop categorization through analogy-making, and it operates all along the continuum we've described, which stretches from very mundane to very sophisticated acts of categorization. And it's this unified mechanism that allows us to understand sentences that run the gamut of zeugmaticity, from complete non-zeugmas (requiring only mundane categorization mediated by very basic analogy-making) to extreme zeugmas (requiring unusually flexible categorization mediated by much more sophisticated analogy-making).

But let's take our leave of zeugmas and return to the larger picture. We claim that cognition takes place thanks to a constant flow of categorizations, and that at the base of it all is found, in contrast to *classification* (which aims to put all things into fixed and rigid mental boxes), the phenomenon of *categorization through analogy-making*, which endows human thinking with its remarkable fluidity.

Thanks to categorization through analogy-making, we have the ability to spot similarities and to exploit these similarities in order to deal with the new and strange. By connecting a freshly encountered situation to others long ago encountered, encoded, and stored in our memory, we are able to make use of our prior experiences to orient ourselves in the present. Analogy-making is the cornerstone of this faculty of our minds, allowing us to exploit the rich storehouse of wisdom rooted in our past — not only labeled concepts such as *dog*, *cat*, *joy*, *resignation*, and *contradiction*, to cite just a random sample, but also unlabeled concepts such as *that time I found myself locked outside my house in bitterly freezing weather because the door slammed shut by accident.* Such concepts, be they concrete or abstract, are selectively mobilized instant by instant, and nearly always without any awareness on our part, and it is this ceaseless activity that allows us to build up mental representations of situations we are in, to have complex feelings about them, and to have run-of-the-mill as well as more exalted thoughts. No thought can be formed that isn't informed by the past; or, more precisely, we think only thanks to analogies that link our present to our past.

The Rapid Inferences that Categories Provide

A term that will be useful to us in this context is *inference.* As is traditional in psychology, we will use the term much more broadly than it is used in the field of artificial intelligence, where it is synonymous with "formal logical deduction", as carried out by so-called "inference engines". By contrast, what we will mean by "making an inference" is simply the introduction of some new mental element into a situation that one is facing. Basically, this means that some facet of a currently active concept is lifted out of dormancy and brought to one's attention. Whether this new element is right or wrong is not the point, nor does it matter whether it follows logically from prior elements. For us, "inference" will simply mean the fact that some new element has been activated in our mind.

Thus if one sees a child crying, one infers that the child is distressed. If one sees someone shouting, one infers that the person is probably angry. If one sees that the table is set, then one infers that a meal may well soon be served. If one sees a door that is closed, one infers that it can be opened. If one sees a chair, one infers that one could sit on it. If one sees a dog, one has the ability to infer (though one does not necessarily do so) that it barks now and then, that it might bite someone, that it has a stomach, a heart, two lungs, and a brain — internal organs that one doesn't strictly perceive but that category membership allows one to infer. Inferences of this sort are a crucial contribution to thought, and they come from categorization through analogy, for we rely ceaselessly on resemblances perceived between the present situation and ones we encountered earlier. If we did not do this at all times, we would be helpless.

Thus, it is not merely for idle fun that one calls a cat-like thing that one encounters "cat", thereby assigning it to a preexisting category in one's memory; it is principally because doing so gives one access to a great deal of extra information, such as the likely fact that it will show pleasure by purring, that it has a propensity to chase mice, that it may well scratch when threatened, tends to land on its feet, has a very autonomous character... These kinds of things, among others, can all be inferred about an entity once it has been assigned to the category *cat*, without any of them having been directly observed about the specific entity in question. Thus our categories keep us well informed at all times, allowing us to bypass the need for direct observation. If we didn't constantly extrapolate our knowledge into new situations — if we refrained from making inferences — then we would be conceptually blind. We would be unable to think or act, doomed to permanent uncertainty and to eternal groping in the dark. In short, in order to perceive the world around us, we depend just as much on categorization through analogy as we do on our eyes or our ears.

Analogy's Champions and Detractors

Some ancient philosophers, including Plato and Aristotle, were fervent defenders of analogy, seeing it as a fertile medium for thinking rather than as just a figure of speech. Nonetheless, these same thinkers felt compelled to point out its limitations. Thus Plato, using a number of analogies — among them one likening a soul to a city, in his famous work *The Republic* — warned that "likeness is a most slippery tribe". And Aristotle, although just as great an admirer of analogies, cast aspersions on many analogies made by his predecessors. Thus we see that even for some of its strongest backers, analogy has a faintly suspicious aroma, as does its cousin, metaphor. In the minds of such doubters, these two figures of speech, when used ill-advisedly, are liable to mislead both those who utter them and those who hear them.

Immanuel Kant and Friedrich Nietzsche had extremely different personalities, philosophies, and views about religion, but they were united in their unswaying belief in analogy. For Kant, analogy was the wellspring of all creativity, and Nietzsche gave a famous definition of truth as "a mobile army of metaphors". However, analogy has certainly not had such good press universally. Indeed, it's been a favorite pastime down through the centuries to berate analogy for its unreliability, its closeness to wild guessing, and the serious traps into which it leads anyone who depends on it. Some philosophers have had quite a field day denouncing analogy and metaphor, describing them as superficial, misleading, and useless forms of thought.

In particular, the empiricists in the seventeenth century and the positivists in the twentieth raked analogy and metaphor over the coals. The English philosophers Thomas Hobbes and John Locke are often quoted in this regard. Hobbes, in *Leviathan*, his best-known work, declares his love for clear words and his scorn for metaphors:

> [T]he light of human minds, is perspicuous words, but by exact definitions first snuffed and purged from ambiguity; ... [M]etaphors, and senseless and ambiguous words, are like *ignes fatui*; and reasoning upon them is wandering amongst innumerable absurdities.

Hobbes leaves no doubt as to his views. Truth is light, words must be cleansed and purged of ambiguity, and metaphors are nothing but will-o'-the-wisps that would lead one to wander in a wacky world. However, if one stops to look at this passage for a moment, one is struck by a certain ironic quality — specifically, the fact that its author condemns metaphors not by using "snuffed and purged definitions" but through the repeated use of metaphors. After all, what kind of phrase is "the light of the mind"? What about "definitions that have been snuffed and purged"? And how about "wandering amongst innumerable absurdities"? What are all these phrases, if not metaphors? Does a mind really contain light? Can definitions actually be cleansed? And are metaphors in truth unpredictably flickering lights hovering above a swamp?

In his protest, Hobbes is a bit like someone who screams in order to praise silence, or like evangelistic television preachers who whip up the masses by speaking of the sins that lead straight to hell while themselves engaging in the very acts of debauchery that they decry. His protest is also reminiscent of a paradoxical phrase that encapsulated the tragedy of the Vietnam war: "We destroyed the village in order to save it." In short, Hobbes undermines his anti-metaphor credo by expressing it metaphorically.

The eleventh-century Benedictine monk Alberic of Monte Cassino never knew anything approaching the fame of Hobbes, but he too wrote a virulent diatribe against the use of metaphors in his book *The Flowers of Rhetoric.* Here is an excerpt:

> Expressing oneself with metaphors has the quality of distracting a person's attention from the specific qualities of the object being described; in one manner or another, this distracting of attention makes the object resemble something different; it dresses it, if one may put it thusly, in a new wedding dress, and in so dressing it, it suggests that a new kind of nobility has been accorded to the object... Were a meal were served in this fashion, it would disgust and nauseate us, and we would discard it... Consider that in one's enthusiasm for giving pleasure through delicious novelty, it is unwise to begin by serving up flapdoodle. Be careful, I repeat, when you invite someone in the hopes of giving pleasure, that you not afflict him with so much malaise that he will vomit from it.

As we glide from "dressing an object in a new wedding dress" to "serving up flapdoodle" and "vomiting from it", we are treated to one metaphor after another in a passage written for no other purpose than to criticize the use of metaphors.

Eight centuries later, Gaston Bachelard, a highly respected French philosopher of science, did not completely avoid the same trap when he wrote: "A science that accepts images is, more than any other, a victim of metaphors. Consequently, the scientific mind must never cease to fight against images, against analogies, against metaphors." But how can science become a "victim", and how can a mind, scientific or otherwise, "constantly fight" against anything, unless they do so metaphorically?

Are Analogies Seductive and Dangerous Sirens?

And so, are analogies like seductive and dangerous siren songs, likely to lead us astray, or are they more like indispensable searchlights, without which we would be plunged in total darkness? If one never trusted a single analogy, how could one understand anything in this world? What, other than one's past, can one rely on in grounding decisions that one makes when facing a new situation? And of course all situations *are* in fact new, from the largest and most abstract ones down to the tiniest and most concrete ones. There isn't a single thought that isn't deeply and multiply anchored in the past.

To use the elevator in an apartment building that one has never been in before, does one not tacitly depend on the analogy with countless elevators that one has used before? And when one examines this analogy, one sees that, despite its seeming blandness, it depends on numerous others. For example, once you've entered the elevator, you have to choose a small button you've never seen before, and you have to press it with a certain finger and a certain force, and you do that without thinking about it whatsoever (or more accurately, without noticing that you are thinking about it). This means that you are unconsciously depending on your prior experiences with thousands of buttons in hundreds of elevators (and also buttons on keyboards, stereo systems, dashboards, etc.), and that you are working out the best way to deal with this new button by relying on an analogy between it and your personal category *button.*

And when, after you've stepped out of the elevator and are just setting foot in the sixth-floor apartment, you see a big dog coming towards you, how do you deal with this situation if not on the basis of your prior experience with dogs, particularly large dogs? And much the same could be said for when you wash your hands in the sink that you've never seen before with soap that you've never touched before — not to mention the bathroom door, the doorknob, the electric switch, the faucet, the towel, all never before seen or touched.

And if you go into a grocery store that you've never seen before and are looking for the sugar or the olives or the paper towels, where do you go? Which aisle, which shelf, and how high up on the shelf? Without any conscious effort, you recall "the" spot where these articles are found in other familiar stores. Of course you're not thinking of just *one* place, but of a collection of various places that you mentally superimpose. You think, "The sugar should be around *here*", where the word "here" refers simultaneously to a collection of small areas in various familiar grocery stores and also to a small area in the new store, and it's "right there" that one looks first of all.

How mundane is the scene of an employee who, requesting an extra day of vacation, says to her boss, "Last year you offered an extra weekend to Katyanna, so I was wondering if you would be able to give me just one extra day next month…" How could one do anything in life if one felt that it was crucial to be constantly on the alert in order to mercilessly squelch any resemblance that came to mind at any level of abstraction or concreteness? And worse yet, once we'd squelched them all, what would we then do? On what basis would we make even the tiniest decision?

Might there be a rigorous proof that all analogies are dubious? Obviously not, because, as we just saw, everyone depends, without thinking, on a dense avalanche of mini-analogies between everyday things, and these mini-analogies follow on the heels of one another all day long, day in, day out — and seldom do such mundane analogies mislead anyone. Indeed, if they did, we would not be here to tell the tale.

Giant Electronic Dunces

How can computers be so terribly stupid, despite being so blindingly fast and having such huge and infallible memories? Contrariwise, how can human beings be so insightful despite being so limited in speed and having such small and fallible memories? Though perhaps hackneyed, these are reasonable and important questions, focusing as they do on the nearly paradoxical quality of human thought.

Indeed, the human mind, next to a computer, appears fraught with defects of every sort, coming off as hopelessly inferior along most dimensions of comparison. For instance, in carrying out pure reasoning tasks, well-polished computer algorithms reach logically valid conclusions virtually instantly, while people tend to fail most of the time. Much the same can be said about large amounts of knowledge. Where people's minds are saturated after only a few pieces of information are presented, a computer can take into account a virtually unlimited amount of information. And of course human memory is notoriously unreliable; whereas computers never forget and never distort, those are activities at which we human beings excel, for better or for worse. Three days, three weeks, three months, or three years after we've seen a movie or read a book, what details of it remain accessible in our minds? And how distorted are they? We might also mention the speed at which processing takes place in computers as opposed to human brains. What might take us minutes, hours, or far longer can be done by a computer in an infinitesimal eyeblink. Just consider simple arithmetical calculations such as "3 + 5" (a bit under a second for a person), or "27 + 92" (perhaps five or ten seconds), or "27 x 92", a calculation that most people could not carry out in their heads. Counting the number of words in a selected passage of text and correcting a multiple-choice exam are activities that we humans can carry out, but only with pathetic slowness compared to computers.

Overall, the comparison is extremely lopsided in favor of computers, for, as we just noted, computers carry out flawless reasoning and calculation way beyond human reach, handle unimaginably larger amounts of information than people can handle, do not forget things over short or long time scales, do not distort what they memorize, and carry out their processing at speeds incomparably greater than that of the human mind. In terms of rationality, size, reliability, and speed, the machines we have designed and built beat us hands down. If we then add to the human side of the ledger our easily distractable attention, the fatigue that often seriously interferes with our capacities, and the imprecision of our sensory organs, we are left straggling in the dust. If one were to draw up a table of numerical specifications, as is standardly done in comparing one computer with another, *Homo sapiens sapiens* would wind up in the recycling bin.

Given all this, how can we explain the fact that, in terms of serious thought, machines lag woefully behind us? Why is machine translation so often inept and awkward? Why are robots so primitive? Why is computer vision restricted to the simplest kinds of tasks? Why is it that today's search engines can instantly search billions of Web sites for passages containing the phrase "in good faith", yet are incapable of spotting Web sites in which the *idea* of good faith (as opposed to the string of alphanumeric characters) is the central theme?

Readers will of course have anticipated the answer — namely, that our advantage is intimately linked to categorization through analogy, a mental mechanism that lies at the very center of human thought but at the furthest fringes of most attempts to realize artificial cognition. It is only thanks to this mental mechanism that human thoughts, despite their slowness and vagueness, are generally reliable, relevant, and insight-giving, whereas computer "thoughts" (if the word even applies at all) are extremely fragile, brittle, and limited, despite their enormous rapidity and precision.

As soon as categorization enters the scene, the competition with computers takes on a new kind of lopsidedness — but this time greatly in favor of humans. The primordial importance of categorization through analogy in helping living organisms survive becomes obvious if one tries to imagine what it would be like to "perceive" the world in a manner entirely devoid of categories — something like how the world must appear to a newborn, for whom each new concept has to be acquired from scratch and with great difficulty. By contrast, seeing the new in terms of the old and familiar allows one to benefit, and at only a slight cognitive cost, from knowledge previously acquired. Thus, if there were two creatures, one of which (an adult human being) perceived the world using categorization through analogy while the other (a computer) had no such mechanism to help it out, their competition in understanding the world around them would be comparable to a race between a person and a robot to climb up to a high roof, with the human allowed to use a preexistent staircase but with the robot required to construct its own staircase from scratch.

Analogy Operating at All Levels

Categorization through analogy drives thinking at all levels, from the smallest to the largest. Consider a conversation in which several hierarchical linguistic levels are continually interacting. First of all, the choice of a specific word will of course determine the sounds that make it up; similarly, when one is typing at one's keyboard, each word chosen determines the letters composing it, so that they come along automatically rather than being chosen one by one. Analogously, words are often determined by larger structures of which they are but pieces. This happens most clearly whenever one uses a stock phrase (such as "so to speak" or "cut to the chase" or "down to the wire" or "when push comes to shove" or "as easy as stealing candy from a baby"), but it also often happens when no such expression is involved, because one is always working under the constraints of the syntactic and semantic patterns of the language one is speaking, as well as those of one's own habitual speech patterns.

And the same principle holds at more global levels of speech as well. Thus when one writes or utters a sentence, many of the words comprising it come along without being chosen one by one, since they are all serving a higher-level goal that has been pre-selected. Thus, much as with letters being constrained by a word, the words are in a sense constrained by higher-level thoughts. And then, moving yet further upwards, we can say that the same holds when one is developing an idea; that is, the sentences one produces to express this idea are once again constrained by a yet higher-level structure, even if there is more freedom at this level than at the letter-choice level. And the same holds at the level of the conversation itself, because its overall topic, its tone, the particular people involved in it, and so forth, all constrain the ideas that will be thought of. Of course at this level, there is much more flexibility than at the level of letters composing words. And so, in summary, a conversation constrains the ideas in it, the ideas constrain the sentences, the sentences constrain the phrases, the phrases constrain the words, and finally, the words constrain their letters.

Our claim that choices on each of these levels are carried out by categorization by analogy runs against the naïve image of categories as corresponding, more or less, to single words. To be sure, some categories are indeed named by words, but others are far larger, residing essentially at the level of an entire conversation.

For example, consider arguments about the size of the military budget. Those who advocate a large budget frequently trot out the same old arguments over and over again, based on the vital need to protect our nation against unnamed threats of all sorts, the intense pressure to develop ever newer technologies, the idea that advances in military technology help to drive the civilian marketplace, and so forth. Such a line of reasoning can be spun out over a long time, while always depending on a well-known, even hackneyed, conceptual skeleton that has been "seasoned to taste", depending on the context, the occasion, and so forth. But whatever the variations on the theme are, it's always the same conceptual skeleton centered on the need for national defense and for advances in technology. The high-level category determining the overall flow of one's argument is defined by this conceptual skeleton.

Conversely, advocates of trimming the military budget will almost invariably cite the enormous importance of other sectors of the economy and the great inefficiencies in the military. Here once again, such arguments can be spun out at great length, but however they run, they will always be centered on the bloatedness of the current military budget and the crying need for funding other sectors of the economy. This is the familiar conceptual skeleton that will guide the overall flow of the argument.

And thus we see that at the top level of the conversation, we are dealing with the very high-level categories *need for a bigger military budget* and *need for a smaller military budget*, and the activation of either category in an advocate's mind will trigger, with a bit of variability, yet also a considerable degree of predictability, the auxiliary ideas that will pop to mind, and these will promptly enlist appropriate stock phrases and well-worn grammatical patterns, which will in their turn call up the standarized words that comprise them — and these words, in the end, will call up, with essentially no maneuvering room, the letters or sounds that make them up.

One can examine any conversation, whether it's a deep or a shallow one, in this fashion, and one will see how analogies, at all the different levels, are in the driver's seat. Here's a rather light-hearted example based on a real event.

One Saturday evening, Glen and Marina Bayh had a few friends over for dinner. The food was savory, the wine and witticisms flowed copiously, and at last, around midnight, people started rising to get their coats. As they were filing out the front door, Larry Miller, one of the guests, said warmly to Glen and Marina, "It was a terrific evening. Haven't had so much fun in a long time. Thanks a lot. Hope to see you again soon. Bye-bye!" On hearing this innocent remark, Jennifer, another guest, commented, "I always have a hard time saying good-bye to them." Larry, puzzled, replied to her, "But all good things come to an end. We had a great time, it finished, and now we're taking off. What's the big deal about saying bye?" Jennifer answered, "Yeah, you're right, but still it sounds weird, because 'Bayh' is their last name. I mean, for them to hear 'Bye' or 'Bye-bye' all the time must be a bit like it would be for you to hear people saying 'Miller, Miller' all the time, no? That would come across as strange to you, wouldn't it?" Larry burst out laughing and said, "I guess I'm just dense! I'd never thought of that!" Right then, Larry's wife Colleen chimed in, saying, "It reminds me of when I was a teen-ager and every time my parents took me and my brother to our grandmother's house, he and I would always whisper to each other, 'Now don't forget — you have to mind your gramma around here!' We always called her 'Gramma', and she was always correcting our English, so this was our private little way of getting back at her, though she never knew it." Everyone could easily relate to Colleen's story, and of course they all understood why her comment wasn't a *non sequitur* coming out of left field but a perfectly apt segue.

What are the key analogies behind the scenes of this down-to-earth interchange? It was launched by an analogy between the sounds of the words "Bayh" and "bye", which spurred Jennifer to invent an analogy between the last names "Bayh" and "Miller", after which the surface-level topic veered off to visits paid decades earlier to the home of a relative of Colleen's, but still driven by the momentum of analogy — this time an analogy between the name "Gramma" and the word "grammar". At a higher level, however, the trigger that sparked the retrieval of this dormant memory in Colleen's mind was the similarity between a fresh episode and a long-ago episode, both of which involved humorous phonetic resemblances between a normal word and someone's name — in one case, that of "bye" to "Bayh" and in the other case, that of "grammar" to "Gramma". So here we're dealing with a similarity between resemblances — which is to say, with an analogy between analogies.

There is nothing unusual about this conversation or this type of analogy-spotting behind the scenes. It's all par for the course. We exhibited it simply to show how a conversation as a whole mobilizes one or two brigades at a very high conceptual level, how those high-level concepts mobilize a few lower-level conceptual regiments, how these in turn mobilize a larger number of conceptual platoons or squadrons more or less at the level of stock phrases, and finally, how these many "smaller" concepts mobilize hundreds of individual soldiers way down at the word level.

Abstract or Concrete?

What lies behind this universality of analogy-making? In order to survive, humans rely upon comparing what's happening to them *now* with what happened to them in the past. They exploit the similarity of past experiences to new situations, letting it guide them at all times in this world. This incessant flow of analogies, made in broad brushstrokes, forms the crux of our thoughts, and our utterances reflect them, although our specific word choices are usually fast forgotten. The concrete meets the abstract when a down-to-earth phrase is applied to describe a down-to-earth situation but where the concepts that the phrase is built from are distant, on a literal level, from the situation. For instance, in an idiomatic utterance such as "Marie is off her rocker" or "Their love affair went down the drain", the thought is at such a high level of abstraction that one seldom will consciously envision someone falling off a rocking chair or water flowing out of a sink or bathtub.

Much the same happens when a new situation reminds you of another situation (or family of situations) that you previously encountered and that is, on its surface, totally different, but that shares an abstract essence with the new one. Thus if one day your child is not allowed to register for a crucial event at school because the relevant Web site's deadline was at 4 o'clock sharp and you logged onto the site at 4:01, this may summon up a long-buried memory from some fifteen years earlier of a time when you missed a plane because after dashing to the gate, you arrived just seconds after the doors had been closed, and no matter what you said, they wouldn't let you board.

Our daily talk is filled with this kind of meeting of concrete and abstract, but we are unaware of it most of the time. Thus if a professor says, "Only a handful of students dropped in on me in my office yesterday", we aren't likely to envision students tumbling out of a giant hand in the sky and landing in the professor's office. And if someone says, "There hasn't been any snow to speak of today", we don't feel inclined to protest, "What do you mean? You just *spoke* of it!" The concreteness of the words and phrases that we constantly use to formulate our thoughts on all sorts of topics is at one and the same time a sign of the great concreteness of our way of thinking and also a sign of our extraordinary propensity to carry out abstraction, allowing us to cast a situation using words that would seem, on their surface, to refer to totally unrelated things.

Thus a Japanese stockbroker, commenting on the unstoppable collapse of the stock market, said, "One should never try to catch a falling knife." Most people understand this image effortlessly, as well as its relevance to the circumstances. And yet if there is a falling knife here, it's certainly not a knife in its most ordinary sense, and the way this "knife" is falling is invisible, comparatively slow, and not spatially localized. It took a considerable act of abstraction for that individual to use that phrase in that context — but that's nowhere close to the end of the story, for the same phrase might just as easily be applied to a politician in the throes of a corruption scandal and whose once-ardent supporters are suddenly nowhere to be seen, or to a skyscraper on fire that it would be folly to enter, or to someone so deeply depressed that even their best friends are caught up in the atmosphere of bleak pessimism, or to someone who has fallen overboard in a storm so violent that no one dares to go to their aid, or to an approaching hurricane that has destroyed a nearby town and from which everyone has been ordered to evacuate immediately, or perhaps, who knows, even to a person who was injured in their kitchen because they tried (of all things) to catch a falling knife. In short, we see that here we are dealing with a full-fledged category, rich and multifaceted.

As soon as one starts thinking about situations to which the phrase "One should never try to catch a falling knife" could apply, they start to pop up from under nearly every stone in sight. At least for a while, one has the impression that one could paint a large fraction of the world in terms of that phrase alone, since the world is rife with huge, irrepressible forces against which one has no power and which would carry one off to one's doom if one were so rash as to try to stop them. Thus we will find this image jumping to our mind willy-nilly, imposing itself on us whether we like it or not, and unless we can somehow stop thinking altogether, we will simply have to let this aggressive metaphor have its way with us until it has had its fill; after all, one shouldn't try to catch a falling knife.

Synopsis of This Book

In the chapters that follow, we don't aim to speak about the brain at a biological level, but about cognition as a psychological phenomenon. We will not speculate about the cerebral or neural processes that underlie the psychological processes we describe, because our goal is not to explain cognition in terms of its biological substrate but to present an unconventional viewpoint concerning what thought itself is. Our discussion will thus take place at this rather abstract level, but even at that level there will be plenty of grist for the mill.

The book's first three chapters constitute our attempt to provide an account of what categories and analogies are. Chapter 1 focuses on categories associated with single words, and it also puts forth some of the key theses of the book. We show how concepts designated by a single word are constantly having their boundaries extended by analogies. We take a careful look at the development of concepts by observing the long progression that starts with the concept of a child's *Mommy* (a specific adult human) and that gradually leads to all sorts of metaphorical uses such as *motherland*, passing en route through such concepts as *birth mother* and *surrogate mother.* We also show that less concrete words, such as "thanks", "much", "to fix", "to open", "but", "and" (and so on) are, no less than nouns, the names of mental categories that are the outcome of a lifelong series of analogies.

In Chapter 2, we study concepts having lexical labels that are longer than single words. We show that hidden behind the scenes of multi-word stock phrases, even ones as long as a proverb or a fable, there lie concepts that are very similar to those designated by isolated words. Thus a phrase such as "Achilles' heel" is the linguistic "hat" worn by a particular category (namely, the category of *serious weaknesses that may lead to someone's undoing*). Æsop's famous fable in which a fox tries to reach some luscious-looking grapes and when he fails, he declares that he didn't want them anyway because they are sour, is a linguistic embodiment of the abstract mental category of situations featuring something that is the object of someone's ardor, but that, having turned out to be out of reach, is subsequently deprecated by the person who desired it. This abstract quality, often concisely called "sour grapes", is potentially recognizable in thousands of situations, and this phrase could thus be used as the verbal label of any such situation, in just the same way as there are myriads of objects meriting the label "bottlecap" and myriads of actions meriting the label "retrieve". And the same can be said of more abstract categories, some of which have to do with the act of communication taking place at the moment, and which are labeled by adverbial phrases, such as "after all", "on the other hand", "as a matter of fact", "that having been said", and so on. In other words, there are situations in our everyday interchanges that call for the label "after all", and when such situations arise, we recognize them (almost always unconsciously) as such, and we apply that label, deftly inserting it into our real-time speech stream. The chapter concludes with a discussion of intelligence as the ability to put one's finger on what counts in any given situation, and how the repertoire of categories that is handed to one by one's native language and culture tailors one's way of doing this.

Chapter 3 deals with categories for which there is no standard linguistic label; people manufacture such categories spontaneously on their own as they deal with new situations in their complex personal worlds. Later on, such categories often give rise to "reminding episodes", where one event recalls another from another time and place, possibly very distant. As an example, when D. noticed an old friend leaning down to pick up a bottlecap in Egypt's renowned Karnak Temple, he was suddenly and spontaneously reminded of a time, some fifteen years earlier, when his one-year-old son was sitting near the edge of the Grand Canyon and, completely oblivious to the spectacular scenery, was intently focused on some ants and leaves on the ground. Despite all the superficial differences that can be found separating any two situations from each other, when such a reminding incident takes place, it reveals that the two situations in question share a conceptual skeleton at a deeper level, and it shows how extremely rich and subtle is our storehouse of non-lexicalized concepts. By analyzing a series of sentences containing such high-frequency phrases as "me too", "next time", and "like that", we show that lurking behind any such phrase, no matter how casual and simple it may seem, there is a non-verbalized category, sometimes simple and sometimes subtle, based on an implicit perception of sameness, which is to say, based on an analogy.

Chapter 4 deals with the way in which, in our interactions with the world around us, we constantly and fluently move about in our repertoire of categories, and yet nearly always without the least awareness of doing so. The chapter focuses particularly on inter-category leaps that involve shifts between levels of abstraction. The flexibility of human cognition relies profoundly on our ability to move up or down the ladder of abstraction, for the simple reason that sometimes it is crucial to make fine distinctions but other times it is crucial to ignore differences and to blur things together in order to find commonalities. For instance, while one is dining, one will take care to distinguish between one's own glass and that of one's neighbor, but afterwards, when one is placing them in the dishwasher, that distinction will be irrelevant. As another example, parents will try to assure that their children get involved in "activities", whether this means acting in plays, doing judo, or playing a musical instrument. *Activity for my child* is a highly abstract category. The most humble of our acts conceals choices of abstraction that are hard for us to recognize accurately, because such acts are central to cognition. We have a very hard time "seeing" our cognitive activity because it is the medium in which we swim. The attempt to put our finger on what counts in any given situation leads us at times to making connections between situations that are enormously different on their surface and at other times to distinguishing between situations that on first glance seem nearly identical. Our constant jockeying back and forth among our categories runs the gamut from the most routine behaviors to the most creative ones.

Chapter 5 is devoted to the role of analogy in very ordinary, everyday situations. It deals with analogies that, because they are essentially invisible, manipulate us. We are unaware of being taken over by an analogical interpretation of a situation. In this sense, the invisible analogy manipulates us because it has simply imposed itself on us, willy-nilly. And it manipulates us also in another sense — namely, it foists new ideas on us, pushing us around. Unsatisfied with being merely an agent that enriches our comprehension of a situation we are facing, the analogy rushes in and structures our entire view of the situation, trying to make us align the newly encountered situation with the familiar old one. For instance, when a small private plane crashed into a building in Manhattan on October 11, 2006, the analogy with the events of September 11, 2001 was irrepressible, leading instantly to speculations of terrorism, even though the building was not seriously damaged; the Dow Jones average even took a noticeable nosedive for a short while. And thus analogies just jump in uninvitedly, thinking and making decisions for us, without our being aware of what is going on.

In Chapter 6, by contrast, we deal with analogies that, in some sense, we ourselves manipulate — analogies that we freshly and deliberately construct when we run into a situation that arouses our interest, sometimes in order to explain it to ourselves or to others, sometimes to argue for our own point of view. This is especially the case for what we have dubbed *caricature analogies.* These are analogies that one dreams up on the spur of the moment in order to convince someone else of an idea in which one believes. They transpose a situation into a new domain while exaggerating it. For instance, a scientist seeking a job abroad wrote to a colleague: "I love my native land, but trying to get research done here is like trying to play soccer with a bowling ball!" Also discussed in this chapter is the way in which political decisions at all levels flow from analogies perceived by decision-makers between current situations and historical events, our main case study being some of the key analogies that shaped the Vietnam war. A few studies of inter-language translation conclude this chapter, focusing on the analogies used by skilled people in order to create coordinated parallelisms between two languages and two cultures on many scales, ranging from the very small to the very large.

Chapters 7 and 8 deal with analogy in scientific thinking. Chapter 7 is concerned with what we call "naïve analogies" — in particular, the kinds of analogies on which nonspecialists tend to base their notions of scientific concepts. We show that notions that one picks up in school, whether in mathematics, physics, or biology, are acquired thanks to appealing and helpful but often overly simple analogies to concepts with which one is already familiar. Thus an elementary arithmetical operation such as division, which supposedly is totally under one's belt by the time one starts middle school, is generally still rooted (even in the case of most university students) in a naïve analogy with the down-to-earth operation of *sharing* (as in the act of distributing 24 candies evenly to 3 children). To be sure, sharing is quite often a perfectly good way to look at division, but the view that it affords of the phenomenon is overly narrow. For example, this naïve view of division makes it very hard for people to devise a word problem that involves a division whose answer is *larger*, rather than *smaller*, than the quantity being divided. This chapter analyzes the implications, both positive and negative, of naïve analogies for education.

Chapter 8 then looks at the extreme other end of the spectrum — namely, how great discoveries are made by insightful scientists. We show how the history of mathematics and physics consists of a series of snowballing analogies. By examining from up close certain great moments in the history of these disciplines, we reveal the crucial role played over and over again by analogies — sometimes very obvious ones, sometimes very hidden ones. In particular, the deep analogies of Albert Einstein play a starring role in this chapter, including a little-known analogy that led to his hypothesis in 1905 that light consists of particles, an idea that was mightily resisted by the entire physics community for nearly two decades. The most carefully examined historical episode is that of Einstein's own slow and gradual process of coming to understand the various levels of meaning of his celebrated equation "$E = mc^2$".

The epilogue to our book is a dialogue — thus it is entitled "Epidialogue" — in which categorization and analogy-making are compared and contrasted along many dimensions, and although at first the two processes may seem very different, at the end of this careful comparison, the spirited debaters conclude that there is no difference between them, and they realize that in fact they are one and the same.

CHAPTER 1

The Evocation of Words

❧ ❧ ❧

How do Words Pop to Mind?

At every moment we are faced with a new situation. Actually, the truth is much more complicated than that. The truth is that, at every moment, we are simultaneously faced with an indefinite number of overlapping and intermingling situations.

In the airport, we are surrounded by strangers whom we casually observe. Some seem interesting to us, others less so. We see ads everywhere. We think vaguely about the cities whose names come blaring out through loudspeakers, yet at the same time we are absorbed in our private thoughts. We wonder if there's time enough to go get a frozen yogurt, we worry about the health problems of an old friend, we are troubled by the headline we read in someone's newspaper about a terrorist attack in the Middle East, we smile to ourselves at a clever piece of wordplay in an ad on a television screen, we are puzzled as to how the little birds flying around and scavenging food survive in such a weird environment... In short, far from being faced with *one* situation, we are faced with a seething multitude of ill-defined situations, none of which comes with a sharp frame delineating it, either spatially or temporally. Our poor besieged brain is constantly grappling with this unpredictable chaos, always trying to make sense of what surrounds it and swarms into it willy-nilly.

And what does "to make sense of" mean? It means the automatic triggering, or unconscious evocation, of certain familiar categories, which, once retrieved from dormancy, help us to find some order in this chaos. To a large extent, this means the spontaneous coming to mind of all sorts of *words*. Without any effort, one finds oneself thinking, "cute little girl", "funny-looking coot", "same dumb ad as at the airport I was at yesterday", "an Amish family", "sandals", "what's she reading?", "who's whistling?", "where is their nest?", "when are we going to board?", "what an annoying ring tone", "how could I have left my cell-phone charger at home?", "and I did it last time, too", "the air-conditioning is on too high in here", and so on.

All these words! No experience is more familiar to us than this ceaseless barrage of words popping up in our mind extremely efficiently and without ever being invited. But where do these words come from, and what kind of invisible mechanism makes them bubble up? What is going on when one merely thinks silently to oneself, "a mother and her daughter"?

It All Starts with Single-member Categories

To be able to attach the label "mother" to some entity without thinking about it, one has to be intimately familiar with the concept *mother*, which is denoted by the word. For most of us, this intimacy with the concept goes all the way back to our earliest childhood, to our first encounters with the notion. For one-year-old Tim, the core of the concept is clearly his own mother — a person who is much bigger than he is, who feeds him, comforts him when he cries, sings him lullabies, picks him up, plays with him in the park, and so forth. Once this first mental category bearing the name "Mommy" has a toehold, Tim will be able to see that in the world around him there are similar phenomena, or as we prefer to put it, analogous phenomena.

We take a momentary break here to explain a typographical convention of our book. When speaking about a word, we will put it in quotation marks ("table"), whereas when speaking about a concept, we will use italics (*table*). This is an important distinction, because whereas a word is a sequence of sounds, a set of printed letters, or a chunk of silent inner language, a concept is an abstract pattern in the brain that stands for some regular, recurrent aspect of the world, and to which any number of different words — for instance, its names in English, French, and so forth, or sometimes no word at all — can be attached. Words and concepts are different things. Although the distinction between them is crucial and often very clear, there will unavoidably be cases in our text where it will be ambiguous and blurry, and in such cases, we'll make a choice between italics and quotation marks that might seem a bit arbitrary. Another source of ambiguity is the fact that here and there we'll use italics for emphasis, just as we'll use quotation marks to suggest a sense of doubt or approximation (which could sometimes be conveyed equally well by the word "so-called"), and of course we will use quotation marks when we are making a quotation. Alas, the world is simply filled with traps, but we hope that the ambiguities are more theoretical than actual. And with that said, we return to our main story.

One day in the park, Tim, aged eighteen months, sees a tot playing in the sandbox and then notices a grown-up near her who is taking care of her. In a flash, Tim makes a little mental leap and thinks to himself more or less the following (although it's far from being fully verbalized): "That person is taking care of *her* just like Mommy takes care of *me*." That key moment marks the birth of the concept *mommy* with a small "m". The lowercase letter is because there are two members of this new category now (and of course using uppercase and lowercase letters is just *our* way of hinting at what's going on in Tim's head, not *his* way). From this point on, it won't take Tim long to notice yet other instances of this concept.

At the outset, Tim's concept of *mommy* still floats between singular and plural, and the analogies in his head will be quite concrete, a comparison always being made to the *first* mommy, which is to say, with Mommy (the one with the capital "M"), but as new instances of the concept *mommy* are superimposed and start to blur in his memory, the mental mapping that Tim will automatically carry out, each time he spots a new grown-up in the park, will start to be made not onto Mommy, but onto the nascent and growing concept of *mommy* — that is, onto a generalized, stereotyped, and even slightly abstract situation, centered on a generic grown-up (*i.e.*, stripped of specific details) and involving a generic child who is near the grown-up and whom the grown-up talks to, smiles at, picks up, comforts, watches out for, and so on.

It's not our goal here to lay out a definitive theory of the growth of the specific concept *mommy*, as our purpose is more general than that. What we are proposing is that the birth of *any* concept takes place more or less as described above. At the outset, there is a concrete situation with concrete components, and thus it is perceived as something unique and cleanly separable from the rest of the world. After a while, though — perhaps a day later, perhaps a year — one runs into another situation that one finds to be similar, and a link is made. From that moment onward, the mental representations of the two situations begin to be connected up, to be blurred together, thus giving rise to a new mental structure that, although it is less specific than either of its two sources (*i.e.*, less detailed), is not fundamentally different from them.

And so the primordial concept *Mommy* and the slightly more sophisticated concept *mommy* act in very similar ways. In particular, both of them are easily mapped onto newly encountered situations "out there", which leads both of them to extending themselves outwards — a snowball effect that will continue all throughout life. It's this idea of concepts extending themselves forever through a long series of spontaneous analogies that we wish to spell out more carefully in the next few sections.

Passing from *Mommy* to *mommy* and then to *mother*

One day, Tim, who sadly has never met his father, is playing in the park, and he runs into a little girl accompanied by a grown-up who is encouraging the girl to play with the other children. He thinks to himself that this grown-up is the *mommy* of the little girl. That is, Tim's mind makes a link between what he's observing and his new concept of *mommy*. This is an act of categorization. Perhaps the new person is not actually the child's mother but the child's father, or perhaps it's her grandmother, or even her older brother or sister, but even so, that doesn't make Tim's mapping of this new person onto the category *mommy* irrational, because his notion of *Mommy*/*mommy* is wider than ours is (not richer, of course, but less discriminating, due to his lack of experience). This simple analogy Tim has made is flawless; it's just that he hasn't taken into account certain details that an adult would have used. If Sue, his mother, explains to him that this person isn't the little girl's *mommy* but her *daddy*, then Tim may well modify his concept of *mommy*, thereby coming into closer alignment with the people around him.

Gradually, as Tim uses the word "mommy" more and more, his initial image — that of his own mother — will start to recede from view, like a root being grown over ever more as time passes. He will overlay his earliest image with traits of other people whom he assigns to this mental category, and the vivid and unique features of his own Mommy will become harder and harder to find in it. Nonetheless, even when Tim is himself a grown-up, there will remain in his concept of *mommy* some residual traces of his primordial concept *Mommy*.

One day, a friendly woman who's come all the way from her home in Canada turns up and treats Tim very sweetly. He hears the word "mommy" used several times to refer to this grown-up, and so for a while he concludes that maybe he has more than one mommy. For Tim this is conceivable, since he has not yet built up a set of expectations that would rule this possibility out. Sometimes his "second mommy" takes him to the park and she, too, chats with the other mommies. But after a week or so, Tim's second mommy vanishes, which quite understandably saddens him. The next day, one of the mommies in the park asks Tim, "Did your grandmother go back home?" Tim doesn't answer, because he doesn't yet know the concept of *grandmother*. So she reformulates her question: "Where's your mommy's mom today, Tim?" But this question makes even less sense to Tim. He knows perfectly well that *he's* the one who has the mommy (he even had two of them in the past few days!), and so his mommy (that is, the remaining one) can't have a mommy. After all, it's *children* who have mommies (and sometimes also daddies) whose purpose is to be sweet to them, to watch over them, and to help them, and Tim knows that his mommy *isn't* a child, and so she doesn't have a mommy. That's obvious! The woman doesn't push her strange question, and Tim goes back to his playing.

And time passes. A few months later, Tim starts to realize that grown-ups are sometimes accompanied by other grown-ups that they refer to as their "mother". Suddenly everything starts to be clear… What children have are *mommies*, and what grown-ups have are *mothers*. That makes sense! And into the bargain, there's even an analogical bond between *mommy* and *mother*. Of course Tim isn't aware of having made an analogy — neither this concept nor the word for it will be known to him for another ten or more years! — but he has nonetheless made one. And as is often the case with analogies, this one helps clarify things for Tim but it also misleads him a little.

We now will skip over the details, simply adding that the two concepts of *mommy* and *mother* gradually merge to create a more complex concept at whose core there is the primordial concept of *Mommy*. This doesn't mean that the primordial image of Sue springs to Tim's mind every time that he hears the word "mother" or even the word "mommy", but merely that the invisible roots are structured in that manner.

As any concept grows in generality, it also becomes more discriminating, which means that at some point it's perfectly possible that some early members of the category might be demoted from membership while new members are being welcomed on board. Thus the dad at the park whom Tim had first taken for a *mommy* is stripped of the label, and although Tim's grandmother stays on as a member of the category *mother*, she winds up in a less central zone than the *mommy* zone, which is reserved for the mothers of small children. And of course as time goes by, Tim will come to understand that his grandmother herself was once a member of the category *mommy* (just as his own Mommy was once a member of the category *small child*), but at present all of that is well beyond his grasp.

The Cloud of Concepts of *Mother*

One might think that the concept of *mother* is very precise — perhaps as precise as that of *prime number.* That would imply that to every question of the form "Is X a mother or not?", there would always be a correct, objective, black-and-white answer. But let's consider this for a moment. If a little girl is playing with two dolls, one bigger and one smaller, and she says that the big one is the small one's mother, is this an example of motherhood? Does the large doll belong to the category *mother*? Or contrariwise, could one state without risk of contradiction that she does *not* belong to that category?

And if we read a certain book in which a certain Sue is described as the mother of a certain Tim, then does this Sue, who is never anything but a made-up character in a book, truly belong to the category *mother*? Does it make any difference that Sue was modeled on a real person, and Tim on her son? Is Sue more of a mother than the doll is? What indeed is Sue? If in the book it states that she is 34 years old, that she has light brown hair, that she weighs 120 pounds, that she is five feet five inches tall, and that she's the mom of a small boy, does that mean that Sue has a body and once gave birth? A doll, at least, is a physical object, but what is Sue, when you come down to it? An abstract thought triggered by some words on a page, by some black marks on a white background. Does this thought even deserve the pronoun "she"?

When Tim gets to be six, if someone tells him that Lassie is Spot's mother, he certainly won't protest, but if he were told that the queen bee is the mother of all the bees in her hive, it's less clear what he would say, and in any case some mental effort would be needed before he could absorb this idea. And if he were told that a drop of water that he just watched dividing into two drops is the mother of the two new drops, he would almost surely find this suggestion very surprising. Everyone knows phrases that use the word "mother" in ways that go far beyond the senses that apply to Lassie, the queen bee, or even the splitting drop of water — for instance, "my motherland", "a mother cell", "the mother lode", "Mother Earth", "Greece is the mother of democracy", and "Necessity is the mother of invention". Are these true instances of the concept of *mother*, genuine cases of maternity? What is the proper way to understand such usages of the word?

Some readers may feel inclined to say that these are all "metaphorical mothers", and indeed, such a viewpoint is not without merit, but we have to point out that there is no sharp boundary that separates "true" mothers from those that are metaphorical, for categories in general don't have sharp boundaries; most of the time, metaphorical and literal meanings overlap so greatly that when one tries to draw a clear boundary, one discovers that things only get blurrier and blurrier.

When he turns seven or eight, Tim will start to be able to handle phrases in which the word "mother" is used with greater fluidity than back in nursery school. He might run into the statement "Mary is the mother of the Lord Jesus" in a religious context. This is a mild extension of the usual meaning, since Mary is imagined as a woman whereas the Lord Jesus is imagined as a divine being, magical and omnipotent in some ways, even if also, in some sense, as a baby like all others. At age seven, though, Tim probably won't have much trouble envisioning Mary giving birth to the Lord Jesus.

On the other hand, having given physical birth to a baby is not a prerequisite for attributing motherhood to an entity, since even if no one ever teaches us this explicitly, we all come to know that motherhood pulls together several different properties, such as that of *female biological parent*, that of *female nurturer*, and that of *female protector*, and these properties do not all need to be present simultaneously. For example, the familiar fact of adoption reminds us that giving birth is only one possible route for becoming a mother.

If at age nine, Tim is reading a book on Egypt or on mythology and runs into the sentence "Isis is the mother of Nature", he'll have to extend his prior conceptions of motherhood at least slightly, because this time, Isis is not a human being but a deity who, in Tim's mind, looks much like a woman but in some sense is not one, and who is capable of giving birth to some rather abstract things, such as Nature, yet without anything emerging from her body. And yet Tim will rather easily absorb this new instance of motherhood, because she looks enough like hundreds of other members of the category *mother* that are already installed in his memory.

Moving right along, Tim will soon handle cases that are even more abstract, such as "Marie Curie is the mother of radioactivity", "The American revolution is the mother of the French revolution", "The American revolution is the mother of the Daughters of the American Revolution", "Judaism is the mother of Christianity", "Alchemy is the mother of chemistry", "Censorship is the mother of metaphor" (Jorge Luis Borges), "Leisure is the mother of philosophy" (Thomas Hobbes), and "Death is the mother of beauty" (a quote from Wallace Stevens, and also the title of a detailed study of the role of metaphor in thought by cognitive scientist Mark Turner).

And we can go yet further, to the idea of Nature as the mother of all living creatures ("Mother Nature"), or the idea of *Mother Superior* in a convent, or the idea of *den mother* for a Cub Scout pack, or the idea of a company that has a *mother company* from which it sprang at some earlier point, or the idea of the *mother board* in a computer, and so forth. A mother in a park, a mother in a soap opera, an adoptive mother, a den mother, a mother doll, a mother bee, a mother cell, a mother board, a mother drop of water, a mother deity, a mother company, the mother lode… Given that some mothers, such as Tim's mommy Sue, are certainly "real mothers", while others, like the mother board, are just as certainly "metaphorical mothers", the goal of drawing a sharp, objective boundary between the two distinct subcategories seems as if it might well be within reach. However, as we have shown with our list of blurry examples, such as the person in a novel, the doll mother, and the adoptive mother, that hope is but a beckoning mirage.

On the Categories and Analogies of Children

The story we've just told illustrates a central theme of our book — namely, that each category (in this book we use this term synonymously with the term "concept") is the outcome of a long series of spontaneous analogies, and that the categorization of the elements in a situation takes place exclusively through analogies, however trivial they might seem to an adult. A crucial part of this thesis is that analogies created between a freshly perceived stimulus (such as the mother of a little girl in the park, as seen by Tim) and a relatively new and sparse mental category that has only one single member (such as Tim's category *Mommy*) are no different from analogies created between a perceived stimulus (once again, take the same woman in the park) and a highly developed mental category to which thousands of analogies have already contributed (think of the very rich category *mother* in the mind of an adult).

This last statement is among the most important in our book, yet on first sight it might seem dubious. Is it really plausible that the very same mechanisms underlie the act whereby a two-year-old spots a Saint Bernard and exclaims "Sheep!" and the act whereby a physicist of great genius discovers a subtle and revelatory mapping between two highly abstract situations? Perhaps it seems implausible at first glance, but we hope to have made it convincing by the end of the book.

In the meantime, to facilitate building a pathway that will get us to this goal, we'll set up some intermediary bridges. Toward this end, it will be useful for us to take a look at a number of statements made by children, for these statements reveal hidden analogies that underlie their word choices. And so, without further ado, here is a small sampler of children's sentences, many of which were collected by developmental psychologist Karine Duvignau in her work with parents who were observing their children at home.

> Camille, age two, proudly announces: "I undressed the banana!"
>
> She talks about the banana as she would talk about a person or a doll, seeing the peel as an article of clothing that she has removed from it. The banana has thus been "laid bare" (a near neighbor of what Camille said).

> Joane, age two, says to her mother: "Come on, Mommy, turn your eyes on!"
>
> Here a little girl speaks to her mother as if she were dealing with an electrical device having an on–off switch.

> Lenni, age two, says about a broken toy: "Gotta nurse the truck!"
>
> Here, as in the case of Camille, we see a personification of an inanimate object. The truck is "sick" and so the child wants to help it "get well".

> Talia, three years old, says: "Dentists patch people's teeth."
>
> This represents the flip side of the coin, where the child speaks of something alive as if it were an inanimate object (as we just saw Joane do as well).

Jules, three years old, exclaims: "They turned off the rain!"

For Jules, rain is like a television set or a lamp that a person or people can turn on or off with a switch.

Danny, aged five, says to his nursery-school teacher: "I want to eat some water."

In this case, Danny was not speaking his native language but one he was just starting to learn, so he reached out and grabbed the nearest word he knew.

Talia, aged six, says to her mother, "Are you going to go scold the neighbors today?"

The night before, the upstairs neighbors had held a very noisy party, and her mother had told Talia that the next morning she would go knock on their door and complain about their noisemaking. In using the word "scold", Talia was unconsciously revealing her egalitarian value system: any person, whether it's an adult or a child, may sometimes have to be scolded.

Tom, aged eight, asks: "Dad, how long does a guinea pig last?"

While it's true that Tom talks here about his guinea pig in a most materialistic manner, the tenderness with which he treats his pet shows unmistakably that his category *entity of limited duration* is much broader in scope than that of most adults.

At the same age, Tom asks his parents, "How do you cook water?"

This question gets uttered when Tom has generously decided to fix some coffee for his parents one morning, but isn't sure how to start. The distinctions between such kitchen-bound concepts as *to heat up*, *to boil*, *to cook*, and *to fix* are not yet very clear in his mind, but since he announces to anyone who'll listen that he aspires to be a chef in a top-flight restaurant someday, it's to be hoped that this blur won't last too long.

Once again Tom, still eight, says to his uncle, "You know, your cigarette is melting."

This is stated when Tom's uncle is so involved in a conversation that he seems unaware that his cigarette is slowly being consumed in the ashtray. Although Tom knows cigarettes are not for consumption by children, here he links them with certain foods that he knows well, such as ice cream and candy, which can melt.

Tom tips over a wineglass, goes to get a sponge, and chirps, "Here, I'll erase it!"

Part of the tablecloth has just been colored dark, much as paper is colored by pencils or a blackboard is colored by chalk, and so to Tom it makes sense that the sponge will act as an eraser, eliminating all traces of the spilled liquid.

Mica, age twelve, asks his mother, "Mom, could you please roll up your hair?"

He wants to take a snapshot of her and what he means is, "Could you please put your hair up in a bun?", but his thought comes out in a more picturesque way.

Very similar examples are provided by Corentin, who says, "You can stop, Mom, your hair's all cooked now" (meaning it's now dry), or Ethan who observes, "I broke the book" (meaning he's torn it), or Tiffany who declares, "I want to get my nails permed" (meaning she wants a manicure), or little Alexia, who asks, "Mom, can you glue my button back on?" (of course meaning "sew it back on"), and last but not least by Joane, who poses the classic conundrum, "Do buses eat gas?"

Impressive Heights of Abstraction by Children

In each of the cases shown above, one can ask if the child actually was making an error. The key question is, what would constitute an error? If Danny knows the word "drink" but it simply doesn't come to mind, and if he realizes that "to eat" isn't really what he means to say, then saying "I want to eat water" would be an error. But if he has the feeling that what he said is perfectly fine, and if he would be surprised to hear the nursery-school teacher correct him, then we'd say that his statement was correct, at least from his own point of view. Most likely Camille, who "undressed the banana", Ethan, who "broke the book", and Alexia, who wanted the button to be "glued back on", had little or no idea of the existence of the verbs "to peel", "to tear", and "to sew". From their viewpoint, what they were saying was correct, because their concepts of *undressing*, *breaking*, and *gluing* were more inclusive than those concepts are in the mind of an adult, and they could thus be applied to situations having a wider range of diversity. For example, Ethan could almost certainly have said, given the proper conditions, "the curtains are broken", "I broke a loaf of bread", or "they broke the house".

On the other hand, it's very unlikely, even in our society, filled as it is to the brim with technological gadgets, that Joane ("Come on, Mommy, turn your eyes on!") would be familiar with the verb "to turn on" and yet unfamiliar with the verb "to open". Likewise, it's extremely unlikely that Jules ("They turned off the rain!") knows the verb "to turn off" but is unaware of the verb "to stop". And so we ask: are these children making errors, or not?

The line between what is and what is not an error is less precise than one might think. What these children are doing is making semantic approximations, stretching their personal concepts in a way that adults would not feel comfortable doing, because the concepts *to turn on*, *to turn off*, *to open*, and *to close* in these children's minds have not yet reached their adult forms — no more than (to switch from verbs to nouns momentarily) the categories *horse* and *cat* had reached their relatively stable adult stages in the mind of little Abby when, at age three, she saw some greyhounds and called them "horses", and soon thereafter saw a chihuahua and called it a "cat". The concepts silently hidden behind these words will continue to develop in the minds of all these children, just as will the category *mother* in Tim's mind.

The utterances made by such children are not terribly different from the semantic approximations of adults who say "I broke my DVD" instead of "I scratched it", or "I broke my head getting out of the car" instead of "I banged it"; it's just that adults' concepts are a little bit more sophisticated than children's. And then (sticking with the verb "to break" for a moment) there are many usages that are often labeled "metaphorical", such as "to break bread", "to break one's fast", "to break one's silence", "to break one's brain", "to break somebody's neck", "to break the ice", "to break wind", "to break ground", "to break the news", "to break someone's heart", "to break a habit", "to break away", "to break a code", "to break the law", "to break a world record", and on and on. Such usages are obviously built upon analogical extensions of the verb "to break" that go way beyond anything that a child does who says "the book is broken".

Our tale of children's usage of verbs hasn't "turned off" yet. Let's look at little Joane's use of the verb "Come on!" ("Come on, Mommy, turn your eyes on!"). This usage is undeniably a correct one, and it reveals a deep understanding of the situation that this two-year-old is in. What does "Come on" mean? Firstly, it's a verb that indicates that the speaker wants some change to come about, and it's directed at another person who the speaker feels would be able to make that change happen. Secondly, it's spoken as a kind of urging — stronger than and less polite than "please", almost reaching the intensity of "I insist". Thirdly, although it's an imperative based on the verb "to come", it has nothing to do with physical motion. In fact, "Come on!" is such a frozen expression that one might even argue that it is no longer a genuine verb but more of an interjection, rather like "Hey!" After all, no one would reply to the exhortation "Come on!" by saying "Okay, I'm coming on!" But grammar aside, we are dealing with a subtle word choice made by a toddler. She clearly had put her finger on the situation's essence — namely, she wanted her mother to open her eyes — and that desire led to an eager hope that she could achieve this goal by whining.

To put it in another way, already at the tender age of two, Joane had understood that there is a certain class of situations in life that match and that evoke the label "Come on!" This mental category of *Come on!* situations had gained a solid toehold in her mind. One of the situations belonging to this category was the current one, with her napping mother. To put it succinctly, then, we are saying that *Come on!* situations constitute a mental category that is every bit as real and as important as categories such as *eyes*, *truck*, and *Mommy*, which refer to physical entities in the world. The acquisition of the abstract category of *Come on!* situations by a two-year-old child is a small cognitive miracle and is thus an excellent challenge for anyone who has the goal of deeply understanding human thought.

We might equally well focus on the choice of the verb "gotta" by Lenni ("Gotta nurse the truck!"). This two-year-old boy has understood the essence of situations that are labeled with the pseudo-word "gotta" — namely, something is needed in a hurry, there's no time to lose, and so on. It's very likely that Lenni thinks that "gotta" is just one single word (which is why we didn't write "got to" or "I've got to" or "we've got to", etc.), and this would suggest that he hasn't fully understood that it is a verb, even if in different circumstances he might say, "You don't gotta do it" or "I gotted to do it", and other variants, which are clearly attempts to use it as a verb. So once again we observe a case of a high degree of abstraction carried out by a human being who belongs to the category *toddler*.

Here are a couple of other childish pearls that do not involve verbs. Six-year-old Talia announced, "Dad, we have to get some deodorant for the refrigerator!" (since it reeked of seafood), and her two-and-a-half-year-old cousin Hannah, having just licked all the chocolate off her Eskimo Pie, exclaimed with delight, "Look, now the ice cream is naked!"

Even with nouns that denote the most ordinary and concrete of objects, there remain many subtleties. Lenni said, "Gotta nurse the truck!", but what truck was he speaking of? There was no truck in the apartment; there was just a broken toy. Was that object really a truck? Well, yes and no. Lenni knows perfectly well that the trucks he sees on highways are hugely bigger than *his* truck, but for him those are distant abstractions; he's never even touched one. By contrast, his little toy truck is a physical object that drives down the invisible highways on the floor of his apartment. In that sense, this toy is, for Lenni, just as central a member of the category *truck* as are those "real" trucks that drive down the "real" highways — indeed, for him, it's probably more central than those are. Ironically, for Lenni, it's *real* trucks that are metaphorical.

Shining Light on the Moon

Earlier we suggested that there is a strong resemblance between the concrete perceptions of a small child and the abstract mental leaps made by a sophisticated physicist. We wish now to illustrate this thesis by means of a concrete example.

In 1610, Galileo Galilei, having just constructed his first telescope, turned it toward the heavens and peered at various celestial bodies. Recall that at that time, the distinction between planets and stars, which today is quite sharp, was still blurry. Certain celestial lights seemed to wander against a backdrop formed by others, but the reason for this movement was not at all clear. Galileo's choice to focus on Jupiter did not mean he knew what it was; it was probably because Jupiter was one of the brightest and thus most inviting objects in the sky to look at.

Galileo's first surprise was that in his telescope Jupiter appeared not as a mere point but as a small circle, which suggested that this "point of light" might well be a solid object with a definite size. Galileo had certainly had the experience of seeing someone with a lantern approaching him. From afar, the lantern seems to be just a dot without size, but then, little by little, the dot acquires a diameter. By analogy with this familiar phenomenon, Galileo could thus imagine that Jupiter, up till then just a dot of light, was in fact a physical object, much like the objects he knew all around him. A second surprise was that against the background of this small white circle he observed some tiny black points, and moreover — a third surprise — these tiny points moved across the circle in a straight line, some taking a few hours, others a few days. Furthermore, whenever one of these points reached the edge of the white circle, it would change color, becoming white against the backdrop of the blackness of space, and would continue moving along the same straight line, then it would slow down, stop, and reverse tracks; when it returned to the edge of the white circle, it would disappear totally, and after a while would reappear on the far side of the white circle.

We won't go into great detail about Galileo's epoch-making discovery; we want, rather, to focus on the way in which the great scientist interpreted what he was observing through his telescope. He decided that Jupiter was a roughly spherical object around which other, smaller objects were rotating perfectly periodically (with periods ranging from about two days to about fifteen days, depending on which dot he was paying attention to). Galileo knew that the Earth was round and that the Moon rotated around it in a periodic fashion, with a period of about thirty days. All these factors added up and suddenly something clicked in his mind. All at once, Galileo was "seeing" a second Earth in the sky, accompanied by several Moons. We put "seeing" in quotation marks to remind readers of the fact that the key moment of "perception" was Galileo's act of interpretation, since the light stimuli arriving at his eyes hadn't changed in the slightest. The analogy between the Moon and a spot of light (or a black point, depending on where the dot was with respect to the white circle of Jupiter itself) was a stroke of genius — a "vision" of a visionary, so to speak.

Not everyone would have seen what Galileo saw, even if they had been given a telescope, even if they had observed the celestial lights over several weeks, and even if they had focused on Jupiter in particular. The reason is that until that moment, the word "Moon" had been applied to only one object, and the fantasy of "pluralizing" that object was well beyond the imagination of anyone alive at that time (and if someone original had the audacity to think such a thought, that was sufficient to bring about their swift demise: it suffices to recall the case of poor Giordano Bruno, who was burned at the stake in Rome in 1600 for his fantasies about worlds like our own spread throughout space). Moreover, Galileo's daring act of pluralization was the fruit of an analogy that might have seemed laughable to most people — after all, it was an analogy between the entire world, on the one hand (since for most people back then, the terms "Earth" and "world" were synonyms), and on the other hand, an infinitesimal dot of light. This analogy, which might seem far-fetched, nonetheless led to the pluralization of the Earth, since it began by taking Jupiter to be another Earth, and it was rapidly followed by the pluralization of the Moon, which naturally led to the lower-casing of the initial "M". The concept of *moon* had been born, and from that moment on it was possible to imagine one or more moons circling around any celestial body, even around moons themselves.

What Galileo envisioned, in hypothesizing that some small objects in the heavens were rotating around a larger object, was a replica, on an unknown scale, of numerous earthbound situations that were familiar to him, in which one or more objects rotated around a central object. Galileo's stroke of genius was to bank seriously on the daring heliocentric hypothesis of Copernicus and to think to himself that the sky, far from being merely a pretty two-dimensional mural whose purpose was solely to make human life more pleasant, was a genuine *place* that is completely independent of humanity, similar to the places he knew on Earth but much vaster, and as such, capable of housing entities having unknown sizes, and capable of being the site of their movement. In fact, Galileo was completely ignorant of the size of Jupiter and its moons; of course he could imagine a sphere roughly the Earth's size, but doing so would be no more than guesswork, since all he had access to was a set of tiny points. For all he knew, Jupiter might be no larger than the town of Padova, in which he was doing his stargazing, or it might be a hundred times larger than the Earth. Galileo's analogy was an analogy created (or rather, perceived) between something vast and concrete (the Earth and the Moon) and something else that was extremely tiny and immaterial (a circle and some points), but which was nonetheless imaginable as another vast and concrete thing.

Is this profound vision of Galileo's all that different from the vision of the child who sees a very small toy as being a member of the category *truck*, whose other members are so enormous that they are almost inconceivable to the child? One thing is certain — namely, that in both cases, there is a very small object that is imagined as being a very large object, and in both cases, the perceiver uses familiar phenomena in order to understand what is not familiar.

And what about the analogy that *we* are drawing between what Galileo did and what the small child does — is this, too, not just a leap between one scale of sizes and another? Isn't the small cognitive leap by the child, which links a silent, odorless plastic toy truck on the floor with a loud, smoke-belching truck on the highway, simply a small-scale version of the sophisticated cognitive leap by Galileo, which linked the Earth under his feet with the imagined, distant Jupiter, and which linked our familiar Moon with the imagined, distant Jovian moons? Could it in fact be the case that the tiny child's act of calling an everyday object by its standard name is a close cousin of the genius's act of creating a new concept that revolutionizes human life? For the time being, we won't press the point, but we've planted the seed. To go further will require that we look more closely at the subtlety of the most ordinary categories.

Analogies in the Corridors and Behind the Scenes

Some years back, the senior author of this book went to Italy for a sabbatical year. When he arrived, he had a decent command of Italian but, like everyone in such a situation, he made plenty of mistakes — sometimes subtle, sometimes not — most of which were based on unconsciously drawn analogies to his native culture and language. The research institute where his office was located was a building in which some three hundred people worked — professors, researchers, students, writers, secretaries, administrators, technicians, cafeteria workers, and so forth. During his first few weeks, he met several dozen people, whose names he instantly forgot but whom he would continually bump into in the wide, austere corridors of the building, each time he ventured out of his small office. What to say to all these friendly folks who instantly recognized the newly arrived foreigner, the *professore americano*, and who greeted him warmly (or at least politely) whenever their paths crossed in the hallways? And what to say to the people he saw every day but whom he had never actually met?

His initial assumption, coming from his native culture, was that the proper thing to say to anyone and everyone was "Ciao!", even if it was someone that he wasn't sure he'd ever seen before. This was an innocent assumption based on the American way of saying "Hi", and perhaps it seemed charming to those who received such spontaneous greetings and who were naturally inclined to humor their sender because of his status as a foreign guest, but *il professore* soon noticed that his monosyllabic choice did not coincide with that of the majority of the native speakers of Italian whom he ran into. To be sure, there were a handful of people who said "Ciao" to him, but these were his closest colleagues whom he knew well. Otherwise, though, the people in the halls tended to say either "Salve" or "Buongiorno" to him. It took him a while to figure out the levels of formality that were linked to these two forms of greeting, but in the end he devised a fairly clear rule of thumb for himself to guide him in his hallway greetings. Basically you say "Ciao" to people with whom you are on a first-name basis; you say "Salve" to people you see from time to time and whom you recognize (or think you recognize); and finally, you say "Buongiorno" to people you're not sure you recognize, and also to people whom you would prefer to keep at arm's length.

Once he had formulated this rule of thumb and had gotten it more or less confirmed by native Italian-speaking confidants (who, in truth, had never really thought about it and who were therefore not all that sure of what they were saying), he tried to put his new insight into practice, which meant that every time he ran into someone in a corridor, he had to make an instant triage: "First-name basis? ⇒ *Ciao.* Know them a little bit? ⇒ *Salve.* Not sure who it is? ⇒ *Buongiorno.*" He rapidly discovered that this was a cognitive challenge that was not in the least trivial. Fortunately, in each of these three greeting-categories, there were one or two individuals who served as prototypes, and using these people as starting points, he began to feel his way in the obscure corridors of acquaintanceship. "Hmm… This fellow who's approaching me, I know him roughly as well as I know that tall curly-haired administrator" — and zing! — he whipped out a "Salve". Around several central individuals constituting the nuclei of the three categories, there started to form mental clouds that spread out as time passed. The strategy worked pretty well, and after a few months, *il professore* was handling the challenge fluently as he strode through the corridors of what, at the outset, had been a mysterious maze.

This is a concrete example of how new categories form — in this case, those of *ciao* situations, *salve* situations, and *buongiorno* situations — thanks to the use of analogies at every step of the way. And it also allows us to stress another key point — namely, that behind the scenes of even such a simple-seeming thing as uttering an interjection, there is a complex cognitive process that depends on subtle categories.

Let's take an example in English that has many points in common with the one just described. On certain occasions one says simply "Thanks" to convey one's gratitude to someone; on other occasions, one says "Thank you" or "Thank you very much" or "Thanks a lot"; indeed, there is a whole range of thanking possibilities, including such familiar phrases as "Many thanks", "Thanks ever so much", "Thanks for everything", "Thanks a million", "How can I ever thank you?", "I can't thank you enough", and so on. Obviously there isn't one exact and perfect choice for each thanking occasion, but on the other hand, certain situations will very naturally evoke just one of these expressions, and some of these expressions would be wildly out of place in certain circumstances. In short, although there isn't a one-to-one correspondence between situations and expressions, a good choice by a native speaker is far from being a random act, far from being a mere toss of dice. When one is a child, one observes thousands of occasions in which adults use one or another of these phrases without thinking about it for a split second, and pretty soon one starts to do just that oneself. Sometimes adults will smile a little, which conveys the sense that one is probably slightly off-target, while other times one can tell, watching others' reactions, that one has hit the bull's-eye. Thus bit by bit, one refines one's feel for the range of applicability of each of these important and frequent phrases. However, one will probably have no memory whatsoever of the many pathways that collectively led one to one's current status of grandmaster in the day-to-day arts of greeting and thanking.

And what holds for these seemingly trivial acts holds as well for the labels that one pins on all aspects of reality, including verbs (as we already illustrated in the case of children), adjectives, adverbs, conjunctions (as we shall shortly see), and so forth.

"Office" or "Study"?

If one pays attention to the words that are spontaneously uttered in the most mundane of conversations, one will run into many surprises that reveal something of the processes underlying these choices (if indeed "choice" is the *mot juste* here, since words generally bubble up so automatically that they do not feel like choices one has made). Here we'll take an example involving Kellie and Dick, two friends who came from Boston to the house of the above-mentioned *professore* a number of years after he had returned to the United States, and who visited for a few days. As it happened, Kellie and Dick both used the term "your office" to designate the standard workplace of their host, while he himself would always call it "my study". After he had put up with this cognitive dissonance for a couple of days, it occurred to him to ask them, "How come the two of you always go around talking about my 'office' when you both know perfectly well that *I* always call it my 'study'?"

This question caught the Bostonians by surprise, but they quickly hit upon an answer to it, and it was almost surely *the* answer. They said, "In our Boston house, the place where we work [they had a small public-relations firm that they ran from their house] is on the third floor — our house's top floor — and we always call it our 'office'. It's the place where we have our computer, printer, and photocopy machine, all our filing cabinets, and all the slides and videos we've made over the course of the three decades we've been doing this. And for you it's the same thing: your work area is on the second floor — the top floor of your house — and it's where you have all the stuff that you rely on for your work: your computer, printer, and photocopy machine, your filing cabinets, your books, and so forth. To us the analogy is blatant, crystal-clear. It just jumps out at us, no need to think at all. So to us, your workplace is your *office*, clear as clear can be. That's the whole story."

After some reflection on the matter, their host answered, "Aha! I think I see what's going on here. When I was a kid in California, my father had what he called his 'study', which was on the second floor — once again the top floor — of our house. It was the spot where he had lots of papers, books, slide rules, filing cabinets, a mechanical calculator, and so forth. Every day I would see him working there, and it left a vivid impression on me. And also, at the university, on the campus, he had an *office*, where he had many more books, and he often worked down there as well, but the difference between his *study* and his *office* was crystal-clear for me. And today, I too have both a *study* at home and an *office* on the campus here in Indiana. But I would never confuse the two of them. So that's how I see things."

And on this note the exchange between friends closed, but there are important lessons that can be drawn from it. First of all, what's clear is that all parties concerned had depended unconsciously on analogies they had made to very familiar situations. These analogies involved slight "slippages" (third floor instead of second floor; slides and videos instead of books; public-relations work instead of academic work; calculator instead of computer; etc.), but at the same time they respected and preserved a more important essence — namely, both sides of each analogy involved the standard daily workplace, which was separated from the rest of the house and was the storage area for professional material, and so forth. In each case, one sees how the choice of the word to apply to the workplace came from an analogy made to one single familiar situation, rather than what one might have thought *a priori*, which is that assigning an entity to a general category like *office* would depend on the fact that the rich and abstract category *office* had been built up from thousands of different examples encountered over the course of a lifetime. And yet no connection to such a general category took place in this case. Each of the three people, although they all had rich and abstract concepts at their disposal, completely ignored them and instead made a concrete and down-to-earth analogy to a single familiar situation. The numberless prototypical instances of the concept *office*, such as executives' offices, dentists' offices, doctors' offices, lawyers' offices, and so forth, had nothing to do with what went on in Kellie's and Dick's minds. All that mattered was that primordial image of *office* from their own house. This is reminiscent of little Tim's primordial concept of *Mommy*. Even though the concept *mother* has been enormously enriched for Tim as an adult, there's no doubt that his own mother has remained over the decades a potential source for analogies; she never got melted down and lost in the abstract concept of *mother*.

As a postscript to this episode, we might add that the Bostonians, during a visit to their friend's home a year later, occasionally used the term "your attic" when referring to his study. Surprised once again by this word choice, he asked them about it, and they explained that they often used the term "attic" in talking about their office in Boston. For them, in this context, the word "attic" had nothing whatsoever to do with a typical messy and dusty attic in a typical house; quite to the contrary, they were thinking of a room at the opposite end of the messiness spectrum — a very clean space in their house, constantly in use on a daily basis. And thus, once again, we see an extremely down-to-earth analogy linking the new place to just one single familiar place rather than to a generic category in which many places are blurred together.

If Kellie and Dick had discovered a truly prototypical attic in their friend's house, full of cobwebs, ancient checkbooks, huge old wooden trunks shipped from abroad, discarded amateurish paintings, and such things, the word "attic" would certainly have sprung to mind because each of them has in their memory not only the concept *our own attic* but also the concept *typical attic*, which allows them to envisage a standard attic, if need be. For example, if Kellie were reading a mystery novel and she came across the sentence, "Trembling, the aged aunt slowly groped her way up the steep and narrow stairway towards the attic to look for the golden statuette, but after three quarters of an hour she hadn't yet come down", the chance is next to zero that this description would evoke in Kellie's mind an image of her Boston house's attic.

This example of the host's *study*, designated first by his visitors as "your office" and later as "your attic", shows how we are guided by unconscious analogies towards labels that seem to pinpoint just what we want to say. It illustrates why there is no boundary line — indeed, no distinction — between categorization and the making of analogies.

The Structure of Categories and of Conceptual Space

The anecdote we've just related shows that a concept (such as those designated by the terms "attic", "truck", "to open", "to melt", "to nurse", "come on!", "ciao", and so forth) can have specific and very distinct instances. Indeed, if we ask you to think of a golfer, you *might* conjure up the image of an anonymous middle-aged lady riding, on a Sunday morning, down some anonymous fairway on a golf cart. But it's more likely that you would conjure up the image of a famous golfer such as Tiger Woods swinging a five-iron, or perhaps you would recall a golf pro you once took lessons from. Instances of the category *golfer* abound, and around each of these specific and concrete instances there is a halo that extends far out. For example, around Tiger Woods, one can imagine seeing him not only making (or missing) a long putt on a tricky green, but also teeing off with great power, hitting out of the rough, and getting out of a sand trap, not to mention appearing in various airport ads and on television, and so forth. Moreover, in this cloud surrounding Tiger Woods, any golf aficionado will surely find a number of Woods' famous predecessors, such as Jack Nicklaus, Arnold Palmer, Sam Snead, Ben Hogan, and others. Anyone familiar with golf will evoke such images without any trouble at all. And so, what does the concept of *golfer* amount to?

It might seem, *a priori*, that asking about the nature of the concept *golfer* is a minuscule question in comparison with the huge question, "What is human thought all about?", but the fact is that it is no smaller. In any case, our modest musings about the concept *golfer* are bringing to the fore the obvious fact that concepts are densely stitched together through relationships of similarity and context. The concept of *golfer* is quite closely linked to that of *minigolfer*, and less closely to such concepts as *tennis player*, *runner*, *bicycle racer*, and so forth. Among these connections, some are very close while others are so distant that they barely exist (for example, there will be practically no relation at all between the notions of *golfer* and *sumo wrestler* in anyone's mind, aside from the fact that both are types of athletes).

The concept of *golfer* is also connected (at various conceptual distances) to a multitude of other concepts, such as *golf course*, *hole*, *fairway*, *tee*, *wood*, *iron*, *putt*, *green*, *par*, *birdie*, *eagle*, *bogey*, *double-bogey*, *hole-in-one*, *hook*, *slice*, *golf cart*, *caddie*, and *tournament*, and also, of course, to a large number of specific people (or more precisely, to the concepts one has formed that represent these people). Despite the large number of golfers of whom any fan has certainly heard the name and has very likely stored it in memory, it's far more probable that a fan will think of Tiger Woods than of some middling player from the 1960's. Thus the distance from the "center" of the concept *golfer* to the concept *Tiger Woods* is quite small, whereas the distance from the center out to the middling player from decades ago is very great, unless, perhaps, the particular player happens to be one's mother or one's uncle or something along those lines.

And thus we come to the idea of a multidimensional space in which concepts exist, somewhat like separate points; however, around each such point there is a halo that accounts for the vague, blurry, and flexible quality of the concept, and this halo becomes ever more tenuous as one moves further out from the core.

The Endless Chunking of Concepts in a Human Mind

We could not make an analogy between one concept and another if those concepts had no internal structure in our mind. The very essence of an analogy is that it maps some mental structure onto another mental structure. We can only understand how a hand is analogous to a foot if we mentally recall the fingers and the toes, for instance, as well as the way the hand is physically attached to an arm and the foot to a leg. These kinds of facts are part of what "hand" means; they are integral to the concept *hand*, and they make it what it is. But how many such facts are there "inside" the concept *hand*? How detailed are the internal structures of our concepts? This is the question to which we wish to turn for a moment.

Consider a rather complex memory in the mind of a certain professor — say, the sabbatical year she spent in Aix-en-Provence. When she recalls that year, of course she doesn't replay its 300-plus days like a movie; rather, she sees just the tiniest part of it, in its barest outlines. It's as if she were looking down at a mountain range from an airplane, but an extensive cloud layer allowed only a handful of the chain's highest peaks to peek through.

If someone asks her about details of the city of Aix, or about some major event that happened during the year, or about the most interesting people she met there, or about the schools that her children went to, and so on, then any of these aspects will become available upon request, but until that happens, they are all hidden under the "cloud cover". And if she decides to shift her focus to the school that her children attended that year, then still just a handful of the school's most salient aspects will come into view. If her focus shifts still further down onto a particular teacher, then a handful of that person's most salient features come into view — and on it goes. The overarching memory — the sabbatical year in Aix — is never seen in its full glory; rather, just a tiny (but very salient) fraction of it is ever made available. However, pieces of it can be focused in on, and in this way, the large memory can be unpacked into its component pieces, and the same can be done to those pieces, in turn.

All our concepts, from the grandest to the humblest, have the same quality of being largely hidden from view but partially unpackable on request, and the unpacking process is repeatable, several levels down. One might at first think that concepts named by simple words, in contrast to a vast and complex event like a sabbatical year that one is recalling, don't have much inner structure, but that's not the case.

Consider the concept of *foot.* When you first think about a foot, you don't think of cuticles or sweat glands or hairs on it or the fancy swirls making up its five toeprints; you think about toes and an ankle and a large vague central mass, and perhaps a sole and a heel. If you then wish to, you can mentally focus in on a toe and "see" bones and joints inside it, as well as the toenail on top and the toeprint on the bottom. And then, if you wish, you can mentally focus in on the toenail, and so it goes.

So far, our discussion might suggest that concepts are structured according to the physical parts that make them up, with unpacking always moving towards smaller and smaller pieces. Of course, that wouldn't make sense for concepts of events or other sorts of abstractions, but even when a concept is of something physical, this needn't be the case. We'll now give an example that makes this very clear; it is the contemporary concept of a *hub* for a given airline. We chose this concept because the word "hub" is monosyllabic, just three letters long, and sounds very down-to-earth, at the opposite end of the spectrum from fancy technical concepts like *photon*, *ketone*, *entropy*, *mitochondrion*, *autocatalysis*, or *diffeomorphism.* And yet when one looks "inside" this concept, one finds that it, too, is complex — indeed, it has much in common with technical terms. To be concrete, what comes to your mind if we say, "Denver is a hub for Frontier Airlines"? Most people will picture in their mind a map of the United States, with a set of black lines radiating into (or out of) a dot representing Denver, as is shown below.

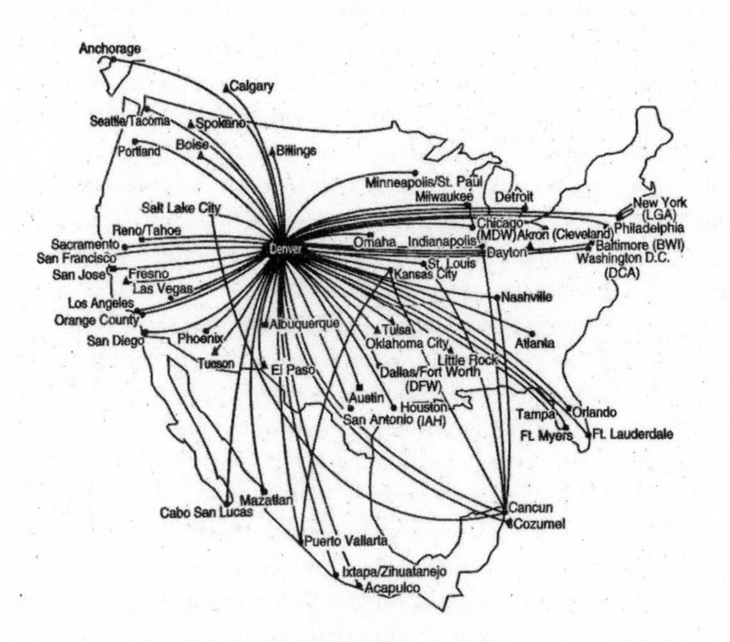

Perhaps they will also think, "Most of Frontier's flights go in and out of Denver", or else "Lots of Frontier planes and gates are found at the Denver airport". This small set of "highest peaks" (*i.e.*, most salient facts) is pretty much all that one needs in most cases where a hub is being talked about. But in fact it leaves out nearly all of what makes up the concept *hub*, and of which most adults in our culture are perfectly aware. The stipulation "in our culture" is crucial, because hundreds of concepts that we take for granted are not part of other cultures or eras. For instance, imagine trying to explain the concept of *hub* to Johann Sebastian Bach, or to Joan of Arc, or to Archimedes, or to Nebuchadnezzar. These were all remarkable individuals in their respective cultures — but how would you go about trying to get across this "simple" concept to any one of them? It would be a rather long story.

To begin with, the word "hub" is the name of a very concrete, visual concept that we learn when we first ride a bicycle and we see the many spokes radiating out of (or into) the wheel's center — its hub. Indeed, it's because of wheels with spokes that airline hubs sport the name "hub", and the concept of *bicycle wheel* is certainly more "primitive" or "elementary" than is that of *airline hub*, not only being learned far earlier in life but also being far simpler to grasp. Let's list some other concepts that are more primitive than *hub* and that are likewise prerequisites to it. There's *airline*, for instance, and *route* and *schedule* and *route map*. And in order to understand the concept of *airline*, you first have to be familiar with the concepts of *airplane* and *company*. And the concept of *route* depends on the concepts of *starting point*, *destination*, *leg*, and *connection*. We won't go on forever, but let's not forget that the *raison d'être* of hubs is economic efficiency — the relentless pressure to cut costs and to reduce the number of different flights — and thus one has to know about the concepts of *trade*, *gain*, *loss*, *profits*, *competition*...

We have only scratched the surface of what goes into the concept *hub*. All of those ingredients are "in there", and they could all, if and when the need arose, be unpacked and revealed. Such unpackings carry one back towards more and more basic, elementary notions — concepts involving motion, vehicles, roundness, acquisition, trading, winning and losing, large and small numbers, and on and on. And note that none of what we have so far spoken of has anything to do with the fact that airports are associated with large cities, or that airports are more than just black dots on maps — indeed, we completely skipped the internal *physical* structure of an airport, with its runways, tarmacs, concourses, gates, jetways, food courts, etc.

The image we've just given of a chain of concepts that depend on other concepts, moving ever downwards in complexity, is reminiscent of the nesting of Russian dolls, and might give the impression that concepts are in fact structured in this boxlike fashion. In truth, however, the phenomenon of concept-building is much subtler and more fluid than that. Concepts are not like nested boxes, with any given concept being rigidly defined in terms of a precise set of previously-acquired concepts, and with concepts always being acquired in a fixed order. Instead, when new concepts are acquired, their arrival often exerts a major impact upon the "more primitive" concepts on which they are based, a bit as if the construction of a house affected the very nature of the bricks with which it was built. Although houses that modify the nature of the bricks of which they are made do not exactly grow on trees, we are all nonetheless familiar with this basic idea, since, for example, children are dependent upon parents in order to exist, but at the same time their existence radically transforms the lives of their parents.

This is true also for concepts. Thus the concept of *hub* depends, without any doubt, on many others, such as *airport*, but at the same time, the concept of *airport* is itself modified by the concept of *hub*. For instance, familiarity with the *hub* idea inevitably brings out the fact that airports are entities that can help airlines to become more streamlined and thus to save money; this notion is certainly not the most obvious fact about airports. Similarly, recalling that airports tend to be transit areas for travelers somewhat reduces the saliency of airports as final destinations. Even if such effects do not cause radical modifications of the concept *airport*, they are undeniably real and demonstrate that the original concept doesn't remain unaffected by the newer one. One can imagine more radical effects of the *hub* concept on that of *airport*, such as novel kinds of architecture aimed at optimizing the design of airports to function as hubs, or the design of new kinds of airport shopping malls specially designed to serve passengers who are making rapid plane-changes and who have only twenty or thirty minutes. And the existence of hubs can change the seemingly obvious correlation between city size and airport size; that is, with hubs, it becomes perfectly conceivable for a relatively small city (such as Charlotte, North Carolina) to have an airport with an enormous volume of air traffic but very few passengers who actually disembark there. We thus see that although there cannot be a "child" concept of *hub* without the prior "parent" concept of *airport*, the child nonetheless changes the identity of the parent.

There are countless examples of this general sort. It happens particularly often in science, where a new idea depends intrinsically on previous ones, but at the same time it casts the old ones in a fresh new light, and often a deeper light. For example, non-Euclidean geometry not only came historically out of Euclidean geometry, but it also allowed a much deeper understanding of Euclidean geometry to emerge. In physics, much the same could be said for relativistic mechanics and quantum mechanics, both of which are "children" of classical mechanics, and together have yielded a far deeper understanding of it.

The same is true for concepts in everyday life. Thus, the relatively new notions of *surrogate mother*, *adoptive mother*, and *single mother* all come out of the concept of *mother*, as does that of a homosexual couple that adopt a child, and each of these new notions modifies the concept of *mother*, showing how a mother need not give birth to a child, need not raise a child, need not be part of a couple, and may even not be a female. In like manner, the concept of *divorce* depends on that of *marriage*, and yet it also has reverse effects on the nature of *marriage* itself (think, for instance, of the effect of prenuptial contracts, and of the fact that today everyone knows, when going into a marriage, that half of all marriages finish in divorce). The notion of *homosexual marriage* clearly depends on the prior concept of *marriage*, and the intensity of the debate over homosexual marriage is in large part due to the fact that opponents claim that the idea not only extends the concept of *marriage* but in fact does the concept serious harm. The concept of *death* both depends on and modifies the concept of *life.* The concept of *fast food* both depends on and modifies the concept of *restaurant.* The concept of *credit card* both depends on and modifies the concept of *money.* The concept of *cell phone* depends on and changes that of *phone.* The concept of *traffic accident* depends on and changes that of *car.* The concept of *airplane* depends on and changes that of *distance.* The concept of *recycling* depends on and changes the concept of *garbage.* The concepts of *rape*, *slavery*, *genocide*, *serial killer*, and others not only depend on but change that of *human being.*

Although the repertoire of human concepts is in a sense hierarchical, in that some concepts are prerequisites to other ones, thus implying a rough temporal order in which various concepts generally are acquired, it is nonetheless extremely different in nature from the precise and rigid way that concepts are built up systematically and strictly hierarchically in mathematics or computer science. In the latter contexts, formal definitions are introduced that make each new concept depend explicitly and in an ironclad fashion on a well-defined set of prior concepts. Ordinary concepts have none of this rigidity or precise dependence. True, a person probably needs some familiarity with such concepts as *wheel*, *spoke*, *takeoff*, *landing*, *leg of a trip*, *jetway*, *concourse*, and *transit area*, for instance, before they can acquire the concept of *hub*, but it's by no means clear what precise role such concepts play in any specific person's notion of what a hub is, nor how deeply such concepts have to have been internalized by someone who feels perfectly comfortable with the sentence "Denver is a hub for Frontier Airlines."

Over the course of our lives, we humans build up concept after concept after concept. This process continues incessantly until we die. This is not the case for many animals, whose conceptual repertoires seem fixed from an early age, and in some cases very limited (think of the conceptual inventory of a frog or a cockroach). And each new concept depends on a number (often very large, as we've just seen in the case of *hub*) of previously existent concepts. But each of those old concepts depended, in its turn, on previous and more primitive concepts. The regress all the way back to babyhood is an extremely long one, indeed. And as we stated earlier, this buildup of concepts over time does not in any way establish a strict and rigid hierarchy. The dependencies are blurry and shaded rather than precise, and there is no strict sense of "higher" or "lower" in the hierarchy, since, as we've shown, dependencies can be reciprocal. New concepts transform the concepts that existed prior to them and that enabled them to come into being; in this way, newer concepts are incorporated inside their "parents", as well as the reverse. Moreover, this continual process of conceptual chunking goes hand in hand with a continual process of conceptual refinement.

Classical Concepts

Until quite recently, philosophers believed that the physical world was divided into natural categories — that is, that each and every thing, by its very nature, belonged eternally to an objective category. These philosophers focused primarily on categories such as *bird*, *table*, *planet*, and so on, whose members were visible entities. In part as a result of these conjectures from long ago, there remains a tendency, even among most contemporary thinkers, to link the notion of *category* with the idea of classifying physical objects, especially objects that we can perceive visually. The idea that situations of someone being *nursed* back to health, for example, or situations of *hoping* for an outcome or of *changing one's mind*, might constitute categories with just as much legitimacy as *table* or *bird* was far from such philosophers' beliefs, let alone the even further-out idea that words such as "and", "but", "so", "nevertheless", "probably" (and so forth) are the names of important categories. If you find it difficult to imagine that a word like "but", which seems so general and perhaps even bland, denotes a category, don't worry; we will come soon enough to this matter, but for the time being we would like to make some observations on the more classical types of categories, since over the millennia certain ideas have become so entrenched in our culture that it is very difficult to overcome them and to start afresh down new pathways. It will thus be helpful for us to make some elementary observations that will paint a picture of concepts that is markedly different from the classical one.

We might begin by asking what a bird is. According to classical philosophers, whose view went essentially unchallenged in philosophy for centuries, until the studies of philosopher Ludwig Wittgenstein, published in the 1950's, and which also reigned supreme in psychology until the pioneering research of Eleanor Rosch two decades later, the category *bird* should have a precise definition consisting of necessary and sufficient conditions for an entity's membership in the category, such as "possesses two feet", "has skin covered with feathers", "has a beak", "lays eggs". (Obviously one could add further or more refined membership criteria for the category *bird*; these few simply constitute a gesture towards the idea.) The set of membership criteria (the defining properties) is said to be the *intension* of the category, while the set of actual entities that meet the criteria (the *members*) is said to be the *extension* of the category. The notions of intension and extension, borrowed from mathematical logic, are thought of as being just as precise and rigorous as that discipline itself, and the use of these terms reveals the ardent desire to render crystal-clear that which at first seems utterly elusive — namely, the abstract essence of all the highly variegated objects that surround us.

A source of problems, however, is the fact that the words used to express the membership criteria are not any more precise than the concept that one is trying to pin down — in this case, *bird*. What, for instance, is a *foot*? And what does "to possess" mean? What does "covered with" mean? And of course, everyone knows that there are all sorts of birds that don't have two feet (perhaps because of an injury or a genetic defect) or that are not covered with feathers (ducklings and chicks, for example). And turning things around, we human beings have two feet, but if we hold a spray of feathers in our hand, this "possession" does not suffice to turn us into birds. And the famous *plume de ma tante* — my ancient aunt's quill pen, which she loved to use to make beautiful calligraphy — would that count as a feather? And if so, would possession thereof make my bipedal old aunt a bird?

At times one gets the impression that the actual goal of ancient philosophers was not to classify specific entities from the material world, such as individual birds, whose variety is bewildering, but rather to characterize the relationships that hold between generic, immaterial abstractions, such as the categories *bee*, *bat*, *egg*, *chick*, *ostrich*, *pigeon*, *dragonfly*, *swallow*, *flying fish*, and so forth. If this is one's goal, then the crucial question would be "Which of these *classes* of entities are birds?" It's clear that one has moved far from the specific and concrete, and has replaced it by an intellectual activity where everything is generic and abstract. This rarefied universe of Platonic concepts, since it lacks annoying exceptions like the plucked or the injured bird, not to mention the old aunt who keeps a quill in her drawer, might appear to be as pure, immutable, and objective as the universes of Euclidean geometry or chess, and this could suggest that in this universe there are a vast number of eternal verities lying in wait to be discovered, much like theorems in geometry. But appearances are deceptive. Even if one considers only abstract categories and pays no attention to their annoyingly problematic instances, one still faces enormous obstacles.

Would a chick's lack of plumage make it lose its membership in the category *bird*? That seems unlikely. Or is there a specific instant, for each chick, when it passes over from the category *chick* to that of *bird*? Would that switchover in status take place at the instant when its skin becomes "covered" with feathers? How many feathers does it take for a chick to be "covered" with them? Or what percentage of the skin's area must be covered for it to count as "covered"? And how does one measure the surface area of a chick, if that is needed in order to decide if we are dealing with a bird or not?

The closer one looks, the more such questions one will find, and the more they are going to seem absurd. And we have only scratched the surface of the issues. Consider the generic idea of a bird that has just died. Is it still a bird? And if so, for how long will this entity remain a member of the category *bird*? Will there be a sharp transitional moment at which the category membership no longer obtains? And let's go backwards in time by a few million years. Where is the boundary line between birds and their predecessors (certain flying dinosaurs)? And to push matters in yet another direction, what about questions such as, "Is a plucked chicken still a bird?" The moment one has created the expression "plucked chicken", the question we posed becomes a legitimate question in the hypothetical formal algebra that governs abstract categories. And with this, we have opened a Pandora's box of questions: "Is a robin whose feet have been cut off still a bird?" (since the first noun phrase is the valid name of a category of entities), or "Is a snake onto which one has grafted some feathers and two eagle's feet a bird?", and so on, without any end in sight.

Even without imagining such radical transformations, one can ask whether sandals are *shoes*, whether olives are *fruits*, whether Big Ben is a *clock*, whether a stereo set is a *piece of furniture*, whether a calendar hanging on one's wall is a *book*, whether a wig is an *article of clothing*, and so forth. People turn out to have highly divided opinions on such questions. In an experiment conducted by the psychologist James Hampton, sinks turned out to be just barely included in the category *kitchen utensils*, while sponges were just barely excluded. Since these close calls are the result of averaging over many subjects in a large experiment, one might imagine that if one were to ask individuals instead, one would find clear-cut and fixed boundaries for each person (even if they would vary from individual to individual). However, even that idea, which runs considerably against the idea of Platonic concepts (which are supposed to be objective, not subjective), turns out to be quite wrong. Many people change their mind if they are asked whether pillows and night-table lamps are articles of furniture and then are asked the same question a few days later. Are these individuals suffering from a pathological state of permanent vacillation, never able to make up their mind about anything? It seems more likely that they are quite ordinary individuals whose categories simply grow blurry toward their edges; if these people were asked about more typical cases, such as whether dogs are *animals*, they would be extremely stable in their judgments about category membership.

Anyone who has taken an interest in the letters of the alphabet will have savored the dazzling richness of a "simple" category like the letter "A", whether capital or small. What geometric shapes belong to the category "A", and what shapes do not? All that one needs to do is take a look at a few handwritten postcards or a collection of typefaces employed in advertising, or for that matter, the figure in the Prologue, in order to see why the boundaries of the twenty-six categories *a*, *b*, *c*, *d*, and so on are impossible to specify exactly. And, to be sure, what holds for the letters of the alphabet holds just as much for other familiar categories, such as *bird*, *bill*, *boss*, *box*, and *brag*.

Summing up, then, the ancient hope of making the categories describing physical objects in the world into precise and rigorous theoretical entities is a vain hope. Such categories are as fleeting and elusive, as blurry and as vague, as clouds. Where are the boundaries of a cloud? How many clouds are there in the sky today? Sometimes, when looking at the sky, one has the impression that such questions have clear and exact answers, and perhaps that's the case on some particular day; however, the next day, the sky will have a radically more complex appearance, and the idea of applying such notions to it as *how many* and *boundary* will simply be a source of smiles.

Concepts Seen in a More Contemporary Fashion

Since the classical view of categories is now generally perceived as a dead end, some contemporary psychologists have tackled the challenge of making the very blurriness and vagueness of categories into a precise science. That is, their goal is to explore those mental nebulas that are our concepts. This has led them to formulating theories of categorization that reject the role of precise membership criteria and instead invoke either the notion of a *prototype* (a generic mental entity found in long-term memory, which summarizes all one's life's experiences with the given category) or else the notion of *the complete set of exemplars* of the given category that one has encountered over one's lifetime. Another influential view involves stored "mental simulators" of experiences one has undergone, which, in response to a fresh stimulus, reactivate certain regions of the brain that were once stimulated by the closest experiences to the current stimulus.

Behind all these efforts lies the appealing idea of non-homogeneous categories — that is, categories having stronger and weaker members — which amounts to distinguishing between more central and less central members. For example, if one times the responses of experimental subjects when they are asked questions of the form

"Is an X a Y?", or if one asks them to write down a list of members of a certain category, or if one gives them a list and asks them to indicate, for each item, its degree of typicality as a member of a specific category, one finds that some very striking trends emerge, and these trends turn out to be stable across all these different ways of testing. Certain members of the category turn out to belong *more* to the given category than others do (recalling how some animals in Orwell's *Animal Farm* were "more equal" than others). For instance, ostriches and penguins turn out to lie close to the outer fringes of the category *bird*, whereas sparrows and pigeons are near its core.

This phenomenon can affect the difficulty one has in understanding a sentence inside a passage that one has been asked to read. Thus, it turns out that the time taken to read and understand a sentence such as "The bird was now just a few yards away" depends on whether, earlier in the passage, there was a reference to an ostrich (an atypical bird) or to a pigeon (a typical bird), in preparatory sentences such as "The ostrich was approaching" or "The pigeon was approaching". The link in memory between *ostrich* and *bird* turns out to be less strong than that between *pigeon* and *bird*, and this tends to impair the understanding of the passage in the first case.

It's important to point out that categorization goes well beyond the intellectual realm of connections among words, which is to say, the names of various categories (such as "sparrow", "ostrich", and "bird"). If, for example, someone were to ask Eleanor "Is a spider an insect?", she might well reply, on the basis of her knowledge from books, "No", and yet if she were to espy a dark blob hanging from the ceiling of her bedroom, it is likely that she would cry out, "Yikes! Get it out of here! I hate insects in my room — they're scary!" If someone were to object to her word choice, Eleanor would say that she knows very well that the "insect" was in fact not an insect but a spider.

Generally speaking, context has a great influence on categorization. The spider in this anecdote was seen as an *insect* in the bedroom, but it would not have been seen as such in the context of a biology test, for instance. And much the same holds in general: a single item in the world belongs to thousands of categories, which can be extremely different from each other, and a good fraction of our mental life consists in placing entities in one category and then in reassigning them to another category. During a basketball game, everyone is aware of the fact that basketballs roll, but it has been experimentally shown that only situations that involve water (such as the loading of a bunch of basketballs on board a ship) evoke the notion that basketballs float.

Context thus changes categorization and can modify how we perceive even the most familiar of items. For example, an object can slip in the blink of an eye from the category *chair* to that of *stool* when a light bulb has just burned out and one needs something to stand on in order to change it. Usually one is unaware of these category shifts because one is mentally immersed in a specific context and such shifts are carried out in a totally unconscious manner. In a given context, just one categorization seems possible to most people. Their lack of awareness of the contextual blinders that they are wearing reinforces the widespread belief in a world in which every object belongs to one and only one Platonic category — its "true" category.

On the other hand, one cannot help but recognize how complex category membership is if one considers the fact that a single entity can easily belong to many diverse categories, such as, for instance:

> 60-kilogram mass, mirror-symmetric object, living entity, biped, mammal, primate, mosquito attractor, arachnophobe, human being, forty-something, book-lover, nature-lover, non-compromiser, non-speaker of Portuguese, romantic, Iowan, blood-type A+, possessor of excellent long-distance vision, insomniac, idealist, vegetarian, member of the bar, mother, mother hen, beloved daughter, sister, big sister, little sister, best friend, sworn enemy, blonde, woman, pedestrian, car driver, cyclist, feminist, wife, twice-married woman, divorcée, neighbor, Dalmatian owner, intermediate-level salsa dancer, breast-cancer survivor, parent of a third-grader, parents' representative…

To be sure, this is but a small excerpt from a much longer list one could draw up, a list having essentially no end, and whose entries would all be terms that anyone and everyone would, without any trouble, recognize as designating various categories.

When Ann had to be hospitalized on an emergency basis and a transfusion was needed, her membership in the category *blood-type A+* dominated all her other category memberships, but in a restaurant she is above all a *vegetarian*, while at work she is a *lawyer*, at home a *mother*, in a PTA meeting a *parents' representative*, and so forth. It may seem useless to point out such obvious facts, but such simple observations carry one well outside the realm of classical categories.

When I Imitate Tweety, Am I a Bird?

Let's come back to the one-word category *bird*, which still has some lessons to teach us. Consider the following candidates for membership in the category:

- a bat;
- an airplane;
- a bronze seagull;
- an eagle in a photograph;
- the shadow of a vulture in the sky;
- Tweety the (cartoon-inhabiting) canary;
- an entire avian species, such as *eagle* or *robin*;
- a chick inside an egg two hours before it hatches;
- a flying dinosaur (or rather, a dinosaur that once flew);
- a pigeon on the screen in a showing of Hitchcock's film *The Birds*;
- the song of a nightingale recorded and played back fifty years after it died;
- a rubber-band-powered wing-flapping plastic object that swoops about in the air.

If you are like the vast majority of humans, you probably felt a keen desire to say "yes" or "no" to each of the candidates in the list above, as if you were taking an exam in school and had to demonstrate the precision of your knowledge, and as if, in each of these cases, there really were a *correct* answer to the question. A sparrow — is it a bird? *Yes!* When you spot a black spot moving unpredictably through the air against a light cloudy background, are you seeing a bird? *Of course!* And when one sees the shadow of a vulture on the ground, is one seeing a bird? *Of course not!* When one hears a loud hooting during the night, is one hearing a bird? *Yes!* And if one hears a recorded hooting (perhaps without being aware that it is recorded)? And what about the case where some person imitates hooting extremely well? And if one dreams about an owl, is there a bird involved? And if one reads a comic book featuring Tweety?

No one ever taught us the boundaries of categories. Our spontaneous sense for their boundaries is an outcome of what we often call "common sense", and no one teaches that in any school. There are no courses on category membership, and even if there were, there would be endless arguments among the students as well as between teachers and students, not to mention the passionate debates that would take place among the teachers themselves. Indeed, expertise doesn't help at all. Here we borrow an anecdote from the psychologist Gregory Murphy, who quoted from a keynote speech once delivered by a world-renowned metallurgist at a conference of world experts in that field: "I'll tell you something. You really don't know what a metal is. And there is a big group of people that don't know what a metal is. Do you know what we call them? *Metallurgists!*"

The recent vehement debates among astronomers over whether Pluto should or should not be deemed a planet (which, as of this writing, it no longer officially is) were due to the blurriness of the concept of *planet*, even in the minds of this planet's greatest specialists, which made the question extremely thorny. For similar reasons, although there is considerable agreement among experts today that it is not correct to refer to our "five senses", since proprioception, thermoception, and nociception (among others) would be left out of such a roll call, there remains a major blur about what our senses really are. Since the experts can't even agree on how many senses we have, let alone on what they all are, they often talk about "our five main senses". And in a similar vein, a standard definition of *life* is still missing, even if biologists, hoping to pin it down for once and for all, are constantly juggling the details of taxonomies that laypersons would have presumed had long ago been cast in concrete. The classification of living organisms has come a long way since Linnæus, and today, many classic terms that he employed in his classification, such as "reptile", "fish", and "algæ", remain present in school texts, but no longer appear in modern phylogenetic classifications. All this goes to show that the blur of categories is not due to some kind of lack of expertise, but is part and parcel of the act of categorization.

How Many Languages do You Speak?

Although psychologists have done a good job in making it clear that no category has precise boundaries, our everyday language and thought are still permeated with residual traces of the classic vision in which category boundaries are as sharp as those of nations (which, to be sure, are often not all that clear, but we'll leave that matter aside). Our intense human desire to avoid ambiguity, to pinpoint the true and to discard the false, to separate the wheat from the chaff, tends to make us seek and believe in very sharp answers to questions that have none.

For instance, people who enjoy studying foreign languages are frequently asked the question, "How many languages do you speak?" Despite how perfectly natural this question might seem, it is based on the tacit idea that the languages of the world fall into two precise bins: languages that person X *does* speak, and languages that X does *not* speak, as if this were a black-and-white matter. But in fact, for each language one has studied, one speaks it to a different degree, depending on many factors, such as when one first studied it, the context in which one studied it, how long it has been since one spoke it, and so forth. When pressed, the questioner may retreat, saying, "All I meant was, 'How many languages can you have an everyday conversation in?' "

But once again, even if this new question sounds reasonable at first, it's just as blurry. For example, it presumes that the category *everyday conversation* is sharp and well-defined. But it might mean a conversation of two minutes about the cost of postage stamps with someone standing next to one in a line in the post office. Or it might mean a half-hour conversation about one's children and family, or about the World Series, or about the sad state of the world economy, with a stranger sitting next to one in an airplane. Then again, it might mean a three-hour conversation ranging over twenty different random topics with seven other people, all native speakers, seated around the table at a lively dinner party. Most people say they speak a language when they have surpassed a far lower threshold than that, but in any case, the threshold for "speaking a language" is not well-defined.

And indeed, the category *language* is itself very blurry. How many languages are spoken in a polyglot land such as India, China, or Italy? In each case, there are many languages and dialects; moreover, what is the precise distinction between a *dialect* and a *language*? The following humorous observation is often attributed to the linguist Max Weinreich: "A language is a dialect with an army", and there is much truth to it, but it still begs the question; after all, what exactly constitutes an army?

In short, the question "How many languages do you speak?" is not a simple question, and has no simple answer — no more so than do the questions "How many sports do you play?", "How many movies do you love?", "How many soups do you know how to make?", "How many big cities have you lived in?", "How many friends do you have?", or "How many things have you done today?"

The Endless Quest for Creative Metaphors

Psychological studies have shown that a mental category, rather than having well-defined and context-independent boundaries, is more like a vast cosmopolitan area such as Paris, which first sees the light of day as a tiny, almost solid, central core (and which, as time passes, will eventually be baptized the "old town", and which shortly after its birth might well have had walls defining its boundary). The "old town" is the original core from a historical standpoint, but the core can move over time and today it may contain modern buildings and roads. After all, both metropolises and categories evolve; it's part of their natural developmental process. Both metropolises and categories exhibit a structure that is the result of repeated acts of extension, and in the case of categories, each new extension is due to some perceived analogy. At every moment in the life of a major metropolis or a "mature" category, there is a crucial, central zone that includes, surrounds, and dominates over the original core, and this zone is considered the town's (or category's) essence. Further out, one finds an urban ring that is not as dense or as historically important, and then there comes a vast suburban ring, which extends far out from the center while growing gradually less and less densely populated, and which has no precise outermost boundary. Nonetheless, one has a pretty clear sense for when one has gone beyond the edge of the metropolitan area, since fields filled with wheat and cattle are evidently no longer part of a city.

In our analogy, the suburban sprawl corresponds to the most recent, fresh, novel, creative usages of the word, which still strike us as metaphorical. And yet over time, these usages, if they resonate with native speakers, will become so widespread and bland that after a while no one will hear them as metaphors any longer. This is essentially what happened to yesterday's suburbs, which today strike us as essential parts of the city, so much so that we have great difficulty imagining how the city ever could have been otherwise.

Seldom if ever reflecting on the literal meaning of what we are saying, we casually speak of such things as:

> the *legs* of a table; the *spine* of a book; a *head* of lettuce; the *tongue* spoken by the islanders; the kisses we *give*; the *window* of opportunity for doing something; the *field* one studies; a *marginal* idea; salaries that *fall* within a certain *bracket*; the *moons* of Jupiter; the *voices* in a fugue; a product of *high* quality; someone's inner *fire*; the familial *cocoon*; a heat *wave*; the *bond* of love; a couple that *splits up*; a relationship that is *foundering*; an athlete who is *worn out*; a team that is *beaten*; a *roaring* wind; a light bulb that is *burned out*; anger that *flares up*; a *handful* of acquaintances; a *circle* of friends; the *friends* of Italian cuisine; someone who *moves* in *high circles*; the *tail* of an airplane; the *burners* on an electric stove; a *ton* of good ideas; the *punch line* of a joke; the *tumbling* reputation of a singer; an idea that one *drops*; a name that one *drops*; the *high point* of a melody; the *crest* of a *fabulous* career; a *slimy* politician; a popular *bodice buster*; a *fleabag* of a hotel; a *rotten* government; a *budding* romance; a wine's exquisite *bouquet*; a belly *button*; a worry *wart*; a traffic *jam*; *laundered* money; an idea that's difficult to *grasp*; the subtle *touch* of a novelist; a *box canyon* in which one is *stuck*; the *block* one lives on; one's *neck* of the woods; a *stream* of insults; the *bed* of a river; the *arrow* of time; an *umbrella* policy; a *haunting* melody; a *skeleton* key…

and of course we could go on forever. The halo of a word gradually moves outwards or, rather, the blurry boundaries of the concept named by a word gradually engulf what were once metaphorical swamps and forests and turn them into apartment buildings, parks, and shopping malls.

Linguist George Lakoff and philosopher Mark Johnson have shown that there are certain systematic tendencies that guide the construction of a number of metaphors in everyday language. Their studies, along with related studies by other researchers, have helped to demonstrate that metaphors, far from being just an elegant rhetorical flourish exploited solely by poets and orators, are the coin of the realm in much of ordinary discourse. For example, time is often characterized linguistically in terms of physical space (*in* three weeks; *at* four o'clock; a *distant* era; the *near* future; *from* now on; a tradition that *goes back* to the seventeenth century), and conversely, space is often represented in terms of time (the first street *after* the traffic light; the road changes name *when* it crosses the river; a star twelve *light-years* distant). Likewise, life is often spoken of in terms of motion or a trip (the *path* of her success; a *sinuous* career; the *dead end* in which they're trapped), with everyday events as places one passes through (I'm *going* to see them tomorrow; I'll *come back* to that point), and happiness and unhappiness are often represented by the concepts of high and low (*raising* someone's morale; to be in *seventh heaven*; to *plunge* into despair; to be very *down*). Abstract notions are often conveyed through comparisons to familiar human activities (her experiment *gave birth* to a new theory; the facts *speak for themselves*; fate *played dirty tricks* on me; life was *cruel* to her; a religion *dictates* certain behaviors; his fatigue *caught up* with him). Complex situations are often cast in terms of a metaphorical fight with a metaphorical adversary (the recession is our *enemy*; our economy has been *weakened* by inflation; corruption must be *fought*; outsourcing *kills* growth; we are *victims* of the stock-market crash; we have *declared war* on the economic crisis; we have *won a battle* against unemployment, etc.). Systematic families of metaphors such as these abound in human languages and they explain, at least in part, the great richness inherent in even our most casual and informal speech.

On the other hand, thousands of words are used metaphorically without belonging to any systematic family of metaphors. Here is a small set of examples:

> they're all *fruitcakes*; you're *nuts*; it's *Greek* to me; while wearing her parental *hat*; he *punted* on the term paper; what a *mousy* person; *watertight* reasoning; today was another *roller-coaster* for the stock market; he *snowed* the committee; my engine is *coughing*; an old *salt*; a *spineless* senator; the company *folded*; a *bubbly* personality; they *creamed* the other team; let the wine *breathe*; to *dress* the salad; a rule of *thumb*; I was such a *chicken*; a *cool* idea; nerves of *steel*; pass the *acid* test; in *round* figures; she's so *square*; you're getting *warmer*; *yellow* journalism; what a *drag*; he just didn't *dig*; *cloverleaf* exchange; *hairpin* turn; make a *hit*; no *soap*; she's really *wired* today; he *swallowed* her story; the old man finally *croaked*; she *drove* me crazy; *carpet* bombing; an *umbrella* clause; a *blanket* excuse; we just *nosed* them out; a *straw* vote; a *blue* mood; we always *horse* around; his *gravelly* voice; they *railroaded* us...

and on and on.

Calling someone "butterfingers", for instance, does not belong to any large, overarching system of metaphors, but the image is very easy to relate to, since butter is slick and slippery, and thus, one imagines, a person whose fingers were covered with butter (or even were *made* of butter) would be completely unable to catch a ball or hold onto anything at all. Therefore, someone who often drops balls that are thrown to them can be easily found in the (metaphorical) halo of the concept of *buttery fingers.* In summary, we often come up with a label for a complex situation by finding a more familiar concrete situation to which it is analogically linked, and then borrowing the standard name of the concrete situation. Such a strategy allows us to create a useful verbal label for a new category of situations.

The act of "metaphorization", whether it is broad and systematic, like the set of metaphors portraying life as a voyage, or narrow and one-of-a-kind, like "butterfingers" and the other phrases cited in the display above, is a crucial aspect of the way in which we naturally extend our categories. The human mind is forever seeking novelty, and it would never be satisfied with a limited and fixed set of metaphors. One might say that human nature is characterized by a constant, intense drive to go beyond all conventional metaphors, which are often labeled "dead metaphors", since when a metaphor is used enough, one no longer hears the original imagery behind it and it loses all its sparkle. Categories are extended successively via metaphors that at first are used over and over again in a vivid, evocative fashion, but then, like dough that first needs to settle before rising, they gradually congeal and become inert, and this very fact sparks a quest for a new extension. Each time a metaphor loses its punch, we push the boundaries further out with new metaphors, always with the goal of understanding more directly and intensely what surrounds us, of adjusting to change, and of adding piquancy and novelty to the way we see familiar things.

Concerning the Literal and the Metaphorical

It might seem tempting to establish precise boundaries for each category, just as we do for cities, and to declare that anything that is found outside of those boundaries is not a member, end of story. In order to retain some flexibility, however, one could grant the title of "honorary member" to certain non-members, as long as they were found within a certain distance of the category's *official* boundaries; in such specially sanctioned cases, one would put the category's name in quotes to indicate that this would be an *official* metaphorical usage. In such a world, then, if someone said, "Ella has a large circle of friends", it could mean only one thing — namely, that Ella's friends were neatly arranged in a big closed curve having a fixed radius; to indicate otherwise, one would have to say, "Ella has a large 'circle' of friends", and in order that one's listeners would realize that the term was not being used literally, one would have to wag one's fingers in a quote-marky fashion or else say, "so to speak" or "quote unquote" or "metaphorically speaking" or something of the sort.

In a world where this linguistic convention held sway, Galileo would not have seen the moons of Jupiter but the "Moons", quote unquote, of Jupiter. And no one would ever come home to the cocoon of their family (since the expression would make no sense, unless the family had acquired one prized cocoon, but even then it would be far too small for a human to fit into) but *so to speak* to the cocoon of their family, or to the *metaphorical* cocoon of their family. One would no longer *give* kisses, but one could *metaphorically* give a kiss to someone or *so to speak* give a kiss to someone. One would never be *under pressure,* but quote-unquote under pressure, and as for the so-called pressure, it too would have to be in quotes, unless one were a diver thirty meters below the surface of the sea. And so on and so forth, without end.

Unfortunately, such a solution would give rise to more problems than it would solve. Firstly, those "precise boundaries of categories" — even of the most common categories — are nonexistent, as we've shown. And secondly, even were we to imagine that categories could be precisely defined, the problem of identifying their so-called "honorary members" would not be solved. Earlier we suggested that some entity located outside the border of a concept would be granted this title provided it were "sufficiently near" the boundary line — but what is the nature of this conceptual distance that would allow us to measure proximity precisely? What kind of yardstick would we use to measure distances? And would there be precise outer limits for the use of quote marks, beyond which even "quote unquote" would not apply? And would all of this be taught to children in courses on categorization and quotation-mark usage?

We could of course imagine introducing second-order quotation marks, which would be used to name entities found in a ring yet further out from the concept's core than the first-order quote-mark ring. One's fingers would soon become indispensable aids to one's mouth in communicating these subtle distinctions. Among the most frequent words and phrases would be "so-called", "in quotes", "so to speak", "metaphorically", and others. In addition, there would be a whole system for expressing the number of quotation marks needed — second-order, third-order, and so on — in other words, oral or manual "roadsigns" telling the distance to the center of the "city". It's "pretty" clear that this "'straitjacket'" would soon "give" "'"royal"'" "'headaches'" to anyone who "wore" it, metaphorically "speaking".

The Categorization/Analogy Continuum

The idea of courses to teach people how to categorize and how to use quotation marks to indicate metaphorical uses of terms seems ridiculous, and for good reason. It's like imagining that in elementary school we should teach children how to walk, eat, and breathe. The reason we don't do that is that our bodies were fashioned by evolution to do such things, and it makes no sense to teach a body what it was designed by nature to do. The same can be said about our brains, which evolved as powerful machines for categorization as well as for quotation-mark deployment. But there is no sharp boundary between pure categorization and quotation-mark deployment, for all the reasons just given. A category has an ancient core, some commercial zones, some residential zones, an outer ring, and then suburbs that slowly and imperceptibly shade off into countryside. It's tempting to say that perceiving something as a member of the "old town" or "downtown" is an act of "pure" categorization, while seeing something as belonging to the outer ring or the suburbs involves a certain amount of quotation-mark deployment — but a bit of thought shows that one passes smoothly and continuously from a concept's core to its fringes, and there are no clean and clear demarcation lines anywhere. All these concentric layers making up a category in its full glory are the result of a spectrum of analogies of different types, collectively made by millions of people over a period ranging from dozens to thousands of years. These analogies form a seamless continuum; they range from the simplest and easiest to make, giving rise to the concept's core (so simple and natural that they are not even seen as analogies by an untrained observer), to more interesting and lively ones, giving rise to the suburbs, and finishing up with extremely far-fetched and unconvincing analogies, giving rise to the remote countryside (that is, objects or situations that hardly anyone would consider as belonging to the category in any sense).

Verbs as Names of Categories

More than once in this chapter we have stated that what holds for nouns, such as "desk", "elephant", "tree", "car", "part", "idea", and "depth", holds just as much for other parts of speech. We already broached this topic in our discussion of some of the charming verb choices such as "nurse the truck" and "patch people's teeth", made by children whose categories *to nurse* and *to patch* didn't coincide totally with those of adults. We'll now go into this idea in greater detail.

It's not so hard to move from nouns to verbs, firstly because many verbs are tightly associated with certain nouns, and vice versa. To start with an obvious example, anyone who can recognize *rain* falling on the ground can also recognize that *it is raining.* The same holds for the category associated with the noun "snow" and the category associated with the verb "to snow"; ditto for "hail" and "to hail". We move effortlessly back and forth between noun and verb, because the words are identical. But even in cases where there is no phonetic resemblance between noun and verb, there are countless cases where the evocation of a particular verb goes hand in hand with the evocation of a particular noun. When you see a dog and hear it make a sudden loud noise, you are simultaneously perceiving a member of the category *dog* and a member of the category of situations where *something is barking.* In much the same manner, given that mouths eat, drink, and speak, we all perceive, many times per day, members of the categories of situations where *something is eating, something is drinking,* and *something is speaking.* In the same vein, the sun *rises* and *shines*, eyes *look* and *see*, birds *fly* and *chirp*, cyclists *ride* and *pedal*, leaves *tremble* and *fall*, and so forth.

Our insistence on the idea that verbs, no less than nouns, are the labels of categories might seem to be merely a fine point of philosophy without any consequence. However, we are insisting on it because the same perceptual mechanisms that allow us to recognize pumpkins, pastries, plows, and pigs also allow us to recognize situations where some marketing, menacing, meowing, or mutating is going on. Once one has had enough experience with situations where menacing is going on, one is able to recognize members of this category, to label them as such, to talk about them with one's friends, to report them to the appropriate authorities, to describe them if called on as a witness in a court, and so forth. One even learns to recognize, from long observation of how people drive their cars, situations where someone is driving in a menacing fashion, occasionally through hearing just a certain telltale squealing of tires. The fact that the verb "to menace" automatically bubbles up to our conscious mind in such situations is in no way different from the fact that a certain noun bubbles up when we look at a canary, a doorknob, or a pair of pants. These evocations of words are the result of categorization. In the case of verbs just as much as that of nouns, the effortless bubbling-up of a word occurs as a result of a vast number of prior experiences with members of the category in question.

If at first glance the collection of all the members of the category *to nurse* seems vaguer and less "real", somehow, than the collection of all members of the category *bridge*, that's simply a prejudice and an illusion. The bridges of the world are not given to us without effort and without blur. Even if all the existing bridges could oblige us by simultaneously lighting up in response to a button-push, there would still be all the bridges from ancient Roman times, ancient Chinese dynasties, and so forth, which have long since disappeared, not to mention all the bridges that are yet to be constructed during this century and all centuries yet to come. And of course, we haven't even touched on the fictitious bridges seen in paintings and films and described in novels. And what about the miniature bridges built by children out of wooden blocks? Or tree trunks fallen over creeks? Or "jetways" (those tunnels on wheels that link an airplane with a gate)? And then there are bridges (or do they count as such? — that's the question) built by ants, for ants, and made out of ants! And what about a toothpick casually placed between two plates, affording a shortcut for a wandering ant? What to say about bridges inside one's mouth, bridges built between distant cultures, bridges between distant ideas? A moment's thought shows that the category *bridge* is highly elusive. At this point, one might even wonder if situations that deserve the slightly abstract verbs "to nurse", "to menace", and "come on!" aren't rather straightforward in comparison with situations that deserve the visual noun "bridge".

Much Ado about *Much*

Let's move on now to such an everyday word so mundane that most people would never think of it as the name of a category or concept. Namely, we'll focus on the word "much". What is the nature of situations that cause this word to spring to one's lips? What do they all have in common? In short, what is this *much* category? Let's take a close look at some examples of this abstraction.

> That's much too little for him. That's a bit too much for me. Much less than that, please. Much the same as the last time. Don't go to too much trouble. How much will that be? Much obliged. I'd always wanted it so much. It's not much, but it's home. I'm very much in agreement with you. Much though I wish I could... Much of the time it doesn't work. Your hint very much helped me. Just as much legitimacy as her rival had. Moths are much like butterflies. As much as I'd like to believe you... So much so that we ran into trouble. She got much the better of him. It didn't do us much good. Her florid writing style is just too much!

What is the shared essence of *much* situations? A *much* situation involves an opposition (usually unconscious) to an imaginary *some* or *somewhat* situation; in other words, a *much* situation involves a mental comparison in which a particular mental knob is "turned up" relative to a milder, more common situation. For example, "I wanted it so much" can only be understood by means of a fleeting comparison with a hypothetical scenario in which the speaker's desire is less intense. In short, the word "much" is evoked in the mind of English speakers when they want to describe an unexpectedly large quantity or large degree of something, whether it's concrete ("too much peanut butter", "not very much air") or not concrete ("much to my displeasure", "much more prestige than it deserves"). As for listeners, when they hear the word used, they understand this intention on the part of the speaker, and consequently, in their heads they turn up a small mental knob in order to reflect the speaker's apparent desire to intensify some part of speech, or even to intensify a phrase or clause.

A *much* situation is thus a situation that resides partly in the objective, outside world and partly in the subjective, inner world of one's expectations about the nature of the outer world. In order to recognize a *much* situation as such, you have to be concentrating not only on something in the world "out there" (such as the amount of soup you're being dished up by someone), or on some internal situation (like being hungry or sleepy), but also on your own expectations in such a situation, or on a typical person's expectations. The exclamation "Hey, they sure didn't give me much soup!" means that, in comparison with one's expectations of the amount of soup typically served in restaurants, this serving is on the low end of the spectrum.

If a speaker didn't feel that some milder contrasting scenario needed to be hinted at (at least subliminally), the word "much" wouldn't pop to mind. "Too much peanut butter", when spoken in a given situation, is aimed at evoking in listeners a hypothetical contrasting situation where the *right* amount of peanut butter was used. It's in this contrast that the phrase's meaning resides. Likewise, "Thank you very much!" is aimed at evoking in a listener, in a subtle fashion, the idea that the speaker *could* have voiced a less ebullient sentiment; it is therefore heard as a desire to convey gratitude more intensely than some other people might do in the same situation, or more intensely than the same speaker might do in a different situation or in a different mood.

We have seen that *much* situations concern the disparity between the external world and an ideal inner world filled to the brim with expectations and norms. Just as one can hope (though always vainly) to pin down what the essence of the category *bird* is, so here we've tried, with the aid of extremely blurry words, at least to hint at what the essence of the category *much* is.

Grammatical Patterns as Defining Mental Categories

As the above list of examples shows, when one is talking, there are certain ready-made syntactic slots into which the word "much" fits very neatly and there fulfills its function. In fact, these syntactic slots themselves constitute another facet of the nature of the word "much". As we grow up and go to school, we encounter the word "much" many thousands of times, and if certain spots where that word sits among other words strike us, on first hearing, as a bit surprising, after a while they become more familiar, then turn into a habit, and in the end they wind up being a reflex that is completely unconsciously integrated into us. Ways of placing the word "much" that at the outset seemed odd and unnatural gradually become so familiar that in the end one no longer sees what could at first have seemed puzzling or confusing about them.

Why do we say "I much appreciate all you've done for me" but not "I appreciate much all you've done for me?" Why do we say "I don't go out much" and sometimes "I don't much go out" but never "I much don't go out"? Why "I'm much in agreement with her" but not "I'm much out of contact with her"? Why "much the same" but not "much the different" or "much the other"? Why "I'm much obliged" but not "I'm much grateful"? Why "much though I'd like to join you" but not "very much though I'd like to join you" or "much although I'd like to join you"? Why is "Many thanks" as common as daisies while "Much thanks" is as rare as orchids? Or is it? A quick Google search revealed a ratio of 200 to 1 in favor of "Many thanks to my friends" as compared to "Much thanks to my friends" — but the fact that the latter exists at all suggests that things might be changing. Here we find ourselves face to face with the blurry and moving contours of the category *appropriate syntactic slots for the word "much"*. Who knows what the just-mentioned ratio will be in five years, ten years, or fifty? Native speakers seldom ask themselves these kinds of questions about word usages, because the patterns are deep parts of their very fiber.

What all this means is that the category *much* — that is, roughly speaking, the full range of situations that evoke the word "much" and a feeling of "muchness" — is a category that possesses not only a *cognitive/emotional* side (while speaking, we feel a need or a desire to emphasize something, to draw a contrast between how things are in fact and how they might have been or may become), but also a *syntactic* side (we sense, as we are building a sentence even while uttering it, various telltale slots where the word could jump right into the sentence with no problem).

A reader might react to this observation by claiming that all we've said is that the word "much" has two facets, one being the concept behind it, and the other being the grammatical roles that the word can play in English, and thus that our claim is merely that "much" has both a semantic and a syntactic side (much as does any word), and that semantics and syntax are independent human mental faculties. Such a stance implies that the mental processes that underlie people's choice of *what* to say and their choice of *how* to say it are autonomous and have nothing in common. But making such a distinction is highly debatable. Could it not be that the mechanisms with which we perceive *grammatical* situations in the world of discourse are cut from the same cloth as those with which we perceive *physical* situations in the world around us?

As a child, one learns to "navigate" (quote unquote!) in the abstract world of grammar just as one learns to navigate in the world of concrete objects and actions. A child starts to use the word "much" in the simplest syntactic contexts at first, such as "too much", "not much", "much more", and so on. These initial cases constitute the core of the category; as such, they are analogous to little Tim's Mommy as the core of his category *mommy*, and to the Moon as the core of Galileo's category *moon.* The child might possibly explore risky avenues such as "a lot much", "many much", "much red", "much here", "much now", "much night", and so forth, but such trial balloons will be popped, sooner or later, by society's cool reaction, and will be given up.

As the years go by, our child will hear, read, understand, and integrate increasingly sophisticated usages, such as "much traffic", "I much prefer the other one", "much to my surprise". These could be likened to the other children's mothers in the mind of little Tim, and to the moons of Jupiter in the mind of Galileo. Each time a new usage is heard (such as "much to my surprise"), that specific case will contribute to a blurry mental cloud of potential usages that are *analogous* to it ("much to her horror", "much to his shame", "much to our disappointment", "much to my parents' delight", etc.) Thus the child will be led to taking further risks by making little explorations at the fringes of these expanding categories — risks such as "much to my knowledge", "much to her happiness", "much to his unfamiliarity", "much to their comfort" — and to the extent that these tentative forays resonate or fall flat with other speakers, they will be reinforced or discouraged.

Children refine their sense for the category of *much* situations (both its semantic and its syntactic aspect) in much the same way as they refine their sense for any other category. And they do all of this on their own, because schools do not teach any such thing and do not need to; children simply become, without any particular effort (let alone a great deal of conscious effort), *much*-ness experts. They will randomly run into the word in poems, in song lyrics, in ancient texts, in slang phrases, and in marginal usages like "it's of a muchness", "thanks muchly", "it cost me much bucks", "too much people here", and without realizing that they are doing so, they themselves will indulge in just this kind of pushing of the linguistic envelope. Bit by bit, this will add up to a personal sense for the limits of the category — the category of *appropriate usages and syntactic slots for the word "much".* For each person, this mental category will stretch out in its own idiosyncratic fashion, but no matter who it is, it will consist of a core surrounded by a "halo". Just as in each person's mind there are prototypical chairs and also quotation-marked "chairs" that flirt with the very edges of the category, so there are prototypical usages of "much" and also edge-flirting usages of the same word.

Words that Name Phenomena in Discourse

A profound aspect of growing up human involves developing an exquisite real-time sensitivity to the many types of expectations that our words set up in the minds of our listeners. In so doing, we acquire a rich set of categories that have to do with these abstract phenomena. Oddly enough, though, some of the most important of these categories are labeled by words that seem boring and bland — "and", "but", "so", "while", and numerous others. Such words may at first strike us as unimportant and even trivial, but that is a most misleading impression. These words denote deep and subtle concepts, and as we shall see, those concepts are grounded in analogies, much as are all other concepts.

Let's look at some examples involving the word "and". No one would be at all surprised if a friend, upon returning from a trip to France, enthused, "I like Paris and I like Parisians." On the other hand, we would certainly be confused if our friend first declared, "I like Paris" and then stated, after a short pause, "I like Parisians." This would give the impression of two ideas that were unrelated to each other, which of course is not the case. Our friend could make it a bit more logical-seeming by adding "also" at the end of the sentence; doing so would acknowledge the fact that listeners want to hear an explicit, sense-making link between the two utterances. Indeed, that's precisely one of the key roles of the word "and" — to set up a natural link between two statements. Thus if our friend declared, "I like Paris and I just bought a pair of pliers", we would be caught off guard by the lack of coherence. A central purpose of the word "and" is to convey to listeners a clear sense of the logical flow that, in the speaker's mind, links one thought to the next one uttered.

The flow of discourse is just as real to human beings as the pathway of a fleeing zebra is real to a pursuing lion. They are both varieties of motion in certain kinds of space; it's just that the space of hunting is physical and the space of discourse is mental. Lions live mostly in the physical world, and although we humans live there too, we also live in the world of language, and a large part of our category system revolves around phenomena that take place in that intangible but no less real world. We perceive and categorize situations that arise in discourse space, and we do so just as swiftly and just as naturally as the pursuing lion, on the savanna, chooses its direction of motion in a split second in chasing its prey.

We all acquire the word "and" and the concept behind it just as we do for other words and concepts — through analogical broadening. Can anyone recall the very first occurrence of the word "and" that they ever heard? Of course not. But as with all other words acquired during early childhood, it was never defined explicitly; rather, its meaning was picked up from context ("Mommy and Daddy", possibly). At first it linked people, we might well suppose. Then it linked people and objects ("Sally and her toy"). Then it linked sequences in time ("I went out and looked"). Then it served to represent causal links ("It fell and broke"). Then it linked combinations of abstract qualities ("hot and cold water"), as well as of relationships ("before and after my haircut") and other abstract attributes ("a hot and healthy meal"). And then many more came, in an avalanche.

Like any category, the category of *and* situations expands gradually and smoothly in each human mind — indeed, so smoothly that after the fact the resulting urban sprawl seems, albeit illusorily, monolithic and uniform, as if it had been constructed all at once, as if there were but one single elementary idea there, which had never needed any generalizing at all. There are no conscious traces left of the many concentric layers of outward expansion of *and*, just as there are no conscious traces left of how we acquired categories that give the impression of being considerably more complex, such as *mother*, *stop*, and *much.* And so this innocent little conjunction, which very few people would think of as standing for a *category*, fits right in with the story of words and concepts that we are here relating.

Contrasting "And" with "But"

Now let us deepen and broaden our discussion by looking at some examples involving the conjunction "but". A totally logic-based view would claim that "and" and "but" mean exactly the same thing except for emotional shadings. However, that's a pretty parochial view of the matter. Let's take a closer look. Were our just-returned friend to say to us, "I like Paris but I like Parisians", we would surely wonder, "What does *that* mean? It makes no sense!" The reason is that hearing the word "but" leads us to expect a *swerve* or a *zigzag* in discourse space, but there was no such sudden switch in direction. Stating that one likes Paris and also Parisians does not challenge common sense, does not violate reasonable expectations, and thus it does not in any way, shape, or form constitute a swerve or zigzag in discourse space. Our hypothetical friend's hypothetical sentence faked us out by *announcing* a swerve but not carrying it out. There is a puzzling inconsistency between the conjunction and the two phrases that it links. Indeed, if such a sentence were to show up in an email message, you might well guess that it was a typo and that your friend had intended to write, "I like Paris but I don't like Parisians." Now that would indeed constitute a zigzag in discourse space.

For effective communication, speakers have to pay close attention to the nature of the flow in the sequence of ideas that they are conveying — in other words, they have to carry out real-time self-monitoring. When motion in the space of discourse continues smoothly along a pathway that has already been established, then the word "and" (or some other cousin word or phrase, such as "moreover", "indeed", "in addition", "on top of that", or "to boot", to list just a few possibilities) is warranted. We'll call situations of this sort "*and* situations". When one recognizes that one is in an *and* situation, one can say "and" and be done with it. By contrast, when motion in the space of discourse makes a sudden, unexpected swerve, then the word "but" (or some other concessive word or phrase, such as "whereas", "however", "actually", "in fact", "although", or "nevertheless", "even so", "still", "yet", "in spite of that" to list just a few possibilities) is warranted. Analogously, these are *but* situations, and of course, when one recognizes that one is in a *but* situation, one can say "but" and be done with it.

What Makes One Say "But" Rather than "And"?

Occasionally one hears sentences like "I don't know what country the florist comes from, but she seems very nice." Why the "but" here? What kind of a zigzag in discourse space is this? Well, first consider how it would sound with "and" instead: "I don't know what country the florist comes from, *and* she seems very nice." It simply sounds like a *non sequitur.* One wonders what these two thoughts are doing in the same sentence. On the other hand, with "but", there is a definite logical flow, although it's a bit subtle to pin down. The feeling being expressed is something like this: "*Despite* my near-total lack of knowledge about her, I would say that she seems affable." "Despite" is a concessive that is a close cousin to "but". The point is that the first part of the sentence is about a hole in one's knowledge, and the second part is about a small but significant counterexample to that tendency. Thus the first part of the sentence suggests a pattern and the second part states an exception to the pattern. Whenever we are about to tell someone a "piece of news" and just before doing so we realize that in some way or other it goes against expectations likely to be set up by what we had just told them moments earlier, we have detected the telltale signs of a *but* situation. The two-clause sentence about the florist has exactly that property, and that's why putting "but" between its clauses makes sense and sounds right to our ears, whereas putting "and" there would make it sound very strange.

Likewise, if someone says "He has big ears, but he's really a nice guy", it doesn't mean (despite the way it sounds on the surface) that the speaker has a stereotype of large-eared people as being unpleasant. Rather, it means something more like, "Although this person is on the negative side of the norm in a certain physical way, he is on the positive side of the norm in terms of his behavior." Once again, we see that the conjunction "but" signals a swerve in discourse space — the person in question is on one side of *one* norm and yet (despite that fact) is on the other side of *another* norm.

The category of swerves that the word "but" denotes is just as real as the category of swerves made by vehicles on roads, though it is more intangible, and the use of the word "but" comes about because as people speak, they are always paying some amount of attention to their trajectory in discourse space and are categorizing its more familiar aspects in real time, just as they are always paying some amount of attention to the scene before their eyes (and the sequence of sounds coming into their ears, etc.) and are categorizing its more familiar aspects in real time.

Sometimes a speaker becomes aware of the real-time linguistic self-monitoring going on as a background process in their brain, and this can affect the flow of speech. It can result in one verbal label being canceled and swiftly replaced by another label. One example is when someone says, "Oh, look at that horse — uhh, I mean *donkey*". The following story involves such a relabeling, but the self-correction involves an event in the speaker's linguistic output stream rather than an object in the environment.

Frank and Anthony, lifelong friends, hadn't seen each other in a long time and were pleasantly catching up on the news of each other's families. Frank wanted to tell Anthony about his daughter, who had been hit by a mysterious illness and, to everyone's relief, recovered from it after a couple of years. One of his sentences went like this: "She got to be an excellent skier during her stay in Montana, and one day on the slopes she just couldn't keep her balance — or rather, *but* one day on the slopes she just couldn't keep her balance..."

As he launched into his sentence, Frank thought he was in an *and* situation, and then suddenly — or rather, "*but* suddenly" — when he started to flesh out the second clause, he clearly heard the abrupt swerve in what he was telling Anthony (it would strongly violate anyone's expectations that a highly accomplished skier will, without any warning, start to fall a lot), and so he quickly spun in his tracks and, changing conjunctions in mid-stream, jumped from "and" to "but", as he realized that from a listener's point of view, the story he was relating involved a kind of zigzag — thus a member of the category of *but* situations rather than of the category of *and* situations.

Further Refinements in Discourse Space

Making the distinction between *and* situations and *but* situations is not a high art, but it is a most useful skill to pick up, and that dichotomy is perfectly adequate in many situations. However, there are numerous subcategories inside the broad categories that we've labeled "*and* situations" and "*but* situations", and people, first as children and later as adults, gradually pick up the finer nuances that will help them to recognize these subcategories and thereby to choose, in real time, the sophisticated connecting word or phrase that best describes the situation in discourse space.

Sticking to just the categories *and* and *but* while making no finer distinctions is rather like making the useful but coarse distinction between the categories of *car* and *truck*, but not venturing into finer details. The *car/truck* distinction is good enough for many purposes. People who are fascinated by motor vehicles, though, are eager for much more detail, and they'll often use a much narrower category than is designated by the generic word "car". In the same way, fluent speakers depend on making finer distinctions than just the coarse "and"/"but" dichotomy. However, just as recognizing whether a vehicle is a Honda or a Hyundai, a coupe or a sedan, automatic or manual, fuel-efficient or gas-guzzling, sporty or family-style, and so forth, takes considerable experience, so deciding whether one finds oneself in a *nonetheless* situation in discourse space, a *however* situation, an *and yet* situation, a *still* situation, an *on the other hand* situation (and so forth) is a subtle skill, since it requires having constructed these subcategories and having a decent mastery of them.

We have no need to delve into the subtleties that underlie such choices. Just as it is not our aim to explain how people distinguish among *studies*, *studios*, *offices*, *dens*, *ateliers*, *cubicles*, and *workplaces*, or among their friends who are *agitated*, *antsy*, *anxious*, *apprehensive*, *concerned*, *disquieted*, *distressed*, *disturbed*, *fidgety*, *frantic*, *frazzled*, *frenetic*, *frenzied*, *jittery*, *nervous*, *perturbed*, *preoccupied*, *troubled*, *uneasy*, *upset*, or *worried*, or between situations calling for "Thanks a million", "Thank you ever so much", "Many thanks", and other expressions of gratitude, so it is not our aim to explain the nature of the nuances that lead a person to choose to say "however" rather than "but" or "nonetheless" or "actually" or "and yet" or "that having been said" or "despite all that". We are concerned not with pinpointing the forces that push for choosing one or the other of these linguistic labels, but simply with the fact that each of these different phrases is the name of a subtly different mental category — a highly characteristic, oft-recurring type of pattern in discourse space to which one can draw analogies.

We might point out here that where English has two most basic conjunctions ("and" and "but"), Russian has three — "и" ("and"), "но" ("but"), and "а" (whose meaning floats somewhere between "and" and "but"). This means that Russian speakers and English speakers have slightly different category systems concerning very basic, extremely frequent phenomena that take place in discourse space. Picking up the subtleties of when to use "а" instead of "и" or "но" takes a long time. It's much the same story as for any set of categories that overlap. We don't want to give a linguistics lesson, so we'll stop here, but the bottom line is that words that to most people seem infinitely far from the most venerable and clichéd examples of categories (such as *chair*, *bird*, and *fruit*) are nonetheless the names of categories, and they are so for the very same reasons, and the categories they name act very much the same.

Ever More Intangible

It might seem logical for a chapter on words to move from the most frequent ones to rarer ones, but we will go against expectations here. We want to finish up by talking about some of the most frequent words of all, which, like "and" and "but", are almost never thought of as being the names of categories. Consider words like "very", "one", and "too", for instance. What category does "very" name? Of course we can't literally point to members of the *very* category the way we can point to members of the category *dog*, say. Still, let's try for a moment. Usain Bolt is a very fast runner. Cairo is a very big city. Neutrinos, they are very small. That's very *you*. There; that's enough to give the feeling. Much like *much*, *very* is a category having to do with norms built up over a lifetime of prior experience. Where Rome is a big city, Cairo is a *very* big city.

We learn to use the word "very" just as we learn to use the word "much" — by hearing examples of its usage and feeling our way around in the world of sentence construction. Does the fact that the crux of the notion *very* has to do with the formation of sentences disqualify it from being a concept? No, not at all. The concept *very* is just as genuine a concept as is *dog*. The concept *very* is all about relative magnitudes, expectations, importances, intensities. All of that is deeply conceptual.

And while we're at it, let's not forget that Albert Einstein was one very smart dude. Yes, no doubt about it, Einstein was *one* smart dude, as opposed to being *several* smart dudes; but why was he not just *a* smart dude? The word "one" can convey more information, it seems, than just the number of items that somebody is talking about. In this case, saying "one smart dude" emphasizes the extreme rarity of a genius of Einstein's caliber; it is a subtle way of squeezing extra information into the sentence via a very unexpected channel. However, the choice of the word "one", as opposed to the word "a", also conveys information about the persona of the speaker (earnest, candid) as well as about the tenor of the conversation (informal, casual). Moreover, using the word "dude" strongly resonates with using the word "one", and vice versa — indeed, when used together, these two words paint a vivid portrait not only of Albert Einstein but of a certain brand of English speakers who are prone to use this kind of phrase.

To put it more explicitly, probably most native speakers of American English have developed a category in their minds that could be labeled "the kind of person who goes around saying 'one smart dude' ". However, the category is not as narrow as this label suggests. To be sure, it would be instantly evoked if one were to hear the above remark about Albert Einstein's intelligence, but its evocation doesn't depend on having heard the specific words "smart" and "dude"; it would also be evoked by remarks like "Doris Day was one cute cookie" or "That's one bright lamp!" We thus see that even bland little words like the numeral "one" intoned in a certain fashion, which might seem very close to content-free, can evoke rich and subtle categories in our minds.

Having just considered "one", let's move along to "too". Of course that word has two quite separate meanings — namely, "also" and "overly much" — so let's focus on just the latter. What are some quintessential members of the *too* category? Well, perhaps the idea that eating a whole fudge cake would be too much. Or the idea that teaching general relativity to elementary-school kids would be too early. We'll let readers invent their own *too* situations. The point is that doing this little exercise will make it vivid for you that there are analogies linking each *too* situation to other *too* situations, and thus to the abstract concept of *too*-ness.

When we considered the concept *much*, we pointed out that part of its richness is how it is used in sentences. Indeed, the realm of discourse is one of the richest domains we humans come into contact with. Just as there are concepts aplenty in the worlds of linear algebra, molecular biology, tennis-playing, and poetry, so there are concepts galore in the worlds of discourse, language, grammar, and so forth, but we seldom think about them. Thus a high-school student might pen a poem in flawless amphibrachic hexameter without ever suspecting that there is a standard name for such a meter. Likewise, we native speakers of English are all past masters in the use of words such as "the" and "a" without ever analyzing how they work. But the Polish linguist Henryk Kałuża wrote a whole book — *The Articles in English* — to teach non-native speakers "the ins and outs" (one of his examples of "the") of our language's definite and indefinite articles. As it turns out, Kałuża's book is all about the *meanings* of these rich words, but nonetheless, some people resist the idea that "the" and "a" have meanings, arguing that they are not "content words" but just grammatical devices. It seems that since these words do not designate tangible objects, some people think they are devoid of meaning (not unlike people who insisted for centuries that zero isn't a number). It seems strange, however, to suggest that the difference between "the president" and "a president" has nothing to do with *meaning*. There is a great deal of content conveyed by the distinction between "the sun's third planet" and "a sun's third planet", between "I married the man in the photo" and "I married a man in a photo", between "the survivor died" and "a survivor died".

Trying to pin down how words like "the" and "a" are used in English is not our purpose here — no more than trying to specify the type of circumstances likely to evoke the word "office" as opposed to the word "study". What we are emphasizing is that this subtle knowledge is picked up over many years thanks to one analogical extension after another, usually carried out without the slightest awareness of the act.

And thus we have moved our discussion from fairly low-frequency words, like "hub", "attic", and "moon", to the very top of pile — the most frequent word in all of English — the definite article "the". In so doing, we have also moved from very visual, concrete phenomena to phenomena that are largely intangible and mental. But what's crucial is that in making this move, we have never left the world of categories. Just as "hub" denotes a category (or perhaps a couple of different categories — the centers of bike wheels as opposed to certain major airports), so "the" denotes a category (or perhaps a few distinct ones, as the world-class "the"-expert Henryk Kałuża would be quick to point out).

Carving Up the World Using a Language's Free Gifts

Any language has an immense repository of labels of categories that people over millennia have found useful, and as we grow up and then pass through adulthood, each of us absorbs, mostly by osmosis, a decent fraction of that repository, though far from all of it. The many thousands of categories that we are handed for free and that we welcome, seemingly effortlessly, into our minds tend to strike us, once we have internalized them, as self-evident givens about the world we live in. The way we carve the world up with words and phrases seems to us *the right way* to view the universe — and yet it is a cliché that each language slices up the world in its own idiosyncratic manner, so that the set of categories handed to speakers of English does not coincide with the set handed to speakers of French, or to those of any other language. In short, "the right way" to see the world depends on where and how one grew up.

A striking example is provided by English and Indonesian. The English words "brother" and "sister" seem to us anglophones to cover the notion of *siblinghood* excellently, as well as to break that concept apart at its obvious natural seams. However, the Indonesian words "kakak" and "adik" also cover the notion of *siblinghood* excellently, but they break it into two subconcepts along an entirely different axis from that of sex: that of age. Thus "kakak" means "elder sibling" while "adik" means "younger sibling". To speakers of Indonesian, this seems the *natural* way to slice up the world; they don't feel a need to be able to say "sister" using just one word any more than anglophones feel a need to be able to say "older sibling" using just one word. It doesn't cross their minds that something is *missing* from their language. Of course Indonesian speakers can say "female kakak or adik", and that effectively means "sister", just as we English speakers can say "older brother or sister", and that effectively means "kakak". Each language can express through a *phrase* what the other language expresses through a *word.* And the French language does an admirably diplomatic job with these concepts, managing to slice the world up in both ways. The male/female dichotomy tends to be the more frequently used one in French ("frère" vs. "sœur"), but the older/younger one exists just as well ("aîné" vs. "cadet"), and thus all possibilities are available. As this shows, slicing the world up at its "natural" joints is not quite so natural as one might think.

Different ways of cutting up the world are far from being exceptional picture-postcard rarities. In order to unearth good examples of the phenomenon, one certainly doesn't need to resort to pairs of languages that are spoken halfway around the world from each other. We can find plenty of them right under our nose, simply by poking about a bit in the languages that are closest to our native tongue, even limiting our search to words and concepts that are unquestionably central.

Thus, nothing seems more obvious to us anglophones than what *time* is. We know *what time it is* right now, we know *how much time* it will take to drive to the airport, and *how many times* we've done so before. These three ideas strike us as being very clearly all about just one central, monolithic, and hugely important concept: the concept known as "time" (in fact, the most frequent noun in the English language). And yet, most strangely, there are languages that don't see those three ideas as being about the same concept at all! If you're a francophone, you know what *heure* it is right now, you know how much *temps* it will take to drive to the airport, and how many *fois* you've done so before. They aren't the same word or even related words, and the three concepts labeled by the words "heure", "temps", and "fois" seem quite distant from each other for French speakers. As if this weren't bad enough, the French word "temps" doesn't denote only a certain subvariety of English's concept of *time* — in addition, a good fraction of the time, it means "weather". Thus speakers of French, in their whimsical fashion, somehow manage to confuse the weather and the time! On the other hand, we speakers of English manage to mix up the hour of the day with the number of occasions on which something has happened! Which mistake is sillier?

The English and French languages certainly don't agree on how the world should be broken up into categories, even for the nouns of the highest frequency that exist, let alone for categories labeled by verbs, adverbs, prepositions, and so forth. For example, those incorrigible French speakers, they irrationally distinguish between two kinds of "in" — namely, "dans" and "en". What could make less sense than that? Whereas we clear-sighted English speakers, we distinguish (most rationally, of course) between two kinds of "de" — namely, "of" and "from". What could make more sense than this?

These kinds of discrepancies are totally typical of how different languages carve the world up differently from each other, and between any given pair of languages there are myriads of such discrepancies. How, then, do people ever communicate at all across language boundaries?

Spaces Filled Up with Concepts

To help answer this question, we would like to offer a simple visual metaphor for thinking about the words of a language (and more generally about lexical expressions) and the concepts that they represent. We begin by suggesting that you imagine a two-dimensional space or a three-dimensional one, as you prefer; next, we are going to start filling that space up, in our imagination, with small patches of color, using a different color for each different language that we are interested in — say green for French, red for English, blue for German, purple for Chinese, and so forth. It is tempting to think of these concept-blobs as something like rocks or jelly beans — odd little shapes having very well-defined edges or boundaries. The truth is far from that, however. While each blob is intensely colored in its center (deep red, deep green, whatever), as one approaches its "boundaries" (which in truth don't exist), it grows lighter in shade — think of pink or chartreuse — and then it simply fades out, passing through lighter and lighter pastel shades as it does so. This image of blobs with hazy contours of course echoes our metaphor likening concepts to very dense cities that gradually turn into suburbs and then fade into countryside.

We will call the space itself, before the insertion of any colored blobs (somewhat like a house without furniture), a "conceptual space" (there are many such, which explains the indefinite article). At the very center of each conceptual space are found the most common kinds of concepts — those for very common tangible objects, intangible ideas, phenomena, properties, and so forth — the concepts whose instances are encountered all the time by people who belong to a particular culture (or subculture) and era, and which those people must be able to categorize quickly and effortlessly in order to survive, or simply to live.

The core items in a typical conceptual space include, quite obviously, the concepts for various entities such as the main parts of the human body; general classes of common animals, such as *bird*, *fish*, *insect*, and a few farm animals; general classes of plants, such as *tree*, *bush*, and *flower*; things to eat and drink; common feelings, such as being *cold* or *hot* or *hungry* or *thirsty* or *sleepy* or *happy* or *sad*; common actions, such as *walking* and *sleeping* and *eating* and *giving* and *taking* and *liking* and *disliking*; common properties, such as *big* and *small*, *near* and *far*, *kind* and *cruel*, *edible* and *inedible*; common relationships, such as *belonging to*, being *inside* or *outside*, being *above* or *below*, being *before* or *after*; common degrees, such as *not at all*, *not much*, *slightly*, *medium*, *very much*, *totally* — and so forth. Every language has words for such notions, because all humans require these concepts in order to live. This list merely scratches the surface of the core of a typical conceptual space, of course, but it gives the general idea. In any case, these concepts, all residing at or very close to the dead center of a typical conceptual space, are quasi-universals that most humans deal with constantly, and they are thus bases that are well covered, and necessarily so, by every language.

The idea of conceptual spaces will help to make more tangible and concrete some ideas about the words and expressions of a given language and the concepts used by its speakers. One of the most important ideas that it helps one to think about is how different languages cover, or fail to cover, certain concepts. Between the conceptual spaces of distant cultures there will be large discrepancies. But what about cultures that are bound together by geography, history, traditions, and so forth? In such cases, the conceptual spaces will be very close to each other.

In what follows, we will focus mostly on contemporary Western cultures, simply because we ourselves feel more competent in that context, and we assume that many of our readers (at least those who are reading this book in one of its two original languages) would also feel more comfortable that way. However, our general points have nothing to do with the specific concepts that we will discuss.

Looking at Two or More Languages within a Conceptual Space

How is a conceptual space filled up with sets of blobs of different colors? For instance, how do the repertoires of concepts possessed by French and English speakers who share essentially the same culture compare?

According to our visual metaphor, regions near the very center of a conceptual space are densely filled in, no matter what language we are speaking about. If, as suggested above, French is represented by green, then there is a green blob near the middle of conceptual space that covers the area occupied by the concept *hand*. And if English is represented by red, then fairly much the same area is covered by a red blob of similar size and shape to the green (French) blob. Each different language will cover that same area of a conceptual space fairly well, so there will be blobs of many different colors right there, all closely overlapping with one other.

Some of the different-colored blobs representing different languages' coverages of a given extremely frequent concept will tend to have pretty much the same shape, but in the case of other blobs there will be discrepancies, some minor and some major. We've already seen a pretty major one, involving *time* corresponding to *heure*, *temps*, and *fois,* and *temps* corresponding to both *time* and *weather.* To provide another case, the red blob representing the extremely frequent concept expressed by the word "big" in English aligns quite well with the green blob for the French word "grand", but by no means perfectly so, since some of the meanings expressed by our "big" are usurped by French's "gros" (for instance, things that are large in thickness or width, as opposed to those that are large in height), and conversely, our word "great" usurps some of the meanings expressed by French's "grand" (those that mean "highly accomplished, world-famous, and deeply influential").

Some even more severe misalignments involve extremely frequent prepositions such as "in" (which in fact is covered in French not just by "dans" and "en", but by many other prepositions, depending on the context), and by similarly frequent and enormously protean verbs such as "to get" (which sometimes is best rendered by "obtenir", other times by "prendre", other times by "chercher", other times by "recevoir", other times by "comprendre", other times by "devenir", other times by "procurer", and on and on). Of course, the story is symmetric; that is, each of the just-mentioned high-frequency French prepositions and verbs is likewise covered by all sorts of different English verbs, depending on the context. There's no clean one-to-one alignment between blobs of different colors, although there's a great deal of overlap.

On the other hand (and quite luckily!), for a very large number of truly important concepts — say, *finger*, *water*, *flower*, *smile*, *weight*, *jump*, *drop*, *think*, *sad*, *cloudy*, *tired*, *without*, *above*, *despite*, *never*, *here*, *slowly*, *and*, *but*, and *because*, to give just a few examples — there is generally quite good agreement between French and English, and, for that matter, among all the languages that we are familiar with.

Thus, the center of this conceptual space is inhabited by red and green blobs that often coincide quite well, and when they don't coincide, then there are all sorts of overlapping blobs, each with its own curious shape. Luckily, though, despite the fact that the green blobs covering a certain concept and the red blobs covering the same concept are often shaped rather differently, the central zone of the overlapping space is extremely densely covered both by red blobs and by green blobs (and also, if we want to throw in other languages, blue blobs and purple blobs, and so forth).

Furthermore, there aren't going to be any gaping holes in the linguistic coverage of concepts residing near the dead center of the conceptual space of some other culture (such as the Nepali or the Navajo culture); there won't be blank zones where a human language totally lacks a lexical item labeling a concept that is universally part of the human condition. Any language spoken by more than a tiny, isolated group will easily be able to talk about, for instance, sleeping poorly, or seeing a friend after a long time, or breaking a stick, or throwing a stone, or walking uphill, or feeling sweaty, or being very tired, or losing one's hair, though each one will have a unique way of doing so.

Rings or Shells in Conceptual Space

Let us now imagine moving outwards from the core towards slightly less frequently encountered concepts, such as, for instance *thanks*, *barn*, *fog*, *purple*, *sincere*, *garden*, *sand*, *star*, *embarrassing*, *roof*, and *although*. If these concepts are of comparable importance to one another within the culture, then their distance from the center will be about the same, and we can say that they constitute a ring (or a shell, if you are envisioning a three-dimensional space). These concepts are still important in the conceptual space — and so, once again, we expect that this region of conceptual space, though not belonging to the most central core, will still be quite densely filled with blobs of every color. On the other hand, let's zoom outwards a considerable distance further from the core of our conceptual space, to a different shell where we will encounter (let us say) the concepts *frowning*, *cantering*, *fingernail-biting*, *tap-dancing*, *welcome home*, *income tax*, *punch line*, *corny joke*, *sappy movie*, *vegetarian*, *backstroke*, *chief executive officer*, *wishful thinking*, *sexual discrimination*, *summit meeting*, and *adverb* (just to give a tiny sampling of the hypothetical shell). This latest outward leap has clearly carried us into more rarefied territory, and so we would not expect all the cultures of the world (and of all different historical epochs) to share all the concepts in this shell, nor would we expect all the earth's languages to have words or phrases to denote all the concepts in this shell.

What is Monolithic is in the Eye of the Beholder

Let's take any shell of this conceptual space. Since languages differ enormously, we can easily find a red blob that no single green blob covers precisely. However, a small *set* of green blobs will collectively do a pretty good job of covering all the territory of the red blob (although they will inevitably also cover areas outside the red blob). And of course, what's sauce for the goose is sauce for the gander, meaning that we can easily find green blobs that no single red blob covers precisely.

To make things concrete, let's take an example. English speakers fluently and effortlessly use the word "pattern" to describe regularities, exact or approximate, that they perceive in the world. However, if they wish to talk about such phenomena in French, they will soon learn, to their frustration, that there is no French word that exactly covers this very clear zone of conceptual space. And thus, depending on details of what they mean, they will have to choose among French words such as "motif", "régularité", "structure", "système" , "style", "tendance", "habitude", "configuration", "disposition", "périodicité", "dessin", "modèle", "schéma", and perhaps others.

At the outset, this lacuna in the French lexicon strikes English speakers as a rude violation of common sense, since the concept of *pattern* strikes them as being self-evident and objective, and therefore something that should be universal to all languages. It seems obvious that there "should" be just one word for all those notions that the

English word unites; after all, it feels like just *one thing* rather than many. But in French and in fact in most other languages, there simply isn't such a word. Nonetheless, other languages manage to cover the zone of conceptual space labeled "pattern" in English pretty completely, although somewhat less efficiently, by using a bunch of smaller blobs each of which corresponds to a limited facet of the notion, or else a set of large blobs that intersect partly with the English one.

For the sake of fairness, we should point out that French, too, has words of quite high frequency that have no counterpart in English — for instance, the adverb "normalement", which certainly looks like it means what we anglophones mean when we say "normally" (and sometimes it indeed does), but which a large part of the time means something rather different. Here are a few examples that show typical uses of the word, and that give a sense for the wide variety of translations it needs in order to be rendered accurately in English:

Normalement, Danny doit être arrivé à la maison maintenant.
Hopefully, Danny's back home by now.

Normalement, on va courir à 7 heures ce soir, non ?
Unless we change our plans, we'll be taking our run at 7 this evening, right?

Normalement, nous devions passer deux semaines en Bretagne.
If there hadn't been a hitch, we would have spent two weeks in Brittany.

French speakers will be just as puzzled by English's lack of a single word for the obvious, monolithic-seeming concept expressed by the word "normalement" as English speakers are puzzled by French's lack of a single word for the obvious, monolithic-seeming concept that is embodied in the word "pattern". What is monolithic is in the eye of the beholder.

In cases such as these, where one language has a single word that covers a set of situations that another language needs a variety of different terms to describe, we are dealing with linguistic richness and poverty. Thus in the case of "pattern", English is richer than French, and in the case of "normalement", French is richer than English. More generally, we can say language A is locally richer than language B if language A has a word (or phrase) denoting a *unified* concept — that is, a concept that native speakers feel hangs together tightly, and that seems to have no natural internal cleavages — and if language B *lacks* any single word covering that same zone of conceptual space. We can thus speak of a local "hole" or "lacuna" in language B's coverage of conceptual space, even though language B manages to cover the zone by resorting to a *set* of words.

On the other hand, when a certain area of conceptual space is finely broken up by a given language, and when speakers of both languages agree that this fine break-up is warranted, then a language that doesn't offer its speakers such a fine break-up has to be considered poorer. Take the English word "time", for instance. To native speakers of English, whereas the word "pattern" feels unitary and monolithic, the word "time" does not have that monolithic feel; native speakers readily and easily see (at least if it's brought to their attention) that there are several very different meanings of "time" (for instance, those corresponding to the French words "heure", "temps", and "fois"). Thus in this case, it's the French language that is richer and the English language that is poorer, for the English lexicon doesn't break that large zone of conceptual space into smaller separate zones, as the French language does. An example where French is weaker is the word "beaucoup", which corresponds to both "much" and "many" in English. For us anglophones, it's obvious that these are separate concepts, one having to do with a large quantity of a substance, the other having to do with a large number of similar items. The French word that blurs this distinction thus seems rather crude. Thus in this case, the English language appears to be richer, and French poorer.

In summary, when language A has a word that strikes its speakers as representing a natural and monolithic concept, and language B has no corresponding word, then language B is poorer and language A is richer, because speakers of language B are forced to cobble different words together in order to cover the zone of conceptual space that language A covers with just one word. Conversely, when language B has a set of words that cut up a zone of conceptual space that is covered by just one word in language A, and when the distinctions offered by language B seem natural to speakers of both languages, then it's language B that is richer and language A that is poorer.

The Need to Stop Subdividing Categories at Some Point

When one studies various languages, one discovers that many concepts that one had at first naïvely taken as monolithic, because of one's native language, are in fact broken up into subconcepts, and often with excellent reason, by other languages. And if one studies enough languages, one often discovers numerous different ways of subdividing one and the same concept. Seeing a concept being broken up into all sorts of subconcepts that one hadn't previously dreamt of suggests that it would in theory be possible to continue carving the world up into tinier and tinier blobs, thus making an ever finer mesh of very small, extremely refined concepts, without any end.

But no language in fact does this, because all languages come from the key human need to have categories that apply at once to a vast number of superficially extremely different and yet deeply extremely similar situations. Such categories help us to survive and to have comfortable lives. To be sure, some language could, in principle, have separate words for red books and green books, or for books printed on butterfly wings, or for orange books of under 99 pages, or for puce-colored books about subtropical botany that contain between 221 and 228 pages (but not 225) and are in (Brazilian) Portuguese and are typeset in 13-point Bodoni — but it's obvious that there comes a point of diminishing returns, and it's nowhere near the absurdly fine distinctions just hinted at. There's no reason for any culture to construct any of these categories, let alone to reify it via a word in its language, although the miracle of language — of every language on earth — is the charming fact that any of those odd and far-fetched categories *could* in theory be invented by someone, if they were needed or desired.

We should also point out that category refinement doesn't always move in the direction of an ever-finer mesh. Sometimes refining one's mental lexicon of categories means broadening through abstraction, in the sense of learning to perceive common threads in situations where people who lack the concept would simply see unrelated phenomena (for example, the commonality linking human mothers with animal mothers, den mothers, and mother companies, or the commonality linking female animals with female plants, or the commonality linking hubs of wheels with airports that are hubs, and so forth). The emergence of this type of *broader* category is also extremely useful for the development of a people or a culture.

There is thus a tension between the desire to make finer distinctions that cover very few cases and the desire to make broader categories that cover many more cases. Earlier, we saw that children's perception of the world is quite coarse-grained relative to the perception of adults (this is why some young children uninhibitedly speak of "patching teeth", "eating water", "undressing bananas", and so forth), and we saw that as children grow older, they acquire more and more refinements in their conceptual systems. This is a universal tendency, but at some point, adults stop refining their lexicon when it comes to ordinary objects, actions, relationships, and situations. Each language and culture has found its natural grain size for such entities, and in a kind of unspoken collective wisdom, it ceases to go beyond that, although of course experts are continually refining their technical vocabularies, and each society, as it makes new discoveries and inventions, collectively creates new concepts and new words for them.

Everyone in every culture is constantly refining their conceptual repertoire by acquiring ever more compound words, idiomatic phrases, proverbs, and new catch phrases that enter the language through books, movies, and advertisements; in addition, everyone is also constantly building up a rich repertoire of concepts that have no verbal labels. In the next two chapters, we will turn our attention to these two key ways in which our conceptual storehouse continues growing as long as we live.

CHAPTER 2

The Evocation of Phrases

❧ ❧ ❧

Categories Vastly Outnumber Words

What is it that links words and categories? To be sure, words are often the verbal counterparts of categories. We can describe and refer to categories with them, but that does not mean that categories should be equated with words — not even with the broader notion of lexical items — for categories are mental entities that do not always possess linguistic labels. Often words are names of categories, often they can be used to describe categories, but sometimes they simply are lacking. All in all, the connection between categories and language is complex. A single word can of course bring a category to mind — "mother", "moon", "chair", "table", "office", "study", "grow", "shrink", "twirl", "careen", "thanks", "ciao", "much", "and", "but", and so on — but the correspondence is somewhat lopsided, because in fact we all know many more categories than we know words.

Coining a word is cognitively costly, and our mental categories are so numerous and constantly changing that it would take an astronomical repertoire of words if we wanted to have exactly one word per category. As a consequence, humans have figured out how to economize with words. Thus, there are many words that have multiple meanings, depending on the context. Such words cover a variety of categories (consider the multitude of meanings of a simple word like "trunk", for instance). Another word-saving device is that many categories have verbal labels that consist of a string of words rather than just one word, and that idea will be the central focus of the present chapter. And then there are myriads of categories that simply have no verbal label at all, and the goal of the next chapter will be to shine a bright light on those.

In sum, whereas Chapter 1 focused on categories whose labels are just one word long, this chapter is concerned with categories whose linguistic labels are more complex; thus compound words, idiomatic phrases, proverbs, and fables are among the scenic spots we shall visit.

Psychology does Not Recapitulate Etymology

No less than indivisible words, compound words designate categories. Thus the word "airplane" is no less the name of a category than are "air" and "plane"; the same goes for "airport", "aircraft", "airfield", "airlift", "airsick", "airworthy", "airhead", "airbag", "airplay", "airtight", and so forth. There are many words whose components are so tightly fused inside them that the individual pieces are seldom if ever noticed, since (in most cases) the wholes are not analyzable in terms of their pieces — for example, "cocktail", "cockpit", "upset", "upstart", "awful", "headline", "withstand", "always", "doughnut", "briefcase", "breakfast", "offhand", "handsome", "cupboard", "haywire", "highjack", "earwig", "bulldozer", "cowlick", "dovetail", and so on.

To be sure, in some of these cases — for instance, "cupboard" and "headline" — a little guesswork provides a plausible story about their origins, but the possibility of doing an intellectual analysis doesn't mean that a fluent speaker conceives of the word — that is, hears it — as a compound word. For example, we don't pronounce "The plates go in the cupboard" as if it were written "The plates go in the cup board", and we don't hear it that way. In fact, we never say "board" when we mean a storage location, even if it once had that meaning. What we say aloud sounds more like "cubberd" than like "cup board", and virtually no one hears either part inside the whole. As for "airport", although we can deliberately slow down and hear "air" and "port" inside it, who ever thinks about the *atmosphere* and about a *harbor* when picking up a friend at the baggage area, or when transferring between planes? Indeed, were someone to call an airport an "atmospheric harbor", it would invite ridicule, if not sheer incomprehension.

Many compound words are positively mysterious if one starts to think about them. Why do we sometimes call a woman's purse a "pocketbook"? It's not a book by any stretch of the imagination, and it certainly doesn't fit in any kind of pocket! Nor is it a book of pockets! And how can we understand the compound word "understand"? Understanding has nothing obvious to do with standing anywhere, let alone underneath something. Then again, in certain compound words, just one of the two components sounds strange and strained, such as the "body" in "nobody".

An analysis of where words come from and how they came to mean what they now mean belongs to the classic discipline of etymology, and often it is truly fascinating, but it does not have much bearing on how words are actually perceived by a native speaker. In that sense, psychology does not recapitulate etymology (to tip our hat to the phrase "ontogeny recapitulates phylogeny"). Many compound words simply act like indivisible wholes; we learn them as wholes as children, and it is as such that we usually hear them.

Thus a toddler learns and uses the word "pacifier" without having any idea of the existence of the verb "pacify" or the suffix "-er" inside it (not to mention the Latin root "pax" and the suffix "ify" found inside the verb "pacify"!). For a toddler's purposes, the sound "pacifier" is simply that concept's arbitrary-seeming label, and the word doesn't need to be broken down or analyzed. As for adults, they seldom need to decompose compound words, either; indeed, doing so would often be more confusing than helpful, as in "eavesdrop" or "wardrobe", for example. Who ever thinks of a *wardrobe* as a place where one *wards* one's *robes*? The standard pronunciation ("war-drobe") would not suggest hearing it or thinking of it that way (and in any case, the verb "to ward" is quite a stretch when one is talking about storing clothes in a closet). As for "eavesdrop", well, that's just as opaque as "understand", "handsome", "cockpit", and "cocktail".

Often we hear one part of a compound word quite clearly inside the word, and the other part less clearly. Thus, people called "gentlemen" are always of the male sex (showing that the second component is heard loud and clear), but they are certainly not always gentle. It is perfectly possible to say, "Would the rowdy gentlemen in the corner of the room please pipe down?" On the other hand, a freshman is always fresh (in the sense of being new), but only about half the set of freshmen in a typical high school are men. A great-grandson and a great-grandmother are unlikely to be particularly great or grand, but the former is sure to be somebody's son, and the latter to be somebody's mother. A restroom is certainly a room, but seldom if ever is it a place to take a rest.

Compound words whose components are still at least blurrily heard in the whole can be a bit tricky when it comes to pluralization, because one isn't sure to what extent one hears their parts resonating inside them. Thus when we sit in a café, do we gaze at the passersby or the passerbys as they stroll before us? And how many teaspoonsful of sugar do we add to our coffee? That is to say, how many teaspoonfuls? Are we thinking of giving our children jacks-in-the-box when Noël rolls around, or contrariwise, jack-in-the-boxes at Yuletide? On the golf course, do we aspire to make holes-in-one, or would we prefer the glory of hole-in-ones? And in golf tournaments, do we beam if we are runners-up or are we disappointed to be mere runner-ups? As married folk, are we fond of our mothers-in-law while finding our father-in-laws rather stuffy? And turning the tables, how do those respected elders feel about their sons-in-law and daughter-in-laws?

Looking at the statistics of the rival plurals for compound nouns of this sort gives one a sense for where those nouns lie along the slippery slope on which the parts slowly "melt", over time, into the whole. But once the parts have truly been absorbed into the whole, then the whole becomes truly a single unit, and no one hears the pieces any longer. Thus "handsome" might as well be spelled "hansim", "nobody" might as well be spelled "gnobuddy", "cupboard" "cubberd", and so on — and of course we have all seen "donut", "hiway", and "hijack", which show the parts as they make their way towards absorption (much like the vestigial "five" and "ten" inside "fifteen" and the vestigial "two" inside "twelve" and "twenty").

Often compound words have drifted so far from their etymological roots that native speakers can easily miss what is right in front of their eyes. Thus in German the word for "nipple" is "Brustwarze", which, broken up into its parts (the two nouns "Brust" and "Warze"), means "breast-wart". Once again in German, the word for "glove" is "Handschuh" ("hand-shoe"), and the French word for "many" is "beaucoup", which, decomposed, is "beau coup" — that is, "beautiful blow". But no native speaker would hear these words in the way that they strike us — namely, as ugly or strange — because over time, they have melted together to make category names that are seamless wholes and which therefore feel completely bland.

How could the native speakers of these languages possibly fail to see (or hear) something that is so blindingly obvious? Is it really possible? Well, yes — it's just as possible for them as it is for us anglophones to fail to see or hear the "dough" and the "nut" inside "doughnut", or the "break" and "fast" inside "breakfast", or the "under" and "stand" inside "understand". And keep in mind that no one flinches at the overtly sexual allusions in the common terms "male plug" and "female plug".

Opening the Door Doesn't Require Taking the Lock Apart

"Front door", "back door", "doorknob", "door knocker", "dog door", "dog dish", "dish towel", "dishwasher", "washing machine", "dining room", "living room", "bedroom", "bathroom", "bathtub", "bath towel", "towel rack", "kitchen table", "tablecloth", "table lamp", "lampshade", "desk chair", "hair dryer", "grand piano", "piano bench", "beer bottle', "bottlecap", "toothbrush", "toothpaste"... Here, without our once setting foot outdoors, are some compound words or phrases that designate familiar household sights. Some are written with a space between their components, and some are not. Fairly often it takes a trip to the dictionary to find out which ones take a space and which do not, and at times the official word handed down from on high runs against the grain or seems totally arbitrary, and moreover the official spelling frequently changes as one traverses the Atlantic or the decades. Indeed, from a psychological as opposed to an etymological point of view, the presence or absence of a space (or sometimes of a hyphen) makes no difference to the typical language user (or language-user), who is unaware of such fine points and will usually just improvise in writing such things down. One's point will be made equally well whether one writes "door knob", "door-knob", or "doorknob".

Although the types of words (and phrases) shown above have visible, hearable inner parts, these expressions are every bit as much the names of mental categories as are "simpler" nouns, such as "chair", "table", and "door". These longer words and phrases are, just like the things that they name, wholes that are built out of parts. And yet, no more than we need to understand a physical tool in order to use it do we need to take apart a compound word or phrase in order to use it. We use our dishwashers and our loudspeakers as wholes or "black boxes", undismantled and unexamined, and much the same holds for their names.

This observation has important consequences. Contrary to what one's intuition might suggest, using a compound noun or phrase rather than a "simple" word does not mean that more cognitive activity is required to understand it, or that the named category resides at a higher level of sophistication. When we hear "living room", for instance, it doesn't mean that first we activate the most general concept of *room* (which includes dining rooms, bedrooms, bathrooms, restrooms, waiting rooms, etc.) and then maneuver inside the abstract space of *room*-ness until we locate the appropriate subvariety. Our concept *living room* enjoys the same status as do "simple" concepts such as *room* or *bed*. In other words, the fact that "living room" is a compound word doesn't cast doubt on its status as the name of a stand-alone mental category. The same holds true for "bottlecap". Understanding this word doesn't require locating it among the subcategories of the concept *cap*, which include polar caps, yarmulkes, dental crowns, and lens covers. Cognitively speaking, *bottlecap* is no less simple a concept than are *cork*, *plug*, and *lid*, which, like bottlecaps, are devices for closing containers of liquid.

Jumping around from language to language helps make this idea clearer and more believable. Thus to express our simple noun "counter", French uses three words — namely, "plan de travail" ("surface of work") — while Italian uses just one — "banco". Our two-word noun "dish towel" is merely the atomic "torchon" in French and the slightly molecular "strofinaccio" ("wiper") in Italian; similarly, our "living room" is merely "salon" in French and "soggiorno" in Italian. Our compound noun "bedroom" is "chambre à coucher" ("room for sleeping") in French but merely "camera" (not a compound noun) in Italian. And our "camera" is "appareil photo" ("photo device") in French and "macchina fotografica" in Italian. And oddly enough, our compound noun "video camera" is simply "caméra" in French and "telecamera" in Italian. The moral here is that what seems like a blatant compound in one language may perfectly well seem atomic — that is to say, unsplittable — in another language. (Speaking of atoms, the indivisible English word "atom" comes from a compound word in the original Greek — "a-tomos" — meaning essentially "without a cut" or "part-less". Thus, as was wittily pointed out by David Moser, the word "atom" is an unsplittable etym in English despite not being so in the original Greek, and contrariwise, physical atoms are now known to be splittable despite what their etymology would suggest.)

In order to understand a compound noun, we do not need to break it down into its parts and then put together their "simpler" meanings in order to figure out what is being spoken of. To be sure, we are all aware that the words "bath" and "room" are found inside "bathroom", and that "tablecloth" means a piece of cloth that one spreads out on a table, but we don't need to take those words apart to understand them — no more than we do with "afternoon", "psychology", or "atom" — unless there is a special context that calls for it, such as explaining their meanings to a foreigner or a child.

By Concealing their Constituents, Acronyms Seem Simple

A widespread linguistic phenomenon that clearly illustrates the universal human tendency to represent complex concepts by short chunks whose parts are clearly "there" and yet are seldom if ever noticed is that of the creation and propagation of acronyms. Among the earliest-known acronyms are in Latin: "SPQR", standing for "Senatus PopulusQue Romanus" ("The Roman Senate and People") and "INRI", standing for "Iesus Nazarenus Rex Iudaeorum" ("Jesus the Nazarene, King of the Jews"). For ages, letter writers have used "P.S." ("post scriptum"), and mathematicians, not to be left behind, have for centuries used the classic Latin abbreviation "QED" ("quod erat demonstrandum" — "which was to be demonstrated") to signal that the end of a proof has been satisfactorily reached. For centuries the British have used "HRH" (His/Her Royal Highness) and "HMS" (His/Her Majesty's Ship), and of course there is the famous old call for help, "SOS" (Save Our Ship).

In the early twentieth century, the tendency to reduce stock phrases down to either the initial letters or the initial syllables of their component words grew more widespread, with such examples as "Nabisco" (National Biscuit Company), "Esso" (Standard Oil), "Texaco" (The Texas Company), "GBS" (George Bernard Shaw), "FDR" (Franklin Delano Roosevelt), "RCA" (Radio Corporation of America), "CBS" (Columbia Broadcasting System), and so on. And as the century progressed, the tendency gradually heated up, and the acronymic world started becoming more and more densely populated, with such well-known denizens as:

TV, LP, UFO, ESP, BLT, LIRR, ILGWU,
SPCA, PTA, YWCA, RBI, HQ, BBC, AA, AAA

Most fluent adult speakers of American English today should be able to say without too much trouble what lurks behind most of these acronyms, although perhaps a few of them will elude solution because, several decades after having been coined, they have run their course and are becoming dated.

A number of twentieth-century American political figures were popularly known by their initials (*e.g.*, JFK, RFK, MLK, and LBJ); indeed, it is said that Richard Nixon was intensely jealous of JFK's having been thus "canonized", and dreamed of becoming canonized as "RMN", although that monicker never caught on.

By the end of the twentieth century, what had a hundred years earlier been just an amusing little novelty had become an unstoppable tsunami, with opaque sets of initials coming at speakers of English left and right. And although our stressing their opacity may make it sound as if we are pointing out a defect, it is precisely that quality, paradoxically, that makes acronyms so catchy and so cognitively important, as we will discuss below.

We give the following sampler of acronyms in various fields as a set of challenges for the reader to unpack into their constituents. Although many will be fairly easy, others will probably be hard, either because they are almost never unpacked or because they are now growing obsolescent or are already obsolete:

computers and information technology: WWW, HTML, CRT, IT, URL, PDF, JPG, PC, CPU, CD-ROM, RAM, SMS, PDA, LED, GPS;

banking and finance: ATM, SEP-IRA, GNP, VAT, NASDAQ, NYSE, IPO;

automobiles: HP, MPH, MPG, RPM, GT, SUV;

companies: GE, GM, IBM, AMOCO, BP, HSBC, AT&T, HP, SAS, TWA;

business: CEO, CFO, CV, PR, HR;

chemistry and biology: TNT, DNA, RNA, ATP, pH;

communication: POB, COD, AM, FM, VHF, TV, HDTV, PBS, NPR, CNN, ABC, NBC, CBC, CD, DVD, WSJ, NYT;

photography: SLR, B&W, ASA, UV;

medicine: MD, DDS, AIDS, HIV, ER, ICU, ALS, CLL, DT's, HMO, STD, MRI, CAT, PET;

entertainment: PG, T&A, HBO, MGM;

labor: AFL-CIO, UAW, IBEW;

government: AEC, HUAC, DOD, DOE, FDA, NSF, CIA, FBI, NIH, NASA, SSN;

the military: GI, AWOL, MIA, MAD, ICBM, NORAD, USAF, USN, ABM, SDI, WMD;

education: GED, BS, BA, MA, MS, MBA, PhD, LLD, SAT, LSAT, MCAT, TOEFL, TA, RA, ABD, MIT, UCLA, USC, UNC, UNLV, UTEP, SUNY, CCNY;

sports: AB, HR, RBI, ERA, TD, KO, TKO, QB, NBA, NFL, NCAA;

organizations: AMA, AAAS, APS, UN, UNICEF, UNESCO, PLO, IRA, MADD, NAACP, NRA, NATO, IMF;

cities and countries: LA, NYC, SF, SLC, DFW, UAR, UAE, USA, UK, USSR (CCCP), PRC, GDR

miscellaneous: WASP, FAQ, LOL, BTW, IMHO, R&R, VIP, PDA, AKA, LSD, RSVP, OED, MOMA, GOP;

and so forth. Our challenge list is, of course, just the TOTI.

The Utility of Acronymic Opacity

The purpose of acronyms, and the reason that they are so popular all around the world these days is, of course, that each one takes a long (sometimes very long) and complex linguistic structure and makes it much simpler and more digestible, by sweeping the parts under a kind of "linguistic rug", or, to change metaphors, by making a black box that carries out its function very efficiently but into which no one ever bothers to peer, or at least not very often. The parts of acronyms are deliberately buried so that listeners and readers won't see them, can't get at them, and thus will not be distracted by mental activity going on at too fine a level of detail. Listeners and readers are meant to focus on a higher, more relevant, more chunked level.

Indeed, the parts of an acronym are hidden by a kind of membrane or "skin", making a concept that might otherwise be off-putting become palatable and even sometimes pleasing by its relatively simple, attractive packaging. Thus for most people, "DNA" is easy to remember, while "deoxyribonucleic acid" seems forbiddingly technical and complex. The fact that "DNA" seems to mean nothing at all whereas "deoxyribonucleic acid" clearly *does* mean something is precisely the advantage of the acronym. It becomes much more word-like and much less like a technical term.

When an utterance uses an acronym instead of the full phrase that it stands for, the number of visible parts in it is smaller than it would have been, as several pieces have been chunked into a single piece, and so the processing by the mind is easier. The principle here is similar to that of checkout lanes marked "10 items or fewer" in grocery stores, where a pack of six bottles of beer counts as just one item, as does a bunch of grapes with 100 grapes, and a bag of sugar containing a million grains of sugar. If each beer bottle were autonomous, if each grape were wrapped in an individual small bag, or if sugar were sold by the grain (heavens forbid!), it would be quite another story. Just as chunking of grocery-store items greatly simplifies the processing, so does linguistic chunking in acronyms. Our short-term or working memory does not get overloaded by too many items.

As an example, consider the following hypothetical announcement, which may seem a little heavy in the acronyms, but compared to much of the bureaucratic email we receive, it is actually pretty tame:

MIT and NIH announce a joint AI/EE PhD program in PDP-based DNA sequencing.

As is, it contains about fifteen "words", but if it is unpacked into more old-fashioned English terminology — "The Massachusetts Institute of Technology and the National Institutes of Health announce a joint artificial-intelligence and electrical-engineering doctor of philosophy program in the sequencing of deoxyribonucleic acid based on parallel distributed processing" — it would be over twice as long. And is it clearer or more confusing in this unpacked version?

An actual bureaucratic email contained the following noun phrase:

the URDGS IT Training and Education Web Markup
and Style Coding STEPS Certificate Series

In this phrase, "URDGS" stood for "University Research Division and Graduate School", "IT" for "Information Technology", and "STEPS" for "Student Technology Education Programs". Thus if one unpacks all acronyms (and does not rephrase in an attempt at increasing clarity), one gets the following:

the University Research Division and Graduate School Information
Technology Training and Education Web Markup and Style Coding
Student Technology Education Programs Certificate Series

This is quite a proverbial mouthful, and it certainly taxes one's linguistic processing capability at or beyond its limits. The phrase with the acronyms is still hard to parse, but it comes closer to being humanly parsable.

Using acronyms is a favorite device of bureaucrats, but it's also popular usage, because if they're used in moderation and with care, they can be very helpful. Because our technological society is growing in complexity in many ways at once, we simply have to have ways of ignoring the details underlying things, whether they are physical or linguistic. A typical teen-ager's cell phone, for instance, has many millions of times more parts than does a grand piano, for instance, and yet because of the way it has been cleverly engineered for user-friendliness, it probably seems far simpler than a piano to the teen-ager. Just as we need to hide the massively complex details inside our fancy gadgets by elegant and user-friendly packaging, so we need to hide the details of many ideas in order to talk about them in a sufficiently compact way that we won't get lost in a mountain of details. And thus acronyms flourish.

Furthermore, acronyms become more and more opaque over time, like metaphors. Just as we speak of "dead metaphors", so we could speak of "dead acronyms". For instance, probably most people today do not realize that the following words first saw the light of day as acronyms:

yuppie ("young upwardly mobile professional")
laser ("light amplification by stimulated emission of radiation")
radar ("radio detection and ranging")
modem ("modulator–demodulator")
snafu ("situation normal all fucked up")
scuba ("self-contained underwater breathing apparatus").

And indeed, who would want to think of, or say, "radio detection and ranging" instead of just "radar", or "light amplification by stimulated emission of radiation" instead of just "laser"? Cognitively, we *want* these membranes to be opaque. Just as we are happy not to see people's veins, intestines, brains, and other internal organs, so we don't want to be constantly reminded of all the infinite details inside the things we deal with on a daily basis. We want our eyes to be closed so that we can see better. In a word, we want to be spared looking at the trees so that we can clearly make out the forest.

Catholic Bachelors who are Jewish Mothers

Few dictionaries would have an entry for the compound noun "Jewish mother". And yet despite this lack, the phrase is the name of a well-known and fairly easily described category. At its core is the notion of an extremely overprotective, constantly worrying, ever-complaining mother, so much so that she wants to know everything about her children's lives, and to control everything in them. Her children are the entire focus of her life and she wants to be the same for them. Clinical psychologists might find Jewish mothers to be interesting case studies, while other people might tell jokes about them, caricaturing the nature of the category that they belong to:

You know she's a Jewish mother if, when you get up at night to go to the bathroom, your bed is already made when you come back to go to sleep.

A Jewish mother considers her fetus to be viable when it has finished medical school.

A Jewish mother calls up the airline and without a word of prelude asks, "Excuse me; when will my son's plane be arriving?"

Simon calls his mother up and says, "Hi, Mom, how are things?" "Oh, they're fine, Simon." "Oops! I'm sorry, ma'am — I must have dialed the wrong number."

The curious thing about this expression, revealing that it names something quite different from what its two sub-words would suggest, is that members of the category *Jewish mother* don't need to be Jewish, nor even mothers. A father, a grandparent, a co-worker in one's office, someone in a bureaucratic hierarchy — all of them can be members of the category *Jewish mother,* as long as they exhibit its more central and crucial features. Consider the following scenario, for instance.

> One of William's co-workers has taken William under his wing. He does all he can to help William rise up the company's ladder, taking it for granted that for William professional ascent is the absolute number-one priority. In fact, he wants William to consider him to be the linchpin of his professional life, so much so that he looks downcast whenever he sees William talking to any other co-workers. Not only does he advise William professionally, but he's taken it upon himself to give William personal counseling. He is convinced he knows what's best for William. In addition to making sure William gets promoted, ever since he found out that William is single, he's gotten into playing the role of matchmaker as well.

Calling William's intrusive and oversolicitous co-worker a "Jewish mother" involves dropping some of the *a priori* expected requirements for membership in the category — specifically, that it should involve a biological mother, that the person should be a woman, that there should be some kind of parental link, and of course that the person should be Jewish. Indeed, the key characteristics of a *Jewish mother* don't devolve from or imply any kind of religious beliefs. Thus a single and childless Catholic man — even a priest — could easily belong to the category *Jewish mother*, and contrariwise, many Jewish mothers are at best weak members of the category *Jewish mother.* What matters most of all for us to see someone as a *Jewish mother* is that the category's most stereotypical characteristics (overprotecting; kvetching; deriving one's main satisfaction from the successes of another person; giving boundlessly and expecting boundless reciprocation thereof) should be present to a sufficient degree, because it is they that most crucially help us to recognize members of the category.

Our ability to make analogies is what allows us to extend this particular category so that it includes all sorts of entities, such as William's co-worker, that share the category's most central characteristics independently of whether the surface-level description of those entities is consistent with the verbal label. When a category is deeply enough rooted in one's mind, its standard verbal label is but a relic reminding one of the early stages of the category's creation, rather than a fence sharply setting off the category's boundaries.

A Modest Sampler of Idioms

So far, we have been looking at rather short phrases. But what about phrases that stretch out a bit longer? Below we offer a sampler of idiomatic verbal phrases, none of which should strike a native speaker of American English as particularly strange:

to be up to one's ears in work, to go in one ear and out the other, to roll out the red carpet, to roll one's sleeves up, to be dressed to the nines, to be in seventh heaven, to be dead as a doornail, to wait until the cows home, to burn the candle at both ends, to swallow one's pride, to eat humble pie, to take it for granted, to kick the bucket, to let the floodgates open, to drop the ball, to catch the drift, to be caught off guard, to get away with murder, to read between the lines, to read the handwriting on the wall, to lick someone's boots, to have the time of one's life, to drop something like a hot potato, to throw someone for a loop, to throw someone into a tizzy, to get a kick out of something, to play it by ear, to bend over backwards, to fly in the face of the evidence, to tie the knot, to get hitched, to open a can of worms, to scrape the bottom of the barrel, to drop a bombshell, to be caught between a rock and a hard place, to paint oneself into a corner, to eat one's words, to let the cat out of the bag, to spill the beans, to be knocking at death's door, to play the field, to make a mountain out of a molehill, to shout at the top of one's lungs, to be scared out of one's wits, to act like there's no tomorrow, to take a rain check, to cry all the way to the bank, to cross swords, to drag someone over the coals, to hit pay dirt, to make hay while the sun shines, to rise and shine, to set one's sights on someone, to make someone's blood boil, to shout something from the rooftops, to lord it over someone, to even the score, to give someone a taste of their own medicine, to turn the tables, to miss the boat, to jump on the bandwagon, to have no truck with someone, to put the cart before the horse, to close the barn door after the horse is out, to while the hours away, to kill time, to spend like a drunken sailor, to get the hell out, to take it out on someone, to go for broke, to even the score, to be in the pink, to be riding high, to be down in the dumps, to throw the baby out with the bathwater, to carry coals to Newcastle, to scatter to the four winds, to open a Pandora's box, to be carrying a torch for someone, to get something for a song, to be whistling Dixie, to need something like a hole in the head, to tell it like it is, to be playing with a stacked deck, to make a long story short, to give someone short shrift, to be feeling one's oats, to sow one's wild oats, to butter someone up, to slip someone a mickey, to laugh on the other side of one's face, to hit the nail on the head, to miss the point, to make the grade, to lose one's marbles, to grasp at straws, to be on pins and needles, to run the gauntlet, to blow one's chances, to shoot one's wad, to keep one's cool, to throw a monkey wrench into the works, to screw things up royally, to look daggers at someone, to look white as a sheet, to be pushing up daisies, to send someone to kingdom come, to knock someone into the middle of next week, to cut the mustard, to cut to the chase, to jump ship, to crack the whip, to go belly-up, to be champing at the bit, to have one's cake and eat it too, to kill two birds with one stone…

This colorful list, illustrating the richness of the English language, names as many categories as it has entries. Suppose, for instance, that you're in a situation where you know something catastrophic might happen at any moment — for example, you have a heart condition that could trigger a sudden heart attack without warning. You might say, "I feel a sword of Damocles hanging above my head." You might also say this if you live in the San Francisco Bay Area, notorious for its seismic activity, and whose residents live in fear of "the big one" (that is, the next big earthquake, whose date is of course completely unknowable). You might also say this if you live with someone who, once in a blue moon, throws a terrible temper tantrum. You can surely think of many other situations that belong to this natural-seeming category — for indeed, that is precisely what "the sword of Damocles" names: a *category*, with all that that entails.

If you are taking your daughter to breakfast in a coffee shop where you know that the server has a tendency to sound gruff but in fact he has always been very nice to you, you may well tell her in advance, "Don't take the server's tone seriously — his bark is worse than his bite." You might also say this about an old car that occasionally makes some strange loud noises when you drive it but that has run smoothly for years without ever giving you the slightest problem. Your spouse might also say this about you if you have occasional fits of pique in which you let steam off vociferously, but the moment it's over you're as good as new and as friendly as can be. And you can surely think of many other situations that belong to the *bark-worse-than-bite* category.

Expressions of this type (long phrases that superficially seem very narrowly focused but that in fact have a very broad coverage) pervade spoken and written language, and one gains mastery of them much as one masters individual words. One gradually extends the category boundaries in just the same way as one does for shorter linguistic expressions — by noticing analogies between a new situation and the existing category. The actual words constituting the category's name — "a sword of Damocles" or "to jump on the bandwagon", for instance — merely hint at the full richness of the associated category, often revealing little if anything about its nature.

Did I Spill the Beans or Let the Cat out of the Bag?

Colorful expressions often denote categories that are quite different from what a literal reading would suggest. Indeed, a literal reading often has nothing at all to do with the expression's meaning. Thus who can explain why the phrase "to spill the beans" involves the action of spilling and, in particular, the spilling of *beans*? Why should *beans*, of all things, symbolize hidden secrets? And why would the act of dumping them out onto some surface be synonymous with revelation? Why couldn't the phrase have been "to tip over the broccoli", "to pour out the peas", "to flip the Brussels sprouts", "to drop the apricots", "to release the acorns", "to liberate the peanuts", "to free the fleas", or even (really stretching things to the limits of plausibility) "to let the cat out of the bag"? Of course there is a good *etymological* reason behind the real phrase, but that doesn't make it *psychologically* more convincing.

And yet every adult native speaker of English takes this phrase for granted. We all know that it means that *a small group of people were sharing some secret and one of them, perhaps deliberately, perhaps accidentally, couldn't resist the temptation of revealing the secret to a non-member of the cabal (most probably by simply blurting it out without any forethought), and suddenly the secret was no longer a secret, to the regret of all its members.* When it is spelled out explicitly this way, one sees how complex and subtle the category really is, and yet there is no hint whatsoever of all this complexity and subtlety in the few words that constitute its concise name.

And then there is another phrase — a cousin phrase — that might at times be considered synonymous with "to spill the beans" — namely, "to let the cat out of the bag". The two expressions both stand for situations in which once-secret information has, to the regret of certain parties, been revealed to a larger public. And yet the two phrases, for all their similarity of meaning, don't apply to exactly the same set of situations. That is, they are names of slightly different categories (whose members have a considerable degree of overlap). Thus when a member of a criminal gang reveals (whether to the police or just to an outsider) the gang's plans for wrongdoing, it's a case of *spilling the beans* (and probably not of *letting the cat out of the bag*), whereas when a married couple tells a few of their close friends very early on that the wife is pregnant, despite having earlier resolved that they would wait a few more weeks before telling anyone, they are *letting the cat out of the bag* (and probably not *spilling the beans*). These are close calls, and some native speakers might disagree (actually, in an informal poll of native speakers of English that we took, almost all fully agreed with our judgment), but what is undeniable is that most of the time, just one of these phrases will pop to mind while the other remains dormant, and the reason is that the evoking situation *fits* one of the two cases more than it fits the other. The subtle difference in flavor between the categories denoted by the two phrases is certainly not a standard piece of conscious knowledge on the part of native speakers (most of whom would be hard put to spell it out), but is simply something that is acquired over time as the phrase is encountered in a wide range of contexts. There is nothing in the phrases themselves that reveals these subtleties in even the slightest degree.

To convince oneself that idioms are often arbitrary, one need only take a look at a few foreign-language idioms, as they are frequently resistant to literal interpretation. Who would have guessed that "to let go of the piece" ("lâcher le morceau") and "to sell the wick" ("vendre la mèche") are the closest French expressions to "to spill the beans" and "to let the cat out of the bag"? And how do French people feel who *have the peach* ("qui ont la pêche")? Well, they are *full of beans* (that is, energy and good health). And what is a French mother doing when she *passes a soap* to her child ("elle lui passe un savon")? Why, she's *giving him what-for*, of course! And French people who proclaim that they'll *see the mason at the foot of the wall* ("c'est au pied du mur qu'on voit le maçon"), well, what they mean is that *the proof of the pudding will be in the eating.* All of this is *clear like some water of rock* ("clair comme de l'eau de roche").

The writer Jean-Loup Chiflet has played with English and French idioms in his books, taking English idioms and translating them "at the foot of the letter" (that is, literally) into French, and vice versa. The results are often very amusing, because as we've just seen, most idioms, if translated literally, make no sense. Thus "Our goose is cooked", familiar to any native anglophone, if rendered as "Notre oie est cuite", will bring a puzzled look to a French face. Likewise, "Il a vu des étoiles" ("He saw stars") and "Personne n'osa faire allusion à l'éléphant dans la pièce" ("No one dared mention the elephant in the room") will cause brows to be scratched. Conversely, literal translations into French of the English sentences "The carrots are cooked" and "He fell into the apples" will be colorful eye-openers ("seront des ouvre-œil colorés").

If our idioms sound opaque to people from other cultures but clear to us, it's because they have, over time, lost their evocative power for us and become *dead* metaphors — labels whose literal meanings are no longer heard by us but that jump out at foreign speakers. To them, such expressions appear at first to be live metaphors, and thus, quite understandably, they hope that a sufficiently dogged effort at making sense of the stream of words will, in the end, result in a flash of illumination.

Indeed, looking at the component words in an idiomatic expression might help someone who is unfamiliar with it, though it's always a bit risky; however, that method is bypassed by native speakers, who retrieve the appropriate abstract category directly from their memory, without proceeding via a literal, piece-by-piece understanding. If it were necessary to figure out every idiom's meaning from the words that make it up, then our understanding of speech, normally very rapid and seemingly effortless, would turn into a complex problem-solving session with no guaranteed results.

Behind the Scenes of Mundane Sentences

As we have seen, mental categories don't limit themselves to what nouns denote; verbs, adjectives, adverbs, conjunctions, and interjections are every bit as much the names of categories as are nouns. So are longer phrases. And for that matter, full sentences (or sentence fragments) that do not seem at all like opaque idiomatic phrases can constitute the names of categories as well. For instance:

> What's up? What's new? Just barely made it. Why does it always happen to *me*? It's your bedtime. Are you out of your mind? Who do you think you are? Just what do you think you're doing? And don't come back. I'll be right with you. Can I help you? How's your meal? The check, please. Will that be all? Anything else? You're more than welcome. Oh, great… that's all I needed. I told you so! Spare me the details. *That's* a likely story! I wasn't born yesterday! Don't give me that. Don't make me laugh. I've really had it. Well, what have we here? And who would *this* be? That's beside the point. There you go again! I've heard that one before. You can say *that* again! *Tell* me about it! Get to the point, would you? Give me a break! I'm no fool. I hope I've made myself clear. So *now* you tell me! Don't get me wrong! Well, I'll be damned! How was *I* supposed to know? Now why didn't *I* think of that? You want it *when*? Go jump in a lake! Have it your way. See if I care! Take my word for it. *That's* putting it mildly! That's no excuse. I wouldn't know. What makes you say that? You've got to be kidding! There's nothing to do about it now. Might as well make the best of it. It's not worth the trouble. Keep it to yourself. Mind your own business. You think you're so smart. So where do we go from here? Don't worry about it. Don't give it a second thought. Oh, you really shouldn't have! It could be worse! What won't they think of next? Shame on you! I don't know what I was thinking. That'll be a hard act to follow. No harm trying! So what? What do you want me to do? So what am *I* — chopped liver? Is *that* all you wanted? All right, are you done now? Haven't I seen you somewhere before? We can't keep on meeting this way.

Each of these sentences (or fragments) names a familiar category — not because it is an idiomatic expression, but simply because it is so commonly used in certain contexts that it has acquired a rich set of implications. These useful little formulas, built from simple words and utterly bland-looking, are in fact the names of important categories, as they pithily encapsulate certain notions that crop up all the time in everyday exchanges. What appears to be a freshly manufactured sentence is in fact a stored phrase that can be called up as a whole by a situation that a speaker is in, and the phrase carries standard connotations that go well beyond the literal sense of the words making it up, in the same way as, for a dog, its master's retrieval of the leash goes far beyond the mere prospect of having the leash imminently attached to its collar — it connotes going outside, taking a walk, smelling things everywhere, encountering other people and dogs, marking one's territory, and eventually returning home.

For instance, the sentence-level categories *It's your bedtime* and *So what's new?* and *Are you out of your mind?* are as crisp, clear, and rich with layers of implicit meanings as, for a dog, is the retrieval of its leash, and as are categories designated by idiomatic phrases.

The category *It's your bedtime* involves, to be sure, the idea that the child being addressed needs to go to bed very soon, but it also involves the idea that one has to sleep well to be alert in school tomorrow, the higher priority of school than of playing video games, the importance in life of good grades, the fact that in family life, parents are the bosses, and the fact that children need more sleep than adults do.

So what's new? conveys much more than just the desire to be informed about recent events. It says that one cares about the life of the other person, that one would like to have a chat, that one is concerned about how the other person is currently doing. When this category has been activated, the range of possible answers is fairly well defined: family, personal projects, professional activities. If someone answered "My shirt" when asked "So what's new?", it would be totally out of line with expectations, and would constitute a joke rather than an answer.

As for *Are you out of your mind?*, this rhetorical question reveals not just a sharp disagreement but a sense of surprise and shock, a fair degree of familiarity with the person addressed, and an aggravation, and it also implicitly asks for some kind of explanation or else a sudden turnaround on the part of the person addressed, and lastly, it warns that there is a potential fight brewing.

Just as a non-native speaker can gradually master the subtle art of choosing different flavors of greetings or thank-you's in another language, so a native speaker slowly acquires the mental categories that are designated by short everyday sentences or fragments like those exhibited above, whose subtlety and complexity are masked by the bland appearance of their constituent words.

Truths Lurking in Proverbs

Sentences of the sort we've just considered fit into daily life in a very frequent fashion, because they involve extremely common categories of experience, some of which are encountered multiple times in a single day: asking others how they are doing, saying how one is doing oneself, expressing disagreement, trying to figure out how much one disagrees with someone else, dealing cordially with people in a business role, suggesting that a conversation is approaching its end, and so forth. On the other hand, proverbs and sayings, although they are also frozen sentences, allude to situations that one may never have personally experienced but that nonetheless allow one to see events in one's own life from a novel and useful slant.

Proverbs are ideal illustrations of our book's thesis — that analogy-making and categorization are just two names of the same phenomenon. When, in a real-life situation, one finds oneself spontaneously coming out with "Once bitten, twice shy", "You can't judge a book by its cover", "A rolling stone gathers no moss", "An ounce of prevention is worth a pound of cure", "The early bird catches the worm", "Better late than never", "The grass is always greener on the other side of the fence", "When it rains, it pours", "If it ain't broke, don't fix it!", and so forth, the two sides of the coin of categorization through analogy-making are equally visible. Let's take a look at a particular example.

> Lucy, aged three, has just built a fence with her wooden blocks on the living-room floor. Jim, a family friend, accidentally bumps into her fence as he crosses the room and knocks over a couple of blocks with his foot. Lucy bursts into tears. A few minutes later, Jim is crossing the living room again, and as he approaches the same area, he conspicuously veers away from Lucy's fence and blurts out, "Once bitten, twice shy."

Now everyone will grant that Jim has come out with an *analogy*: he has implicitly mapped what just transpired in the living room onto a mythical situation in which an anonymous person, having been bitten one time by one dog (or some other animal), makes extra-sure never to go anywhere near any dog ever again. Obviously, the person is Jim, and the traumatizing bite is (*i.e.*, maps onto) Lucy's tears after the toppling of a block or two. The avoidance of all dogs henceforth maps onto Jim's pointed gesture of going far out of his way in order not to knock anything down the second time. What maps onto the fear of the bitten person? Clearly it's a more abstract concern than that of being hurt by a dog's teeth — it's the empathetic desire not to see Lucy in tears again. *Voilà* — there's the analogy, spelled out in full.

And yet we can just as easily characterize Jim's quoting of the proverb as an act of *categorization*, because he sees what has just transpired as a member of the public category *Once bitten, twice shy*. In quoting these four words, Jim has declared that this event belongs to that standard category. The very existence of the proverb in the mental lexicon of an English-speaking person amounts to the existence of such a category in their mind, and the triggering of the proverb by a particular situation reveals that at least for them (and hopefully for others), the situation is a member of that category. No less than public-category labels like "chair", "gentleman", "pacifier", "spill the beans", "go up in smoke", and "take matters into one's own hands", proverbs and sayings are the public labels of public categories — categories that most adult speakers know and share.

The act of recognizing in a given situation a case of a familiar proverb can cast new light on the situation. It provides a fresh, abstract, and non-obvious viewpoint, going well beyond the situation's superficial details. Since proverbs are the labels of rather subtle and complicated categories, slapping a proverb onto a situation is a way of bringing out aspects that otherwise might remain hidden. The use of a proverb as a label is a way of making sense — albeit perhaps a biased type of sense — of what one is seeing. Applying a proverb to a freshly-encountered situation results in a kind of insight that comes from filtering what one sees through the lens of the proverb, rather than from a purely logical analysis. In summary, a proverb is a convenient, concise label for a vast set of highly different situations — past, present, future, hypothetical — that are all linked to each other by analogy.

The experience-based (rather than purely logical) character of proverbs means that different people will see different proverbs (and hence will take different perspectives) in a given situation. For example, in France they say "L'habit ne fait pas le moine" ("Clothes do not make a monk"), while in England they say "Clothes make the man." Indeed, as Blaise Pascal once observed, "A truth becomes a falsity once it crosses the Pyrenees" (and probably he should have added "or the Channel"). As they say, "One man's meat is another man's poison", and this is certainly the case for proverbs. Thus, the sad tale of a nonconformist youth who was exiled and went on to become a famous poet but could never return home again (it could be the story of Dante) might be perceived by person A as teaching the important life lesson "To thine own self be true", while person B might see the selfsame story as exemplifying the wisdom of "When in Rome, do as the Romans do." A's selected proverb thus casts the story of the banished poet as a generic lesson that one should blithely ignore the masses and fearlessly step over the line in the sand, while B's selected proverb casts the same story as a lesson that one should respectfully follow the majority and cautiously toe the line.

The preceding examples are not in the least exceptional; there are enough pairs of mutually contradictory proverbs to make one's head spin:

Strike while the iron is hot...	but then again,	*Look before you leap.*
Good things come in small packages...	but then again,	*The bigger, the better.*
Nothing ventured, nothing gained...	but then again,	*Better safe than sorry.*
Two's company, three's a crowd...	but then again,	*The more, the merrier.*
Half a loaf is better than none...	but then again,	*Do it well or not at all.*
Absence makes the heart grow fonder...	but then again,	*Out of sight, out of mind.*
A penny saved is a penny earned...	but then again,	*Money is the root of all evil.*
Many hands make light work...	but then again,	*Too many cooks spoil the broth.*
Opposites attract...	but then again,	*Birds of a feather flock together.*
Don't judge a book by its cover...	but then again,	*Where there's smoke, there's fire.*
The pen is mightier than the sword...	but then again,	*Actions speak louder than words.*
It's never too late to learn...	but then again,	*You can't teach an old dog new tricks.*
He who hesitates is lost...	but then again,	*Fools rush in where angels fear to tread.*
Practice makes perfect...	but then again,	*All work and no play makes Jack a dull boy.*

We are tempted to add to this list one bilingual example — namely,

Pierre qui roule n'amasse pas mousse… but then again, *A rolling stone gathers no moss.*

Oddly enough, even if dictionaries compiled in Britain tend to agree with the French-language interpretation of this international proverb (namely, that by constantly moving about one never acquires any deep roots or anything of value), we have informally observed that most Americans hear this proverb in the opposite fashion. That is, they consider the gathering of moss to be an obviously bad thing to happen to a person (or a stone), and so from their point of view, the proverb exhorts people to stay constantly on the move in order to avoid acquiring a nasty crust. The irony is that although the English and French proverbs say the same thing on a word-by-word level, their interpretations are often quite opposite, and for Americans the meaning tends to be roughly, "Keep on rolling so you won't stagnate." Pascal might have said, "A truth becomes a falsity once it crosses the Atlantic."

But back to the main list… The fact that each line features a pair of proverbs that assert contradictory things shows that what counts is not a proverb's truth, but its ability to cast light on a situation, allowing it to be seen as more than simply a recitation of events. *Don't judge a book by its cover* and *Where there's smoke, there's fire* are categories that help one to highlight, on the one hand, the importance of not being distracted by cheap attention-getting tricks and of looking below the surface of things, and on the other hand, the importance of not ignoring what's right in front of one's eyes and of paying attention to salient clues. These two opposite stances, embodied in short and familiar phrases, can, if they form part of one's lexicon, be used to pin pithy labels on, and thus concisely categorize, novel situations that are very complex, thereby implicitly conveying entire attitudes about them.

The categories denoted by proverbs are not statements any more than other categories are statements. Thus the category *Don't judge a book by its cover* is not, despite its surface appearance, a statement (indeed, one shouldn't judge a book by its cover!) — no more than the category *table* or *bird* is a statement. It is a point of view that can be adopted on various situations. Just as the category *bird* is a platform for making inferences (if something is a bird, probably it flies, sings, has feathers, lives in a nest…) rather than a statement, so saying "Don't judge a book by its cover" is a sign of recognition that one is in a situation where prudence is called for in judgment, and where one should make sure to look well below the surface and to use one's critical faculties. And it's important to remember that this categorization of a situation, just like others, can be an inappropriate one. Just as one can assume that a small glass container filled with fine white grains contains salt rather than sugar, soon discovering one's mistake, so one can sometimes categorize a situation as belonging to the category *Don't judge a book by its cover*, only to realize later that this was an ill-advised judgment. In some cases, books *are* in fact perfectly represented and appropriately judged by their covers, and in some life situations, making a snap judgment based solely on surface-level cues can in fact be crucial. A person who intones "Don't judge a book by its cover" has not necessarily put their finger on the crux of the situation that they have so labeled. It may well be a *Where there's smoke, there's fire* situation instead.

A Stolen Cell Phone can "Be" a Dog Bite

Obviously, "Once bitten, twice shy" goes way beyond the idea that someone who has suffered a dog (or snake) bite will henceforth steer clear of all dogs (or snakes) at all times. Although the proverb is ostensibly about animal-bite victims, it is really about any number of other situations whose details are completely unforeseeable. What counts is that those other situations should share a conceptual skeleton with the micro-event conjured up by the four words in the proverb. Thus, we could easily see any of the following situations as meriting the label:

After marrying, A. had two children, and then her marriage started falling apart. She found out that her husband had been cheating on her and lying to her for years. It ended up in a very painful divorce. Ever since then, A. has been suspicious of all men, no matter how gentle and kind they are.

B. and C. are from China and live in San Francisco. One day, their son was the victim of racial taunts from a classmate in his public school. The next day, his parents pulled him out of that school and enrolled him in a very expensive private school.

While walking down a steep staircase, D. slipped and fell down several stairs. Although his fall had no serious effects, when he got back to his feet, he was trembling, because he knew he could easily have broken several bones. For the next two days, everywhere he walked, D. took exceeding care. While going up and down his stairs at home he went at a snail's pace, and the mere idea of riding his bicycle struck him as the height of insanity.

After her apartment was burglarized, E., who till then had paid no attention to safety matters, all at once bought a fancy burglar alarm as well as the most expensive safety locks, and she promptly installed the locks on all her doors and windows, including her basement windows, which were so small that for anyone to break in through them would have been well-nigh inconceivable.

F.'s cell phone was stolen in broad daylight by a mugger in the middle of the street in a somewhat dangerous part of town. Ever since then, whenever he uses his cell phone, F. is constantly on the highest level of alert, looking all around himself with great nervousness, even when he is in swanky hotels or ritzy restaurants.

As this shows, "Once bitten, twice shy" conveys the idea that when some event leads to negative consequences, some people develop a hypersensitive avoidance strategy, even at the price of missing out on potentially excellent opportunities, in order not to re-encounter any situation that is even vaguely reminiscent of the triggering one, no matter how little risk it would seem to pose objectively. More succinctly, in the wake of a painful event, people tend to be skittish about events that remind them, however superficially, of the original event.

This idea, having to do with the aftermath of a traumatic event, is not self-evident. The idea that an emotional shock can have lasting negative consequences — that there can be "wounds to the soul" — became acceptable only in the last hundred and some years as a psychological notion. Trauma, originally thought of solely as *physical* damage to a living being, was extended to the realm of *psychic* damage when it became part of the received wisdom that deep emotional shock can cause long-lasting repercussions, which suck the victim into a vortex of changes at many levels, sometimes reversible, sometimes not.

The Irrepressibility of Analogical Associations

Several languages, including Turkish, Italian, Spanish, German, and French, have proverbs about the irrepressibility of seeing certain analogies. Thus in French one says, "Il ne faut pas parler de corde dans la maison d'un pendu", and it has a very rare English counterpart, "One mustn't speak of rope in a hanged man's house", and, even more obscure, "One mustn't say 'Hang up your fish' in a hanged man's house". The idea expressed by such proverbs is of course that people cannot help making analogical associations at the drop of a hat, and that everyone should be sensitive to this fact. Thus, even if one innocently wishes to allude to a piece of rope that was used to tie a package, or to say that some fish should be hung out to dry, it would be boorish to do so in the presence of the family of someone who had been hanged. The hanging would be vividly present in the uttered words, no matter how the thought was phrased. And so in certain circumstances, certain things cannot be said or even hinted at.

This proverb tips its hat to the fluidity of human cognition, but of course it doesn't tell the whole story. Indeed, the spontaneous retrieval of proverbs, triggered by situations one encounters (as described in the book *Dynamic Memory* by cognitive scientist Roger Schank), shows that our everyday perception goes far beyond just seeing the hanging of a loved one in a mention of rope. When a proverb comes to mind in but a fraction of a second, a link has been discovered between two situations that would seem, on first glance, to have nothing whatsoever in common. For example, in the story where Jim, as he widely skirts Lucy's rebuilt wooden-block fence, suddenly blurts out, "Once bitten, twice shy", the connections exist only at a deeply semantic level. There was no dog, no bite, and no physical pain; instead there was an accidental kick, a falling block, and some psychic pain witnessed (in other words, not Jim's own psychic pain, but vicariously-experienced anguish). Rather than fearing a deliberate external attack bringing about his own physical pain, Jim was concerned about accidentally causing someone else mental anguish. And yet the analogy seemed obvious, even trivial, to him — a throwaway remark, a mere bagatelle, nothing to write home about — hardly a mental feat to be proud of. And for all the other *Once bitten, twice shy* situations given in our list above, one could make similar comments. There is no dog, but there is an "abstract dog"; there is no bite, but an "abstract bite"; there is often no physical pain, but just something that maps onto it. At the core of each event, however, there is a person who overreacts, sometimes wildly so, to an unpleasant situation. That is the crucial shared core.

The worldwide category *Once bitten, twice shy* pops up in many different verbal incarnations in various cultures around the globe, all of them superficially different, but tied to one another by their shared conceptual skeleton. It is interesting to notice how simple and down-to-earth each culture's quintessential situation is, a fact that makes the proverb's message seem very plausible, no matter what language it is in. Thus, for instance, in Romania people say, "Someone who gets burned while eating will blow even on yogurt." In Afghanistan, "Someone bitten by a snake fears even a rope." In China, "Bitten by a snake, frightened of tiny lizards." And of course in English-speaking countries, "Once bitten, twice shy." And thus this same category, whatever its surface-level linguistic guise may be, has a good chance of being evoked whenever (1) an event gave rise to negative consequences, and (2) a superficially similar event was subsequently avoided, no matter how unlikely it was to have negative consequences.

In France, the image is of boiling water scalding a cat, followed by the cat's shunning of all water, even cold. The fact that cold water cannot scald shows that the desire to avoid it is irrational, and thus that the caution is overdone. Likewise, while a snakebite is painful and harmful, neither a rope (superficially resembling a snake) nor a tiny lizard (a distant biological cousin) presents the slightest risk of harm.

Novel "proverbs" along the same lines can be created at will, which serve to label exactly the same category, or very close categories. The reader may find it amusing to play this game, giving birth to alternative versions of "Once bitten, twice shy". Here are a few sample pseudo-proverbs, just to set the ball rolling:

> Mugging victims flinch at their own shadows.
> Once fearless on ice, now fearful on driest dirt.
> Broke a bicuspid on a bone, balks at biting butter.
> Assaulted by one's enemy, afraid of one's best friend.
> Struck once by a stone, the cur now cringes at cotton.
> Robbed in the red-light district, terrified in a teahouse.
> A woman betrayed shuns even the most virtuous of men.
> Caught cold one winter; now dons sweaters each summer.
> One who's been through bankruptcy spurns the surest of deals.
> Little fingers smashed in doors will ever steer clear of doorknobs.

All the pithy phrases we've considered, whether taken from real cultures or invented by us, bring to mind and apply to situations centered on a traumatic event. In contrast to so many idioms that are impenetrable on the basis of just their component words, such as "to see red", "to sing the blues", "to be yellow", "to be in the pink", "to be in the black", "to spill the beans", "to shoot one's wad", "to fly off the handle", "to go on a wild goose chase", "to go Dutch", "to be in Dutch", "to say uncle", or "to be a Dutch uncle", a proverb has the twofold virtue of naming a category transparently and doing so in a catchy fashion. Indeed, unlike the preceding idioms, which, even if an etymologist could explain their origins, will still strike foreigners as being just as opaque and arbitrary as compound words such as "cocktail", "understand", and "handsome", proverbs readily conjure up easily visualizable scenarios — "All that glitters is not gold", "A leopard cannot change its spots", "A rolling stone gathers no moss" — and this tightens and strengthens the link between the category and its linguistic label.

The Proper Scope of a Proverb

How broadly does a proverb apply? How wide is the scope of situations that a given proverb can be said to cover, without one feeling that one is stretching things uncomfortably? As we have seen in the foregoing, the mental categories associated with proverbs have members that on the surface are extremely different. This means that such categories are very broad, and that they bring together situations whose common gist is located only at a high level of abstraction.

The French proverb "Qui vole un œuf vole un bœuf" has a relatively little-known counterpart in English: "He who will steal an egg will steal an ox." There is also a proverb in Arabic that says "He who will steal an egg will steal a camel." Someone might argue that these two proverbs express very different ideas, a camel and an ox being rather different beasts. Of course this takes things at a ridiculously literal level. In hearing either proverb, we are meant to understand something far more general than the notion that a male human being who has stolen an egg will one day also steal either an ox or a camel. We are supposed to infer, through our natural tendency to generalize outwards, that any person, male or female, who steals something smallish stands a good chance of going on and committing more serious acts of thievery later on. A schoolchild who swipes a candy bar may well steal Picassos as a grownup, or perhaps "Paper-clip filcher at five, hardened bank robber at twenty-five." But the intended lesson hidden behind the proverb's surface is probably considerably broader than that, since thievery is not really the point here — the targeted idea is bad deeds of any sort, including cheating on tests, engaging in fights, and so forth. The crux of the proverb is that bad deeds on a small scale can be but the initial step on a slippery slope leading towards subsequent bad deeds that resemble them but on a much larger scale.

Aside from the idea of scaling up the initial bad deed, it is also possible that as the bad deed grows in size over time, it also changes in nature, moving from an insult to an assault, from an assault to an assassination. The kid who steals a pencil from another kid's locker in school and then as an adult becomes a hired killer would thus be covered by "He who will steal an egg will steal an ox."

But we are not done yet, for who says that our proverb covers only crimes? Why not let the category flex a bit more, allowing it to cover all kinds of negative behavior, criminal or not? For example, being fresh to one's parents as a kid could lead to habitual aggressive language when one is grown up, or telling little white lies as a kid could lead to telling whoppers to one's spouse, or saying "Darn!" as a kid could be a prelude to swearing like a sailor when one is big. All of these cases would then be covered by "He who will steal an egg will steal an ox." Or would they? Where are the implicit, unspelled-out boundaries of this proverb's category?

Suppose we allow the scope to become more encompassing yet. We could, for instance, drop the idea that the behavior in question has to be negative. In that case, the proverb's meaning becomes roughly, "Small acts are a prelude to larger acts." This might mean, for instance, that a child who drops a penny in a beggar's cap stands a good chance of going on to head up a charitable organization when grown up, or that someone who starts a musical instrument when young will turn into a concert artist.

On the other hand, we suspect that most people would say that we've gone way overboard here — that expanding the scope of the proverb so that it applies to positive as well as negative actions, and not even caring about any similarity existing between the earlier and later acts that it is centered on, is not faithful to what it genuinely means. It's like taking the word "chair" to stand not only for all the standard chairs that people have deliberately designed over the millennia, but also for countless other physical objects, since a person can sit on just about anything. At that point, the word "chair" has lost most of its useful meaning. All this suggests that there is an optimal level of generalization of the proverb that does not dilute its meaning to the point of absurdity.

It is certainly too narrow to hear it as applying solely to acts of thievery, because the key idea seems to be some kind of slippery slope leading from small "sins" to larger sins of roughly the same type. To hear only an allusion to thievery in the proverb would be very limiting. Presumably, the proverb's purpose is to put people on guard concerning all sorts of negative actions early in life, so that they might try to prevent those actions from growing out of control as time goes by. "Nip bad acts in the bud!" would be the crux of the advice being given.

If, however, the scope is extended to actions without negative import, then the idea of being on guard against them no longer makes any sense. We don't need to be on guard against good deeds, don't need to nip them in the bud. To be sure, we can easily imagine a slippery slope leading from small good deeds to large ones — but that misses the proverb's point. In so doing, we will be sacrificing much of the "bite" of the proverb. Such a sacrifice might be seen as a standard consequence of the nature of abstraction, since by definition, "to abstract" means to abandon the less important aspects of what one is dealing with, but if a series of acts of abstraction is carried out without any attention being paid to intent, sooner or later the gist will simply be lost. Indeed, a small sin of abstraction may lead to a large sin of abstraction.

In the case of our proverb "He who will steal an egg will steal an ox", we could take things one level further in abstraction, not just ignoring the idea of magnification (from egg to ox), but also ignoring the sameness of the verbs in the two clauses and even any semantic relation between them; this would lead us to conclude that the proverb means that *one thing leads to another.* This extreme level of abstraction includes all situations in which there are causes and effects, but what good does such an extreme leap upwards do anyone? The richness of the original proverb is lost, and in fact, when carried to this stratospheric level of abstraction, "He who will steal an egg will steal an ox" winds up being no different from "Once bitten, twice shy."

Although jumping up the ladder of abstraction rung by rung may in some cases be a sign of intelligence and fluid thinking, if it is taken too far, it becomes a vacuous and frivolous game, and playing such a game with a proverb reveals an impoverished and superficial understanding of it. Indeed, in the end, an excess of abstraction winds up being similar to an excess of literality, because seeing any two things as analogous is no more insightful than not being able to see any analogies at all.

It would thus seem that there is an optimal level of abstraction, and that if we stop before reaching that level, we will exclude a host of situations that fit the proverb like a glove, such as the pint-sized swearer in nursery school who many years later turns into a volcanic spouse, and contrariwise, if we go beyond the optimal level of abstraction, we will let in a flood of irrelevant situations, such as the kid who at age six made three dollars selling lemonade at the corner and went on to become a billionaire in the soft-drink business at age sixty.

Jumping up to such a rarefied level of abstraction as a quest for some ultimate meaning of a proverb is reminiscent of a person who would label every object in sight a "thing", a "thingy", a "deal", or a "whatchamacallit", and thus would be prone to coming out with such abstract observations as "The thingy is sitting on the deal in the whatchamacallit", whereas most of us would find it clearer and more useful to say, "The pen is sitting on the desk in the living room." The greater specificity of the latter sentence strikes us as obviously preferable to the ambiguity of the former, but it's a matter of taste.

This recalls the cartoon figures called "smurfs", who have a highly abstract and concise lexicon, in which all nouns are covered by the super-abstract term "smurf", and other parts of speech are treated somewhat similarly. For example, they might enthusiastically announce, "We're off to smurf a smurf tonight!", and even if one didn't fully get the meaning of this utterance, it would be hard not to be caught up in the general feeling of excitement. And if the smurfs have a stock of proverbs, then it would contain such pearls of wisdom as "A smurfing smurf smurfs no smurf." Perhaps for the smurfs themselves, this phrase would be filled with insight, but for us its meaning remains elusive.

The problem with having only such an abstract mental lexicon — such a rarefied set of concepts — is the paucity of distinctions that it allows to be made, somewhat like the severe paucity of oxygen at rarefied altitudes. Abstraction has its virtues, which we will point out at an appropriate moment, but if one cannot draw distinctions, then thinking becomes as difficult as breathing at the top of Mount Everest.

From Eggs to Acorns, From Oxen to Oaks

Even if we forget about people who steal eggs and oxen, the notion that things can become bigger and better over time is everywhere to be found around us, for after all, grownups were once children; today's multinational giants were once fledgling outfits; Steve Jobs and Steve Wozniak made the first Apple computer in a garage before going on to found their legendary firm; Sergey Brin and Larry Page incorporated Google in a humble dwelling in Menlo Park; Albert Einstein first learned to read and write before developing his theories of relativity; popes were once priests in little churches; conquerors of Everest climbed small hills before moving on to the big time; major acts of philanthropy were preceded by minuscule acts of charity; every great friendship was once just a tentative affinity; virtuoso instrumentalists were once musical novices; every chess master had to learn the rules at some point; powerful ideas gave rise to modest fruit before resulting in huge advances… All of this is far from egg-stealers turning into ox-stealers, but it nonetheless deserves a proverb or two, along with the rich category that any proverb covertly symbolizes.

How might "He who will steal an egg will steal an ox" be converted into a more upbeat thought? The key question is to pinpoint the core idea that one wishes to capture in this category, since various possibilities, related but distinct, might be imagined. The most general version would simply be the idea that things on a large scale are the fruits, in some fashion or other, of a process of "amplification".

Thus, amplification can come from putting together a number of small items: a big thing is the "sum" of many little things. There are numerous proverbs that capture this idea in some form or other. For example: "Many a mickle makes a muckle" (whatever those items might be!); "Many drops make a shower"; "United we stand; divided we fall" (which paints both an optimistic and a pessimistic picture); "E pluribus unum". All of them get across the idea that individually insignificant entities, when put together in large numbers, can give rise to entities of great magnitude and strength.

However, in none of these is time's flow explicitly involved. If we are looking for an optimistic counterpart to the pessimistic egg-thief-to-ox-thief metamorphosis, then our goal is a proverb highlighting the slow but steady process of evolution or growth of a single good thing over a very long time. Thus we have the famous proverb "Mighty oaks from little acorns grow", and a more elaborated version of this thought, from eighteenth-century English writer David Everett: "Large streams from little fountains flow, tall oaks from little acorns grow."

A related notion of amplification over time involves putting together many small acts (or objects), one by one, over a very long period of time in order to accomplish a grand goal — in short, temporal accumulation: "A long journey starts with a single step"; "Rome wasn't built in a day"; "Little and often fills the purse"; "Drop by drop fills the tub" (or, seeing things in a time-reversed fashion, "Drop by drop the sea is drained", which, in French, is a genuine proverb); and finally, in a more destructive vein, "Little strokes fell great oaks", a homily found in *Poor Richard's Almanac* by Benjamin Franklin.

A clear semantic reversal of "He who will steal an egg will steal an ox" is illustrated by the fund-raising style of American universities, which might be summarized as "A little giver will one day a grander giver be", or, more poetically, "Mighty donors from humble tippers grow." Or then again, "Someone who donates a book today may donate a library tomorrow", or even "Butter up a million, smile and snag a billion." Finally, to give it a more antique tone, we might rephrase it thus: "A giver of eggs may one day give an ox." Or a camel. Or a dromedary. Or a dormitory.

In Memory Retrieval, We Are All Virtuosos

So far in this chapter, we have been discussing acts of categorization that result in the retrieval of composite lexical entities or ready-made phrases, including compound words, idiomatic expressions, ready-made sentences, and proverbs. In particular, our discussion of people's effortless understanding of proverbs they hear and their fluent insertion of proverbs into conversations was intended to bring to light some of the memory-access processes that we all engage in automatically and ceaselessly, processes that transpire in but fractions of a second and with truly impressive precision. This fluid fashion of tapping into deep, dormant reserves of memory is a variety of virtuosity, and far from being limited to a few gifted and highly trained individuals, it comes along free with the possession of a normal human brain.

Telling one's loudly protesting children in the back seat that it's important to fasten their seat belts, one comes out spontaneously with the phrase "Better safe than sorry" without having had any prior intention of quoting a proverb. Hearing from friends who returned from a weekend vacation to find that their teen-age daughter had thrown a wild party in the house during their absence, one finds oneself thinking, "When the cat's away, the mice will play!" Advising a friend who's applied for several long-shot jobs of which one has suddenly come through and needs an instant reply or it will be lost forever, one blurts out, quite off the cuff, "A bird in the hand is worth two in the bush!" (We momentarily interrup the natural flow to remind readers that in Chapter 1 we announced our intention to distinguish linguistic expressions from the concepts they denote by using quotation marks or italics, respectively. In this paragraph, though, all the examples are delicately poised between the two; indeed, they are all cases of the category named by the phrase "neither fish nor fowl". But we had to make a choice, and so, throughout the paragraph, somewhat arbitrarily, we opted for quote marks.)

Just as words like "bottle", "table", and "chair" strike us as being objectively *there,* staring us in the face, when we are in the presence of certain visual stimuli, so the various proverbs that we have just cited above (and many others as well, needless to say) can, on occasions like those just cited, simply materialize out of thin air in our minds, as if handed to us on a silver platter — and when this kind of effortless, instantaneous retrieval happens, they simply feel *right*, every bit as right as calling the objects in front of us "a bottle", "a table", and "a chair". On such occasions, the members of the abstract category seem to us to be objectively *there* and objectively *real* — just as real and almost as visible and tangible to us as are the material objects before our eyes, even though, of course, they are not visible or tangible in the way physical objects are. No less than there are seat belts and children in the back seat, there is a *Better safe than sorry* situation inside the car. No less than there is a frightful mess of half-empty beer bottles and a bevy of reveling teen-agers in the living room, there is a *When the cat's away, the mice will play* situation in the house. No less than there is a job offer dangling in cyberspace and a threat of losing it forever, there is an *A bird in the hand is worth two in the bush* situation floating in the air. As these examples demonstrate beyond any doubt, categories go far beyond what is labeled by single words.

Fables

The further we've gone in this chapter, the longer the labels of the categories concerned have grown. First we discussed words such as "pacifier", "understand", "handsome", "cockpit", "cupboard", "wardrobe", "pocketbook", and "eavesdrop"; these words are made of components that are no longer felt as such by those who use them fluently, so that the compounds have a unique flavor all their own, reminiscent of a process of sedimentation in which the initial constituents have all melted together, or of a rich, tasty sauce that is so subtle that only an expert chef can figure out what went into it. Then we looked at compounds whose components seem more transparently present, such as "bedroom", "airport", "bottlecap", and "Jewish mother". Next, we moved to prefabricated idiomatic phrases like "to drop the ball", "to catch the drift", "to be caught off guard", and "to jump on the bandwagon". Then we moved on to short ready-made sentences such as "What's up?" and "It could be worse", and most recently, to proverbs such as "Better safe than sorry". At every stage of the game, our point has been firstly that such expressions are the names of categories in our minds, and secondly that, thanks to analogical perception, we feel the presence of instances of those categories no less than we feel the presence of instances of categories whose linguistic labels seem far more atomic, like "table", "chair", "moon", "circle", "office", "study", "think", "spend", "much", "and", "but", and "hub" (and whose presence we also detect through analogical perception). This is a quintessential theme of this book.

So how about fables, now — short fanciful tales that wind up stating a moral? Might those, too, be the names of categories? We shall answer in the affirmative — not just sometimes, but in all cases, as long as the fable is clearly understood. Reading a fable allows one to construct a category that is succinctly summarized by its moral; the fable itself is just one member — a very typical member — of the category, among a myriad potential members. After the fable has been understood, then, as is the case for any category, new situations will from time to time be encountered that, thanks to an analogy perceived, are seen to have a common essence with the fable, and will thus add to the category's richness.

From then on, a fable will act much like a word. It becomes a label that jumps to mind when someone who has incorporated it in their memory runs into a situation that "matches" or "fits" the fable — not in a word-for-word fashion, obviously (fables are seldom memorized), but by an abstract alignment with its moral, or with its title, or just with a blurry memory of its basic plot. If a flat surface comes into view off of which a person is seen eating food, this is very likely to trigger the word "table"; if a few children come into view who are playing hopscotch, this is very likely to trigger the word "jump"; in much the same way, there are certain combinations of actors and events in which they participate that are very likely to trigger the retrieval of certain fables (or of fable-labels, at least). Our claim — a very serious one — is that fables jump to mind as situation-identification labels no less than do proverbs, idioms, compound words, and "simple" words, and any situation that evokes the memory of a particular fable will be perceived as a very real member of the category that lurks behind the scenes.

Scorning What is Out of Reach

Æsop is remembered for the fables he wrote in the sixth century B.C., of which one of the most famous is "The Fox and the Grapes". It was so successful that it passed down through the æons and was even adapted by a number of later authors who, despite changing its form, retained its content. The Roman fabulist Phædrus included it in his collection of Æsop's fables in the first century A.D., and in the seventeenth century, the French poet–fabulists Isaac de Benserade and Jean de La Fontaine did the same. Here are the versions by these four authors, all translated into English:

"The Fox and the Grapes", by Æsop (sixth century B.C.):

> A famished fox observed some grapes dangling above him, on a very high trellis. They were ripe and the rascal would very gladly have absconded with them. But jump though he might, the trellis was simply so high that he couldn't reach them. Seeing that all his efforts were futile, the fox strutted away with his head held high, declaring, "I could grab those grapes in a trice if I had the slightest interest in them, but they look so green that it's simply not worth the trouble."

"The Fox and the Grapes", by Phædrus (first century A.D.):

> Driven by hunger, a fox was lusting after some grapes on a high vine, and he jumped with all his might to reach them. But he failed, and as he walked away, he remarked, "They aren't yet ripe, and I don't want to eat sour grapes."

"The Fox and the Grapes" (after Æsop), by Isaac de Benserade (1612–1691):

We can't have all we seek, alas, as shows this little scene.
A picture-perfect bunch of purple grapes was dangling high.
To snag them, up jumped Fox, but missed, despite his valiant try;
He jumped and jumped until he sighed, "Those grapes are far too green."

Our Fox, to put it frankly, felt despair and rage and pain.
Perhaps more calmly, later, he would muse in tones forlorn,
"Those grapes were ripe for plucking — but we mortals always scorn
The things we strive for valiantly but never do obtain!"

"The Fox and the Grapes" (after Æsop), by Jean de La Fontaine (1621–1695):

A certain fox from Normandy (though others say the South)
Was at death's door from hunger, when he spied, upon a vine,
A tempting bunch of grapes that he would fain place in his mouth,
All covered with a lovely skin so red it looked like wine.

The plucky fox would happily have made of them a meal,
But due to the chance happenstance that they were up too high,
He snapped, "Who'd want such sour grapes? They're food fit for a heel!"
He just was letting off some steam, for foxes never lie!

And now our readers, having read these four versions of the classic fable, are invited to supply the missing conclusion to the following brief anecdote:

> Professor C. had campaigned very hard to be elected Head of the Department, but Professor A. won the election hands down. At that point, Professor C. declared to everyone in the Department…

Here we have no animals, no vine, no trellis, no grapes, and no hunger, but rather some humans, a university, a department, an election, and a lust for power. And yet the reader most surely guessed the kind of sentence that Professor C. might have uttered. For example:

> "I ran for office solely out of my selfless dedication to our department's welfare. However, having lost the election, I am at peace with my conscience and luckily I will not have to do a thankless job."
>
> "Whew! Now, at last, I'll be able to dedicate myself to my true passion — research."
>
> "This responsibility would have eaten me alive. I'll be much happier being able to devote myself to my family, and to watch my kids grow up."
>
> "This department is just a bunch of prima donnas. Lucky for me that I escaped the nightmare of trying to run it!"

What amazing psychological insight allows us to come up with these conclusions to a story that has so little to do with Æsop's fable? Well, of course, there is no miracle here. The story of Professor C. is clearly understood as belonging to the same category as the fable itself — namely, the category of *things that one once craved deeply but that one failed to obtain, and that one therefore disparages.* This category of situations is well known to many people in our culture who have never read Æsop's fable itself. The familiar expression "sour grapes" is a very standard label for such situations.

Curiously, although a roughly equivalent expression exists in French — "les raisins sont trop verts" ("the grapes are too green") — it does not enjoy anything like the popularity of its English counterpart. This phrase, borrowed from La Fontaine's rhyming version of the fable, appeared in the 1832 edition of the official dictionary of the French language published by the French Academy, and it has remained in the dictionary ever since then. But even a speaker of French who has never run into the expression is quite likely to have observed that people often deprecate things that they have failed to obtain; such a person has thus already constructed the category without being aware of it.

If one hasn't already created the category, then reading the various versions of the story of the fox and the grapes will naturally and easily lead one to manufacture it. Once the fable has been understood, the category thus created has a decent chance of being evoked on occasions when a failure to obtain something cherished is followed in short order by a revised estimate of how desirable the original goal was.

We'll now take a brief look at some flagrant cases of the category of *sour grapes* — short scenarios that should very easily trigger the category, especially in the mind of a native speaker of English, for whom the category comes pre-equipped with a familiar, standard label.

A. didn't want his son to go to the local high school and tried to get him into an elite private school whose admission standards were very high. When his son was not accepted, A. declared to everyone in hearing range that he was actually very glad it worked out this way, because now his son would get to live in an environment of great social diversity, rather than finding himself cut off from reality and surrounded only by arrogant and superficial people.

B. wasn't able to purchase last-minute airplane tickets to Hawaii and thus had to give up his elaborate vacation plans. But he said to his friends that he was in fact relieved, not disappointed, because all the best spots in the islands are always hugely overcrowded during vacation periods, and that ruins all the fun of going there.

C.'s great dream was to become an actor, but after suffering a number of rejections, he finally said he had dropped that goal and would look for a more conventional kind of career. He added that the unhealthy atmosphere in the world of acting would have corrupted him, and that he would be far happier leading a more balanced life far from the glare of the footlights.

D. has just been dumped by her boyfriend, whom she'd always described to her friends as "Mister Right". Now, though, she's telling all her friends that their breakup has taken a great weight off her shoulders, and that she can finally breathe again; deep down, she'd always known that their love affair was doomed, but she just hadn't been able to take the step of breaking off with him herself, because she hadn't wanted to hurt him.

E. learned that her favorite rock band was going to give a concert in her town. As fast as she could, she scrambled to get a ticket, but unfortunately she was too late; they were already all sold out. E. said to her friends, "The auditorium is so huge that no one will really be able to see anything at all; you're probably better off watching the concert on television."

All the situations in the list above belong to a single category whose members, though very different, all share the same core — namely, the moral of the fable of the fox and the grapes. Each of these scenarios exemplifies, in its own way, the notion of *failure followed by belittling of the original goal*, and they are all located quite close to the core of the category *sour grapes*, which comes from the fable itself. Although the resemblances among all these scenarios probably strike you as glaringly obvious and thus of no interest, it's that very fact that is so remarkable. We all tend to pay so little attention to the surface level in these stories that it is very easy to slip into the belief that seeing the *sour grapes* concept in all these diverse contexts is utterly mechanical and trivial; the truth, however, is that spotting this pattern beneath the surface is anything but a mechanical act. No search engine today is anywhere near being able to spot the deeper aspects of an anecdote like this, and to detect the *sour-grapes*-ness of all sorts of situations. Indeed, making these kinds of seemingly trivial perceptions has been a stumbling block for many years for researchers in artificial intelligence. Rapid spotting of this kind of essence is (at least so far!) a uniquely human capacity, and computers can only dream with impatience of that far-off day when they, too, will at last be able to perceive that two situations so different on their surface level are nonetheless "exactly the same thing". In the meantime, though, they all pooh-pooh the interest of such a goal...

How to Reduce Cognitive Dissonance in a Fox

Æsop's fox-and-grapes fable, more than two millennia old, insightfully anticipated some rather recent ideas. From the 1950's onwards, thanks to the pioneering work of social psychologist Leon Festinger, the notions of cognitive dissonance and its reduction have been part of psychology, and they are direct descendants of the fable, which, in expositions of the theory, is often given as a quintessential example. The basic idea of the contemporary theories is that the presence of conflicting cognitive states in an individual results in a state of inner tension that the individual tries to reduce by modifying one or another of their conflicting internal states. Thus, the fox is in a state of cognitive dissonance, since his desire to eat the grapes conflicts with his inability to reach them. He thus modifies one of the two causes of the conflict by denying that he wants to eat them. Since they are sour (so he says), they are no longer desirable, so his failure to reach them is no longer upsetting.

Much as the concept *once bitten, twice shy* contains the essence of the modern psychological notion that a traumatic experience leaves lasting after-effects in its wake, so the sour-grapes fable contains the essence of the notion of reduction of cognitive dissonance, and more generally, the notion of *rationalization*, where a painful situation is rendered less painful by the unconscious generation, after the fact, of some kind of arbitrary and often unlikely justification.

The blatant nature of the fox's lie makes the fable an ideal core member of the *sour grapes* category, and allows one to understand the structure of all *sour grapes* situations. The genius of Æsop was to have come up with such a simple, appealing situation in which dissonance is reduced. For this reason, his fable not only has survived many centuries but it also anticipated developments in modern psychology.

To see how the sour-grapes fable relates to the notion of cognitive dissonance in its full generality, one can cast the notion of *disparagement of an unrealized yearning*, which is the fable's crux, as a special case of the more general notion of *regaining a peaceful frame of mind by distorting one's perception of a troubling situation*, which is what the reduction of cognitive dissonance is all about. Equipped with this new category, we will far more easily and more rapidly recognize situations in which people spontaneously invent novel justifications, sometimes rather bizarre ones, in order to reconcile themselves with disappointing outcomes. And this new, more general category will start expanding in an individual's mind as that person encounters unexpected situations that have varying degrees of similarity to the most central members of the category.

Let's now take a look at a sampler of situations that might fit into the new, broader category.

F. has reserved a table for two in a fancy restaurant highly recommended by friends. However, he and his date are caught in a traffic jam on his way, and their reservation is canceled. F. says, "There are terrific restaurants everywhere around here; let's go find one ourselves. It'll be much more romantic that way."

G. has a tradition of buying slashed-price theater tickets from a special agency. Tonight is the last night of a play that's received rave reviews from the critics, but it is sold out, and G. has to give up his plan. He muses, "That's the first time this has happened to me in all these years of using this strategy. That's a pretty darn good track record!"

H. is drooling over a certain *à la carte* dish in a restaurant. When orders are being taken, the server has disappointing news for H.: they've just run out of her dish. "Oh, well," says H. with a philosophical, on-the-rebound chuckle, "this way I'll save myself hundreds of calories and some cholesterol to boot." And she orders a lighter, healthier dish from the menu, one that her eye hadn't been so drawn to when she was first scanning the menu.

I. has just learned that her deeply-desired request for a transfer within her company has been denied. "All right, then — so be it!" says I. "I'm not going to let it bother me; I'll just quit and get another job in another firm. And my chances to make headway in my career will be a lot better than if I had stayed in this stodgy old place."

J. wasn't admitted by the art school he'd applied to. He says that only people who pull strings ever get admitted there; that's how everything is in today's corrupted society. The thought of so much rampant injustice everywhere in the world makes him sick.

K., after several years of marriage, is taking stock. He still feels great affection for his wife, but physical passion and spiritual intimacy are largely things of the past. "Everything has a way of eroding with time," K. thinks to himself, "and so it is with our marriage. But even if what I feel isn't as intense as it once was, our love has grown ever so much deeper."

L. just barely lost an election to represent his district in the state legislature. Months of sacrifice and day-and-night work have gone up in smoke. But L. says to himself that failures are part of the learning curve of politics; through this defeat he is becoming broader and deeper.

All these ingenious rationalizations do the job, in one way or another, of reducing some kind of tension created by the gap between hopes and reality. But do we easily see these as cases of *sour grapes*? Probably not, and this is in part because the device of dissonance reduction is a defense mechanism, and it's in our own best interest not to be aware of how we protect our delicate psyches by deluding ourselves with defense mechanisms. If everyone saw through all defense mechanisms — their own as well as other people's — it would be a loss in many ways. In any case, the successful survival of the dissonance-reduction "trick" over eons reveals that we all have a certain kind of blind spot concerning it.

Sour grapes is certainly not the sole way of categorizing the string of anecdotes shown above. One could also see, in each of these (mis)adventures, a kind of instinctive wisdom in reacting to their various disappointments, an optimism that focuses on the positive and minimizes the negative. If we wished to categorize these little vignettes in different ways, we could focus on this aspect of the people's reactions, and could perfectly reasonably cast them as instances of the category *seeing the silver lining* (presumably at the edges of a storm cloud). Then again, some of them fit the category *walking on the sunny side of the street* or the category *counting your blessings* or the category *thanking one's lucky stars* or the category *being thankful for what you have.* And then again, some of them might arguably best be placed in the clichéd but nonetheless perfectly valid category *seeing the glass as half-full instead of half-empty.*

As is the case for any categorization, filing a situation under the "sour grapes" label is the result of a judgment call. We opened Chapter 1 with the idea that one is never confronted with a single isolated situation but is always in the midst of a vast multitude of situations, and that there is never just one single valid point of view to take but always a variety of reasonable points of view. For instance, in order to label a situation as "sour grapes", one has to recognize that someone saw a rare opportunity, tried to seize the moment, failed and was disappointed, and in the end came out with some dismissive comments that would never have been thought up had the original goal been obtained. The man who cheerily said that it's more romantic to chance upon a new restaurant while strolling about would doubtless have found dining at the posh restaurant that his friends had recommended to be boundlessly romantic had his reservation not been canceled. The great benefits of losing in politics would not very likely have sprung to the mind of the politician if he had just barely won the election instead of just barely lost it; in that case, he might more likely have thought to himself, "The world is my oyster!"

One way of verbalizing the essence of *sour grapes* situations is as follows: they are situations in which *disappointment turns a person into an intellectual opportunist* — that is, into someone who tries to paint a failure in rosy colors. The behaviors that are often called "seeing the silver lining" or "counting one's blessings" are somewhat different from this pattern. They involve a person searching for small positive aspects that lurk unseen in a mostly troubling situation. In contrast to *sour-grapes* situations, which involve the expedient distortion of one or more beliefs, *seeing-the-silver-lining* situations are ones in which the protagonist, though upset, does not distort any beliefs but instead is *selective* in terms of which beliefs to focus on.

The fox-and-grapes fable is (by definition!) a prototype of the *sour grapes* category. However, it is a very poor member of the *silver lining* or *blessing-counting* categories, because the fox does not foreground any positive aspects of the frustrating situation that he finds himself in. On the other hand, this fable is a good member of the *bad faith* category (that is, situations whose protagonist lacks honesty and sincerity). In this category are found many situations that have nothing to do with the reduction of cognitive dissonance. Some simple examples would be:

A person who would file false reports after having had an automobile accident;

Politicians who would distort facts about the current economic situation in order to boost their chances of re-election;

A kid brother who would scream, "She started it!" when he knows there is no truth at all to his claim.

Any situation permits a host of diverse categorizations. The category that winds up being selected will determine the perspective that colors how the perceiver interprets the various facts that constitute the situation. The range of situations that have been explored in the preceding sections shows the impossibility of thinking of categories as having fixed boundaries and interpretations as being unique. To the contrary, categories evoked by fables are, like all categories, overarching frameworks that guide interpretations. The mental act of categorization shines a particular light on a situation. Thus the fox-and-grapes fable gives a first, basic sense for situations that clearly involve *sour-grapes*-ness, and later it helps us recognize this quality when it is less obvious. In the end, the fable enriches us with a sense for the various creative ways that people manage to find comfort in situations that in fact bother them.

Lacunæ in a Conceptual Space

We now take up once again the theme of *conceptual spaces,* which we introduced at the end of Chapter 1 to describe the relations between languages and concepts. Each word or expression of a given language is thought of as a colored blob occupying a portion of a conceptual space (and each color is thought of as representing a language). The center of a conceptual space consists of the set of concepts most frequently used in a given culture. All the different languages that share the same culture cover the core of the conceptual space in different fashions, using blobs of different sizes and shapes, and of course different colors.

We also introduced the metaphor of a ring or shell of concepts, meaning those concepts that share approximately the same frequency or importance. We had built up the image of concentric rings or shells that, given a particular color, are filled up with blobs of that color, with each blob having its own unique shape and size, and representing a specific concept. At the core of a conceptual space, each color does an essentially perfect job of filling the space up, and as one moves outwards to rings that lie near the core, each color continues to do an excellent job.

If we keep on going out further, however, sooner or later we come to areas of conceptual space where single-word lexical items almost never suffice, and where each language has a quite different way of covering those zones. For instance, English has the phrase "it's nothing to write home about" (meaning "what happened isn't particularly thrilling or memorable", and if someone were to ask (quite reasonably), "How does French cover that zone?", the answer would be that it's not by reference to hypothetical postcards or letters that were never written or sent to one's family, but in a radically different fashion. The French get this same idea across by recourse to the colorful (although rather nebulous) phrase "ça ne casse pas trois pattes à un canard" ("it doesn't break three legs of a duck"). In some sense, the French phrase and the English phrase mean just the same thing, but they nonetheless convey the meaning in very different ways, since the concrete images that might be conjured up in the minds of speakers or listeners involve extremely different scenarios.

These differences between idiomatic expressions can be of any size. For instance, the closest French counterpart of the idiom "to be flat on one's back" (meaning "to be very sick") is "être cloué au lit" ("to be nailed to one's bed"), which conveys a similar but much more painful scenario. So we get discrepancies between different languages' ways of filling up a conceptual space not only in the sense that the blobs in the same part of space are shaped differently, but also in the way in which idiomatic phrases get across their messages.

Eventually, if we move out far enough, genuine *holes* will start to turn up — patches in conceptual space that one language covers very neatly with a single blob, but that the other simply doesn't cover with any standard word or phrase, no matter how voluminous is its repository of lexical items. A typical example was given in the Prologue, where we mentioned the lack in Mandarin of a generic verb meaning "to play" applicable to any musical instrument.

Thus, given any idiomatic phrase in English (and there are untold thousands of them), it makes reasonable sense to ask, "How does one say this in French?", because fairly often there is a nearly-perfect counterpart phrase, but sometimes the answer is not what one wants to hear; indeed, sometimes the blunt truth is: "There is no standard way to say that in French." Sometimes the answer is, in effect, "In the zone of conceptual space that is pinpointed and highlighted by that phrase in English, there is unfortunately a gaping hole in the French lexicon." Of course the French language can always *describe* the idea, but in these kinds of cases it cannot do so by means of a standard lexical item known to all or most native speakers. We hasten to point out that exactly the same phenomenon of unexpected lacunæ is encountered also by French speakers seeking to say things in English.

And the farther out one moves from the center of the shared conceptual space, the more often one will encounter these kinds of regions that, although easily and naturally accessible in one language, are simply uncovered by another language. Eventually, each language, as it approaches its own outer reaches, offers only spotty coverage, growing ever spottier as one gets further out. If at this point you are envisioning something like a nebula or galaxy whose core is densely packed with stars but whose fringes are populated more sparsely, and which eventually tails off totally, yielding to the utter blackness of the cosmos, then you have in mind exactly the image that we wish to convey.

Eventually, then, every language simply gives out, and from a certain point onwards the conceptual space is simply empty, uninhabited. What does this imply? It implies that if someone wants to talk about things in that remote zone of conceptual space, they can't just quote one standard building block, but instead must take a number of standard building blocks and string them together, thereby constructing a pathway that leads to the desired zone. In short, they must concoct new phrases or sentences. And if no single phrase or sentence will suffice, then a paragraph may be required. And if no single paragraph will suffice, then an article or a story may be required. In this fashion, arbitrarily remote spots in the black depths of conceptual space *will* be reachable by any language.

The Genius of Each Language

Here we are not primarily concerned with extremely remote, nearly empty areas of conceptual space. Instead, we wish to focus on little local pockets of conceptual space that are covered by one language's lexicon while being uncovered by another's. Are there any implications when some language hands to all of its speakers a ready recipe for picking out a small spot somewhere in conceptual space, while another language does not do so at all?

Let's take an example. American English has the picturesque idiom, "That's the tail wagging the dog!" Adult speakers of American English know what this means, which is to say, they readily recognize situations to which it applies and they can use it themselves in such cases, and they can also easily understand what is meant if someone else applies it to some situation.

In order to convey the meaning of this idiom, a speaker of American English cannot simply translate it word for word in the hopes that a French speaker will just "get it", suddenly becoming enlightened. That strategy won't work. One might try to get the concept across by giving an abstract description of the idea behind this idiom, and although doing so could be a good first step, it might be more helpful to provide a few quintessential examples of *tail-wagging-the-dog* situations, either by retrieving them from memory or by inventing new ones on the spot. Thus our imaginary American could recall or invent the story of seven-year-old Priscilla, a spoiled girl whose parents were eagerly planning a short vacation to New Orleans and were planning to take her along, but she didn't want to go at all, so she threw such a violent temper tantrum that her folks totally dropped their plans and submissively stayed home. Hearing about this, friends of the family tsk-tsked and said, "That little *enfant terrible* has her parents wrapped around her little finger. Talk about the *tail* wagging the *dog*!"

In order to convey the idea that *tail-wagging-the-dog* situations are not limited to those in which spoiled children have temper tantrums and foil their parents' vacation plans, our American could then recount the story of the grand new city hall that was being designed to beautify the central square of Waggington. After the first sketches had been submitted by the architect, the town council complained that there was no provision for parking. The idea was sent back to the architect, who responded with a new plan that included a parking area, but when this was submitted to the town council, it was again rejected because, they claimed, this time, that there wasn't *enough* parking. After a couple more iterations of this, with the building growing smaller each time and the parking lot coming to dominate the entire design, one outraged citizen wrote a letter to the local paper that said, "So the need to park a bunch of cars is dictating the appearance of our new city hall? Well, if *that* ain't the tail waggin' the dog!"

As a brief third example, let's mention the story of a runner who had to stop running each day when his kneecap started to hurt. Thus his kneecap dictated to him how many miles he would run. Another excellent case of the tail wagging the dog!

After a few such stories, the gist of *tail-wagging-the-dog*-ness would hopefully have been gotten across pretty effectively; from there on out, the French speaker would hopefully be able to use the American idiom appropriately, although at the outset there might be some need for fine-tuning to clarify where the idiom is eminently applicable and where it is less so, though of course the borderlines are blurry, so that native speakers won't always agree. The French speaker might even start, at about this stage of the game, to feel a frustrating sense of French's "vacuum" in this part of conceptual space, not unlike the slight sense of vacuum created by the lack of a familiar phrase corresponding to English's "sour grapes".

Here, we would like to even up the score by giving English speakers the chance to experience the just-described feeling of vacuum, and to do so we will cite a typical French idiomatic phrase, often attributed to the philosopher Denis Diderot, that has no good English counterpart (and of course this one isn't unique; there are hundreds of others) — namely, "avoir l'esprit d'escalier". What does this mean? Well, translated literally (in the manner of Jean-Loup Chiflet's books), it means "to have the spirit of staircase", but as an idiom it basically means "to come up with the ideal retort to an annoying remark right after one has left the party and is heading down the stairs". In other words, to put it a bit more pithily, "to have staircase wit". Although it is a frustrating thing to find the perfect parry only when it no longer counts, it is also a fairly widespread phenomenon in life, and so you would think that the famously rich English language would offer its speakers a stock expression that gets efficiently at this notion, but no. That's just the way the cookie crumbles.

This contrast between language A, which has a blob where language B has none, is what we mean by the phrase "the genius of language A"; it is the special ability of language A to get at certain concepts that no other language gets at as easily — and complementarily, it is also the set of weaknesses that language A has in expressing certain things that, in some other languages, are as easy as falling off a log. Perhaps a language's unique set of frailties doesn't merit the positive-sounding word "genius". The phrase "lexical coverage" might be a bit more accurate, but in its staid neutrality it fails to suggest the special flavor of the idiosyncratic subtleties and the evolutionary potential of each different language.

Out near its fringes, each language has its own unique set of little blobs that fill up certain small zones of conceptual space that are covered by no other language. When Language A features a blob that elegantly fits into an area that was previously uninhabited, then speakers of Language B may want to follow suit and fill in the same zone, either by coming up with a brand-new phrase or by literally borrowing Language A's appealing phrase (oftentimes, however, unintentionally changing the boundaries of applicability of the phrase, so that in Language B it no longer means exactly what it did in Language A).

Thus we English speakers occasionally have *déjà vu* experiences that give us a *frisson*, we try to avoid *faux pas* (they make us feel so *gauche*), we indulge in *hors d'œuvres*, *soupe du jour*, apple pie *à la mode*, and even *sorbet*, and once in a while we wear *décolletés* (as long as they're not too *risqué*), we sometimes take in *avant-garde* films, read an article about *coups d'état* caused by *fin-de-siècle* decadence while *en route* to a secret *rendezvous* whose *raison d'être* is to engage in a *tête-à-tête*, enjoy ogling a *femme fatale* who's *petite* but very *chic* and all decked out in *haute couture* duds, we always seek the *mot juste par excellence*, have an *idée fixe* of one day having *carte blanche* to hobnob with the *crème de la crème*, and of course if we are *nouveaux riches*, we seek out *objets d'art* (not likely to be made of *papier mâché*) to decorate our *pied-à-terre* while indulging ourselves in *dernier cri* technology. *Ooh la la!*

The French, meanwhile, leave their *break* (station wagon) in the *parking* (the parking lot), in order to go play *foot* and *flipper* (soccer and pinball), listen to *jazz* and *rock* on their *hi-fi*, place their *rosbif* and *pop-corn* in their *caddie* (shopping cart), and later that day they go to their *dressing* (clothes closet) in order to find a *smoking*, a *pull*, and a pair of *baskets* (a tuxedo, a sweater, and tennis shoes) to wear to a *rallye* (a high-society *surprise-party*), and last but not least, they read *magazines* about *le marketing* in order to be *smart* and they use *shampooing* in order to have a *look* that is very *sexy* in order to get a *job* very *cool*.

As is clear, some of these words have retained their original meanings, while others have somewhat drifted from their moorings. Indeed, we should keep in mind that these terms have been imported precisely in order to fill a gap in the receiving language. The new word fills the lacuna, even if the shape that it takes on may not exactly match the shape covered by the original blob in the source language. For instance, when speaking of "a hamburger", English speakers do not necessarily envision the ground beef as being found inside a bun (though of course it is a strong possibility), whereas for French speakers, the bun is an integral and necessary part of the concept (indeed, the bun even has to be circular!). What was missing in the French language was a phrase to denote ground beef between slices of bread, rather than a phrase to denote that kind of meat alone, since the expression "steak haché" (which already had an English flavor to it) was already available.

Moreover, unless a borrowed word or phrase has been so deeply integrated that its origin has been totally forgotten, it will generally exude a tone that conjures up something of the other culture, or at least a stereotyped vision of the other culture, and in itself that already means that a bit of drift has taken place. For instance, in English, the term "pied-à-terre" has a somewhat fancy or rich connotation to it, while in French that need not be part of the image at all.

Amusingly, some borrowings are the result of a series of cross-Channel bounces, where, for instance, old French becomes English and then bounces back home to become new French, or vice versa. An example is provided by the French word "budget", which of course is a wholesale import from English, but the last laugh is on the anglophones, for it was they who, many centuries ago, acquired the word "budget" by importing (and distorting) the French word "bougette", meaning a small purse worn on one's belt. Another example with a similar story is the French word "étiquette" (meaning "label") which, in crossing the Channel, lost its first syllable and thus became "ticket", after which, decked out in its new guise and sporting a new meaning, it returned home, where it became a close cousin and occasional rival of the word "billet". Interestingly enough, there are dozens of such ricochet stories.

The upshot of such cross-cultural, interlingual borrowing processes is to enlarge both "galaxies" in conceptual space, adding blobs at various spots on their fringes, pushing them ever further outwards.

The Sapir–Whorf Effect

There are cases where one language pleasingly fills in some small zone, yet for some reason others do not follow suit. In such cases, it can be argued, speakers of that language benefit from the extra concept thus provided for free by their language. Let's take an example from American culture. There is an ancient disreputable business practice related to the timeless con game played with three shells on a table, in which an unsuspecting customer is lured by an attractive offer but then is told that that particular item is unfortunately out of stock or slightly outmoded, or that for some reason they are not eligible to buy it; then, in its place, another item, far more expensive, is aggressively pushed on the customer.

Variations on this theme are legion. For instance, a family seeking to buy a car is shown a model that they gush over. The wily dealer, quickly picking up on their strong interest, initially tells them that their down payment will be just $2,000. Delighted, the family eagerly says they want to buy, but then, when it comes to signing the contract, they are told that for some technical reason that they don't fully understand, the amount will "unfortunately" have to be "just a little bit higher" — and sooner than they can count to three, it has slid from $2,000 all the way up to $6,000.

People who rent cars will also be familiar with very tempting offers that give the impression that one can rent a car for a nominal sum, but when one shows up at the agency, one invariably discovers that the conditions for such a rate are very restrictive, and so in the end one winds up paying at least twice the rate quoted in the ad.

Such disreputable techniques, which often work like a charm, bear the evocative name, as our readers surely know, of "bait-and-switch". The category is broader than might be supposed. For instance, here is a case that in some ways is the flip side of the coin, yet it too counts as an excellent member. During a financial slump, an elegant old house has been on the market for some months with no takers, but one day, buyer A shows up and offers $1,000,000 for it. Shortly thereafter, by coincidence, Buyer B arrives and ups the ante to $1,050,000. The seller is ready to let B have it for the higher offer, but then along comes buyer C, who raises the stakes all the way to $1,200,000. On hearing this, both A and B immediately drop out of the bidding and out of sight, angry to have been displaced after weeks of negotiation. And now, buyer C, having gotten rid of the competition, is much freer to maneuver than before. After having some inspections made, C suddenly declares, "Oh, what a shame — I can't stick to my previous offer, because the inspectors found some serious problems; nonetheless, I'm willing to offer $900,000." At this point, the seller has lost much precious time and is growing desperate, so the house winds up going to buyer C, but for far less than it is actually worth. This is a classic bait-and-switch maneuver, despite the fact that this time the actor doing the bait-and-switch was not the seller but a buyer.

The fact that this term exists in English and is daily used by thousands of people means that the idea in great generality (for instance, including the "flipped" case we just gave) is readily accessible and immediately understandable. At first, the existence of this term may not seem of much consequence, since anyone can understand the idea if told a couple of stories of this sort, but in fact the term's existence can help the concept to spread quickly and it also lends a sense of legitimacy to the concept (approaching a sense of total objectivity), since so many people know it. For instance, the existence of the concept and its standard name may well catalyze the writing of laws that seek to squelch the many-headed hydra of this phenomenon. By contrast, a culture in which there is no standard name for this disreputable technique will be less likely to enact laws that prevent it, because the notion is not "in the air"; it's not a recognized regularity in the world that most people are explicitly aware of, even in its more common forms, let alone in its more exotic variants.

Thus we see a genuine power that comes along with providing a concept with a name: it allows speakers to spread knowledge of it around easily and quickly, and that in turn allows it to enter public discourse on many levels, and to exert influences both on individuals and on society as a whole. The effect whereby the existence of a term in a given language allows its speakers certain advantages is known as the *Sapir–Whorf effect,* and although the idea has occasionally been advanced in extreme forms that have lent it a bad name, the fundamental premise is perfectly clear and there can be no denying that it exists.

What is Intelligence?

These considerations about thinking and concepts lead one naturally to wonder whether human intelligence might not reside, at least in part, in the number of concepts one has and the intricacy of the network that weaves them together. After all, we human beings are formed by the culture in which we grow up, which hands us vast numbers of conceptual tools. Does it then follow that our level of intelligence is determined by the repertoire of concepts that we inherit from our culture?

Indeed, what is the nature of the elusive quality called "intelligence"? Countless theories have been proposed. A search through dictionaries, encyclopedias, textbooks, and the Web will yield dozens of definitions rather quickly, many of them overlapping considerably, although occasionally one will turn up that has very little overlap with the others. The most frequently occurring themes are (in no particular order):

- the ability to acquire and use knowledge;
- the ability to reason;
- the ability to solve problems;
- the ability to plan;
- the ability to achieve goals;
- the ability to remember important information;
- the ability to adapt to new situations;
- the ability to understand complex ideas;
- the ability to think abstractly;
- the ability to learn and apply skills;
- the ability to profit from experience;
- the ability to perceive and recognize;
- the ability to create products of value;
- the ability to attain what one seeks;
- the ability to think rationally;
- the ability to improve.

Among the many characterizations of intelligence that we ourselves have run into, although each one undeniably touches on some qualities of the phenomenon, none quite strikes the bull's-eye. They all hover near it, but they all fail to pinpoint intelligence's core; they don't get to the heart of the matter, let alone hit the nail on the head. Never quite managing to put their finger on its essence, they merely skirt the crux, flirt with the nub, and miss the gist, curiously unable to zero in on the kernel of the phenomenon of intelligence.

Readers may well be anticipating what our own conception of intelligence is, but before we state it explicitly, we thought it would be of interest to quote here a provocative sentence that we uncovered about, of all things, military strategists, since the author of this sentence, in describing the quality that defines a great military leader, came up with a phrase that is very similar to the words that we would use to characterize intelligence:

> What distinguishes the great commanders — Napoleon, von Moltke, Grant, Patton, Zhukov — from the more ordinary leaders is the ability to see the essence of a situation at a glance, and strike directly at the enemy's greatest weakness.

Oddly enough, the author of this sentence is an individual identified merely as "Admiral Ghent" in a military role-playing game. The quality that Admiral Ghent most admires is the ability to pinpoint the gist of a situation in a flash — the ability to sort the wheat from the chaff, the ability to get quickly at what matters and to ignore the rest. Well, this is what we would take as our definition of intelligence.

Intelligence, to our mind, is the art of rapid and reliable gist-finding, crux-spotting, bull's-eye-hitting, nub-striking, essence-pinpointing. It is the art of, when one is facing a new situation, swiftly and surely homing in on an insightful precedent (or family of precedents) stored in the recesses of one's memory. That, no more and no less, is what it means to isolate the crux of a new situation. And this is nothing but the ability to find close analogues, which is to say, the ability to come up with strong and useful analogies.

Trekking High and Trekking Low on the Slopes of Mount Analogy

The final chapter of our book is devoted to showing how analogy-making of a high order has given rise over millennia to the great ideas of mathematics and physics. But of course, the majestically soaring peak of Mount Analogy is by no means the entire mountain. Up there at the top, one finds analogies of great abstraction, while on the lower slopes one finds more concrete resemblances, which, although doubtless less scenic and striking, still result from the same cognitive mechanisms, merely applied in humbler and more familiar contexts.

For instance, in the previous chapter, we encountered, as they were meandering on the low foothills of Mount Analogy, two-year-old Camille, who "undressed" her banana, and eight-year-old Tom, who saw his uncle's cigarette "melting". These juvenile strollers were unwittingly demonstrating their keen intelligence when they retrieved those analogues from their personal stock of experiences, putting their finger on the crux of the matter at hand. Camille's idea of "undressing" her banana is quite a bright one, coming right to the point; indeed, it's the flip side of the quip someone twenty years older might make, after dancing all night: "When I got home, I had to peel all my clothes off my body!" As for Tom's idea of a cigarette "melting" in an ashtray, it's the flip side of what an adult might say upon finding that ten boxes of very expensive candy they'd bought for friends had all melted: "All those luscious chocolates went up in smoke."

When we effortlessly call something we heard a moment ago "a sound" rather than "a noise", it is because we've just made a mapping between a fresh mental structure, representing the sonic event, and a prior mental structure that we'd built up as a result of thousands of prior occasions — and we unconsciously chose *that* dormant structure because that mapping struck our brain as the best analogy in town. It's not as if we were ever formally taught the distinction between *sounds* and *noises*; indeed, we'd be hard pressed to explain what that elusive distinction is, but no matter: when we hear something, just one of those categories tends to be activated (*i.e.,* to spring to mind).

It's rather miraculous that we are all so good at unconsciously making these kinds of instant judgment calls among our many thousands of concepts, given that we were never taught formal criteria for them. What, indeed, is the difference between a *hill* and a *mountain*, or a *country* and a *nation*, or an *enemy* and an *adversary*, or a *sign* and a *symbol*, or a *piece* and a *part*, or an *idea* and a *thought*, or a *shop* and a *store*, or *picking* and *choosing*, or *falling* and *dropping*, or *throwing* and *tossing*, or *putting* and *placing*, or *smiling* and *grinning*, or *big* and *large*, or *sick* and *ill*, or *pretty* and *lovely*, or *delicate* and *fragile*, or *however* and *nevertheless*? No one would dream of trying to teach such distinctions in school.

The ceaseless activity of making mappings between freshly minted mental structures (new percepts) and older mental structures (old concepts) — the activity of pinpointing highly relevant concepts in novel situations — constitutes the analogical fabric of thought, and the unceasing flurry of analogies that we come up with is a mirror of our intelligence. Thus when we reflexively make the fine discrimination of calling a very small object "teeny-weeny" (as opposed to "tiny", "teeny", "teeny-tiny", "teensy", and "teensy-weensy"), or when we unconsciously distinguish between cases of *clutching*, *clasping*, and *clinging*, or when we casually describe part of a city as a "district" (as opposed to "area", "zone", "region", "spot", "place", or "neighborhood"), we are unwittingly displaying our great finesse at the art of rapid retrieval of apposite analogues from our enormous storehouse of experience. In truth, far from being an unthinking activity, the art of super-rapid right-on retrieval is the core of thinking.

When a woman toting two bags nonchalantly saunters out of a butcher shop into the street in front of a car in which you are a passenger, the chances are virtually nil that you will exclaim, "Watch out for that biped!" or "Watch out for that female!" or "Watch out for that redhead!" or "Watch out for that customer!" or "Watch out for that carnivore!" To be sure, in different circumstances, the bag-laden damoiselle might well be perceived primarily as a *biped*, a *female*, a *redhead*, a *primate*, a *shlepper*, a *lady*, a *dress-wearer*, a *customer*, or a *carnivore* — but in this circumstance, she is most importantly a member of the category *pedestrian.* "Pedestrian" may not be the word we utter, but instantly recognizing that she is playing this role is a quintessential act of thinking.

Much the same could be said about rapidly spotting, in highly diverse situations, the telltale signature of the protean concept *mess.* Here we give a handful of typical members of the category (and we urge readers to come up with others):

- a spoiled child's bedroom, with toys strewn all over the place;
- a toolshed in which no one has set foot in decades;
- a plate of spaghetti accidentally dropped onto a white rug;
- a shoe with chewing gum stuck in the grooves on its sole;
- a china shop after a bull has been let loose for a half hour in it;
- books replaced at random on a shelf by someone who has just dusted the shelf;
- a complex algebraic expression that doesn't yield at all to attempts to simplify it;
- a musical manuscript covered with crossouts and revisions everywhere;
- the discovery of a pile of important bills that one had forgotten to pay;
- having hired a close friend's son who turns out to be totally incompetent;
- commitments made to two colleagues to meet them at exactly the same time;
- losing one's passport the day before one has to set off on an international trip;
- the decades-long strife in the Middle East;
- a romantic triangle.

No courses are needed by any speaker of English to learn the many subtleties of this concept; in fact, for a school to offer such a course sounds like an utter absurdity. Every adult will understand these cases of *mess*-ness without expending any effort.

We humans excel at making fluid mappings between new situations and old concepts lying dormant in our memory, although we seldom if ever focus consciously on the many thousands of such mappings that we carry out each day. Just as consummate dancers are constantly demonstrating their virtuosity at making rapid-fire maneuvers in physical space, so consummate speakers of a language are constantly demonstrating their virtuosity at making rapid-fire maneuvers in conceptual space, where a "maneuver" consists in darting into just the appropriate nook in one's vast stock of experiences and from it delicately plucking a highly apposite memory, overlapping in a deep and important way with the situation at hand.

Does Having More Concepts Mean One is Smarter?

If intelligence truly comes down to the ability to pinpoint the essence of situations, then it would seem that the larger and the more fine-grained the repertoire of concepts one has at one's disposal, the more intelligent one will be. After all, each of us grows up in some culture, and that culture provides its members with a myriad of useful conceptual tools. Thus it might seem that one's intelligence will be determined by the set of conceptual tools one inherits from one's culture. The question then becomes whether someone who grows up in a culture that is endowed with more conceptual tools will be more intelligent — that is, more capable of rapidly putting their finger on the nubs of situations they face — than someone whose culture is lacking such concepts.

We who live in today's highly technological, intensely commercial, advertising-drenched world are awash in a lush semantic sea rife with untold thousands of concepts that people of, say, two centuries ago lacked totally, and those tools pervade, and help to determine, our moment-to-moment thoughts. Consider, for instance, the following picturesque phrase that we encountered not long ago:

an ego the size of a Macy's Thanksgiving Day Parade balloon

To understand this phrase, one has to be familiar with Sigmund Freud's notion of an *ego*, with the notion of *department stores*, with the notion of *Thanksgiving* as well as the idea of vast long parades that march down large boulevards in big cities on holidays. In addition, one needs to know something about the gas called "helium" — at least the fact that a balloon filled with it will rise into the air. And lastly, one has to be familiar with the specific cartoon-character-inspired lighter-than-air balloons that are regularly featured in the Macy's Parade each year, and with how huge they loom above the massive crowds that line the wide avenues of Manhattan each Thanksgiving Day.

No one could possibly have dreamed of a phrase of this sort 200 years ago. And yet today it is a very clear, run-of-the-mill phrase that most American adults would have no trouble understanding. But this is only one tiny example. Below are listed some concepts — just a minuscule subset of the concepts that our culture abounds in — the possession of which would seem to give us a substantial leg up on people from previous generations or centuries:

Positive and negative feedback, vicious circle, self-fulfilling prophecy, famous for being famous, backlash, supply and demand, market forces, the subconscious, subliminal imagery, Freudian slip, Œdipus complex, defense mechanism, sour grapes, passive-aggressive behavior, peer pressure, racial profiling, ethnic stereotype, status symbol, zero-sum game, catch-22, gestalt, chemical bond, catalyst, photosynthesis, DNA, virus, genetic code, dominant and recessive genes, immune system, auto-immune disease, natural selection, food chain, endangered species, ecological niche, exponential growth, population explosion, contraception, noise pollution, toxic waste, crop rotation, cross-fertilization, cloning, chain reaction, chain store, chain letter, email, spam, phishing, six degrees of separation, Internet, Web-surfing, uploading and downloading, video game, viral video, virtual reality, chat room, cybersecurity, data mining, artificial intelligence, IQ, robotics, morphing, time reversal, slow motion, time-lapse photography, instant replay, zooming in and out, galaxy, black hole, atom, superconductivity, radioactivity, nuclear fission, antimatter, sound wave, wavelength, X-ray, ultrasound, magnetic-resonance imagery, laser, laser surgery, heart transplant, defibrillator, space station, weightlessness, bungee jumping, home run, switch hitter, slam-dunk, Hail Mary pass, sudden-death playoff, make an end run around someone, ultramarathon, pole dancing, speed dating, multitasking, brainstorming, namedropping, channel-surfing, soap opera, chick flick, remake, rerun, subtitles, sound bite, buzzword, musical chairs, telephone tag, the game of Telephone, upping the ante, playing chicken, bumper cars, SUVs, automatic transmission, oil change, radar trap, whiplash, backseat driver, oil spill, superglue, megachurch, placebo, politically correct language, slippery slope, pushing the envelope, stock-market crash, recycling, biodegradability, assembly line, black box, wind-chill factor, frequent-flyer miles, hub airport, fast food, soft drink, food court, VIP lounge, moving sidewalk, shuttle bus, cell-phone lot, genocide, propaganda, paparazzi, culture shock, hunger strike, generation gap, quality time, Murphy's law, roller coaster, in-joke, outsource, downsize, upgrade, bell-shaped curve, fractal shape, breast implant, Barbie doll, trophy wife, surrogate mother, first lady, worst-case scenario, prenuptial agreement, gentrification, paradigm shift, affirmative action, gridlock, veganism, karaoke, power lunch, brown-bag lunch, blue-chip company, yellow journalism, purple prose, greenhouse effect, orange alert, red tape, white noise, gray matter, black list…

Not only does our culture provide us with such potent concepts, it also encourages us to analogically extend them both playfully and seriously, which gives rise to a snowballing of the number of concepts. Thus over the years, the concept *alcoholic* has given rise to many spinoff terms such as "workoholic", "chocoholic", "shopoholic", and "sexoholic". Here we have linguistic playfulness marching hand in hand with conceptual playfulness. The ancient concept of *marathon* has likewise in recent times engendered countless variations on its theme, such as "dance-athon", "juggle-athon", "cookathon", "jazzathon", and so forth. In a more serious vein, the concept of *racism* has spawned many variations, including *sexism*, *ageism*, *speciesism*, and *weightism*, and today there are words for yet other forms of discrimination that previously had had no identity and that were therefore difficult to pick out from all the background noise.

One doesn't need, however, to engage in the act of coining catchy new words to benefit from the great richness of concepts of this sort. One can simply use conceptual broadening in the way it has always been done since time immemorial. Thus these days one often hears such sentences as "they had to make an end run around the President", "the two missile-rattling countries played chicken for several months", "we're just not on the same wavelength", "there's a huge gridlock in congress", "and as for the President's stance on tax cuts, well, that's still a bit of a black hole...", "there was a chain reaction crash on the freeway involving 80 cars", "those universities are playing musical presidents". In short, the concepts that our culture hands us are constantly being stretched outwards by analogy, increasing their reach and their power.

Given that such a list of contemporary concepts that are "in the air" could be extended for many pages, and that most adults can effortlessly apply many if not most of these abstract and insight-providing concepts to novel situations that they run across, does this mean that as culture marches forward in time, people are inevitably becoming ever more intelligent, ever more capable of rapidly pinpointing the cruxes of the situations they face, and of doing so with ever greater precision?

As evidence in favor of this idea, many people have pointed to what is now called the "Flynn Effect", after James R. Flynn, a political philosopher who in the 1980s drew attention to the fact that all around the world, scores on IQ tests were slowly but steadily rising, at the rate of roughly five points every twenty years. This unexpected observation has been confirmed many times in many countries. What could possibly account for such a striking effect, if not the notion that human intelligence is in fact steadily on the rise? And what could possibly lie behind the steady drumbeat of rising global intelligence if not the constant proliferation of new concepts coming from all across the vast spectrum of different human activities?

Are we to conclude that because our culture has handed us so many rich concepts on a silver platter, it follows that a random individual today might spontaneously come out with off-the-cuff remarks whose perspicacity would astonish Albert Einstein, James Clerk Maxwell, Alexander Pushkin, or Mark Twain, not to mention Shakespeare, Galileo, Newton, Dante, Archimedes, and so many other geniuses? Without doubt, the answer is "yes". Since people and cultures develop through the construction of new categories that sometimes are highly idiosyncratic and sometimes are shared by vast numbers of people, an individual who seems quite ordinary in today's society might well have a great intellectual advantage in various domains over people from earlier generations, simply because human beings, rather than storing their acquired ideas or abilities in their genetic material and passing it to their progeny at birth (Lamarck's vision of evolution was long ago discredited), store it in their personal concepts and in their shared tools and culture. Each person's repertoire of categories is the medium through which they filter and perceive their environment, as they attempt to pinpoint the most central aspects of situations that they come into contact with. And since our conceptual repertoires today are far richer than those of earlier eras, a random person today might well be able to astonish brilliant minds of previous ages by doing nothing more than making observations that to us seem routine and lacking in originality.

Does this mean that geniuses of bygone times would do poorly on contemporary IQ tests? And if so, what would that imply concerning their actual intelligence? It's hard to say what scores would have been obtained by historical figures on IQ tests, but the more interesting question is whether the intelligence level of geniuses from long ago has been reached and even surpassed by average people in today's world. We believe the answer to this question is "no", because the great gift of those exceptional individuals was that of being able to home in on what really mattered in situations that no one had ever understood before, by constructing original and important analogies that were built on whatever repertoire of categories they happened to have at their disposition. This is always a deeply rare gift, no matter in what era it arises.

Different Styles of Ascending Mount Analogy

Imagine some mountaineers who come across a very tall, sheer rock face for the first time. At the outset, only the very best climbers in the world can scale it, and even they do so only with extreme difficulty. But some of those highly talented climbers leave behind pitons in the rock face, which allow less experienced climbers to do some of the ascents. And then those climbers in turn leave behind yet more pitons, and after a few years the once-barren face is full of pitons, and now almost anyone with a modicum of experience in rock-climbing can scale what once was nearly unclimbable. It would be absurd, however, to conclude from this that today's climbers are superior to yesterday's. Only thanks to the great climbers who made it up *without* pitons or prior known routes can the average climbers of today negotiate the once-formidable cliff. The excellent climbers who blazed the first trails up the sheer face had numerous abilities, such as the skill of spotting promising routes, the intuitive sense of where it would be advantageous to place pitons, and the skill of knowing how to drive pitons into the rock so that they will remain reliable for future climbers.

We who are alive today are the beneficiaries of countless thousands of conceptual pitons that have been driven into the metaphorical cliffs of highly abstruse situations. We can easily climb up steep slopes of abstraction that would have seemed impossible a few generations ago, for we have inherited a vast set of concepts that were created by ingenious forebears and that are easy to use. And the set of concepts available to us is constantly expanding. Does all this easily accessible power, however, wind up making us smarter and more creative than our forebears?

Think of today's electronic music keyboards, which come with a host of built-in rhythmic accompaniments for many types of music. Does having a raft of such canned accompaniments turn the instrument's user into a deeply creative musician? Does having a slew of highly variegated typefaces at one's fingertips make one into a great graphic designer? Do the myriad bells and whistles supplied by PowerPoint turn all users of that software into world-class presenters of complex ideas? Certainly not! Likewise, the fact that we can easily put our finger on scads of situation-essences by exploiting standard labels that have been handed to us by our culture does not mean that we could do so in a trackless, uncharted wilderness where no one has gone before.

Concepts have a special property that distinguishes them from physical tools: as opposed to being just an external device, a concept becomes an integral part of the person who acquires it. The mathematician Henri Poincaré is said to have stated, "When a dog eats the flesh of a goose, it turns into the flesh of a dog." He was referring to how we internalize knowledge we acquire, and how it differs for that reason from mere tools, which remain separate from us, much as a piton is totally separate from a mountain climber. Merely having a library filled with books about, say, mathematics, fashion, or word origins does not make one a mathematician, a fashion designer, or an etymologist. What counts, rather, is the degree to which the concepts in those books are internalized by a person, thus enriching their conceptual space and turning them into a thinker able to make new categorizations and analogies. In contrast to the image suggested by our mountain-climbing metaphor, conceptual pitons are not just tools, but devices that enrich and transform people, allowing them to make deeper, more insightful, and more precise categorizations. These mental pitons are no longer just inert objects in an external cliff, but become parts of the person using them. They cannot be easily removed in the same way that one can take a piton out of a rock, because to remove a concept is to take away some of the person who owns it.

How would Albert Einstein contribute to contemporary physics, were he a young physicist today? What would Alexander Pushkin bring to today's poetry? What would Shakespeare or Dante write if they were alive today? What would Henri Poincaré give to mathematics, and Sigmund Freud to cognitive science? What analogies would they discover lurking implicitly in today's concepts? What depths could they perceive in the world around them, by using the tools of their new conceptual universe to interpret the surface appearances that they would encounter all around them?

Sailing Off into Outer Conceptual Space

In this chapter and the preceding one, we have presented an image of any particular language's repertoire of lexical items as forming a "lexical galaxy" in conceptual space. We want, however, to convey a polyglottal image — thus, the idea that different languages overlap strongly at the center of conceptual space, and that as one drifts outwards towards the fringes (where concepts are more and more complex and thus rarer and rarer), each language's coverage becomes not only sparser but also more idiosyncratic. The particular lexical galaxy associated with any specific language defines that language's "genius". And lying further out beyond each galaxy there is empty space — the sheer blackness of the untracked conceptual cosmos.

But things are not as bleak as that sounds. The fact is that a very large proportion of the concepts belonging to any person have no linguistic labels and yet are just as real as ones that have standard labels, such as "hand", "pattern", "green", "dogmatic", "twiddle", "sashay", "but", "indeed", "living room", "Jewish mother", "play it by ear", "sour grapes", "tail wagging the dog", "esprit d'escalier", and "bait and switch". This idea that so many of our concepts, often ones that we care deeply about, entirely lack names was saluted by American poet Tony Hoagland in the following poem.

There Is No Word

There isn't a word for walking out of the grocery store
with a gallon jug of milk in a plastic sack
that should have been bagged in double layers

— so that before you are even out the door
you feel the weight of the jug dragging
the bag down, stretching the thin

plastic handles longer and longer
and you know it's only a matter of time until
the bottom suddenly splits.

There is no single, unimpeachable word
for that vague sensation of something
moving away from you

as it exceeds its elastic capacity
— which is too bad, because that is the word
I would like to use to describe standing on the street

chatting with an old friend
as the awareness grows in me that he is
no longer a friend, but only an acquaintance,

a person with whom I never made the effort —
until this moment, when as we say goodbye
I think we share a feeling of relief,

a recognition that we have reached
the end of a pretense,
though to tell the truth

what I already am thinking about
is my gratitude for language —
how it will stretch just so much and no farther;

how there are some holes it will not cover up;
how it will move, if not inside, then
around the circumference of almost anything —

how, over the years, it has given me
back all the hours and days, all the
plodding love and faith, all the

misunderstandings and secrets
I have willingly poured into it.

CHAPTER 3

A Vast Ocean of Invisible Analogies

❧ ❧ ❧

The Rarity of the Word "Analogy" in Everyday Language

A central thesis of this book is that analogy-making defines each instant of thought, and is in fact the driving force behind all thought. Each mental category we have is the outcome of a long series of analogies that build bridges between entities (objects, actions, situations) distant from each other in both time and space. These analogies imbue the category with a halo lending it a suppleness that is crucial for the survival and well-being of the living being to whom it belongs. Making analogies allows us to think and act in situations never before encountered, furnishes us with vast harvests of new categories, enriches those categories while ceaselessly extending them over the course of our lives, guides our understanding of future situations by registering, at appropriate levels of abstraction, what happened to us just now, and enables us to make unpredictable and powerful mental leaps.

And yet, for all this, the word "analogy" is seldom heard in ordinary speech. Its rarity conveys the impression that analogies are unusual delicacies, like caviar or asparagus tips, or precious gems, like rubies or emeralds. The word "analogy" tends to come to mind only when we see someone explicitly link two entities that at first glance strike us as deeply unlike each other, and hearing the word makes us anticipate a feeling of surprise, delight, or revelation, such as when someone suggests a mental link between two entities as remote and unrelated-seeming as, say, asparagus tips and analogies.

If a politician were to compare Saddam Hussein to Adolf Hitler, everyone would call this act an analogy (not necessarily an excellent one), especially if a number of connections were explicitly pointed out, such as the way the two leaders seized power, the way they governed, or their invasions of bordering countries. The category *analogy* would light up in an instant if a physicist were to suggest that the molecules in a gas are constantly bashing into each other like a myriad of billiard balls banging against each other on an enormous pool table, or if a biologist were to describe the way that two strands of DNA can come apart and then rejoin each other as being like a zipper on a jacket. And if a journalist described a fan who hovered around a movie star as a satellite in orbit around a planet, labeling this an "analogy" would seem very natural.

All the above are fine analogies, but they reinforce the prejudice that analogies must always be spicy, picturesque, and unexpected, like those in the sampler below:

z... a
d... w
a... z+1
abc... xyz
abd... wyz
wing... fin
song... drug
to die... to part
sexism... racism
division... sharing
God... Santa Claus
to be born... to arrive
an animal heart... a pump
the atom... the Solar system
giving birth... running a marathon
creating a work of art... giving birth
leukemia... ivy creeping all over a house
an upside-down wineglass... the Eiffel Tower
global warming... the warming in a greenhouse
a moon orbiting a planet... a planet orbiting a star
a suicide bomber... a wasp that perishes when it stings
an animal circulatory system... a national highway system
immune system protecting a body... army protecting a country
a concept growing inside a brain... a metropolis spreading in a valley
an insect hovering around a streetlight... a moon in orbit around a planet
a chain reaction... dogs barking making other dogs in the neighborhood bark
the next-to-last letter of the roman alphabet... the second letter of the roman alphabet
humans surrounded by analogies they don't notice... fish surrounded by water they don't feel

Each of these one-liners is at least a bit provocative, thus matching the stereotype of analogies, but in truth, most analogies are unprovocative, yet are analogies no less.

The Swarm of Resemblances Buzzing in our Heads

The last line of our sampler puts it clearly. Like fish swimming in a medium of which they are unaware but that allows them to dart nimbly from one spot to another in the vast briny depths, we human beings float, without being aware of it, in a sea of tiny, medium-sized, and large analogies, running the gamut from dull to dazzling. And as is the case for fish, it's only thanks to this omnipresent, unfelt medium that we can dart nimbly from one spot to another in the vast ocean of ideas.

In this chapter, we will concentrate on analogies that, unlike the stereotype, lack spice and do not grab attention, but are different from those dealt with in the preceding chapters. In those chapters, we showed how simple words and common expressions — lexical items — are constantly jumping to our consciousness thanks to "little" analogies that are found unconsciously and ultra-rapidly. These "analogettes" constitute the most basic and crucial acts of categorization in our lives. Their *raison d'être* is to allow us to relate instantly and easily to the most standard situations that we face, and also to allow us to talk with others about them. However, the analogies that spring up at every moment in our heads are not limited to those that slap linguistic labels on things.

When we go beyond the activation of categories having pre-existent verbal labels, we enter the realm of *non-lexicalized categories.* By this we don't mean that it is impossible to describe such categories using words — in general, this can be done perfectly well, and that's a blessing, since otherwise we would be unable to discuss such categories in this chapter! All we mean is that there is no previously existing label, whether it be a single word or a phrase, that bubbles up from memory. There is a lexical gap, in short, like the vividly painted verbal vacuum in Tony Hoagland's poem. Now if people had to rely on a special *alternative* set of cognitive mechanisms every time they ran up against a lexical gap, then the centrality of categorization's role in cognition would be cast in grave doubt. However, the existence of non-lexicalized categories reinforces our thesis that categorization through analogy-making is the universal fabric of cognition.

Categories that We Construct on the Fly and Juggle with

We rely constantly on concepts that have no name. Words and concepts are two different things; indeed, linguists classically distinguish between a linguistic label and the thing to which the label refers. This distinction overlaps the distinction made by psychologists between the *mental lexicon* (our storehouse of labels) and *semantic memory* (our storehouse of concepts). If one were to fail to make this distinction, then there would be no meaning to a phrase such as "the meaning of a word". The distinction between a label and a category is crucial, and hopefully is clear. Although many situations trigger concepts designated by standard words or phrases, we also face many situations for which we have no ready verbal label. However, this doesn't mean that such situations are less categorizable than ones for which a standard word or phrase exists.

Every day, without reflection, we construct a fair number of fresh new concepts, most of which we wind up never thinking about again because they are applicable only in a specific, one-of-a-kind context. The psychologist Lawrence Barsalou launched the study of such categories, which arise when one suddenly finds oneself driven to attain some unfamiliar new goal. He gave them the name of "ad-hoc" categories (meaning "spontaneous" or "improvised") because they are created on the fly in the service of the new goal. For instance, the category *possible Christmas presents for one's twelve-year-old* is an ad-hoc category that might count among its members such items as a backpack, a compact disk, a pair of running shoes, a video game, a trip to an amusement park, a dinner in a favorite restaurant, a flight in a balloon, and so forth, despite the fact that, without knowledge of this special connection among them, the items in this list might seem to have nothing in common.

If one keeps one's eyes peeled, one can discover collections uniting some rather bizarre bedfellows on signs that list prohibited activities and entities in public places. For example, in Palmer Square in Princeton, New Jersey, a sign declares, "No skateboarding, rollerblading, bicycle stunts, horseplay, littering"; in a park in Manhattan's Lower East Side, one is warned: "No bike riding, pigeon feeding, dogs"; and on a beach near Bloomington, Indiana, "No glass, pets, or alcohol".

It's possible to enumerate all sorts of ad-hoc categories, such as those in the following hopefully provocative list (which includes a number of cases suggested by Lawrence Barsalou himself):

> Items to save when one's house is burning down; people one would like to drop in on when visiting one's home town; people to inform when one's father dies; foods that are compatible with one's diet; things one can pack into a small suitcase; shoes that won't hurt one's blistered foot; items to pack for a picnic; objects one might stand on to change a lightbulb; things one could use to put under a table leg to make it stop wobbling; places in someone's apartment where one can lay down one's handbag; restaurants to which one might invite a vegan friend; clothes to wear to a "seventies" theme party; people to ask for advice about a good moving company; potential sellers of gypsy jazz guitars; activities to engage in on a camping trip; tourist activities on a trip to Beijing; relatively uncrowded spots to sunbathe in on Memorial Day weekend.

(We might add that the list itself is an excellent example of an ad-hoc category!) Any of the categories in this list could have become suddenly relevant or even important to anyone at some point in their life or even on multiple occasions; it all depends on what one is trying to do. The existence of ad-hoc categories reinforces this book's thesis that categorization, through analogy-making, lies at the core of thought. Indeed, this phenomenon shows that whenever one is pursuing some goal (trying to escape from a burning house while grabbing the most important items, organizing a picnic, changing a lightbulb, planning one's vacation, choosing one's evening attire, and so forth), the lack of appropriate pre-existing categories in one's mind is compensated for by the spontaneous creation of a new category.

And the category created on the fly is by no means a luxury, because there will be no chance of attaining one's goal without it. For example, take the case of the earlier-mentioned category *possible Christmas presents for one's twelve-year-old.* For the parent who conceives this goal, the brand-new category becomes a powerful filter through which the environment is perceived. As is the case for any category, this one will become ever more refined in the mind of its creator as time passes and more and more interactions take place with the environment.

Some Other Non-lexicalized but Stable Categories

This is far from being the end of the story, for in our heads, there are a vast number of other categories, just as non-lexicalized as ad-hoc categories but more durable, and of which we are generally unaware in daily life. They are not among the categories covered by the genius of one's native language (though some might be lexicalized by certain languages), but they exist in the minds of many people and can be appropriately activated when the proper situation arises. And it's quite possible for such categories to emerge out of ad-hoc categories that one has used many times. For example, ad-hoc categories such as *activities typical of camping trips*, *objects that could be useful for a picnic*, *objects that can comfortably be carried around in a small suitcase* are categories that have a tendency to become very stable in the minds of those people who are fans of camping, of picnics, or of traveling. More generally, and just as in the case of lexicalized categories, this kind of category can be evoked to help one confront a new situation — that is, to understand it, to think about it, and to make decisions about it. Such categories could be listed with no limit in sight, but the sampler that follows at least gets the basic idea across:

> People who were once household names but who've been largely forgotten, and about whom, when one reads they have just died, one thinks, "Oh, hadn't so-and-so died long ago?"; things one could swipe from a friend's house without feeling in the least guilty (*e.g.*, a paper clip or a rubber band); the "cousin" category of things that one could borrow from a friend's house without asking permission, intending to return them very soon (*e.g.*, a pen or a pair of scissors); the last item in the bowl (*e.g.*, the poor little cherry tomato that everybody is eyeing but that nobody dares to take); people who, when they take the train, always want to have a seat facing forward; items that are in themselves cheap but whose auxiliary items are devilishly expensive (printers, certain kinds of coffee machines, cell phones, razors with replaceable blades); one's former romantic partners with whom one is still friends; people whom one might have married; the children one might have had with such potential mates; the clothes one wears when one is feeling thin; items in one's house that have been passed down from generation to generation; dishes that taste better reheated than when originally fixed; friends whom one thinks of as family members; those very old friends with whom one no longer has the least thing in common; friends' children whom one watched as they grew from babyhood, and who are now all grown up; once brand-new technologies that have been rendered obsolete by recent advances (*e.g.*, floppy disks, photographic film, audio cassettes, tape recorders, fax machines, etc.); our great personal plans that have not yet been carried out; the things one almost never remembers to purchase when one goes to the grocery store (salt, flour, toothpaste, shaving cream, etc.); people who made a major career switch in mid-life; main courses that one can eat with one's fingers without being frowned at (French fries, chicken drumsticks, slices of pizza); rich people who live in a very modest fashion; people who have the same first and last names as a celebrity; people whose last names are also common first names; occasions where someone says to you "I'll be right back" and then takes ages…

Does each of these categories exist in everyone's mind? There's little reason to think so. The fact that none of them has a standard lexical name is a cue that they don't crop up very often in real life, and for that reason, it seems unlikely that these categories would be universal. On the other hand, it is probable that most people have come across, at least fleetingly, a number of these categories, although perhaps without being consciously aware of doing so.

Who among us has not, at some point or other, swiped a paper clip or a Post-it from a friend without asking, or a piece of paper, or perhaps a piece of chewing gum or a candy that was lying around? Clearly it would be unthinkable to swipe a friend's pen or tie clip, let alone a pretty decorative item on the mantelpiece. And surely, now or then, you must have been annoyed at yourself when you realized that you forgot to pick up some salt or some napkins during the grocery store run you just got back from. How many of us feel perfectly fine about helping ourselves to French fries with our fingers, but wouldn't dream of eating string beans that way? How often have you heard, "Hey, *somebody's* gotta eat it, don't let it go to waste!" in connection with the last olive in the bowl, the last slice of cheese, or the last piece of cake, which nobody dared to take for fear of looking discourteous? In some languages there is a standard phrase for this phenomenon — in Spanish, it's "el pedazo de la vergüenza" (more often just "el de la vergüenza"), and in Italian, "il pezzo della vergogna" (both translatable as "the morsel of shame") — but in English no such phrase seems to exist, at least as of yet.

As soon as one starts paying attention to categories of this sort, one realizes that many of them had already been created and were present in the recesses of one's memory, ready to bubble up when needed, whereas others, though not already present, could easily be manufactured on the spot. Although such categories are usually too trifling or too esoteric to merit anointing with standard lexical labels, they nonetheless provide excellent evidence for the constant churning of categories in our minds.

Yes, There's a There There!

We turn now to analogies of a special type that people perceive not only effortlessly but wordlessly, and that will most likely seem so elementary and simple-minded that many readers will at first probably balk at calling them analogies at all.

> A man casually tells his daughter, who happens to be accompanying him one evening as he is taking his usual commuter train home, "Yesterday some teen-age girl was sitting right *there* [so saying, he points to the seat across the aisle from them], and she blabbed so much on her cell phone that one couldn't get one moment of shut-eye."

What could be more natural than saying "sitting right *there*" and pointing at a specific spot? To be sure, the young woman hadn't really been *there* — far from it! In fact, it would be a good exercise to list as many differences as possible between the two *there*'s, and then to think of a number of other circumstances in which someone might have said, "sitting right *there*", and in which his daughter would have understood perfectly easily just what he meant. (For example, they might have been traveling in a bus or an airplane rather than a commuter train, or he might have pointed one seat ahead or two seats behind…) The chatty young woman had obviously not been sitting *there* in the strict sense of the word, but although the man was not telling his daughter the truth on a *literal* level, he was nonetheless telling her the truth on a different level — an *analogical* level — and that's how we communicate all the time: with a minimum of effort and a maximum of very simple analogies.

This may strike you as an example of a *lexicalized* category — namely, the category denoted by the word "there". But the word was accompanied by a gesture, which was crucial. The category that was retrieved in his daughter's mind was triggered by the combination of a word and a gesture. Rather than the vast and vague concept that would be evoked by the word "There!" in the absence of any context, the man triggered in his daughter's mind an infinitesimal subset of the full set of *there*'s that exist in theory — namely, just seats in a train (or perhaps a bus, a plane, a boat, etc.) that are across an aisle from where one is sitting (or across from where someone whom one knows is sitting, or was sitting, or would have been sitting, and so forth). By deploying a word–gesture combination, the man tried to indicate to his daughter the *analogous* spot, within the current frame of reference, to the spot where the chatterbox had been sitting. The spontaneous creation of this new concept — at once very general, since it could work in so many contexts, and also very specific, since it is so precise and so concrete — allowed the daughter to imagine very vividly the situation that her father had experienced during his train ride the other day: communication via finger-pointing analogies can be extremely efficient.

Her Hero Shows up in Her Office

In a situation that is "roughly of the same sort", a young professor has been invited to give a talk at a prestigious research institute. To her astonishment, she sees, in the middle of the front row, an elderly professor whom she has long admired, and who she would never have dreamed would come to hear her speak. During her talk, he listens with clear interest, and at the end he asks a simple but incisive question, and even does so with a sense of humor. The lecturer is thrilled. When she returns to her own university, she meets with her graduate students in her office, and says, "It was fantastic! Professor X was sitting right *there*!" And so saying, she points to the empty space between two of her students, straight in front of her. And she's quite right, because in a sense Professor X had indeed been *there*, even *exactly* there — but in another sense of course he hadn't in the least been *there.* Nonetheless, the implicit analogy easily wins the competition against nitpicky logic and petty-minded precision.

Someone might say that what the urban commuter and the young professor did was not merely "roughly of the same sort" but *exactly the same.* True enough — and yet, in order to see that what they did was "exactly the same thing", one has to ignore almost all the details of the two situations in order to extract from them one single shared essence.

Here and There

Here's yet another situation of "roughly the same sort". Two participants at the annual meeting of the Cognitive Science Society run into each other unexpectedly at a certain press's display, and one of them exclaims, "Didn't we run into each other right here last time?" It matters little that the event "last time" took place some five years earlier at an anthropology conference in another city on another continent, and at a rival press's display. Even so, it was indeed "here".

And then, returning momentarily to our meta-analogies, would it not seem that this is "exactly the same analogy" once again? In this case, the unspoken background behind the word "here" is a specific display in the publishers' showroom at a specific meeting, but the category blurs outwards from this specific publisher and meeting to include other publishers, other rooms, other meetings, other cities, other years, and so on. The conference participant unconsciously, and perfectly reasonably, assumes that her colleague will understand the implicit context, which plays the role of the hand gesture in our previous examples. The meaning of "here" in her sentence is thus just a tiny subset of all the possible meanings of "here" that might exist in theory.

The Slippery Slope from Shallow Analogies to Deep Analogies

The following two anecdotes will help shed light on the subtleties of *there* situations. A music lover arrives in a great city in whose largest cemetery his favorite composer is buried. Early one morning, as a gesture of homage, he makes his way to the cemetery, but to his dismay, he finds the entrance locked. He decides to walk around the cemetery to see if there is another entry. After 45 minutes, he finds himself right back where he started, but now he happily discovers that in the meantime the main entrance has been opened, so he can make his pilgrimage. That evening, he returns to his hotel around midnight, and, to his shock, he discovers that the front door is locked, and a circuit of the building reveals that there is no other way to get in. Luckily, another client shows up just then and opens the door with a night key, letting him in as well. Tired and relieved, he goes up to his room and sits down on the bed. As he looks at the door, he notices that on it there's a small map of the hotel showing how to escape in case of a fire. He places his finger on the map and says, "Here I am at the locked main door!" Then, sliding his finger on the map, he runs it all around the hotel, retracing in his mind the circuit he just made on foot. Right at the halfway point he smiles, for this finger-circle reminds him of his walk around the cemetery that morning. As he finishes up his circular gesture he says, "And when I'd gone all the way around the cemetery, I wound up *here*!" His single circle has done double duty for him.

And now we come to a middle-school science teacher, who starts out by drawing in the middle of the board a yellow circle, then adds some smaller objects rotating around it. All her students recognize this as the solar system, a topic that they just covered in class. Then she says, "Today you'll see that this same picture works for atoms, too. And so *here* [so saying, she touches one of the planets and makes a large circular gesture that applies to all the planets at once] some electrons are in orbit, and *there* [pointing at the yellow sun-dot in the middle] is what is called the nucleus." This orbital analogy, which uses scientific terms, might seem to be more sophisticated than the tourist's down-to-earth analogy, but is that really the case? Both analogies merely map a *large* circular gesture onto a *smaller* one using a single diagram, after all.

Far be it from us to suggest that the act of saying "Right *there*!" while pointing with one's finger would be deserving of a Nobel Prize in physics, and yet such a banal act is remarkably close to the profound analogy that links the atom and the solar system. That discovery was made collectively, around the turn of the twentieth century, by brilliant scientists, both experimentalists and theoreticians, from many countries; among them were Hantaro Nagaoka, Jean Perrin, Arthur Haas, Ernest Rutherford, John Nicholson, and Niels Bohr. The images at the heart of this analogy were extremely elusive at that time, and it took remarkable intellectual daring, supported by a large number of empirical findings, to come up with such bold ideas. And yet only a few decades later, the educational system had fully integrated these once-revolutionary ideas, and it is in this sense that understanding the analogy between the solar system and the atom's structure is not all that different from understanding analogies that we all make, day in and day out, totally off the cuff, when we say "here" or "there".

Analogies and Banalogies: Their Utility and Their Subtlety

It's a common thing for people to convey their understanding of a situation that someone else just described by nodding and saying, "Exactly! That's what always happens!", or "I've often seen that before", or else "The same thing has happened to me a bunch of times." The blandness of such comments masks their subtlety.

Above all, these kinds of frequent and banal-sounding utterances are intended to convey the idea that in spite of the novelty, uniqueness, and complexity of the situation just described, there is nonetheless in it an essence that one is familiar with, and that although no single word or phrase that one knows captures that essence, one has already lived through such an experience, either personally or vicariously. One is saying, in effect, "Of course what you just recounted was a unique, one-of-a-kind event, but even so, I've been there myself. I can recall a number of events sharing the exact conceptual skeleton of your story, and so I understand deeply — in fact, perfectly — what you went through."

We'll now take a look at some specific examples of this phenomenon, all taken from real-life conversations, and on their surface so bland that few people would pay any attention to them, and yet much richness lurks in them.

The Quintessential Banalogy: "Me too!"

Paul and Tom are attending a conference. They are having a lively conversation in the bar of their hotel. An hour passes and Paul says, "I'm going to pay for my beer." Tom replies, "Me too."

Tom's minimalist answer hardly seems to be overflowing with cognitive complexity. No one would believe that it harbors deep mysteries that it would take an Einstein to make sense of. And yet, Tom's act of uttering "Me too" and Paul's act of understanding it did in fact take considerable cognitive agility.

First of all, Tom didn't mean he would do *exactly* the same thing as Paul — namely, pay for Paul's beer. That idea wouldn't occur to anyone. What he did mean, of course, was that he would do something analogous. But what? A natural thought is that Tom meant he would pay for his own beer. That's a reasonable interpretation, but he hadn't had a beer; actually, he'd sipped a Coke while munching some peanuts. Did his "me too" thus mean that he would pay for *those* items? Actually, no — he wasn't intending to *pay* for anything. Since he was an invited speaker, his expenses were covered by the conference's budget. His intent was thus to put the peanuts and Coke on his room's account. *That's* what Tom meant when he said, "Me too."

Tom's tiny remark thus turned out to be surprisingly complex, and indeed, in the "geometry" of situations, there are seldom if ever truly parallel lines. In this case, what was "parallel" to Paul's beer was not *one* thing but a *pair* of things (provided one accepts a large number of peanuts as just "one thing"). Moreover, one item in this pair was not a beer and the other wasn't even a drink at all. Mapping such a pair of entities onto a beer hence requires a non-negligible amount of *conceptual slippage* (letting one concept play the role of another). And the same can be said for the mapping of Tom's hotel account onto the change in Paul's pocket (or the bills in his wallet, or perhaps his credit card). Tom's casual "me too" really means, if one looks closely, "I understand the intention you just described concerning the situation you're currently in, and it's my intention to do the analogous thing in the corresponding situation that I find myself in."

If Paul hadn't said "I'm going to pay for my beer" but something very general, such as, "I intend momentarily to take such actions as are necessary in order to cancel my debt toward a certain purveyor of consumable goods and services that responded favorably to a request that I initiated", then Tom's reply "Me too" would have been literally correct. However, Paul didn't utter any such thing; no one ever says anything like that in everyday life. Speaking in legalese would not be helpful at all, because to communicate smoothly, we all make small analogies and we know that others will understand them despite their imprecision and sloppiness. By contrast, a complex and "precise" legalistic remark such as we just put into Paul's mouth, although it might seem very general and applicable to many circumstances, is very hard to understand.

Procrustes' Ill-proportioned Bed

It may seem to you that Tom and Paul are merely solving a proportional analogy puzzle — namely, "*paying for a beer* : *Paul's world* :: X : *Tom's world*". And that's true, in a sense, since it's possible to force any analogy into the classical schema "$A : B :: C : D$", as long as one sees B and D as two large situations taken as wholes, and A and C as small aspects or constituents of the "worlds" B and D. The analogy would of course be the fact that A's role inside world B is "the same as" C's role inside world D.

Take the conversation between Paul and Tom, for example. What, in fact, are *Paul's world* and *Tom's world*? Neither notion is in the least well-defined. Both men were in a bar, in a hotel, attending a meeting, in the same city, in the same state, and so forth. How many of these common attributes would be relevant to Tom's me-too? On a smaller scale, the fact that Paul generously helped himself to the peanuts that Tom ordered somewhat blurs the boundaries between their separate worlds inside the bar. And on a much larger scale, what we are calling "Tom's world" might or might not include his airplane ticket, the title of his talk at the conference, the broken air-conditioner in his hotel room, the glitch in the lapel microphone when he started to give his talk, the stool he was sitting on in the hotel bar, and so forth. In a nutshell, the boundaries of the situations designated by the simple-seeming phrases "Paul's world" and "Tom's world" in the proportional analogy just cited are neither sharp nor explicit.

It's not our intention to mystify the cognitive process that took place inside Tom's head, but it strikes us as unhelpful to pretend that one can formalize the situation by using a symbolic notation that carries connotations of mathematical precision. All analogies can be cast after the fact as proportional analogies, but only in very few cases is such a formulation at all natural. The proportional notation is but a set of hints that ignore all the psychological complexity that we just described; this lack of psychological relevance undermines the interest of casting analogies in proportional form.

A dogmatic insistence on the uniformity and "proportionality" of all analogies recalls the legend of the Greek demigod Procrustes, a mythical giant who grabbed all innocent passers-by and forced them to lie down on his bed. Then, depending on the sizes of their bodies, he would either violently stretch them until their body had grown to be as long as the bed, or else he would chop their feet off in order to keep their body from sticking out beyond either end of the bed. In this way, Procrustes always reached his goal of making each "guest" fit the Procrustean bed perfectly, but at quite a sacrifice.

The Subtlety of the Blandest Analogies

Sometimes one feels like saying "me too" to indicate empathy or agreement, or to convey the sense that one sees the world through the same eyes; other times one says it simply not to be impolite. Thus the phrase "me too" can come bubbling up to one's lips even before one has a clear idea of what one means by it. For instance, store clerks in airports have a habit of saying, "Have a nice flight" to passengers on the way to their planes, and in return they quite often hear, "You too!" Much the same thing happens to servers in restaurants who say "Enjoy your meal!"

Sometimes a me-too is ambiguous. Thus, if a friend says, "I'd like it if you could come over for dinner one of these days", and you reply "Me too", are you looking forward to going to your friend's house or do you hope your friend will come to your place? It can mean both things at once, in fact, and we may not even be conscious of what we really mean by such a statement. If we are, then we may realize that for the other person there is a potential ambiguity, and we may feel obliged to be more specific for purposes of crystal-clarity.

We'll now present a series of examples in which the slippages between categories are unlikely to be noticed, yet some of them involve an impressive amount of mental fluidity, since in every case, conceptual slippages take place in the passage from the first person's frame of reference to the second person's frame of reference, and what slips and what remains fixed is subtle and depends on many unspoken factors.

A: "I see the moon." B: "Me too."
C: "I have a bad headache." D: "Me too."

In the first example, A and B could be standing next to each other, or hundreds of miles apart, talking by phone, but it's always *the* moon that both see — not one moon each. In the second, by contrast, there are different heads and different headaches.

E: "I had a dog when I was a little girl." F: "Me too."
G: "I had a crush on you when I was a little girl." H: "Me too."

In the first example, we can imagine F being a man (in which case he was of course never a little girl). It's also possible, though unlikely, that F had some other animal, perhaps a cat or even a less common pet. In the second, it's clear that H had a crush on G. The mental slippage here effortlessly flips the direction of Cupid's arrow.

I: "I'm sorry I didn't control myself last night." J: "Me too."
K: "I can forgive myself just this once, because of my accident." L: "Me too."

In the first example, one understands that I and J had an argument, and that each is telling the other one that they are sorry for their own behavior. In the second, by contrast, K feels that a recent accident constitutes an excuse for having committed some peccadillo, and L simply agrees with K that this is a reasonable point of view.

M: "I forgot my wife's birthday this year." N: "Me too."
O: "Oops! I forgot my present for my wife in your car." P: "Me too."

In the first example, we understand that both M and N are men and that both of them were negligent toward their respective wives this past year. Thus, "my wife" shifts in meaning from "M's wife" to "N's wife" when the speaker changes. On the other hand, in the second example, we can safely infer that both men arrived together in P's car (rather than each of them driving up in the other one's car!), and so P's remark means, "We *both* left our presents for your wife in my car. How can we be so dumb?"

Q: "I once dropped my wedding ring into the wastebasket." R: "Me too."

When R says, "Me too", it's obviously not because R once dropped Q's wedding ring into the garbage. R is speaking of R's own wedding ring. Even if R is not now married, and even if R was married not just one time but several times in the past, the noun phrase "my wedding ring" is understood as meaning "the ring that I wore during a certain period to signify that I was then married". As for "the wastebasket", it certainly doesn't mean The Wastebasket (as if there were only one single wastebasket in the entire world, like The Moon before Galileo's act of pluralization); in fact, R's "wastebasket" could have been a garbage disposal, a sandy beach, a bowl of soup, a lake, a drain, a mailbox — who knows what! So many diverse scenarios could be seen as matching Q's story. And if, perchance, the memory that came rushing up from one's subconscious was of one's teen-aged niece accidentally dropping a piece of costume jewelry from a chairlift into a bank of powder snow below her, perhaps one wouldn't actually pronounce the words "Me too", but such an act of conscious censorship doesn't prevent the memory from having bubbled up, and it doesn't mean that one wasn't at least *tempted* to say "Me too". Even in that extreme case, we're still dealing with a *me-too* situation, albeit a marginal member of the category.

Edward Finds Himself on the Spot

One Thursday afternoon, while Edward was waiting to pick up his son from school, he was talking with Stephanie, the mother of one of his son's friends. Stephanie spontaneously said to him, "I think your wife is very pretty — she has a gorgeous face." Edward instantly put his mind in gear, trying to find a worthy reply to this compliment bestowed upon his wife. As he wanted to keep it in the same vein, he found himself in search of a me-too. But what me-too could it be?

The first idea that came to mind was based on the category *mate*: "Since Stephanie just complimented *my* mate, maybe I ought to do the same for hers." However, the idea of telling Stephanie that her husband was very handsome made Edward somewhat uncomfortable. Appreciating the faces of other men was not something that came naturally to him, and furthermore, the general way that women have of savoring other women's attractiveness seemed to him very different from the way a man might react to another man's good looks. And on top of all that, such a me-too would have an artificial, nearly mechanical quality, because it would be so formulaic, copying in such a direct fashion the original remark, that its sincerity would instantly be cast in doubt, no matter how genuine the feeling behind it might be. It would simply seem that whenever Edward received any sort of compliment, he felt instantly obliged to send another one back, parrot-like. Such a character weakness is not something one wants to convey to other people. And so, having weighed all these factors in his mind in an eyeblink, Edward suppressed his first tendency to say, "Me too — I think your husband is very handsome; his face is extremely attractive."

Next, Edward thought of a me-too that was based not on the category *mate* but on the category *female member of the couple*. "Since Stephanie just complimented the female in *my* marriage, I could turn around and compliment the female in *her* marriage — namely, herself." But he couldn't bring himself to do this either, because he would have felt uncomfortable telling Stephanie that he found her very pretty; such a remark could easily be taken the wrong way. And so he dropped the idea of saying to Stephanie, "Me too — I think that *you're* very lovely."

At this point, Edward realized that what troubled him about the me-too that he had just given up on was that he and Stephanie were of the opposite sex, whereas his wife and Stephanie were of the same sex. Therefore, wouldn't the best me-too respect the fact that the admirer and the admiree belong to the same sex? In other words, if complimenting a *female* is more important than complimenting a *mate*, it might also be the case that the compliment should *come* from a female rather than from a male. This was a novel idea! It meant that he could tell Stephanie that *his wife* found her to be lovely. Although his mind was now racing to find something to say before he started to look impolite, Edward still wasn't satisfied with this option, because it would presume to be giving the viewpoint of a third person, whereas Stephanie had voiced her *own* opinion. And so, Edward dropped the idea of telling Stephanie, "My wife admires *your* face, too, Stephanie."

In the end, the analogy that he settled for was the most literal one possible, the most down-to-earth and the most elementary of all analogies, and yet the one that turned out to be the most charming, in his view. Stephanie had told him his wife was lovely, and so, with a smile, Edward graciously replied, "Me too! I think she's very lovely!"

Killing Two Birds with One Stone

As we have seen, understanding the meaning of a me-too remark can require fluid and subtle analogical thinking, despite the surface-level banality of the remark. There are many other expressions in any language that act in pretty much the same manner, in that they involve analogies that at first seem trivial, until one looks at them more closely. We'll now take a look at just such a remark.

> Carol has just finished making out a check for her friend Peter when she blurts out, "Drat! One more signature all bollixed up! I'm sorry — I'll write you another one." Peter smiles at her and asks, "How long have you two been married now?" Carol replies, "Oh, a bit over six months…" Consolingly, Peter says, "Don't feel bad — that happens to me every January."

Now just *what* is it that happens to Peter each January? It's surely not the case that he bollixes up his own signature by using his maiden name instead of his married name whenever January rolls around. Quite obviously, he is referring to the phenomenon whereby one writes the old year on a document, rather than the new year.

The pronoun "that", used so casually by Peter, has a subtle double-valuedness. Into it he squeezed both the concept of *signing a check with one's maiden name instead of one's married name* and also the concept of *dating a document with the old year instead of the new year*, thus killing two birds with one stone. Peter's usage of "that" brought into existence a new abstract category in his mind, having both types of event as members. Earlier, we saw similar phenomena involving subcategories of the vast categories *there* and *here*; now we are witnessing the spontaneous creation of a subcategory of the inconceivably vast category denoted by the pronoun "that". This category, which includes just about everything one could imagine, has essentially zero utility without a context drastically restricting it. But the restricted version of "that" created by Peter's comment denoted a subtle, abstract, and interesting category — a regularity in the world that appears in many different guises in addition to the two Peter was thinking of.

A summary of the essence of this subcategory of *that* might be: *A sudden and unusual change of status requires one to modify the way one writes a certain item, but for some time thereafter, the old habit keeps on taking over and as a result, one frequently writes the item in the outmoded way.* Despite its verbose, quasi-legalistic description, this is a category that we all understand and are familiar with from various episodes in our lives. Below we give a number of scenarios, some that we've observed, others that we dreamt up, all of which share the essence of this category.

Marsha switched jobs and now works in another part of the city. The first few months after the change she found herself repeatedly heading off to her old workplace.

Pluto is no longer officially considered to be a planet, but most people still name it (or say that there are "nine planets"), when asked to describe the solar system.

Mike died but even after several months, his grieving widow, despite her best efforts, keeps on saying "we" and "us".

A Dutch couple who lived in Paris for many years has returned to Amsterdam, and the wife, talking about her husband, who has just taken off on a business trip to New York, says, "Whenever possible, he uses Air France." However, what she meant to say was, "Whenever possible, he uses KLM" (the Dutch national airline).

Richard, who just got a new car, is looking for it in the parking lot, but he can't find it because the "search image" that he has in his mind is that of his old car. He doesn't even see his new car when he's looking straight at it!

Several weeks after school is back in session, a mother is asked the name of her son's school, and she gives the name of his school the year before.

George still says, "When I turn fifty years old", and yet his fiftieth birthday took place six months ago.

To call home, Larry dials the phone number that he used to have before he moved.

Julie forgets that her cousin gave birth to a second baby last year, and she asks her cousin how her son is doing as if there were only one.

A newly promoted lieutenant is caught off guard every time he hears anyone address him as "Captain Green".

Lisa studied elementary education in college, and by coincidence her first teaching internship was in the classroom of her fourth-grade teacher, Mr. Long. He told her to call him "Marty", and she tried, but often she blurted out "Mr. Long" by reflex.

A curious thing about the foregoing list is that nary a one of its members belongs strictly to the category defined above in italics, because if you read the fine print, you will see that it involves someone making an error in *writing*. And yet, the range of situations remains quite faithful to the word "that", as used by Peter in his phrase "*That* happens to me every January." Poor absent-minded Marsha or Captain Green could say to Carol, to console her, "Don't feel bad — *that* happens to me all the time!"

The above list of scenarios greatly extends the category centered on Carol's slip in signing and extended by Peter's throwaway pronoun "that" — and yet it does so without in any way diluting the category's essence. To the contrary, this broadening of the category, this fleshing-out of a kind of "sphere", has made the category's essence more vivid — namely, it is those habits that persist despite a sudden change that makes them invalid, or put otherwise, it consists of those situations in which one is trapped by a reflex that one has trouble reprogramming.

Someone might well ask, "Was such a conceptual sphere born only when Peter made his consoling remark to Carol?" We would reply that when Peter heard Carol's lament, he understood its literal sense perfectly, but at the same time he extracted from it a more general idea — a more abstract notion that could apply to anyone, at any time, in any place. Peter's phrase "that happens to me" was an act of unconscious pluralization, an offhand act of abstraction that broadened a "mental dot" (consisting of a single event) into a small "mental sphere" (a nascent concept).

Indeed, any sentence uttered by a human is implicitly surrounded by one or more "variations on a theme" (since no situation has only one essence) that come without bidding to the minds of listeners. That is, once a situation has been described, it naturally invites analogies to be made that will generalize it and render its essence (or rather, one of its essences) ever clearer. Such a generalized situation, created as an automatic by-product of a person's comprehension of an uttered sentence, amounts to a new abstract category; as such, it invites simple mental slippages that will link it by analogy to yet other situations, differing in minor or even major ways from the "charter members", and these situations will in turn become new members of the category. This process of repeated category extension based on analogy-making allows the fledgling category to expand very far, depending, of course, on the particular analogous situations that happen to be encountered as time goes by. And as the concept continues to grow outwards, the key idea that lies behind it becomes ever sharper in the mind of the person who is building it up, like a knife growing sharper with each new use.

A Piano in His Bed?

David, a jazz musician who plays piano, trumpet, and guitar, one day heard that one of his old friends from music-school days, a bassoonist, had fallen off the roof of her parents' house when she was fixing it up with her father, and the doctors said that she would never walk again. And yet the very next day after the accident, she managed to sit up in her hospital bed and ask for her bassoon. On hearing this, David exclaimed, "Unbelievable! *I* would never have been able to do that."

Just *what* did he mean he would be unable to do? Was he imagining himself in the hospital, one day after having fallen from his parents' roof while repairing it with his father, sitting up and asking for one of his three instruments (and we won't even mention the idea of a bassoon!) to be brought to him? It seems very doubtful. Most likely, he simply wanted to communicate his admiration for the astonishing pluck of his former classmate as she struggled with the meaning of the tragic accident. And yet it's not possible to completely suppress the image of David in his hospital bed shortly after some sort of traumatic accident, probably some kind of fall, asking for his trumpet (of his three instruments, the one that most resembles a bassoon).

It's unlikely that David had a very precise idea of what he meant by his throwaway remark, but probably what he had in mind, and wanted to evoke in his listeners' minds, was located somewhere between an extremely rarefied abstraction (any human act that reveals surprising pluckiness) and a nearly exact replay of the event (in which David himself would have fallen off his friend's roof while working with her father, heard that he would be paralyzed for life, and the very next day asked for a musical instrument in his hospital bed). Such a midway scenario, featuring a set of middle-sized conceptual slippages, might be one in which David, shortly after being hit by an unexpected lfe-changing event, resists the temptation towards self-pity and rage and instead focuses on some passionate interest, perhaps musical, perhaps not, that he has had for many years.

Damn that Dam

One evening in Cairo toward the end of his first visit to Egypt, Nick took a taxi to go to a restaurant. Being naturally chatty and curious, Nick struck up a conversation with the taxi driver, who said that he was from the southern half of the country and belonged to the ancient tribe of the Nubians. On hearing this, Nick, who had recently visited the huge dam built by a Russian–Egyptian consortium on the Nile in the city of Aswan, where many Nubians live, asked the driver, "What effect did the building of the dam have on the Nubians?" The driver replied, "It was not good at all for us." Nick said compassionately, "Yeah… It always turns out that way."

The key question is: What was Nick actually talking about? Was he alluding to the previous dam built on the Nile at Aswan some seventy years earlier by a British–Egyptian consortium? Or to some other dams on the Nile in Egypt, located near cities where Nubians live? Or to other tribes that live in Egypt, or that live on the Nile in general? Or to all large-scale acts of construction undertaken by the Egyptian government that had infringed on minority groups? Or to any acts undertaken by any governments at all that infringed on minority groups? Or to any acts of construction that brought suffering on any group of people? Or was his remark the benevolent expression of a generalized sadness concerning the dwindling fate of native peoples scattered all around today's industrial world? Or did he mean that the weak are always victims no matter what happens? Most likely Nick himself didn't really know what he meant, and was simply leaving the door open to most of these possibilities (and possibly others). Nothing comes through unambiguously about either the scale or the nature of the category that Nick concocted in his off-the-cuff remark. Many interpretations are plausible, but all of them rely on analogy and go far beyond the specific case of the Aswan Dam and the Nubians, and each one corresponds to a blurry-edged category.

Banalogies by the Bucketful

The last few sections focused on the phenomenon of "banalogies", in order to give some sense of their great frequency in everyday thinking. The diversity of the phrases that people use to convey such banal analogies hints at how common the phenomenon is, but on the other hand, this very richness hides one key difficulty for would-be collectors. Indeed, these simple analogies can slip by even the most attentive of ears, because the analogies are not spelled out explicitly and the sentences that express them tend to be bland and boring-sounding.

Just as an ordinary-looking figure wearing bland clothing does not stand out in a large crowd, especially when one is paying attention to the more striking people and activities all around, so me-too analogies don't draw any attention to themselves in the midst of a sparkling conversation. Although they are understood with no problem, they tend to go by unnoticed. However, one helpful hint for people interested in studying the phenomenon is that they are often conveyed by phrases such as "I won't let *that* happen again!" or "There you go again!" or "I'll watch out from now on!" or "Next time this happens" or "Exactly the same thing happened to me!" or "I have that habit, too" or "I won't ever fall for *that* again" or "That's what you said last time" or "Don't let them treat you that way ever again!" — and so on. Though bland, these are giveaway phrases. Another helpful hint is the word "exact" or "exactly", as in the phrase "The exact same thing happened to me the other day!" The person who comes out with such a phrase is generally alluding to two situations that share a conceptual skeleton despite having many differences at the superficial level. Thus, for instance:

> A man tries to read in bed but his wife says that even the extremely small battery-operated light that he attaches to his book keeps her from sleeping. Exasperated, he complains about her hypersensitivity to light. Not missing a beat, she replies, "You should talk! You do exactly the same thing in the morning. If I get up early and try to dress as quietly as a mouse, you always say that the noise, no matter how tiny it is, wakes you up. *You're* the one who's hypersensitive!"
>
> A woman says that she had to cut her vacation short and head home immediately because during a huge thunderstorm her apartment had gotten flooded in two feet of water. Her friend says that he'd had "just the same experience" a few years earlier when he was in the middle of an important meeting at work and was called home because the neighbors' house was on fire.

This reveals that when we listen to others tell stories, we all strip off each story's surface particulars instantly, automatically, and unconsciously, until we feel we have arrived at the story's true essence, and we then take that skeletal essence as the genuine core of what we've just heard. It strikes us as being "what really matters here". But this conceptual skeleton, this abstract core, this pure gist, is too stripped-down, too "naked" for us, and so we immediately try to wrap it into more familiar clothing by launching a search in our memory. And it is when a concrete personal experience comes bubbling up from our memory that we can most easily relate, in a concrete and vivid fashion, to what we have just been told. This explains why people can quite honestly exclaim, "Exactly the same thing happened to me!"; they are concentrating so hard on the shared conceptual skeletons of two stories that for them the differences, no matter how great they are, seem to evaporate. At the gist level, indeed, the two stories *are* exactly the same — they are just one single story.

How We Try to Understand Our Own and Others' Experiences

Our natural inclination to relate to stories told by other people by converting them into first-person experiences dredged out of our dormant memories — this propensity to make analogies that link us with other people, or, more generally, to interpret any new situation in terms of another similar situation that comes to mind — is omnipresent, because doing so fulfills a deep psychological need.

Virginia receives her semi-annual royalty statement from her publisher. Eagerly she scans its pages of data. When she comes across the check — that's the bottom line — she's quite delighted. How come? Because the previous two times the amount had been rather low, and she had begun worrying that this trend was going to continue. The favorable comparison makes her happy. Once again, we have a trivial analogy — I compare what I just got with what I got a short while ago. One would be hard pressed to find a simpler mental act than that, but that doesn't mean that such elementary comparisons don't take place all the time, day in, day out, in the life of every human being.

But back to Virginia for a moment. What is her next thought? It's the following: "What about Susan? How much did *she* get this time?" Yet another trivial analogy! Now Virginia is comparing herself with a close friend, also a novelist, and with whom Virginia has always felt a vague sense of rivalry of which she is a bit ashamed, but what can she do about that? Wondering about the size of her friend's royalties is a knee-jerk reflex, and she can't suppress it. It's a perfectly normal psychological pressure pushing for a mapping to be made. Obviously she knows that no oracle is going to supply her with the unknown figure, but that doesn't in the least keep her from wondering about it.

At last, Virginia takes a closer look at her own royalty statement, comparing the amounts that her various novels have brought her this time. Once again, she's making a series of mini-analogies — I compare the income due to my most recent tale *Carnival after Doomsday* with the incomes due to my previous books *Symphony in Ugly Minor*, *Hike of the Hellbound*, and *The Tyranny of Well-behaved Moppets.*

There will no doubt be some who will protest that we are not talking about analogies here, but just about simple comparisons between numbers. But in fact it's a good deal more than just that. These figures are all members of the category *royalty amounts*, and they all apply to the same novelist, and moreover, they all belong to the same biannual statement. It would never have crossed Virginia's mind to compare her income from *Carnival after Doomsday* with a random figure, such as the price of her hairdryer or her county taxes ten years ago, let alone the temperature in Beijing or the number of lions in the local zoo. To be sure, she's comparing one numerical figure with another one, but she's doing so because the two figures have a tight conceptual connection and because making this comparison will afford her some kind of insight into her life. It's undeniably an analogy — an analogy between the royalties brought to her by two of her books — a trivial analogy, admittedly, but no less an analogy for its simplicity or naturalness.

Swimming in a Sea of Analogies

Mark is reading the newspaper. He sees that the swimmer Michael Phelps, shortly after winning his umpteenth gold medal in the Olympics, has just said, "I was hoping to break the world record in this race, but okay, I guess a gold medal isn't too bad." Mark asks himself, "Is that guy Phelps arrogant, or what?!" And in order to think about it more clearly, he wonders, "Well, what would *I* have said if I'd been in his shoes?" Comparison, analogy — no doubt about it. And more generally, in order to relate more deeply to the article he's reading, Mark imagines, as would any of us, what it would be like to be a world-class swimmer at the tender age of 23, what it would feel like to be there and to participate in all these events, to be madly churning down the final lap and to see one's own hand touching the wall ahead of all others, to throw one's arms in the air in jubilation, to receive congratulations from one's teammates, to hear loud rounds of cheering, and so forth.

This is how we human beings understand such an event — we try to mentally simulate it, inserting ourselves into it by likening it to events that we have known in our lives. Perhaps Mark himself once won a medal long ago; in that case, the memory of that event will jump to mind instantly. Perhaps he never participated in competitive sports but once swam very fast in a friend's swimming pool, and his friend voiced amazement; he will recall it vividly. Perhaps he was once warmly congratulated by a bunch of his schoolmates; then that memory will come to mind. Perhaps one time in school he was called up to the stage to receive some award, and this lovely souvenir comes back to him. That's how it goes.

And what if this Mark were Mark Spitz, the American Olympic swimmer who won seven gold medals in the 1972 Olympics in Munich? What would Mark Spitz have been thinking as he was watching Michael Phelps on television during the 2008 Olympics in Beijing? It's hard to imagine that he wouldn't have been making scads of analogies. Not surprisingly, in an interview, Mark Spitz said, "Phelps is pretty much my double. He reminds me of myself."

And what if Michael Phelps were Jewish (as is Spitz), and if he had grown up in the town of Sacramento (as Spitz did)? Well, the analogical link between him and Spitz would have been all the stronger, and that would have made the experience even more intense for Spitz. On the other hand, if the sensational athlete in Beijing had been a *woman* who had a chance at winning eight gold medals, the analogy would have been less compelling to the mind of male swimmer Mark Spitz. And if this woman had been Indonesian and if she was shooting for eight gold medals not in swimming but in archery, then Mark Spitz might well have had little or no interest in her quest.

Now why have we taken the trouble to dream up a long set of variations on the theme of Mark Spitz who, at the age of 58, is watching his 23-year-old quasi-double Michael Phelps on television? Our goal was merely to point out once again that human minds are constantly swimming in a sea of comparisons that mostly go unnoticed, and which are all mini-analogies whose experienced intensity is a function of the strength of the analogy. It's a simple connection: the tighter the analogy, the more strongly it tugs. And what earthly purpose does this nonstop deluge of analogies serve? The flood of analogies sweeping through our brains at all moments is part and parcel of the human condition, and they are manufactured because their presence helps us to put our finger on the essence of the new situations that we confront. Our insatiable compulsion to make comparisons between the brand-new and the previously seen is a necessary prerequisite for staying afloat in a world that is so complex and unpredictable.

Let's take one last look at the Phelps/Spitz comparison. If this analogy were really all that pointless and vacuous, why would Mark Spitz be thinking about it at all? Why would he be glued to his TV in order to see what will happen to his quasi-double? Why would he say with some nostalgia, "Back in 1972, they didn't have a 50-meter race as they do now; if it had existed back then, I would probably have won eight gold medals"? And why would journalists from all over the world have gone into a feeding frenzy comparing in great detail the performances of the two Olympians, day after day? Every time someone makes a comparison, no matter how simple-minded or trivial it is, one feels compelled either to reject it or to deepen it. Analogies are addictive!

Are Analogies Always Filled with Surprises?

It might be objected that the comparison "Michael Phelps is like Mark Spitz" does not belong to the same family as the analogies listed in our pyramid of analogies at this chapter's start ("a song is like a drug"; "sexism is like racism"; "dying is like parting"; "wings are like fins"; "an animal's heart is like a pump"; and so forth). Those in the pyramid would be *genuine* analogies because each of them reveals something new and unsuspected when one runs into it, whereas "Michael Phelps is like Mark Spitz" is merely the flattest, most anemic of resemblances, having no interest or consequences, almost as if someone said to a friend, "I'm analogous to you, because we both have a head, two arms, and two legs."

Well, yes — that is, in fact, quite a fine analogy between two people, which, no matter how trivial it may be, can still be perfectly useful. For example, if your ankle is giving you some trouble, and I have already had some ankle problems myself, my advice might be helpful to you. Or if you don't know how to get an eyelash out of your eye and I have a very reliable trick involving pulling my eyelid down over my eye with my finger, I could teach you that trick and save you some suffering. Or even simpler: If we've just taken a hike together and I'm feeling ravenous, I might well suspect that you're hungry too.

Analogies far simpler than "Michael Phelps is like Mark Spitz" — extremely banal analogies like "I am like you" or "this human being is like other human beings" — pervade our thoughts. At all moments, we depend intimately on such analogies, though in an entirely unconscious fashion. Thus, we see someone in the New York subway take out at a map of New York, unfold it, and study it; we relate, because we, too, have taken out, unfolded, and studied maps of New York (and of Paris and Madrid and Tokyo...; and guide books, and instruction manuals...) hundreds of times. We see someone scratch her elbow; we relate, because we, too, have scratched our elbow (and our knee, and our neck...), thousands of times. We see someone yawn; we relate, because we, too, have yawned tens of thousands of times. These kinds of analogies are certainly not deeply insightful, but they are nonetheless deep, because they lie at the roots of our understanding of other beings — it would be no exaggeration to describe them as the cornerstones of compassion and empathy — and because they determine our style of relating to the world.

Let's think about the very down-to-earth analogies that link one grocery store to another. The concept *grocery store* carries a great deal of knowledge within it, such as where bananas are likely to be found. Such knowledge is acquired through the making of analogies and, when needed, it is triggered by analogy. When we say to ourselves, "The bananas ought to be somewhere over there", the word "there" designates certain familiar aisles in certain familiar grocery stores, yet at the same time it also designates some never-before-seen aisles in the unfamiliar grocery store in which one finds oneself for the first time. If I know where to find the bananas in my usual grocery store, then a "bananalogy" will no doubt help me to find bananas in an unfamiliar grocery store, even one in a foreign country. To be sure, this idea is so lackluster that it hardly feels like an analogy, let alone like a thought of interest or consequence. And yet, for all its lack of luster, it is useful in helping me guess where I can find bananas in a new store.

A crucial aspect of categorization is that it allows us, through analogies that we note, to make guesses or to draw conclusions. These analogies, whatever domain they are in, are based on very familiar categories, such as *person*, *swimmer*, *athlete*, *Olympic champion*, *swimming legend*, *grocery store*, *aisle*, *banana*, and so on. Without such categories, all thought would come to a crashing halt. Indeed, everyone, at every moment, is betting their very life on the validity of an enormous number of trivial, unconscious analogies whose existence they never suspect at all. Every act of thinking, no matter how small, relies on such analogies, and the tighter the analogy, the more unavoidable the conclusions it leads to would seem to be.

Creatures that Live Thanks to the Efficient Triggering of Memories

George has just heard the sad news that his best friend's father has died of a heart attack. Reflexively, he recalls the unexpected death of his own father several years earlier. And again involuntarily, the recent sudden death of the woman who lived on the same floor of his apartment building flashes to mind. He recalls the time six years earlier, when he himself had to be taken to the emergency room because he was experiencing irregular heartbeats. He recalls the only time when he saw his best friend weeping, and how much that sight had moved him. He remembers how profoundly the death of an old aunt had devastated her husband, and he tries to put himself in the place of his friend's mother at this terrible time... In short, a flurry of analogies comes rushing helter-skelter into George's head.

That evening, George calls up his thesis advisor and says, "My best friend's father died last night, and I'm going to have to be away for a few days for the funeral." His advisor wistfully replies, "Ah, I understand... You know, our old cat died last week. My wife and I are very sad." This is yet another analogy, and you might well find it in bad taste — and yet if *your* cat, a cat you'd had for nearly twenty years, had just died, and if someone had phoned you to tell you that a close friend of theirs had just died, wouldn't your cat's demise inevitably spring to mind? Sensitivity to the other person's feelings would almost certainly keep you from mentioning it, but it would not prevent such remindings from occurring silently in the privacy of your head. Moreover, this whole paragraph is tacitly relying on yet another analogy — namely, the comparison between how *you* would act in this situation and how George's thesis advisor acted. Although that comparison is just a routine, mundane act of alignment, it is nonetheless important to you, because it's helping you figure out how you feel about a situation you've just encountered. Are analogies not, indeed, irresistible and unsuppressible?

In a very large exotic airport at four in the morning when it's swelteringly hot and you're hemmed in by several hundred noisy people who, all anxious to get through customs (or better yet, to sneak through) as fast as possible, have established mysteriously fluid line-like filaments that go way beyond the boundaries of the familiar category labeled "line", and in which people left and right are cutting in and elbowing others, and in which quite a few people can be seen heading down passages that look illegal (hard to say where they go), and where you don't know a single word of the language (or are there several languages here?), and where you have in your three suitcases (one of which hasn't yet shown up despite over an hour's wait, and another of which arrived without its handle, and the third of which lost one of its wheels) a wide variety of quite valuable objects — well, in this unfamiliar situation, it won't be so easy to lean on one's rich and usually trustworthy repertoire of familiar and comforting categories as one tries to decide how to behave among all these elbowers, line-cutters, and customs agents!

What is the conceptually closest situation from your past, and at what level of detail? How do you strip this complex situation down to its most essential details to see right to its heart, in the way that a local would do effortlessly? For natives, knowing what to do in this situation is the most natural thing in the world; they're at home, and it's simple. For a visiting traveler, though, even if this scenario were to evoke a few situations that have been experienced first-hand, it probably wouldn't bring to mind any category that would seem very promising and thus very comforting — no helpful haven of a memory would spring to mind. In this kind of situation, one has to be satisfied with behaviors suggested by memories that are less strongly analogous and which, for that reason, cannot give such precise and reliable tips. It's too bad, but this, too, is a frequent part of life.

If no memory whatsoever bubbled up in your mind as you read the description of the frenzied scene in the airport, we would be surprised. Human nature is such that we cannot help digging down unconsciously into the many layers of our memory in order to understand situations that others tell us about (and all the more so for situations that we encounter directly). We are creatures that live thanks to the efficient triggering of memories.

Whenever we hear of the divorce of a close friend, or of a fire in the home of some neighbors, or of a burglary in the neighborhood, or of a colleague's car accident, or of a flat tire far out in the boondocks, or of a wedding ring lost and then miraculously recovered, or of an incredibly long line to get through security at the airport, or of a just-barely-missed airplane, or of someone who brazenly cut in line at the theater, or of friends getting lost at midnight in the middle of Slovenia, or of a twenty-dollar bill found on the sidewalk, or of a painful immunization required to get a visa, or of a moving reunion with someone after a separation of twenty-five years, or of a very serious cancer in remission, or of two people who randomly bumped into each other in an exotic land neither had visited before, or of a small child saved from drowning by her mother, or of a tortoise that turns up again in someone's yard after not having been seen for three years — in short, whenever we hear about virtually any event at all that has some interest to us — then one or more specific memories just come floating up out of the subterranean murk without ever having been invited, and those memories afford us a personal perspective on the given event.

On the other hand, if we hear about a very bland event — a bill that was paid, a pizza that was eaten, a telephone that rang, a trip that was made to the grocery store, some distant relative's flu, an old car getting sold, the construction of a building somewhere in the suburbs, and on and on — then what gets evoked in our head is much less specific and detailed than in the cases that pique our curiosity. After all, we've all experienced or witnessed many cases of the flu, seen hundreds of buildings under construction, not to mention eaten pizzas or watched them being eaten. Our countless experiences with pizza-eating have been stored in memory and have piled up one on top of the other to the point where they've simply blurred together and made a composite and rather vague image that is the category of *pizza consumption*, and it's this image that is generally evoked, rather than a highly specific pizza-eating anecdote, when someone tells us that they just had a pizza at their local pizzeria.

Every story that we hear, whether it fascinates us or bores us, consists of many small components put together in a unique way, such as the story of the pizza that our aunt ate at the airport recently and that nearly made her miss her plane. The anecdote as a whole may well remind us of a specific story that happened to us — for instance, the time when we nearly missed a train because we spent a few moments getting a soft drink from a vending machine — but its tiny component, the humble pizza, doesn't evoke any anecdote at all on its own. Whereas a one-sentence pizza-eating *anecdote* is easily capable of coaxing explicit memories out of your unconscious storehouse, most if not all of the *words* making it up do no such thing. Look, for instance, at the various words in this very paragraph and ask yourself if any of them — for example, "ate", "but", "humble", or "words" — summoned an entire anecdote to mind.

We would be lost if every single word (or stock phrase or idiomatic expression) in a long tale dredged its own anecdote up out of long-term memory. Were that to happen, we would be engulfed in a tidal wave of random stories and we would drown in the catastrophic mental confusion that would ensue. But luckily no such thing takes place, because only once in a blue moon does a mere word in a story (a "cameo actor", so to speak) trigger the retrieval of an anecdote.

Danny and Dick, Canyon and Karnak: A Canonical Reminding

Reminding is a profound mystery, as the following case shows.

> Doug and Carol arrive with their son Danny, fifteen months old, at the Grand Canyon's North Rim. While his parents are captivated by the huge chasm, Danny is riveted by a few ants and a leaf on the sandy ground, fifty feet from the canyon's edge. For a moment Doug is surprised, but then he realizes that such a young child is unable to appreciate entities of dimensions greater than ten or twenty feet, let alone miles (and the Grand Canyon is many miles wide). Although his infant son's reaction now makes perfect sense, Doug cannot suppress a smile at the irony of the situation.
>
> Fast-forward roughly fifteen years. Doug and his two children get off their cruise ship on the Nile in the city of Luxor. They are with their friends Kellie and Dick, and the whole group sets off on foot for the famed Temple of Karnak. While the other visitors are soon absorbed by the splendor of the great columns that surround them and by the erudition of their guide, Dick is irresistibly drawn to a few bottlecaps he spots lying in the dirt, and he leans down with joy to pick them up, thereby augmenting a modest collection that he'd started when they landed in Egypt just a few days earlier.
>
> This act, reflecting Dick's fascination with rusty knickknacks on the ground as opposed to the splendor of the ancient ruins soaring above, reminds Doug of something far back in his past: the time when his tiny son was engrossed by a handful of insects scuttling about on the ground rather than by the awesome sights surrounding him.

Below we present a more lyrical viewpoint on these two parallel stories, coming from our friend Kellie Gutman, who recounted them in the form of verse. The poem about her husband Dick was written originally to commemorate their cruise down the Nile, and a couple of years later, when we asked her, Kellie graciously indulged us by writing a twin poem about Danny at the Grand Canyon, adhering to precisely the same poetic constraints. For the French version of our book, we translated these two poems, once again obeying all the original poetic constraints, and in Chapter 6, where we discuss the role of analogy in translation, we will come back to these poems.

Arizona Ants

When Doug and Carol and their son,
a toddling Danny, all of one,
left Indiana on a trip
out West, they knew they couldn't skip
the Grandest Canyon known, bar none.
They piled into their clipper ship

with no set plan, but just a notion
to shoot for the Pacific Ocean,
with detours here and there, thus reaping
the benefits of highways sweeping
across the landscape. Gentle motion
would lull their infant into sleeping,

while Mom and Dad drank in the views
of Colorado's Rockies, whose
steep craggy peaks filled them with awe.
They drove through reservations, saw
the Navajo, and got to choose
some turquoise stones without a flaw.

At last, the North Rim: strange striations
with shades evoking exclamations —
unless you're Danny… Then you treasure
the leaves and bugs! While grownups measure
the grandeur of vast rock formations,
you play with ants — a simpler pleasure.

Karnak Caps

The bottlecaps were all around
us, rusted remnants that once crowned
the Cokes and Fantas near the shop
that sold falafels, where we'd stop.
In Alexandria, the ground
was paved with flattened tops from pop.

We left for Cairo next, departing
the same day Dick announced, "I'm starting
an Egypt bottlecap collection!"
Each specimen, upon inspection,
was added to his pile for carting
back home. He had a wide selection:

in every bar, for every beer,
his pointing outstretched hand made clear
the cap was what he had in mind.
On dusty streets, he searched behind
the soda stands (a new frontier!),
and often came back with a find.

In Karnak's heat, our guide expounded
on gods and temples, while surrounded
by columns far too grand to measure.
We contemplated them with pleasure,
but as we gazed on high, dumbfounded,
Dick stooped to pluck a humbler treasure.

As described here, Doug's reminding may not seem particularly striking, but you have been handed the two analogous scenarios on a silver platter. It would be quite simple, *ex post facto*, to display in a diagram the identical conceptual skeletons of the two situations, or to encode each separate story into a concise sequence of formal expressions exhibiting a perfect one-to-one correspondence. But such a diagram or chart would not do justice to what goes on in the human mind when such remindings take place out of the blue. Such a display would be reminiscent once again of Procrustes, who always obtains the result he desires (in this case, his heart is set on creating a flawless one-to-one matchup between two sequences of formal expressions) but only at a great sacrifice (a complete neglect of the complex mental processing that underlies the discovery of the analogy), which makes the whole exercise of little interest. An *ex post facto* diagram might be an elegant summary of the result of Doug's reminding, but it wouldn't cast the slightest light on how the reminding took place inside his head.

There is, after all, a deep scientific mystery here. Over fifteen years had elapsed since the Grand Canyon visit, and in all that time, Doug had only very rarely — a handful of times at most — thought about that fleeting moment of fascination, on his son's part, for the ants and the leaf. While they were on their Nile cruise, the furthest thing from Doug's mind was this memory. How, then, could such a distant, shadowy memory have been so rapidly and easily brought back to life in Doug's mind?

The Enigma of Encoding

As we pointed out at the start of Chapter 1, everyday situations are not handed to us in pre-wrapped packages — that is, with precise boundaries that surgically carve them away from the rest of the world. Rather, we filter our environment, dealing with only part of it, and in a biased manner. Each person somehow "decides" (that is, encodes) what a situation includes and doesn't include, and what its key ingredients are. This of course takes place on the fly and not in the least consciously: at every moment of our life, we are as busy as a beaver, encoding situations in memory along dimensions that will determine which future events will be able to remind us of these situations.

How was the small event at the Grand Canyon originally perceived by Doug and stored in his memory? The privileged candidates for the encoding were, first of all, the most local and visually salient aspects of the scene — thus, the huge canyon, the little boy sitting on the sand, the sandy soil, the ants, the leaf, and so forth. Next, on a more global level, the encoding might have included the reasons for the trip, the route that had been followed, the other national parks that had been visited, the car they were driving, the year, the season, the weather that afternoon, the ever-present car seat for the toddler, and various other details.

Other more abstract aspects of the situation were certainly encoded as well — namely, the relative sizes of the key entities in the story, and their wildly different levels of importance in the culture (the fame of the canyon as opposed to the insignificance of the ants and leaves). Last but not least, the feelings of surprise and irony were crucial aspects of this scene, and had to be included in the memory package as well. However, had Doug merely tagged this complex event with a very general label like "ironic" or "surprising", that index on its own couldn't possibly have triggered the reminding, since untold thousands of other ironic and surprising events were stored in his memory, and none of those were brought to his mind by the Karnak bottlecap event.

Indeed, any event that Doug saw, no matter how ironic and surprising it might be, wouldn't stand a chance of re-evoking his memory of Danny at the Grand Canyon unless it shared other key features with that old story. However, local features alone wouldn't do the trick, either. The commonplace event of seeing somebody in a random place stoop over to pick up a small object on the ground would not suffice to get Doug to recall his son playing with ants at the Grand Canyon; if such a reminding took place, it would seem weird and senseless. Equally critical for the triggering of the old story is that the second event's encoding should explicitly include the contrast between an insignificant object and a grandiose setting, which gives rise to a sense of absurdity. We thus see that the most abstract aspects of this situation mingled inseparably with some of its most specific aspects, and that the reminding was due to similarities at several levels at once. In short, the encoding of a situation, if it is to facilitate appropriate remindings in the future, must be both abstract and concrete. While the encoding has to retain some salient details, it must also include salient abstract characteristics.

A Trivial Side Show that is More Fascinating than the Main Event

To be sure, we don't want to suggest that an event has just *one* abstract structure. Speaking, *a posteriori*, of "the" abstract structure of an event gives the impression that there was and is only one correct way to perceive it and to have a memory of it get triggered, which is not the case; any of various distinct conceptual skeletons might get attached to a given situation, depending on the frame of mind one is in when it is encountered.

The wide variety of potential encodings of *Danny at the Grand Canyon* means that many extremely different situations, possibly encountered much later in life, could induce that old memory to bubble up from dormancy. Such situations would thereby become members of the category *Danny at the Grand Canyon*, or, to propose a more generic name for it, *trivial side show more fascinating than the main event.* Let's look at a few scenarios that might trigger such remindings:

> A French family is taking a vacation in Italy's famed Cinque Terre region, on the Ligurian coast, where five small towns hug a rugged coastline. After finding a bed-and-breakfast place in the countryside, the parents are eager to set off and explore the towns, but the children, delighted by their discovery of a family of grasshoppers in the garden, refuse to budge.
>
> A four-year-old girl has just received a lovely Christmas present, but she's paying no attention to it, concentrating instead on the sparkling wrapping paper.
>
> In the cathedral-like main room of Cairo's Egyptian Museum, a boy is totally ignoring the tiny three-inch ivory statue of Kheops, which is the focus of all the guides and crowds; rather, his eye is caught by the cracked paint covering the vast room's high walls, in which he thinks he can make out the shapes of gigantic dragons and other fantastic monsters.
>
> A young mother is more absorbed in the photos of her baby than in the baby itself.
>
> A neurologist gives a talk to which she brings her ten-year-old niece, visiting for a few days. After the talk, the little girl says, "I liked your speech a lot, but I have just one question about it." The neurologist is delighted and even wonders if her niece is some kind of prodigy. Then the girl goes on, "When you thanked a bunch of people at the end and you mentioned someone named Janet, were you talking about my mom?"
>
> A graphic artist is reading a famous novel but is paying less attention to the plot than to the typesetting and page layout.

> Two intensely motivated scholars spend all their days and nights in a tiny cramped office working on a specialized treatise, when all of Paris, with its magnificent monuments, museums, cafés, and restaurants, is out there, just begging to be explored.
>
> Two co-authors are frittering away every day and evening of their Parisian vacation visiting museums and monuments and munching on *pâtisseries*, when instead they could find a secluded spot and sit down and jointly create a masterpiece.
>
> A mosquito hovering about Albert Einstein's body sees it as nothing more than a warm object filled with liquid sustenance.
>
> Doug and Carol don't in the least appreciate the exquisite marvels of the tiny ants and delicate leaves on the sandy soil, but instead, aping the mob of awestruck philistines around them, they merely gawk at some random hole in the ground.

Each of these scenarios has its own unique kind of resemblance to the story of Danny at the Grand Canyon. Although a certain essence is shared by all of them, there are a few, especially the first one, that are extremely similar to the original episode, despite differing in many superficial aspects, while others, especially those toward the end, enjoy a far more abstract connection. Readers will have different personal judgments about how forced or natural each analogy is, depending on how they encode these scenarios. There is no objective way of ranking these episodes in order of their degree of resemblance to the episode of Danny in the Grand Canyon, because similarity is not only subjective (encodings can vary wildly from one person to the other) but also multidimensional (many aspects of a situation play roles in its encoding). Still, these anecdotes will allow us to illustrate the variety of encodings that can analogically trigger memories of the episode of Danny at the Grand Canyon.

Thus, a relatively superficial encoding such as "children who are more interested in playing with insects than in savoring a famous tourist spot" will suffice to connect the episode of the family taking their vacation in the Cinque Terre with the Grand Canyon episode. On the other hand, the analogy involving the little girl who is more interested in the wrapping paper than in her gift is based on a rather more abstract similarity, because her story involves no insects and no famous tourist destination. This time, the shared gist is the lack of interest paid to something that is supposedly fascinating to any child (a gift) in contrast with the preferential treatment afforded something that is presumably devoid of interest (wrapping paper).

The third episode brings in the new idea of size inversion. The phenomenon that is supposedly universally interesting — the statuette — is now tiny, while the supposedly boring phenomenon — the cracks in the paint on the walls — is huge. For the two episodes to be analogous, this difference has to be transcended by an encoding that stresses the contrast between an entity that is conventionally considered important and another that is conventionally considered trivial, no matter what their sizes are.

As for the mother who is absorbed in the photos of her baby while paying little attention to the baby itself, the irony here comes from the fact that paying more attention to the photos than to the baby seems to suggest more interest in a copy than in the original, more interest in inanimate objects than in a living being, more interest in the past than in the present, and more interest in what is frozen than in what is changing; these priorities constitute stark contrasts with conventional priorities, thus abstractly linking this scenario with that of Danny at the Grand Canyon.

The conceptual skeleton implicitly abstracted from the story of the child who attends her aunt's lecture matches that of the scenario of Danny at the Grand Canyon provided that one's perception goes up a notch in abstraction. Here there are no physical entities or places on which to base the mapping. The key thing is the contrast between a child's superficial (indeed, *childish*) interest in a person's name mentioned only fleetingly during a technical talk and the abstruse neurological ideas that all members of the audience were, at least theoretically, deeply engrossed in.

The case of the graphic artist who is more interested in the typesetting than in the novel is similar in some ways to the case of the child who prefers the wrapping paper to the gift, but there is a difference of scale. The child is neither an expert in nor a huge fan of wrapping paper, while the graphic artist is a specialist in letterforms and words and their placement on the page. This anecdote is related to the two that follow it (those involving the two scholars and the two co-authors), which form a matched pair of episodes that bring out particularly clearly the subjective nature of people's points of view on a situation (Paris, in this case). The connection to the episode of Danny at the Grand Canyon is clearer when one reads these three stories at a yet higher level of abstraction than would be suggested by this section's title ("A Trivial Side Show that is More Fascinating than the Main Event"), because here it would seem that the linking idea is *each person sees things in their own fashion.*

It's not so much Danny's fascination with a trivial side show instead of with a spectacular geographical spot known the world over that links that episode by analogy to the two Paris episodes in our list; what matters is simply that to Danny, an ant and a leaf are every bit as interesting as the Grand Canyon is to adults. From Danny's point of view, the ants and leaves have every bit as much going for them as the Grand Canyon does. The notions of *trivial side show* and *main event,* which are central to the notion of *a trivial side show that is more fascinating than the main event,* have to be relativized and seen simply as natural consequences of a particular individual's idiosyncratic point of view. Indeed, for some people, the stereotypical attractions of Paris — monuments, museums, cafés, restaurants — constitute the core meaning of that city, whereas for other people, completely different kinds of things dominate their perception of Paris. Likewise, ants and leaves are just as interesting to little Danny as the Grand Canyon is to adults; indeed, from his point of view, the meaningless pattern of red and orange hues that his parents are ogling doesn't hold a candle to the exquisite ants and leaves. It's at this abstract level of encoding of *Danny at the Grand Canyon* and of the two Parisian anecdotes that the identity of all three episodes emerges most clearly and naturally.

As for the final two scenarios in our list, they too involve the same basic idea — what is central and crucial for one being can be peripheral and trivial for others. The mosquito doesn't understand Einstein's genius one whit and is limited to its far more down-to-earth refueling interests; in this regard, this particular mosquito is no different from any other member of its species, and nobody is surprised or amused by its irreverent behavior when it lands on Albert Einstein. For the mosquito, the crux of the situation is that this nameless object is a warm vessel filled with fuel, and we all understand that that is the only way the mosquito is capable of relating to Einstein.

This scenario in fact leads easily to a new set of variations on a theme. Thus one can imagine a dog that goes along one night into the countryside with some astronomy students but is indifferent to the beauty of the starry sky. The intrinsic inability of the mosquito or the dog to appreciate what is exceptional in the situation right in front of their eyes brings to mind a familiar proverb: "It's casting pearls before swine." (Note that, through a deft analogical sleight of hand, Einstein has just been transmuted into a set of pearls, and a mosquito into a herd of pigs.) Indeed, such an encoding of the situation, once it's been made, casts the original story in a new light — namely, taking Danny at age one to the Grand Canyon is a member of the category *casting pearls before swine.* At first, this analogy may seem harsh, but on reflection, it makes perfect sense and clearly it is not singling out Danny for criticism in any way. Once the idea is out on the table, it seems a quite natural one, and yet it took many years and the writing of this book for us to discover this proverb-based way of perceiving the event. This shows how difficult it can be to transcend one's first spontaneous encoding of an event.

The final scenario in the list implicitly takes the point of view of a hypothetical Danny who has the same interests as a typical one-year-old but who is also miraculously endowed with the critical faculties of an adult and who sees a huge hole in the ground as singularly uninviting to the gaze, especially given that botanical and entomological marvels are right there in front of his eyes. It's of interest to note that even if this final anecdote shares all the superficial attributes of the original episode of Danny at the Grand Canyon (Danny himself, his parents Doug and Carol, the ants, the Grand Canyon, Doug and Carol's fascination for the Grand Canyon, and so forth), it is also the most distant from the original episode, in the sense that in order to see the analogy between the two stories, one has to encode them in a "relativist" fashion, based on the idea that the received wisdoms of a given society might well strike some of the society's members as absurd. Only in this fashion can one overcome the natural resistance to letting actors play opposite roles to those they played in the initial anecdote.

As is hopefully clear from our analysis, the way an event is *unconsciously interpreted* (the choice of cues that will enable remindings to take place far in the future) should not be confused with the much larger set of *perceived facts* about the situation. What matters is the *perspective taken* on the perceived facts: which dimensions are used in encoding the event, and the amount of abstraction along each dimension — how the event is *distilled*, in short. Thus if, in the Grand Canyon episode, one saw only ants and leaves and a toddler looking at them, then one would be stuck at an extreme level of literality, and the range of possible events that might remind one of this episode will be reduced to practically nothing. If one opens up one's point of view to the slightly broader idea that there are some insects and a child who finds them interesting, that encoding will allow slightly more distant memories to be triggered, such as that of the family on vacation in the Cinque Terre, but even so, the set of possible retrievals remains very sparse.

When one goes yet further and perceives a person who is concentrating on something trivial and ordinary instead of something grand and extraordinary that is right there before their eyes, then one has taken a large step in the abstraction of one's encoding. At that point, the set of potential remindings becomes much wider (it's only at this level of abstraction that the episode of Dick at Karnak is seen as analogous). And nothing keeps us from making yet more abstract perceptions of the situation, letting us see Danny, for example, as a person whose passions are at odds with all the standard clichés. This perception might suggest an analogy likening little Danny to sad, outcast poets who, from their youth onwards, are eternally misunderstood by society. This analogy may strike some readers as an absurdity, and in some ways it is one (here we are touching on the previous chapter's ideas about the gradual impoverishment of the essence of an event as one glides ever higher in the abstraction spectrum), but be that as it may, it illustrates the wide variety of possible encodings of a single situation, and it is these encodings, unconsciously carried out, that will determine the remindings that might, in principle, occur at unpredictable moments of the foggy future.

Categories Belonging Solely to One Lonely Soul

One's personal cognitive style goes way beyond just how one encodes events, for it depends on the entire repertoire of concepts built up over one's lifetime. At the heart of this personal repertoire lie many thousands of concepts that, since they possess linguistic labels, are shared throughout the culture, such as those designated by the *nouns* "ant", "bottlecap", "flaw", "mind"… ; by the *adjectives* "strange", "wide", "gentle", "dusty"…; by the *adverbs* "then", "often" "not", "once"…; by the *prepositions* "through", "upon", "across", "behind"…; by the *idiomatic expressions* "bar none", "all around", "at last", "on high"…; by the *proper nouns* "North Rim", "Karnak", "Pacific Ocean", "Alexandria"… — and so on. And yet, although such a collection of lexicalized categories is very impressive, it is merely the foundation on which is erected the huge idiosyncratic conceptual edifice that develops over the course of each person's unique life.

Every human being, at all times and usually without being aware of doing so, launches new abstract categories based on situations that, in many cases, were noticed only by that person. Often these categories have no publicly shared name, and usually they have no name for the individual either, but that doesn't keep the individual from forming them. Their existence will allow their possessor to interpret subsequent events in their own fashion — otherwise put, to view the world in a unique personal way.

A non-lexicalized category is born when a striking episode is encoded and committed to memory (such as Danny at the Grand Canyon). Then, as we have seen, this buried episode will automatically drift up to the surface if and when one encounters another situation whose freshly-minted encoding overlaps sufficiently with the long-ago encoding of the first episode. Although such idiosyncratic categories are deprived of lexical labels and are grounded in unique, private experiences, they are no less solidly anchored in our memory than the thousands of public, lexicalized categories are. For each of us, they constitute our personal secret garden of categories.

Remindings Can Give Rise to New Categories

As a rule, our personal categories grow slowly and steadily on their own — unobtrusively, imperceptibly. Thus we are usually unaware of a new category when it is just born. Think, for instance, of Peter's casual remark to Carol, "That happens to me every January." Was he thinking to himself, "Wow, I'm building a new personal category to be exploited in the future!" That seems rather unlikely. And when Doug's mind first noticed the Dick–Danny similarity (as Dick leaned down to pick up a bottlecap at Karnak), he wasn't aware of having created a new category either. All he felt happening inside his head was a linking-up of two situations that were different but that had an essence in common — something that happens all the time. Only when he consciously tried to analyze this case of reminding did the new category come into focus as such, and at that point its essence started to become clear.

Such unlabeled and personal categories have much in common with lexicalized categories — those labeled by words, expressions, proverbs, fables, and so on — to which we devoted our first two chapters. They are different, however, in the way they are constructed. One distinction is that they are built on events experienced in private, as opposed to events picked up from exposure to one's culture. Another is that they are idiosyncratic, in the sense that they might wind up never being shared with anyone else. Nonetheless, their usefulness in helping us to spot the essence of certain situations in a pithy, catchy fashion means that if need be, they can be described and transmitted in words to other people — and when this is done, many listeners will find analogous episodes in their lives springing to mind, unbidden. To illustrate these ideas, we'll describe a couple of categories that grew in a couple of people's secret gardens.

The Secret Agent Dashing through the Tunnel

The source of this category in Patrick's memory was a comic book he read when he was a youngster. In it, an aspiring secret agent was required to undergo a series of grueling tests, and at some point he found himself in a very long corridor, not knowing what to do. All of a sudden, the roof started descending and he realized he would have to sprint all the way to the far end in order not to be crushed. He ran like the wind, and as he approached the opening at the end, he said to himself, "These evil monsters required a world-record time, but I'll show them! I'll make it out." However, the final few yards of the tunnel were covered in a thick tarry muck, which slowed him down drastically. In the last fraction of a second, instead of reaching the opening and emerging into freedom, he was mercilessly crushed into a lifeless pancake.

Every time Patrick is hoping to finish some project by making a crash effort and some last-minute complication crops up to threaten his success, this gripping image flashes before Patrick's mind's eye. It could be a sudden loss of access to the Web just as the deadline to submit a grant proposal is approaching. Or it could be the sudden cold-sweat realization, just as his taxi pulls up to the international terminal, that in his frantic rush to pack his bag, he forgot to put his passport into his pocket.

When Patrick told Nadine about this personal category of his, she was immediately reminded of her first date ever. She'd spent a long time getting ready for it and walked out the door at the last minute, already knowing that she would be a bit late. However, her cat slipped out the door as well, and she had to spend a quarter of an hour coaxing it back into the house, and as a result she was disastrously late. Then another occasion came to mind — the day of a friend's wedding. She searched in many stores before finding an appropriate gown to wear, and then, back home, dressing in a great hurry, she tore the brand-new gown and it took her a half hour to mend it before she could leave for the wedding. Once again, she was disastrously late. A third case came to her mind as well, which involved a theater outing. She'd left her house late, had run like mad to catch the subway, and arrived on the platform just as the train's doors closed, so she wasn't allowed into the theater until the first intermission.

The ease with which Nadine was able to retrieve episodes from her memory that shared the conceptual skeleton of Patrick's personal category, while at the same time differing from it in scads of surface-level details, showed that over the course of her life, she, too, had constructed a similar unlabeled category, despite never having read that comic book. The reason she was able to find, among her innumerable memories, a few scenarios roughly describable by the phrase "an unpredictable last-minute glitch drastically reduces one's chances of achieving an already very urgent goal" is that when she encoded and memorized these episodes of her life, some quite abstract aspects were unconsciously taken into account. How else could one hope to explain the much later spontaneous bubblings-up of those old memories?

"God is a Sniper"

The word "sniper" acquired world renown during the war in which the former country of Yugoslavia was splintered into many fragments. Indeed, the main avenue of the Bosnian capital Sarajevo was nicknamed "Sniper Alley" for some years, since people walking along it would frequently shout out, "Watch out for snipers!"

Ilana came up with the phrase "God is a sniper" as a label for a personal category she had made; it was her way of indicating that fate can be as cruelly random as a sniper. The reason for bringing God into the picture, aside from indulging in a bit of irreverence, is that fate was thus anthropomorphized — a very common type of analogy. Every time Ilana hears that someone innocent has been hit out of the blue by an incomprehensible stroke of terrible luck, this troubling phrase comes to her mind.

Thus, someone recently told Ilana about a relative who had a loving and tightly-knit family, sweet children who were blossoming, and much success in her career — but who, only a few weeks after being diagnosed with cancer, abruptly succumbed, leaving friends and family in profound grief. The phrase "God is a sniper" popped instantly to Ilana's mind. For her, this category has gradually grown in importance, to the point where now it plays a central role in her personal philosophy of life. The niche that Ilana has carved out in her mind to accommodate the whims of a cruel shooter lurking in the heavens helps her to deal with life's unexplainable tragedies.

The Crucial Role of Emotions in the Evocation of Dormant Memories

So far, we have not stressed the role of emotions, but if one looks back over the various examples that we have given in this chapter, one will see that emotions are present in a nearly universal fashion. They include jealousy, anguish, sadness, irony, and anxiety. Emotions play important roles inside conceptual skeletons, allowing the retrieval of ancient memories by analogy, as the following example shows.

When Doug was in elementary school, he was fascinated by numbers and the operations that combine them in various ways. One day, his father, a physicist, told him how one could put a number to a power, and taught him the notation of exponents (as in "x^1", "y^2", etc.). Enchanted, Doug started putting together a table of integers raised to various powers, and with delight he noticed little patterns in it. One morning a few weeks later, while walking by his father's desk at home, he saw a physics article that he couldn't make head or tail of, but in which a certain mathematical notation jumped out at him — namely, *subscripts* (as in "x_1", "y_2", etc.). His curiosity was instantly piqued. What wondrous kind of calculation could a subscript, looking so similar to an exponent, stand for? To his frustration, he had to wait for his dad to come home in the evening. The moment he walked in, Doug asked him to explain the enticing new mathematical operation. However, his father merely said, "Oh, those are called 'subscripts'. It's just a way of making new names for variables, since we only have a small supply of letters of the alphabet. Subscripts don't stand for any kind of numerical calculation." Doug's avid hopes were suddenly dashed. He'd eagerly waited all day long, only to be let down by his dad's revelation of the mathematical emptiness of this notation so parallel to that for powers, a concept that had fascinated him.

This story is centered on a reminding (namely, subscripts reminded young Doug of exponents), but not a reminding mediated by emotions. It was a *visual* reminding. However, many years later this entire episode was echoed in a most unexpected way, and the act of witnessing it sparked a sudden reminding filled with complex emotions.

One-year-old Monica was playing on the floor of the playroom in her family's house. That evening she had come across a Dustbuster, and her parents, Carol and Doug, were amusedly watching her push the device's ON/OFF button over and over again. Each time, the buzzing noise delighted her. She was having a grand time. All at once she spotted a second button at the device's other end. Her parents could see her putting two and two together, and in a flash she started pushing on button #2 to see what kind of noise it would bring about. She pushed and pushed but no sound came out. She kept on trying, but still nothing happened. Thinking he'd be a helpful dad, Doug went over, gently took the Dustbuster from her, and, flicking the second button, showed her that a little compartment opened up in which there was a bag containing dust and other grime that the mini-vacuum-cleaner had sucked up. As Monica watched this, an expression of disappointment crossed her face. Far from charming her with this revelation, her dad had deprived her of a hoped-for pleasure by showing her that the very promising resemblance between the second button and the first button, that source of exquisite buzzing, was of no interest.

Here again we have a young child's reminding mediated by a visual analogy — the similarity of one button on a gadget to another button elsewhere on the same gadget. But that minimal similarity would hardly suggest an intimate link between this story of a toddler who is delighted by loud noises and a story about a different child who is in love with number patterns. Even when you toss in the extra fact that the two Dougs in the two stories are one and the same person, although at a remove of forty years, this additional common element still constitutes only the weakest of links between the stories — hardly the stuff of a deep reminding.

But the two stories have much more in common. Indeed, when 48-year-old Doug saw the sudden letdown on his daughter's face as she realized that button #2 had no interest, a sudden feeling of *déjà vu* swept through him as the long-buried memory of his own letdown at the words spoken by *his* father surged up out of nowhere. That event, in a pithy encoding, had been lying inert in his brain for four decades, like a book gathering dust on a shelf in a remote corner of an enormous library. It was certainly not served to Doug on a silver platter, as it has been to the readers of this chapter.

Overture

Doug (age eight)	⟺	Monica (age one)
Fascinated by math ideas	⟺	Intrigued by droll noises
Putting a number to a power	⟺	Finding a Dustbuster on the floor

Infatuation

Making a table of powers	⟺	Pushing a colorful button
Thrilled by unexpected number patterns	⟺	Delighted by unexpected buzzing sounds
All aflush with excitement	⟺	All aflush with excitement

Discovery

Random discovery: subscripts	⟺	Random discovery: another button
Visual analogy: exponents ≈ subscripts	⟺	Visual analogy: button #1 ≈ button #2
Anticipation of great new math patterns	⟺	Anticipation of a great new kind of buzzing

Deflation

Father (Robert), who explains that…	⟺	Father (Doug), who explains that…
Subscripts have no mathematical interest!	⟺	Button #2 has no sonic interest!
Doug's balloon is popped…	⟺	Monica's balloon is popped…

The out-of-the-blue resurrection of this memory was stunning to the adult Doug, for in many ways the episodes are wildly different. But they are also deeply similar. At the crux of each is a very simple visual analogy suggesting a new source of delight to a hopeful child. For young Doug, it was the seductive analogy between exponents and subscripts, while for little Monica, it was the seductive analogy between two buttons.

But above and beyond the fact that both of these stories revolve around misleading visual analogies, what imbues this reminding incident with its depth and interest is the poignant ending that the two stories share: a well-meaning father who, having no inkling of the distress he's about to cause, disillusions his hopeful child by revealing that the child's visual analogy, though seemingly a gold nugget, is in fact just a piece of fool's gold. All in all, there are a large number of elements common to both stories, the most crucial one being a hopeful child's sudden experience of keen disappointment.

Much like the case of Danny playing with ants on the ground and Dick picking up bottlecaps from the ground, this case shows that abstract remindings aren't triggered solely by a close matchup between the most abstract cores of the two events, but that matchups at several levels of abstraction are often needed to bring them about. Obviously many details are irrelevant, such as the colors of the buttons on the Dustbuster, the ages of the children, the houses where the events took place, and so on, but one aspect that is not irrelevant is the fact that in both stories, it was a parent — specifically, the father — who disillusioned a child. Although this is a superficial aspect of both stories (as is the dirt in the Danny and Dick stories), it certainly contributes to their striking parallelism. And there's no doubt that the analogy is rendered yet more intriguing by the fact that little Doug, the "disillusion*ee*" of the first story, grew up to be the disillusion*er* in the second story. To anyone who hears both stories and sees their many analogical connections, this role-reversal has to be one of the most central and ironic aspects of the whole thing, and it no doubt strengthens the analogy, even though the younger and older Dougs play opposite roles in their respective stories.

In sum, several emotions — an initial fascination, the pleasant surprise of a simple visual analogy, the high hopes that ensue from this discovery, and finally a sudden deflation — all play critical roles in this reminding. This illustrates a general tendency — namely, that remindings that take place at a deep level are often dependent on emotional aspects of the two episodes they link together. This is because very often the most central aspects of an event are the strong emotions that it churns up.

Events are Encoded Not by Rote but by Distillation

So far, we've been using the term "encoding" as if what we meant by it were self-evident. Our physiology restricts our perception to the standard sensorial modalities (vision, hearing, etc.) and to the features afforded by those modalities (colors, movements, and shapes, for example); our perception is also constrained by the resolving power of our senses (the world would appear very different to us if our visual system could directly perceive microbes). Our psychology also limits our perception, allowing us to recognize and memorize events only in terms of certain modalities of encoding. We perceive through our sensory organs, to be sure, but no less through our concepts; in other words, we perceive not just physiologically but also intellectually. There is thus an unbreakable link between perceiving and conceiving. On the one hand, our conceptions depend on our senses, since our concepts would be quite different if our senses were different, but on the other hand, our perceptions depend on our repertoire of concepts, because the latter are the filters through which any stimulus in our environment reaches our consciousness.

It may not be obvious why we need to encode our experiences at all — that is, why we need to reduce them to a tiny fraction of their entirety. To see why encoding is necessary, imagine trying to memorize an event without any simplification taking place; the result might be called a "total rote recording" or "perception without concepts". An experience would be captured in its entirety in our neurons, much as a film can be stored on a DVD. In the case of Danny and the Grand Canyon, having such a "total rote recording" would mean that the entire scene had been "filmed" in Doug's brain while he was experiencing it, and then that, some twenty years later, when he observed Dick stooping to pick up a bottle cap, this specific film had been reactivated in his brain by a mental search algorithm running through all filmed scenes in his memory.

This idea of "perception without concepts" can be summarily rejected, because no mental process based solely on visual cues would be able to connect one-year-old Danny with the mature adult Dick, or to see the link between ants and a bottlecap, not to mention the link between the enormous geological concavity of the Grand Canyon and the architectural and archeological masterwork of the Temple at Karnak. A search for purely *visual* resemblances, based on techniques having to do with alignment of images, could not possibly lead to such a reminding. And so far we have only alluded to the *physical entities* in the scene, while completely leaving out the *actions* (manipulating small items on the ground), the *context* (a two-week-long pleasure trip), the *relationships* (the contrasting sizes of the entities involved in the situations), the *emotions* (a feeling of irony), and so forth.

If conceptual encoding did not take place, it would be impossible to retrieve events stored in one's memory. Just think how indispensable it is for users of Web sites on which photos or videos are shared that linguistic labels ("tags") are attached to each item. There are many additional arguments for the necessity of encoding in the act of committing experiences to memory, based on such phenomena as selective forgetting, partial or distorted recovery of memories, and the reconstruction of memories.

No purely image-based search process, no matter how sophisticated, would be able to "see" the connection between Dick at Karnak and Danny at the Grand Canyon, or between Doug's disappointment about subscripts and Monica's disappointment about button #2, because such events' connections are not visual. The moral is that we do not store in our memory a collection of "objective" events through which we run, seeking perceptual resemblances, whenever a new event happens to us; rather, events that befall us get encoded — that is, perceived, distilled, and stored — in terms of prior concepts that we have acquired. Remindings are possible because certain aspects of long-ago situations were noticed and stored at that time, yielding encodings of those situations. But which aspects get paid attention to, and how are they encoded?

Do Our Brains Instantly Pinpoint Timeless Essences?

The activation of certain memories from one's earlier life is not a mental game that we humans engage in merely because we find it intellectually pleasurable to connect present and past through the spotting of similarities. In general, the automatic behaviors that we engage in play a crucial role in ensuring our survival. Being reminded of a past event is not a luxury add-on that might optionally take place after we have understood a new event; rather, such a reminding is deeply implicated in the very act of understanding the new situation. This idea was stressed early on by cognitive scientist Roger Schank, one of the pioneers in the study of reminding.

Let us restate this more concretely. Doug was reminded of his son Danny at the Grand Canyon when he watched his friend Dick at Karnak because his perception of the fresh new Karnak situation activated certain concepts that had been encoded in the course of "processing" the Grand Canyon event many years earlier. Doug knew he was in a very special sacred place when he was at Karnak. He noticed that a companion had leaned down to pay attention to something small and was momentarily ignoring the guide. His emerging sense of irony in this situation led him to "replay" the scenario of Danny at the Grand Canyon, as it existed in his memory.

Does this mean that the category *trivial side show more fascinating than the main event* was necessarily created at the moment the first event was encoded? Does it mean that Doug, when he experienced the second event and encoded it, automatically reactivated the category's founding member because the two had been encoded *identically*? This is the idea on which Schank's theory of the mechanisms underlying reminding is based. However, is that the only way to explain the bubbling-up of a dormant memory?

Does a successful reminding presume that both events were encoded at such a high level of abstraction that the two situations are simply specific cases of the abstraction? Can one be reminded of a past situation only if, already in the moment when one was experiencing it, one succeeded in putting one's finger on such an abstract "timeless essence" that later, many years down the pike, when another situation sharing that same timeless essence comes along, one will be all prepared to activate it? If this were the case — let's call it "instant pinpointing of timeless essences" — then Doug, when he observed Danny's fascination with the ants and the leaves, would have instantly created, albeit unconsciously and without an explicit linguistic label, the abstract category *trivial side show more fascinating than the main event.* His sudden recollection, as he watched Dick at Karnak, of his son at the Grand Canyon would have been due to the creation of this category some fifteen years earlier. If one were to believe the hypothesis of "instant pinpointing of timeless essences", then such a reminding could have taken place only if the conceptual skeleton *trivial side show more fascinating than the main event* (we stress that it's not the sequence of English words that we are talking about, but the abstract idea that it denotes) had been created at the moment of encoding the first scene, and then, fifteen years later, rediscovered in another scene.

To put it mildly, this is most implausible. The instant spotting of a shared timeless essence is not the only mechanism that could, in theory, give rise to a reminding sometime down the pike. The problem is not that it's hard to perceive a deep and precise conceptual skeleton, for in fact we all perceive such skeletons all the time, when the categories involved are those of our day-in-day-out thinking — that is, categories with which we have some degree of expertise. The problem is that when the categories are unfamiliar to us, we cannot see nearly as deeply into situations involving them.

We effortlessly and unconsciously perceive certain shared aspects of situations that are very different — for example, despite their enormous differences, we recognize all sorts of *elephant-in-the-room* situations, *once-bitten-twice-shy* situations, *you're-pushing-your-luck* situations, *can't-see-the-forest-for-the-trees* situations, *if-it-ain't-broke-don't-fix-it* situations, *killing-two-birds-with-one-stone* situations, *a-bird-in-the-hand-is-worth-two-in-the-bush* situations, *you-can't-have-your-cake-and-eat-it-too* situations, *you-could-hear-a-pin-drop* situations, *people-who-live-in-glass-houses-shouldn't-throw-stones* situations, and so forth — and the recognition of these highly abstract qualities allows us to spot commonalities in situations that have nothing at all in common on their surfaces. In a word, we are all experts at perceiving conceptual skeletons when the categories involved are ones in which… we are experts!

And so, could it be that Doug unconsciously perceived the conceptual skeleton *trivial side show more fascinating than the main event* while watching Danny, and used it to encode that event? Could it be that this made him an expert with this concept? Could it be that thanks to his expertise with this concept he was able to recall that event many years later, while visiting Karnak? Well, it's possible in theory, but it's most unlikely.

Spotting the essence of something that one is already an expert in is one thing, but spotting the essence of something novel and unfamiliar with is quite another thing. To put it more pithily, abstraction is one thing, but *deep* abstraction is quite another thing. The hypothesis that we can instantly spot "timeless essences" leads, in certain cases, to absurd conclusions. When experiencing an event, we cannot peer into the future and clairvoyantly guess exactly which highly abstract encoding to construct so as to allow that memory to be triggered by events that will take place many years down the pike.

More concretely put, when Doug was eight years old and he asked his father what the meaning of subscripts was and his father's answer gave rise to a great feeling of loss, the abstract conceptual skeleton that this event would wind up sharing, forty years later, with the episode of Monica and the Dustbuster could not possibly have been clear to him at that early point in his life. What, then, *did* the eight-year-old child encode at that time? Alas, one would need both a time machine and a mind-reading machine to give a precise answer, but there is no doubt that *some* abstraction was involved in the encoding. The crux of perception — even for an eight-year-old or a two-year-old — is the act of abstracting; abstraction is the principle that allows us to create new categories and to extend them throughout the course of our lives. We are forever extending our categories because the strictest form of literality ("total rote recording", as we dubbed it above) does not allow *any* resemblances to be noticed, and thus excludes all thinking. In the case of eight-year-old Doug, there had to be at least enough abstraction in his perception to allow the connection with the Dustbuster scenario forty years later to be perceived, even though that scenario was, in so many ways, vastly different from it.

First Encodings Can Go Only So Far

In a course one of us taught, a student heard the anecdote of newlywed Carol signing with her maiden name and her friend Peter who "did the same thing every January"; this reminded the student of a time four years earlier, when she had changed jobs. In her new job, she was supposed to answer the phone and say, "Hello, this is Company X", but in her first few weeks she would often say the name of her previous employer, upon which she felt very silly. Does the fact that this student hit on that memory mean that four years earlier, she had constructed the highly abstract category *situations in which one is trapped by a habit that one is unable to update*? It's hard to imagine what would have driven her to create such a high abstraction; it's unlikely that she felt that someday she would need to be an expert in the spotting of situations of that sort. One can get along quite well in life without having constructed such a narrow category. Although we can be pretty confident that this student didn't encode her mildly embarrassing phone-answering gaffes at such a rarefied level of abstraction, her encoding nonetheless involved a fair amount of generalization, because hearing the story about Carol's maiden-name-signing gaffe brought her own old gaffes back to her mind swiftly and effortlessly.

Among the likely aspects of her encoding were the fact that she blurted out the wrong thing, that what inadvertently came out of her mouth was a relic of her recent past, that she wasn't in control of what she said, that she was embarrassed by her slip, and so on. And so it seems probable that her retrieval of that four-year-old memory, triggered by the anecdote of Carol's signing error, didn't involve just one single and concise abstract conceptual skeleton that perfectly matched both the old and the new situations, but rather, that it depended on a number of separate, abstract, and important aspects of her old job situation, each of which had contributed to how she had perceived that situation's essence at the time it happened. In other words, rather than one single perfectly-fitting conceptual skeleton — the magic key to memory retrieval — having been created when she made her telephonic errors, a number of smaller and independent concepts were built, and hearing the story of Carol reactivated enough of those concepts to remind her of having said, inappropriately, "Hello, this is company X". In sum, although there is always some degree of abstraction in the act of encoding, there is not always exactly *one* highly abstract conceptual skeleton shared by the triggering event and the event retrieved from memory.

For a current event to trigger the recall of a far-off event that one hasn't thought of in many years requires strong resemblances. For the long-buried memory to be triggered by what one is currently experiencing means that each side of the connection has to "give" a little — that is, some dimensions of the way we perceive the new and the old situation have to have sufficient flexibility. Even if the encodings of the events are far from reaching the maximal level of abstraction, they will go far beyond the most literal details of the experienced situations, because such literality confines remindings to the level of the very mundane. For example, in the case of the children in the Cinque Terre who were fascinated by grasshoppers, it takes only a tiny amount of abstraction to link this situation with Danny at the Grand Canyon, since both situations involve children, insects, and famous tourist spots. One feels that the stories are so close to being carbon copies of each other that such a reminding verges on the trivial. Luckily, the human mind goes way beyond this low level of reminding in the mental leaps that it carries out among countless stored memories.

In conclusion, is there always some amount of abstraction involved in the encoding of memories? To be sure. But does being reminded by a current event of a past event always depend on the two events exactly sharing a conceptual skeleton in one's mind? By no means.

The Humble List that Aspired to Become a Magnificent Category

A local chef has just finished sautéing a fresh fish caught on a small fishing boat based in a nearby village. The side dish consists of steamed vegetables grown on a local organic farm. Meanwhile, elsewhere on the same planet, a microwave oven has just heated up a frozen dinner featuring fish grown in a factory and stuffed to the gills (literally!) with bone meal, accompanied by genetically modified vegetables that were grown in chemical fertilizer laced with pesticides.

Which of these two dishes appeals more? It's probably not too hard to choose, but unfortunately, the first is expensive to produce and certainly is not amenable to mass production, and in addition it requires people who are passionate about what they do. Today it's nearly impossible to make a living in such an authentic, old-fashioned way. And so, is there any way to successfully combine *that which allures but is unprofitable*, on the one hand, with *that which is cheaply manufactured though unappetizing*, on the other? Well, a good strategy would seem to be that of mass-producing something that emanates down-home appeal — if this is not a contradiction in terms.

Below we list some cases where marketing trickery can give rise to a false impression that tempts customers to buy without realizing what lies beneath the surface. Naïve would-be buyers, convinced that they have hit on genuine authenticity, fall for slick mass-produced articles — a successful ploy resulting from cold calculations in the business office.

Snails labeled "escargots de Bourgogne" seldom hail from Bourgogne. Today, this classic dish, so redolent to so many of France, is mostly imported from Eastern Europe, Turkey, or China, and when one is eating a snail, the chances are very slim that it actually grew in the shell in which it is found.

In a now-defunct chain of American bookstores, certain prominent sections used to be dedicated to "Local Authors"; this gave the impression that that very bookstore's staff had played a role in the selection of the books found in that section. But in fact, the choice of which local authors to showcase was made far away, in the chain's national headquarters, without any input at all from the local store.

Ads for certain little clay figurines representing historical figures or quaint folk icons proudly proclaim, "Hand-painted", which projects a sweet old-fashioned postcard-like image of how these statuettes came to be. But the truth is that "Hand-painted" usually means "Made in China". The statuettes of the little Danish mermaid or of the Napoleonic soldier seem far more exotic to the person who paints them than they are to the person who buys them.

These days, on the Venetian island of Burano, known for centuries for its intricate lacework tablecloths, blouses, scarves, doilies, and so forth, nearly all such items are in fact made in China and are exported to Italy. The appearance of authenticity is preserved, however, by the elderly women selling them, who sit in the tiny shops, wearing lace clothing and working away on lovely lace items.

In a town along the Nile in Egypt, a child comes up to tourists and offers to sell them "antiques" that she claims she found, when in fact they are mass-produced objects made for tourists and artificially aged in sand and water. In a small shop in the same town, a young boy is making scratches on a metal tray decorated by a machine, in an attempt to give the impression that he himself made all the decorations on it, and thus that all the trays have been locally decorated.

In a Christmas market in the main square of a small Austrian town, a "Corsican peasant" dressed in a traditional Corsican costume is selling "Corsican salamis", but his only connection with Corsica is that he likes to vacation there. Next spring he'll don a costume from the Auvergne and bald-facedly hawk "Auvergne cheese".

In Cabourg, on the coast of Normandy, a *crêperie* on the main street is owned by a Parisian couple. On weekends, they rise very early and leave for Cabourg a few hours before the tourists do; symmetrically, a few hours after sales are over, having cleaned up and waited for the traffic jams to clear up, they head back home to Paris.

The recent upswing in popularity of sushi bars in Paris has coincided with a downswing in the popularity of Chinese restaurants. As a consequence, Chinese restaurateurs have opened up Japanese eateries. For unsavvy Caucasian customers, nothing tips them off that the Asian people waiting on them are as out of place as a flotilla of Greek servers would be in a French restaurant in the middle of Tokyo.

The items in this list all clearly share some quality, but that quality has no standard name. What, if anything, is the difference between them and the members of a category that does have a standard name? We would suggest that the items in this list implicitly define the boundaries of a category that is just as reasonable and intellectually appealing as any lexicalized one, such as *forgeries*, which is a category clearly related to this one but more general than it.

The Humble Category that Aspired to Acquire a Label

Our list implicitly defines a new concept whose only distinction from lexicalized categories is that it is not yet lexicalized. And so, grafting "faux" onto "authenticity", we suggest the term "fauxthenticity". This act of explicit labeling will help to anchor the concept in memory and will increase the likelihood of its being further extended in the future. Indeed, instances of fauxthenticity are easily found, sharing the conceptual skeleton that emerges from reading the previous examples, while also extending it in new directions.

On envelopes one receives these days, it's common to see one's name and address in what looks like handwriting, when in fact it has simply been computer-printed, using an intentionally informal-looking cursive font. There are even certain fonts that have randomizing algorithms in them that allow each letter token to be slightly different from other tokens of the same type, thus giving the impression that every letterform has been uniquely penned by a human hand.

In the automated telephone trees that one encounters whenever one calls any large business, a new trick has been added fairly recently — that of having the recorded voice insert hesitant pauses, or make slight mistakes, or even sound surprised by an afterthought, as if the "person" had just remembered something they should have said earlier. "Uhh… oh, yeah. So now, could you, um, just tell me your confirmation code one more time? Thanks a lot!"

In Italy, companies that send out automated emails to customers have largely abandoned the respectful second-person pronoun "Lei", which is the traditional way of addressing people one does not know (corresponding to the French "vous"). They now use the informal pronoun "tu", which can lead to absurdities, such as a "personalized" form letter that opens with "Egregio Professore, ecco la tua nuova carta Bancomat", which comes across more or less as "Hey there, distinguished professor, old pal, here's your new credit card!"

Sometimes one reads that a particular article is made of "genuine leatherette" or even "genuine artificial leather", and oddly enough, no irony is intended. The idea is presumably that there is an industrial standard for imitating leather (or other natural products), and that meeting this standard constitutes a kind of authenticity.

We hope that some readers will be inspired to look around for further examples of *fauxthenticity,* as they are quite widespread, just waiting to be spotted. Much as we said about the category *bait-and-switch* toward the end of Chapter 2, people who have explicitly constructed the category *fauxthenticity* have a much clearer sense of the phenomenon than those who have not.

Analogies and Categories in Canine Minds

We've looked at a wide variety of situations in this chapter, in order to show that people constantly construct rich new abstract categories that have no labels, sparked by the perception of unconsciously perceived resemblances between situations. But what about animals less reflective than we are? Well, they do much the same, though they are limited by their mental level, which depends on the species they belong to. Take dogs, for instance. What concepts does a typical dog construct, in the course of its lifetime? Below we give a list of a number of concepts that we have observed in dogs we have personally known. (We've taken the liberty of using English words to express these categories.) Each of these concepts gets formed as a result of a long series of analogies made, day in and day out, over many years.

humans; male humans and female humans; adult humans; children; babies; friends and strangers; letter carriers; veterinarians…

dogs; puppies; my best dog-friend; birds; squirrels; cats…

water; my food; human food; my little treats; bones…

delicious; hot; cold; hard; soft; open; closed; nice; mean…

in; on; under; next to; in front of; behind…

day; night; pain; hands; mouth; eyes; feet; paws; orders; threats; my doghouse; my yard; other people's houses; other people's yards; the kennel; dog-doors; games I play with people; toys I can play with; toys I'm not allowed to play with; balls; frisbees; sticks; branches; doors; chairs; tables; leashes; rain; snow; trees; lakes; swimming pools; dog-dishes; cars; sidewalks; streets; staircases; little toys that look like dogs; robot dogs; stuffed animals; my master's voice; my master's voice on the telephone; thunder; fake barking (heard on the radio or the television, etc.)…

eating; drinking; playing; fighting; walking; staying; going out; going in; jumping; swimming; sitting; waiting; lying down; seeking; fetching; catching; going up; going down…

what's good to eat and what's not; what's good to drink and what's not; objects that I'm allowed to chew on and objects that I mustn't chew on; loud harmless noises versus loud noises that could mean danger; places to swim; places where I can "do my duty" and places where I mustn't; places where I'm allowed to sleep and places where I'm not; places where I'm allowed to eat anything that I find versus places where I mustn't do so…

When a thinking being lacks linguistic labels for the phenomena that it encounters, the sharp distinction that many people believe they see between categorization and the making (or spotting) of analogies becomes well-nigh impossible to make. A dog, not possessing linguistic labels, has instead a set of experiences with inanimate objects, animate agents, actions, and situations, and in order to survive and live comfortably in the world, it depends on its ability to see new phenomena in terms of situations that it has already been in. Thus categorization for a dog is clearly the creation of analogical bridges to prior knowledge.

The fact that every dog can reliably recognize other dogs, birds, cars, trees, balls, and leashes is hardly astonishing to us humans. But for a dog to be able to reliably distinguish places where it can "do its duty" from places where it shouldn't is more impressive, because this category seems quite a bit subtler. Given that analogies are distributed all along a broad spectrum, ranging from very simple to very subtle, it seems reasonable to ask what degree of sophistication these nonverbal animals can attain in making analogies. To this end, we asked some of our dog-loving friends if they could recall some interesting analogies made by their pets, and in return we received a good number of fascinating anecdotes. Here we reproduce four of the stories that we found particularly striking.

One evening when my parents were keeping Char [a Labrador] for a few weeks, my mother said, "He needs a bath tomorrow." Just before going to bed, they called Char but he didn't come. A long search finally wound up in the basement, where Char was patiently waiting next to the big sink in which I'd always given him his baths during previous visits. Not only had he understood the word "bath", but he'd also remembered the unusual tub we'd used in that house, as well as the place where it was located.

Fenway [a dachshund] recognizes any and all suitcases, and every time we get out some suitcases from our attic, she starts looking sad, as she's worried we'll leave without her (which is often the case). If she sees one of the suitcases has been packed and shut, she'll jump onto it and lie down, hoping this will convince us to take her along. She also recognizes the small duffel bag in which she does her own traveling, and she knows what it means when we get it out of the closet — namely, that we're all going on a trip. Every time we've taken Fenway to California, the moment we start packing for the return trip, she jumps right into her duffel bag, making sure we don't leave her behind.

Fenway had just had a small growth removed surgically from one of her rear legs, and in order to do the surgery the vet had had to shave off the hair surrounding the growth. As soon as Fenway came back home, she went off in search of her little stuffed moose and started chewing away on one of its hind legs, until she had created a small patch of "bare flesh" that looked just like the bare patch on her own leg.

The first time we took Fenway to a big "dachshundfest" in a park, she spotted, far off, a white dog with long unruly curly hair, which looked very much like her pal Scruffy. In a flash she was off and running, but when she reached the other dog, the two eyed each other up and down a bit confusedly. A moment later Fenway realized this wasn't Scruffy, upon which she flipped right around and ran straight back to us across the park.

This last anecdote shows that dogs, just like people, can make mistaken analogies, because sometimes a resemblance, even a very strong one, is misleading. A perfect stranger can have the most striking resemblance to one of our best friends, and then one can't help wondering whether the powerful resemblance is a reflection of a deep similarity of two souls, or is only superficial. At least one can't suppress such wonderings if one belongs to the genus *homo sapiens*…

Who Does This Young Icelandic Professor Remind You of?

John has gone to Iceland to give a talk, and his wife Rebecca has come along. After two days of tourism, they arrive at the university, where they meet some friendly people, among whom is a young professor named Thor. John is struck by Thor's chiseled face and he keeps asking himself, "Who does he remind me of so much? I know it's someone I know very well!" All at once it comes to him that he's thinking of his friend Scott in California. This makes John feel at ease with Thor, because his friendship with Scott is very strong. After the talk, John and Rebecca are taken out for dinner by Thor and colleagues, and they spend several pleasant hours together.

The next morning, John asks his wife, "Was there anyone yesterday that reminded you of someone you know?" Rebecca replies, "Thor, maybe? Is that who you mean?" "Exactly!" says John. "And what struck you about him?" "Well, I'd say he's very charming, and he also looks like Scott in California." John replies, "I agree. It seems that we see him with the same eyes!" Rebecca adds, "Maybe it's the corners of his mouth when he smiles. It suggests a kind of gentleness that he shares with Scott." "Exactly. Also their husky voices, wouldn't you say?" "That too," says Rebecca. John continues, "Yesterday at dinner I felt as if I knew him well, so I said some things that otherwise I would never have said. It was just like talking with an old friend, and I'm glad you're confirming my impression. To me, the fact that we agree shows that the connection between them is something *real*, not just a personal flight of fancy."

Rebecca's confirmation of what John noticed reminds us that a good analogy is something that one can share with others. Such a feeling of objectivity reinforces the intuitive idea that an analogy connects two entities *in the external world* — in this case, Scott (a lawyer in California) and Thor (a professor in Iceland). Certainly the conversation quoted above gives this impression, but we should remember that John's analogy between Scott and Thor was created (or observed) by him *in Scott's absence*, and Rebecca's analogy was created (or observed) in the absence of *both* individuals. Rebecca was relying only on memories stored in her head, and so her link was necessarily between two *mental representations* of people, rather than between two people in front of her. However, she found it much more natural to think and say that her analogy linked the *sources* of her mental representations — namely, Scott and Thor — than to think and say that she had constructed or discovered a connection between two neural patterns inside her brain. After all, compared to "Thor reminds me of Scott", a sentence such as "I just constructed a mental bridge between my mental representations of Thor and Scott" would sound absurdly pedantic, as well as very weird.

Mental Entities and the Connections Between Them

This shorter way of saying things (and thus of thinking) is preferred by everyone. Instead of saying that we've created a mental link between two mental entities, we humans prefer to project the two ends of our analogical bridge *outside* of our heads, and in this way our analogy seems to become an external bridge — a soaring metaphorical rainbow at whose ends are two entities located in places that may be very far apart, like Berkeley and Reykjavík. And when someone else confirms such a personal analogy, the appearance of objectivity suggested by the agreement in viewpoints reinforces the naïve image of an analogy as being like a rainbow high in the sky — a celestial arc leaping between objective, external entities. But if one thinks about it, one realizes that the arc linking the two entities is not a *rain*bow but, so to speak, a *brain*bow. And if it is *objective* — that is, if two or more people see the same analogy — it's because there can be, in two different brains, two "parallel" brainbows — that is, two brainbows connecting internal representations sparked by the same external sources.

To make this abstraction more concrete, let's take a bridge that exists in the head of each of your authors. At one end of this bridge is a mental image of Mark Twain's face, and at the other end is a mental image of the face of Norwegian composer Edvard Grieg. It's quite possible that "the same" bridge was independently constructed by and in numerous other heads and has lived happily in them for a long time. Indeed, we have little doubt that anyone who looks at images of these two gentlemen will rapidly construct, in their head, this "same" mental bridge, this "same" brainbow.

Mark Twain

Edvard Grieg

It is very tempting — indeed, it is indispensable for efficient interpersonal communication — to use a kind of shorthand to describe several analogous analogical bridges built in different heads in a shorthand fashion, according to which there is simply *one analogy*, an *objective* analogy, between two entities belonging to the world "out there". Just as John and Rebecca shared an "objective" link between Thor and Scott, so can many people share the analogy linking Twain and Grieg (or more precisely, linking their faces). Suppose that you have just created a personal "brainbow" between your images of Twain and Grieg. It is very plausible that your mental images of these two men might start to mix and blur, thus resulting, in the end, in a new mental entity that we might baptize "Twain/Grieg" — the name not of a *person* but of a *category*.

Analogy-making and Categorization: Two Sides of the Same Coin

If you were now to run across some photos of Albert Einstein, you might well think, "An excellent example of the category *Twain/Grieg*!" As a result, your mental entity *Twain/Grieg* would slightly change, taking into account this third member. The outcome would be a more general category, and as such it would deserve a new label, just as when a small company grows large, it no longer belongs solely to the two people who founded it many years earlier. As a matter of fact, your category initially based on the facial resemblances of Mark Twain and Edvard Grieg (and then Albert Einstein), as it came to include more similar-looking people, could, in view of its growing generality, change its name, perhaps adopting the acronymic label "TGE", or it could even lose its label completely.

To round out this story, let's suppose that after building in your mind the category in which are blurred the faces of Twain, Grieg, and Einstein, you ran across a photo of the famed humanitarian doctor Albert Schweitzer. Would assigning Schweitzer to this growing mental category be *the making of an analogy*, or would it be an act of *categorization*? It would be both. What happens inside the head of a person looking at the picture of Schweitzer is *the construction of a mental bridge* linking a fresh *new* mental representation (triggered by seeing the picture of Schweitzer) with an *older* mental structure whose existence was collectively due to having seen and fused the faces of Twain, Grieg, and Einstein. Calling such a mental bridge-building operation "an act of categorization" and calling it "the making of an analogy" are equally valid choices.

Albert Einstein

Albert Schweitzer

If the *TGE* category keeps on growing by accepting more members, thus making more of an anonymous blur, one may start to forget who its founding members were, and at that point and for that reason, one will probably be more inclined to use the term "categorization" than the term "analogy-making". But no matter; in both cases all that's going on is the recognition of a correspondence between a newly-minted mental structure and an older one — in short, the construction of an analogical bridge.

The Debate Dies Down

We would like to consider one last case — the seemingly trivial case of the recognition of a cup as a cup. Suppose you are at a friend's house and want to fix yourself a cup of tea. You go into the kitchen, open a couple of cupboards, and at some point you think, "Aha, here's a cup." Have you just made an analogy? If, like most people, you're inclined to answer, "Obviously not — this was a *categorization*, not an analogy!", we would understand the intuition, but we would propose another point of view. Indeed, there is an equally compelling "analogy" scenario, in which you would have just constructed inside your head a mental entity that represents the object seen in your friend's cupboard. In this scenario, you would have created a mental link between that mental representation and a pre-existing mental structure in your head — namely, your concept named "cup". In short, you would have created a bridge linking two mental entities inside your head. And as we just noted, in examining the assignment of Albert Schweitzer to the *TGE* category, the response to the question "Is this an act of *analogy-making* or of *categorization*?" is once again that both labels are correct.

If you are uncomfortable with the idea that calling a cup a cup is a case of analogy-making, then try to pinpoint the crucial difference between building a bridge linking your brand-new Schweitzer-photo percept to your prior *TGE* concept and building a bridge linking your brand-new percept of a certain ceramic object in the cupboard to your prior *cup* concept. If there is any noteworthy difference between these two actions, it can only be in the difference between the concepts *TGE* and *cup*. The former is a relatively fresh new concept in which there still remain fairly clear residues of its three founding members, whereas the latter is an old concept in which there remains no such residual trace. (Who remembers the primordial cups that initiated their concept *cup*?) Apart from this distinction, the two bridges have the same nature.

The moral of this fable is that recognizing a cup's "cupness" is no less a case of analogy-making than is recognizing a new instance of the concept *sour grapes* or of the concept *fauxthenticity*. We hope that this thesis, although it runs against the grain of most people's intuitions, has now become familiar and resonates with *your* intuition. It is a unifying viewpoint on human thought, placing categorization and analogy-making, fused into one thing, at the center. And armed with this perspective, we now turn our gaze to what this implies about the mechanisms of thinking.

CHAPTER 4

Abstraction and Inter-category Sliding

❧ ❧ ❧

X is Not Always X

It's 3:30 in the Parisian afternoon, and Emmanuel and Doug are taking a break to go down to the corner café Le Duc d'Enghien, where they're planning, as usual, on having, well, a *café*. Despite the scorching temperature, Doug is in the mood for *un crème* (a coffee with cream), while Emmanuel is wavering between a Coke and *un diabolo menthe* (a mint-flavored cold drink); finally he settles on the latter. After a few minutes' chat, the co-authors cross the street to the *pâtisserie*, where, as per their daily routine, they get some pastries. Emmanuel chooses *une tartelette aux fruits rouges* (a berry tart) and Doug goes for a popsicle; then they head back up to the office to resume their writing.

There's little of great moment in the foregoing, except that Emmanuel's coffee wasn't a coffee and Doug's pastry wasn't a pastry. And yet no one would accuse them of lying. Even if Doug had ordered a lemon tea instead of his *crème*, it would have seemed perfectly fine to say that the co-authors had gone out for a coffee. And so, what exactly is meant by "a coffee"?

We can distinguish at least four types of context — four levels of abstraction, in this case — in which the term is understood differently and yet always with ease. First of all, there are situations where "having a coffee" means "chatting while eating or drinking something light". In such a context, the word "coffee" is so open and abstract that it covers any type of drink, or a sandwich, or an ice cream, or for that matter nothing at all, as long as the establishment doesn't object. We'll call this "coffee_4".

Next, there are situations in a restaurant where, after the meal, the server asks the customers, "Who'll have a coffee?" A reply such as, "A tea for me, please" would seem perfectly in order here, and no one would think that it contradicted the question that was asked, whereas asking for a cognac, or worse yet, some more wine or another order of fries, would seem totally incongruous. In other words, there is still some abstraction involved, but it is not as great. We'll call this "coffee_3".

Then we come to a situation in a café where two regulars are greeted with the customary question, "And what might your coffees be this morning, ladies?" Here the server is expecting answers such as "A *crème* and a decaf, please", or "Two *macchiatos*, please", since those belong to the category explicitly mentioned. Here the word "coffee" is taken more narrowly, but despite the approach towards literality, there are still numerous ways for it to be realized (and so we'll label this case "coffee_2").

Lastly, there's the type of situation where someone walks into a Paris café and says, "*Un café, s'il vous plaît*", or "A coffee, please." Here it's clear that it couldn't possibly mean a tea or an ice cream (etc.); the drink being requested is a straight espresso, without cream. This is the default interpretation for the word "coffee" in Paris cafés, and that's what we'll mean by "coffee_1". We've hit the rock-bottom level of abstraction in our spectrum of coffees, and so we'll temporarily draw our exploration of "cafégories" to a close.

What has this exercise shown us? That the members of a category change with context; that we effortlessly understand the nature of the context that we are in; that a single word in a given language can denote numerous different categories; and that these categories can have different levels of abstraction.

Road Map of This Chapter

Our first three chapters were an attempt to give an answer to the question "What is a category?" by examining various types of categories, including those covered by a single word (Chapter 1), those covered by a composite lexical entity (Chapter 2), and those that have no lexical label at all (Chapter 3). With this chapter, we open a new phase in our book, in which we analyze how categorization works. In particular, we will look carefully at "leaps" or "slippages" between categories (and in the interest of less repetition, we'll use both terms).

Our goal is to reveal the fundamental importance of slippages between categories in the act of thinking, and specifically slippages that carry one up or down a vast range of abstraction. First we'll say what we mean by "abstraction" and then we'll show that the fate of objects and situations in this world is to be shunted around, in a manner that is both facile and unconscious, from one category to another. Just as Molière's Monsieur Jourdain spoke prose without realizing it, so we are all experts in making leaps from one category to another without realizing it. We are all constantly practicing the art of mentally shifting objects and situations from one category to another.

To show that a good part of what we call "flexibility" and "creativity" is tied to the eminently human faculty of extending categories and making leaps between them, we'll take a close look at a particular phenomenon that gives insight into the processes underlying the development of concepts — specifically, the linguistic phenomenon called *marking*, which is extremely widespread in language, although people seldom notice it at all; in fact, few even know it exists. The idea is that a single word of a language can designate both a narrower and a broader category, where the narrower one is wholly contained inside the broader one, as was illustrated above by the word "coffee". Although marking can occasionally hinder communication and lead to confusion, it is mostly a useful tool, imbuing language with greater fluidity by allowing several categories to be labeled simultaneously by a single term and by taking advantage of our mind's natural sensitivity to context.

Next we will scrutinize a process at the core of human thought, and which we already introduced in Chapter 1: the development of concepts through category extension. As we saw in that chapter, when categories are born, they are tiny — often they have just one member — and then cores and halos begin to form. Categories grow by welcoming new members, which sometimes are central and other times lie way out at the fringes, at the city limits. The act of welcoming such unexpected members into the fold requires either "pushing the envelope" or else the creation of new categories. In any case, analogy is the motor that drives all such extensions. We will analyze the process at the root of this human ability to understand situations in terms of pre-existing concepts, and at the same time to modify those concepts under the influence of new situations.

We will then turn to another fundamental question, closely linked to the previous one: What makes an expert? This question is important because the concept of *expertise* applies not only to a specialist's knowledge of some narrow domain but also to an average person's ability, developed over a lifetime, to deal with their daily environment. More specifically, we shall see that being an expert doesn't mean just that one has acquired more categories than other people have, but also that one has organized them in such a way as to facilitate useful categorizations at different levels of abstraction, and in such a way as to allow one to glide smoothly, when under contextual pressure, from one category to another.

This indeed is one of the wellsprings of creativity — namely, the ability to make certain crucial leaps that at first seem surprising but that come to make eminent sense after the fact. Even if conceptual slippage is the most ordinary mental phenomenon, there are contexts where it is subtle, rare, and anything but straightforward. Discovery via conceptual slippage gave rise to many of the greatest ideas in history, including scientific discoveries (which we will discuss in more detail in Chapter 8). As the above outline shows, the present chapter has many ambitions; we shall take them one at a time, starting with the notion of abstraction.

What is Abstraction, and What is its Purpose?

Abstraction comes in different varieties. Here we will focus on the variety that we will call "generalizing abstraction". We will say that category A is *more abstract* than category B if B is a subcategory of A — that is, if anything that belongs to category B also belongs to category A. For example, $\textit{coffee}_4$ is more abstract than $\textit{coffee}_3$, because all $\textit{coffee}_3$'s (basically, after-dinner drinks including tea but not wine) are also $\textit{coffee}_4$'s (basically any light thing ordered at all), and moreover a *diabolo menthe* and a Coke belong to $\textit{coffee}_4$ but not to $\textit{coffee}_3$. For much the same reason, $\textit{coffee}_3$ is more abstract than $\textit{coffee}_2$, which in turn is more abstract than $\textit{coffee}_1$.

The notion of abstraction (from now on, we'll omit the modifier "generalizing", since that's the only type of abstraction that we'll be concerned with) applies to classical categories of things occurring in nature (thus the category *bird* is more abstract than the category *sparrow*) as well as to categories of human-made objects (the category *furniture* is more abstract than the category *chair*), and also, of course, to categories of actions (*moving* is more abstract than *walking*) and to categories named by adjectives (*red* is more abstract than *scarlet*, and *colored* is more abstract than *red*). It also applies to categories named by idiomatic phrases, proverbs, or fables (thus, *little misdeeds lead to big misdeeds* is a more abstract category than *little thefts lead to big thefts*, and *better safe than sorry* is a more abstract category than *an ounce of prevention is worth a pound of cure*), and last but not least, it applies to categories that are not lexicalized at all, like those that were examined in the previous chapter.

In short, one category is more abstract than another when it includes the latter category as a special case. The existence in our minds of categories enjoying several different levels of abstraction makes it possible to take different perspectives on a single entity. Sometimes, for instance, a particular entity will be seen as a *sparrow*, and other times as a *bird*.

If we lacked the ability to abstract, our lives would resemble that of Ireneo Funes, the main character in Jorge Luis Borges' short story "Funes, the Memorious", for whom a fall from a horse had the devastating consequence that "Funes not only remembered every leaf on every tree of every wood, but… he was almost incapable of general, platonic ideas. It was not only difficult for him to understand that the generic term *dog* embraced so many unlike specimens of differing sizes and differing forms; he was disturbed by the fact that a dog at three-fourteen (seen in profile) should have the same name as the dog at three-fifteen (seen from the front)." In contrast to Funes, the standard human mind has not only the ability but the proclivity to abstract, in order to deal with the world's vast diversity. It pulls together into a single category items that it sees as similar, and it further organizes its categories according to their levels of generality. We see the dog at 3:15 as being the same as the dog at 3:14, we consider that *dogs* and *cats* are the same as each other in that they are all *animals*, and so on.

The Good Side of Abstraction

Any situation can be categorized in an essentially limitless number of different fashions. Thus one can use many different words to label a situation, and also many different idioms or proverbs. An old piano might be a *musical instrument* to a music teacher, a *piece of furniture* to movers, a *dust trap* to the person who does the weekly dusting, and a *status symbol* to those who proudly display it in the middle of their living room. A tomato might be a *fruit* in a botany course and a *vegetable* in a cooking course. And there are times when a given situation can spontaneously evoke completely opposite and thus contradictory concepts in the minds of different observers; the very same situation can bring to mind the category *all that glitters is not gold* for one person and the category *where there's smoke there's fire* for another.

A given situation can be labeled at many different levels of abstraction because sometimes we wish to *make distinctions* and other times we wish to *see commonalities.* While dining, one naturally wishes to keep track of which glass is one's own and which is one's neighbor's, but while washing them, one will blithely ignore that difference. A refrigerator and a piano have very different purposes, but for movers, they are both simply *big heavy objects.* When one is bringing up children, one wants all of them to be involved in *activities*; for one child this might mean acting, for another it might mean judo, and taking flute lessons, for a third. Indeed, what better way to distinguish two things than to assign them to different categories? For example, the distinction between an eagle and a swallow, or between a barn swallow and a cliff swallow, depends on the existence of distinct categories in the mind of the categorizer. But on the other hand, the barn swallow and the cliff swallow can both be seen as *swallows*, and the swallow and the eagle can both be seen as *birds.* This idea of highlighting a commonality uniting two things, just as valid and useful as the idea of drawing a distinction between them, depends on the existence of a common category to which they both belong.

Our ability to categorize things in many different ways determines how adaptable we are. Indeed, we often shift our perceptions of mundane situations with great speed and fluency, although such reperceptions tend to seem so bland that they usually go unnoticed. And yet such cases reveal the remarkable suppleness of everyday human intelligence, as a small example will now show.

A Mini-saga of Dizzyingly Fast Category Shifts

You open the cupboard and pull out a glass. To do this, you had to recognize that the object was indeed a glass. This seems as simple as simple gets: any object has its intrinsic conceptual box, these boxes are called "categories", and categorization is simply the placing of each object in its proper box, end of story. But perhaps it's not quite the whole story, after all...

> You don't know me. In fact, I don't know myself either; I'm unconscious. But whatever. Here's my story. I was produced on July 11, 2005 in a French factory. From my birth onwards, I have been categorized left and right. So in no particular order, here's what I have been: *artifact, industrial product, commodity produced in the European Union, consumer article, fragile object, glass, item of dishware, drinking glass, water glass, transparent object, recyclable object.* When I was being shipped to the store, I became a *piece of freight* and also a *piece of merchandise,* and while I was sitting on the shelf, I was an *item for sale.* Since my designer seems not to have been super-inspired that day, I remained on the shelf for several months, and the clerks variously reclassified me as an *unsold object*, a *casting error*, an *unsellable object*, and then a *dust-gatherer*, at which point I was declared a *discounted item.* My drastically slashed price finally allowed me to find an owner, and thus I became a *purchase.* Mr. Martin, who certainly is no great shakes in the creativity department, is nonetheless forever shunting me back and forth between categories, and he does so without realizing it in the least. When he's thirsty, he never confuses me

with the plates, bowls, cups, or mugs (let alone the silverware!); indeed, for him I'm not even a piece of dishware or a glass — all I am is a *glass for cold drinks.* I can thus relish being the host for water, soft drinks, and milk, but I'm never given the chance to welcome wine into my person — that role is granted only to a certain special elite that occupies the shelf just above mine. One time, though, a confused guest actually promoted me to the swanky status of *wineglass*, and as such I did a rather commendable job, if I don't say so myself, even though I'm not as sophisticated as my cupboard-neighbors. Usually, after having done my duty, I wind up in the dishwasher, and during that brief stay, no one cares that I'm a *glass*; I'm just *dishware* and that's that. My peripatetic life has occasionally given me the chance to be categorized in some rather extreme fashions. Thus the lady of the house has more than once employed me as a *spider carrier* and quite often as a *knickknack holder.* One time when the family went on vacation, I did duty as a *toothbrush holder* for an entire month, and the next year I was recruited to serve as a *sugar bowl.* I've also done stints as a *home for tadpoles* (this after the Martin children had been playing in the woods), and as a *vase* (a couple of times when the kids had picked some wildflowers for their mother). I was once even a *piece of construction material*, when the kids decided to make a tower using me and some of my peers. Alas, they forgot that I was a *fragile object* and things came to an abrupt and unhappy end when the tower fell down. Luckily, though, since I'm a *recyclable object*, a new life is awaiting me just around the corner, rich in undreamt-of new categorizations.

Categorization pervades every facet of our existence and is never fixed, even in the most mundane of circumstances. Our mini-saga has just demonstrated this, as did our various portraits in Chapter 1 of the 60-kilogram mosquito-attracting mirror-symmetric insomniac object known as "Ann", and as do innumerable other examples. Being moved from one categorical "box" to another, often by being slid up or down the rungs of an abstraction ladder, is the inevitable fate of all objects, actions, and situations.

Some readers might nonetheless feel tempted to think that, despite the incessant bouncing back and forth of our story's unconscious narrator from one category to another, it still has just one true and permanent identy — namely, it is a *glass.* But to think that way is to fall into the trap of Plato's "objectivist" vision, according to which objects have one and only one true identity. That is a naïve vision.

It's nonetheless true that psychological studies have identified certain types of "default" categorizations, usually called *basic-level* categories. For example, people tend to find it more natural to call an object a "chair" than an "armchair" or a "piece of furniture", and likewise they prefer "glass" over "water glass" or "piece of dishware", and this experimentally confirmed intuition might reinforce one's intuition that each object really does have a "true identity" in terms of its *genuine* category. But at any moment, an entity is what its categorization says it is, and that's all. Some objects are of course more *glass*-like while others are more *chair*-like, and they become *glasses* in those contexts when "glass" is the word that they tend to evoke in human minds, but in other contexts they become members of other categories. Thus, as we just saw, a *glass*-like object can become *dishware*, *artifact*, *commodity*, *spider carrier*, *knickknack holder*, and so forth.

(Incidentally, when we write "The glass was categorized as a knickknack holder", we are in fact using sloppy language and we should, in principle, say something more like this: "The entity that in many contexts is categorized as a *glass* has, in this case, been categorized as a *knickknack holder.*" That, however, would be heavy and pedantic, so we refrain from such precision.)

The fact that "our friend the glass" has a clearly dominant category in our minds (the category *glass*, obviously) may make it harder to accept the idea that it doesn't have one single fixed identity. The same could be said of most artifacts (objects made by people with a certain narrow purpose in mind). That narrow purpose will dominate our perception of the object's identity, making us feel that it is indeed the object's true and sole identity.

But playing the game of "musical categories", as we did above, shows that things are not that simple. To be more concrete, let us come back to the just-mentioned case of Ann and ask: Is she first and foremost a human being? A woman? A lawyer? A living being? A mother? An animal? Everyone would agree that Ann is "all of the above", and, depending on the state of mind of the person (or the mosquito) perceiving her, she will be more *one* than *another* of them. Who or what, though, would decide whether Ann is intrinsically more a *living being*, more a *human being*, or more a *woman*? And why would there have to be a "winner" among these diverse viewpoints?

And the crustacean swimming in the aquarium in the restaurant — is it more a member of the category *lobster* or more a member of the category *food*? And what about at the moment when it has just been thrown into the pot of boiling water? And what about at the moment when it arrives on a plate for a diner to consume? And what about after the diner has consumed it? And the cow that is grazing in a meadow, whose short-lived future is already clearly readable in the crystal ball of its human owners, and small pieces of whose cooked flesh will soon be served on plates — is it *a cow* or is it *future meat*? And what about when it is sitting on the plate next to a side of fries? And is the last half-cookie on the tray more a *cookie* or more a *morsel of shame*? Our point should be clear: such debates are not going to wind up yielding precise answers, because the questions make no more sense than did ancient questions about whether various familiar objects (doors, walls, lakes, mountains, the sun, the moon) were "intrinsically" masculine or feminine.

Certain cases of ambiguous identity are particularly helpful in making this point, because they show that one has no way of deciding once and for all what a given object most deeply is, or even if the very notion of *true identity* even applies at all. For example, is a concave piece of granite found on a hike and placed on a table in the hiker's living room as a receptacle for ashes and cigarette butts more a *rock* or more an *ashtray*? Is a piece of cow skeleton found in a field and now used to keep papers from flying away more a *bone* or more a *paperweight*? Is the little four-legged piece of plastic in a dollhouse more a *table* or more a *toy*? Is the remnant of a tree in a garden on which one often spreads a cloth and eats meals more a *stump* or more a *table*?

Is the flexible dangling "snake" made of little linked rings of metal, and tugged in order to turn a lamp on and off, more a *chain* or more a *switch*? Are smoke signals more *smoke* or more *signals*? Are the waves carrying today's news to your television set more *oscillating electric and magnetic fields* or are they more *images*? Does a dentist's patient sit on a *chair* or lie on an *operating table*? Is a hanger that has been twisted so that it can open a locked car door a *hanger* or a *car-opening tool*? Is a paper clip that has been straightened out more a *toothpick* or a *paper clip*?

Are the first few measures of a Mozart symphony as played in electronic tones on a cell phone more *Mozart* or more a mere *ringtone*? Is a receptacle held out by a beggar in order to collect money more a *hat* or more a *purse*? Is the City of the Dead in Cairo, Egypt, where vast throngs of people live in and among tombs, more a *cemetery* or more a *city*? Is a fenced field that is covered with the rusting carcasses of old automobiles and in which a farmer has put a couple of dozen cows out to graze more a *junkyard* or more a *pasture*?

Is a gold-plated leaf attached with a pin to a woman's sweater more a *leaf* or more a *brooch*? Is a bedsheet that has been hemmed at the top so it can be hung in front of a window still a *sheet*, or has it taken on the fresh new identity of *curtain*, not unlike the way a naturalized citizen of the United States has taken on a new identity? Is an umbrella fixed to a table and used to shade people from the sun still an *umbrella*, or has it metamorphosed into a *parasol*? Is a mechanical pencil that has no more lead but still has plenty of eraser still a *pencil*, or is it an *ex-pencil*, or has it been demoted to the status of *glorified eraser*? Is a set of a few dozen shipwrecked people who have lain down on the sand of a desert island to form the letters "SOS" more a *crowd* or more a *cry for help*?

Is a decorative glass pyramid more a *pyramid*, more an *artwork*, or more a *piece of glass*? Is the dark red spot that just appeared on the tablecloth some *spilled wine*, or is it a *stain*? Are the tresses of a young girl more *pigtails* or *hair*? Is the vehicle that was totaled and has now been squashed down into a dense cubic foot of metal still a *car* or is it now merely *scrap metal*? Is the device with which a hanging is carried out a *rope* or a *noose*? Is an extended third finger a *finger* or a *vulgar gesture*? Are the people sitting in the concert hall and listening to classical music an *audience* or are they *the board of directors of Crocodile Computer Company*? Are the 500 animals running in collective panic across a field a *herd of cattle* or are they a *stampede*?

And finally, is the retired but still-proud Concorde that was placed on a large pedestal at the Charles de Gaulle airport so as to evoke the glory of French aviation more a member of the category *airplane*, or of the category *Concorde*, or of the category *ex-Concorde*, or of the category *statue*, or of the category *symbol*?

You can surely enrich the aforegoing list of category dilemmas with examples of your own, but more important, for our purposes, than the question "Is this object primarily an X or not?" is the following question: "Why can't we suppress the inner voice that protests: 'Everything you've said is true, I concede, but in the final analysis the damn thing really is a *glass*, isn't it?' " In other words, why is there such a powerful tendency, even in the most reflective of people, to cling to the belief that any entity has a "true identity", and why do people so valiantly resist the thought that entities are no more and no less than what one's perspective on them makes them be? This nagging question will be taken up again in Chapter 7.

The Telltale Trace of Marking

In language, one sees category membership shifting in a striking fashion thanks to the phenomenon of *marking*, which allows an entity to shift its category membership without changing its lexical label, but simply by changing the level of abstraction that applies to that label. Marking is the phenomenon whereby a word is used sometimes as the name of a general category and other times as the name of a subcategory of the general one. When such a word is used in its broader sense, it is said to have its *unmarked* meaning, while the narrower sense is the *marked* meaning. Marking is worth careful scrutiny, for above and beyond being an intriguing aspect of language, it is the linguistic trace of a mechanism that goes well beyond words, and on which our cognitive system depends totally: the mechanism of conceptual growth.

The Declaration of Independence famously states, "We hold these truths to be self-evident, that all men are created equal, that they are endowed by their Creator with certain unalienable Rights, that among these are Life, Liberty and the pursuit of Happiness. — That to secure these rights, Governments are instituted among Men, deriving their just powers from the consent of the governed..." We cannot be exactly sure what the men who drafted this noble-sounding document meant by the term "all men" or by the capitalized term "Men", since they did not extend the right to vote to women, much less to slaves of either sex, but we will charitably assume that "men" was meant to cover not just males but females as well. Accordingly, we assume that the category they had in mind was man_2. If one takes the word "men" in the (unmarked) sense of man_2, then women are indeed *men* (of a specific sort) — an idea that we still hear echoed in many contexts, even today.

If Man descends from the apes, does it not follow that Woman does as well? The first man on the moon might conceivably have been a woman. If one hears the cry "man overboard", one isn't intended to draw any conclusions as to the sex of the unfortunate individual. The phrase "man's inhumanity to man" presumably includes man's inhumanity to woman as well as woman's inhumanity to man, and last but not least, woman's inhumanity to woman. Although we would probably tend to assume that the richest man in the world, the chairman of the board, a first baseman, a garbage man, a snowman, and a college freshman are all members of the male sex, it could nonetheless happen, at least in principle, that any or all of them could be females. And as for Aristotle, who famously declared "All men are mortal", well, he couldn't possibly have denied the truth of the statement "Madonna is mortal", because, after all, the latter would flow by ironclad syllogistic reasoning from his premise taken together with the self-evident truth "Madonna is a man."

In other situations, however, things are quite the reverse, which is to say, there are clear-cut cases where the word "man" (or its plural, "men") unambiguously *excludes* all females. This is the *marked* case of the word, and we will call this category "man_1". Thus, for instance, when one speaks of the "men's baseball team" one is pretty sure that there are no women on it. When someone speaks of the men's rest room or the men's clothing section, one understands that these are not intended for females. And of course there are countless cases where the words "man" and "woman" are used in a single sentence in parallel positions, which means that they are being contrasted, so one understands that the "man" being referred to has to be a *male* man, not a *female* man. For example, "a woman and a man are staring at each other silently", or "a crowd made up of 500 men and 500 women". If someone heard the word "men" in the latter sentence in its unmarked (inclusive) interpretation, whereby *women are men*, it would be a very strange way of stating that there were, in fact, no males present at all. (Imagine someone saying, "There are 500 mammals and 500 cows standing in that field." At first you would think that there must be 1000 beasts altogether, but then you realize that the 500 mammals coincide with the 500 cows.)

Although these last examples may strike you as being so weird that you would never expect anyone to come out with anything like them, we have in fact heard sentences quite like them in ordinary conversations. For instance, a French friend of ours who was teaching a course said to us (in French and utter seriousness), "In my class there are thirty-four *étudiants*, but only seven of them are *étudiants.*" In her sentence, the word "étudiant" changed horses midstream, so to speak, for in the first clause it meant "students" in a generic sense, while in the second clause, just a few words later, it meant "male students" (a female student being an *étudiante*, although she is of course also an *étudiant*). When we laughed at this, our friend at first defended her sentence as perfectly logical, reasonable, unstrained, and unfunny, but after a couple of moments of thought, she too was starting to be quite amused by her off-handed remark. An analogous example in English might be if someone were to say, "There were thirty-four actors at the party last night, but actually, only seven of them were actors" (meaning that twenty-seven of the actors were female). After all, the Screen Actors' Guild certainly includes both males and females.

Marking can sometimes create genuine ambiguities. For instance, if the order for a group of people in a café is "Three coffees, a macchiato, a double espresso, and a cappuccino", two conflicting interpretations exist, depending on how the term "coffees" is understood. If the server takes it from an inclusive or unmarked point of view ("coffee" standing for the concept $coffee_2$), then the terms "macchiato", "double espresso", and "cappuccino" serve an explanatory role, in which case it's clear that only three beverages have been ordered *in toto.* Contrariwise, from the marked point of view (where "coffee" means $coffee_1$), this is an order for three "default" coffees (that is, American-style coffees, not fancy Italian-style ones) plus three Italian-style drinks. You can try this out on your own and check whether the phrase is indeed heard differently by different people.

The ambiguity inherent in marking can be cleverly exploited. Thus, here is a very short horror story: "The last man on earth was being held in a tiny little hut. All at once there was a knock at the door." The reader shudders, imagining the dire fate that is about to be doled out to the last surviving specimen of the human race, but then relief comes in the next sentence: "It was the last woman on earth."

As this example shows, marking sometimes leads to ambiguities that few people notice but that, once pointed out, can make us smile. Here are some more examples:

"What moos but is not a cow?" This riddle puzzles people (at least children) until they realize that "cow" can be taken to include just females (*cow*$_1$), at which point the answer "a bull" pops to mind. Although a bull is a fine member of the more abstract category *cow*$_2$, it is not a member of the less abstract, marked category *cow*$_1$. At that lower level of abstraction, the words "bull" and "cow" are diametric opposites.

"Who lived in caves 30,000 years ago, raised crops, used tools, and wore clothes, but wasn't a caveman?" (Hint: recall the opening of the Declaration of Independence.)

A. says, "I'm bushed; these last two days I haven't slept a wink." B. replies, "What's your problem? Over the past *hundred* days, I never slept a wink and I'm in perfect shape!" Explanation? The term "day" sometimes means *day*$_2$, which is twenty-four hours long and includes nights, while other times it means *day*$_1$, which is half as long and contrasts with "night"; in this case, day and night are as opposite as night and day.

"Passengers with children or babies may board now." Babies are *children*$_2$ but here they are being contrasted with *children*$_1$. In another context, however, such as a dinner party, it would be very confusing to hear someone say, "I have a baby but I don't have any children." Babies *are* children and yet babies are *not* children. And incidentally, are babies members of the category *passenger*?

On the one hand, "Human beings, like all other animals, need food to survive." But on the other hand, "Human beings, in contrast to animals, have developed language, culture, science, and literature..." Thus humans are *animals*$_2$ but not *animals*$_1$, to which in fact they are seen as forming a natural contrast, much as pepper does with salt, or dogs with cats.

Then there are portraits. One would certainly not object if a friend, looking at one's passport photo, were to say, "That's a very nice portrait of you!" On the other hand, one would not object either if a friend walking through one's house were to say, "Those photos of you and your husband in the hallway are great, but do you have any portraits?" A photo can be a kind of portrait (*portrait*$_2$), but it can also be contrasted with a portrait (*portrait*$_1$).

If you're very tall, are you tall? Of course! Anyone who is "very tall" is necessarily *tall*$_2$. And yet, of course not! On a scale of heights where "tall" runs from, say, 5′10″ to, say, 6′2″, then anyone of height 7′2″ is way out of the "tall" range. Thus "very tall" contrasts with *tall*$_1$, although it is a special case of *tall*$_2$. Moreover, a child can perfectly well say, "I'm 4′3″ tall", thus showing that *all* heights count as cases of tallness.

A man and his two children are talking about movies. The son says, "You know, in all the big adventure films, such as *Harry Potter*, *Star Wars*, *Spiderman*, and *Batman*, the hero is a guy." His sister chimes in, "Yeah, that's true — think of *Lord of the Rings*, *The Matrix*, *Daredevil*, and *Indiana Jones*." Then the father tosses in his own two bits' worth: "Hey, guys, you're right — the hero is *always* a guy, but the heroine is *never* a guy!" And thus we have *hero*$_2$, which subsumes the two contrasting concepts, which are *hero*$_1$ and *heroine*. For that matter, we also have *guy*$_2$ and *guy*$_1$...

Marking is not limited to nouns; it can also come about in the case of verbs, and it can wind up in usages that seem extremely strange when placed under the microscope. For example, the verb "to grow" has, over time, acquired a broader meaning than the one most people spontaneously think of. Thus changes that go in either direction — toward the smaller as well as the larger — are often described by the word "grow". This claim may sound so silly that native speakers of English might deny it at first — until they are shown a sentence such as, "As soon as Alice had drunk the vial of potion, she started to grow smaller and smaller", at which point they will admit, "Oh well, I guess we *do* say that, after all..." We frequently use the verb "to grow" in its unmarked sense without thinking in the least about how it contradicts the marked sense.

It seems that what matters is not *size* but simply the fact that things are *changing in time.* For instance, it's quite normal to say, "The bottle grew lighter and lighter as the water evaporated", "The ticket line had grown a lot shorter", "The average intelligence quotient has grown lower and lower over the decades", and so forth. No native speaker of English would bat an eyelash at any of them. We thus see that although "to grow" often means "to become larger" (this is its marked sense, $grow_1$), the same verb can also simply mean "to change over time" (this is its unmarked sense, $grow_2$). However, it is overwhelmingly the idea of $grow_1$, not $grow_2$, that tends to come to mind if a native speaker is asked "What does 'to grow' mean?" For this reason, certain perfectly standard uses of "grow" can make one smile, because on the surface they seem to involve the contradictory notion of "increasing while decreasing".

How can one explain this paradoxical property of language, whereby two words can, in one context, be each other's violently clashing opposites, while in another context, the one merely denotes a subset of the other? Why is it that we would use the very same word to denote two different levels on a ladder of abstraction? Why do languages insist on being so miserly with their words, when it would seem so very simple if, for each different category, there were a different word? The answer is summed up by one word: "adaptation".

The Virtues of Marking

Marking is actually a well-adapted and useful way of exploiting ambiguity in order to maintain flexibility, allowing people to use a word in a variety of contexts. Indeed, although precision is crucial in communication, it's equally important that precision should not entail a stifling rigidity, preventing one from understanding familiar words in unanticipated situations. Marking allows precision (the designation of a very specific category) to coexist with flexibility (the looseness of interpretation that comes from the freedom of finding the appropriate level of abstraction).

As we will see (note that this "we" is broader than we_1, which consists of just your two authors, since it includes our readers as well, hence this "we" means we_2) in the next few paragraphs, if we (note that this "we" is even broader since it includes all of humanity, hence it means we_3) couldn't categorize things simultaneously at different levels of abstraction, it would lead to some unfortunate consequences:

> Gyro Gearloose is extremely proud of his latest invention: a car that obeys spoken commands. No longer does he need to pilot his vehicle; all he needs to do is tell it what he wants it to do.
>
> As they are approaching an intersection, Gyro says to his car, "Go straight at the crossing, but first make sure that no car is coming on either side; if there is one, then slow down and let it pass first."
>
> At the intersection, Gyro's car doesn't slow down in the least, and thus it gets sideswiped… by a truck.

And so, in the final analysis, are trucks cars? In this light-hearted anecdote, one sees the classic signature of marking, since trucks sometimes certainly *are* cars, yet at other times they certainly are *not* cars.

When Megan's father says "Now watch out for cars!" as Megan is setting out for school each morning, he doesn't expect — and Megan knows this very well — that she will blithely step out in front of the first onrushing truck that she sees approaching. What he means by "cars" is anything moving that might possibly constitute a danger to Megan along her way to school, so it includes big trucks and also pickups, motorcycles, motorbikes, bikes, trikes, and scooters — and if, perchance, some unexpected kind of moving entity came along, it too would be naturally understood as squeezing in under the rubric of "cars", even if neither father nor daughter had ever anticipated any such entity when the warning was issued. Thus "car" as an umbrella term would cover a tank in a military parade, a horse-drawn carriage, and a group of teen-aged roller-bladers. All of those possibilities, far-fetched though they may be, were implicitly part of Megan's father's idea when he told his daughter to "watch out for cars" — perhaps lying out toward the fringes of the category of "car", but nonetheless conceivable as members of the category when encountered in the street.

In other situations, however, "car" is more restricted in its meaning. It doesn't include bikes or roller-bladers, but it does include trucks and motorcycles of a certain horsepower — this on highways in Europe where signs designate what kinds of "cars" can travel down them. When Mr. Martin goes to the car dealer looking for a good deal, he totally excludes in advance the idea of trucks and pickups from his category *car*, and also anything that has fewer than (or, for that matter, more than) four wheels.

The riddle of these highway categories doesn't end here. Thus: are pickups trucks? Are SUVs trucks, or are they station wagons, or are they vans? Are SUVs cars? Are motorbikes and motorscooters motorcycles? Are roller blades roller skates? All these categories are marked categories, and thus they can take on wider or narrower senses depending on the context, which in certain situations leads to an affirmative answer, and in others to a negative answer. The fact that a single lexical item denotes categories at different levels of abstraction allows one to select the appropriate level as a function of the situation, and thus to deal with things in an appropriate manner. And so Megan is spared the sad lot doled out to Gyro Gearloose's invention, because she, like other humans, has the ability to adapt her level of abstraction of categorization to the context that she finds herself in.

The fact that one single word or phrase can be attached to a number of related categories, all residing at different levels of abstraction, encourages adaptation to the context. The construction of such categories is carried out by a process of category extension that tries to combine the two crucial features that we pointed out above: namely, categorization allows people to *make distinctions* and also to *see commonalities.*

And thus, what might seem at first merely to be a phenomenon of interest solely to some specialized linguists and philosophers turns out to be at the heart of the development of concepts, for, as we shall now see, marking provides a kind of linguistic pedigree of a category's history over time, all the way from its babyhood to its adult state. The reason that a single term is so often used to denote different categories is that there are abstraction relations between categories, and the understanding of these relationships develops at the same time as the categories themselves develop.

How Did They Bump into Each Other?

Below is a pair of father–son exchanges that clearly show how the phenomenon of marking is correlated with the development of concepts in a human mind. These two short conversations took place when little Mica, aged five, was taken by his parents to Egypt. Here are Mica and his Papa talking during their vacation:

> "Papa, what's the difference between a camel and a dromedary?"
> "A camel has two humps and a dromedary has just one."
> "But Papa, what do they bang into to get them?"
>
> "Papa, how do divers breathe when they're under water?"
> "They have bottles on their backs."
> "But Papa, why do they need to drink when they're under water?"

These small exchanges might be seen as merely amusing demonstrations of a child's naïveté. Adults who read them usually don't even see what Mica could have been thinking at first, and then when they do, they burst out laughing. And indeed, who wouldn't find the image amusing of strange beasts wandering around the vast desert, banging into random objects (each other? cliffs? exotic trees?), and thus getting humps, bumps, or lumps? (The conversation took place in French, and all of those rhyming notions are blended together in the French word "bosse".) And is the image of undersea divers swimming around with bottles of milk, orange juice, beer, or other drinks strapped to their backs any less amusing?

But something more than just naïve charm can be found in these snippets — namely, a revelation of how categories are born, in part through marking. In these dialogues we see major differences between Mica's categories and his father's. Where Mica had just one concept for the word "bosse" so far, his father had several. Their mutual incomprehension came from the fact that though they were using the same words, those words denoted different categories.

Several varieties of humpy, bumpy, lumpy things existed for Mica's father. His most abstract category for the word "bosse" corresponded roughly to the idea of a gentle rise off of a flat surface, and it allowed him to see camels' humps, speed bumps on roads, lumps from mosquito bites, and so forth, all as manifestations of one single, general *bosse* idea (and this even includes "math bumps" — a linguistic relic from the nineteenth-century pseudoscience of phrenology, but despite the notion's lack of scientific validity, in France people still speak casually and metaphorically of someone endowed with mathematical ability as having "la bosse des maths").

For Mica, however, the category denoted by "bosse" was far narrower. For him, the presence of a hump, bump, or lump meant merely that a human being or an animal had banged into something — as surely as the presence of smoke somewhere means that there is a fire nearby. And so the question that leapt to Mica's mind becomes totally natural and obvious, since he was simply trying to get to the bottom of a fact that he had just heard.

Similar remarks hold regarding the two speakers' understandings of the bottles strapped onto divers' backs. Mica's father's life experience had given him a very wide and general concept of *bottle*, and when he said the word, he had in mind, and intended to evoke in Mica's mind, a certain subcategory of that wide category of *bottle*, but again, it was not the one that Mica envisioned, because in Mica's limited experience, a bottle always contained some kind of drink.

We can rewrite these two snippets from Mica's point of view, showing explicitly how he heard what his Papa said:

> "Papa, what's the difference between a camel and a dromedary?"
> "Camels have *two* humps (because they've banged into *two* things), and dromedaries have just *one* hump (because they've banged into just *one* thing)."
> "But Papa, what kinds of things do they bang into that give them humps?"

> "Papa, how do divers breathe when they're under water?"
> "They have bottles (which are full of drinks) on their backs."
> "But Papa, why do they need to drink when they're under water?"

Despite appearances, it's not the gulf between a child's and an adult's mental mechanisms that makes the difference here. It simply depends on the repertoire of categories one has built up. The existence of a marked category reveals a good deal of experience in a domain, because the speaker has to have constructed *both* a wide category *and* a narrow one, which share the same linguistic label.

As an afterthought, it is amusing to point out that Mica's father, who had little knowledge of desert beasts, actually replied slightly incorrectly to his son; once again, it has to do with marking. The truth of the matter is that even the word "camel" is a marked term. Officially speaking, it denotes both a wide category including both one-humped and two-humped beasts ($camel_2$), and a narrow category including only two-humped beasts ($camel_1$). In other words, whereas the narrow category $camel_1$ is in contradistinction to the category *dromedary* (much as car_1 contrasts with *truck*), the wide category $camel_2$ is a superordinate of (*i.e.*, contains) the category *dromedary* (much as car_2 includes *truck*).

Mica's father could thus have given his son a different reply, based on the wider sense of the word "camel", as follows:

> "Camels sometimes have one hump and sometimes two. When they have only one, people call them 'dromedaries'."

Such a strange reply, though technically correct, would have left Mica rather confused, and would still not have explained how these curious beasts managed to bang into various things. It would, however, convey the extra information that the two species of animals belong together, as well as what makes them different (*camel* ⇒ two humps; *dromedary* ⇒ one hump). Zoologists tend to use the word "camel" in the inclusive, abstract fashion, but non-specialists tend to do the opposite — namely, they prefer stressing the oppositeness of camels and dromedaries.

Compared to children, adults typically have a higher level of expertise with concepts such as *hump* and *bottle*, just as zoologists typically have a higher level of expertise than random adults do with concepts such as *camel.* The possession of a higher-level, more abstract concept allows experts and, more generally, experienced people to distinguish the essence of a concept from certain traits that are more contingent. Whereas Mica, at five years of age, had only a single concept of *bosse* — a lump resulting from a collision — which led him to imagine a collision as the *raison d'être* of any lump he heard about, his father, much older, had found a deeper idea in the concept of *bosse*, allowing him to distinguish numerous subcategories of the notion, as well as one generic or "umbrella" category that covered all the varieties, thus uniting disparate phenomena that share the central idea of some kind of protrusion.

To examine more deeply this process of extraction of the quintessence of a concept through the operation of marking, we'll turn to an example of marking that has come up only rather recently in our society.

How a Concept's Essence Emerges

If you possess a computer, you are very likely to possess two desks: the desk that is shown on your screen, and the desk on which your computer sits. As will surprise no one, this terminological coincidence is not a coincidence. One of these types of desk — the screen-based one — is a metaphor, or an analogue, based on the other one. People who regularly use computers, which means nearly all of us today, have long since internalized the metaphor and seldom hear it as a metaphor based on something known earlier. The idea of a "desktop" on a screen is simply a dead metaphor, no longer (or very rarely) evoking any prior notion, just as the expression "table leg" is a dead metaphor that was rooted in the legs of humans (as well as the legs of animals — non-human animals, that is).

Much like the concept *hump* for Mica, the concept of a solid desk — a piece of furniture — was, for adults who grew up before the era of personal computers, a category with an old town, a downtown, and suburbs, like so many other categories. To make this vivid, we can cite a dictionary definition dating back to the pre-computer age. In particular, the following enormous and admirable vintage-1932 dictionary:

FUNK & WAGNALLS

New Standard Dictionary

[Reg. U. S. Pat. Off.]

OF THE

English Language

UPON ORIGINAL PLANS

DESIGNED TO GIVE, IN COMPLETE AND ACCURATE STATEMENT, IN THE LIGHT OF THE MOST RECENT ADVANCES IN KNOWLEDGE, IN THE READIEST FORM FOR POPULAR USE, THE ORTHOGRAPHY, PRONUNCIATION, MEANING, AND ETYMOLOGY OF ALL THE WORDS, AND THE MEANING OF IDIOMATIC PHRASES, IN THE SPEECH AND LITERATURE OF THE ENGLISH-SPEAKING PEOPLES, TOGETHER WITH PROPER NAMES OF ALL KINDS, THE WHOLE ARRANGED IN ONE ALPHABETICAL ORDER

PREPARED BY

MORE THAN THREE HUNDRED AND EIGHTY SPECIALISTS AND OTHER SCHOLARS

defined the word "desk" as follows:

> desk, *n.* **1.** A table specially adapted for writing or studying, often having a sloping top serving as a cover to a repository beneath; by metonymy, position at a desk; the occupation of a clerk: as, from the desk to the bar. **2.** A table or stand to hold that from which one publicly reads or preaches: sometimes, by extension, applied to the entire pulpit or to the clerical profession in general. **3.** A case or box holding writing materials, and having on the top, or when opened, a sloping surface to write upon.

And in exactly the same year, 1932, the *Dictionnaire de l'Académie française* defined the word "bureau" (French for "desk") as follows (the original was in French, of course):

> A piece of furniture having drawers in which one can store papers and having horizontal surfaces on which one can write or draw. By extension, a table on which one does written or other work.

These definitions are based on the idea of a desk as a piece of furniture. That was the "downtown area" of the vintage-1932 concept of *desk.* To be sure, even back then, there was already a good deal of conceptual urban sprawl in various directions. But back in 1932, no one could have dreamt of the kind of desk that we information addicts now spend most of our workdays working "upon".

Let's give the name "hard-desk" to the concept of the 1932-style piece of furniture. It has a physical existence, and it is heavy and rather awkward to move around. Today's screen-based version of the concept — we'll call it "soft-desk" — is immaterial, or in any case it is material only in a highly indirect fashion; it is transportable, instantly copyable, easily sharable, and fits handily on a flash drive, carryable in one's pocket.

One might think that, although one of these categories gave birth to the other one (*hard-desk* being the "mother" of *soft-desk*), the two categories would subsequently have become fully independent of each other, and that each would have followed its own developmental pathway without regard for the other, as is often the case in nature for mother and child, and as is also often the case with words that engender other words. Take, for example, the word "brand", a close cousin to "burned". Originally it meant simply a flaming stick, but at some point it acquired a second meaning, generalized and abstracted from the first meaning — namely, the kind of mark made with such a stick on the hide of an animal or the skin of a criminal in order to label them forever. At a later point, this second meaning was further generalized and abstracted to the idea of a publicly recognizable symbol permanently identifying any entity, and thus eventually it took on its current overwhelmingly dominant meaning of the name of a company that manufactures goods — a far stretch indeed from a burning stick! Clearly these three very different concepts (a flaming stick; a mark on an animal; a company's name), all associated with the noun "brand", diverged long ago and simply went their separate ways. Of course, this etymological story is unlikely to have much to do with how these concepts are represented in the mind of a person who grows up with them.

What we just said about brands does not, however, hold for desks, for *hard-desk* and *soft-desk* have clearly retained their deskness, which means that they are both work spaces. Indeed, any time we want to prepare or edit some document, it would be perfectly reasonable to consider which of the two types of desk might be preferable. If we want to write a handwritten letter with a pen and paper, well then, *hard-desk* will be our choice; if we want to produce a professional-looking printed document, then *soft-desk* will prevail. But interestingly enough, in many contexts, we can talk about "the desk" without it being relevant whether we mean *hard-desk* or *soft-desk.* Thus, *hard-desk* and *soft-desk* are sub-categories of *general-desk*, which could be defined as any kind of workspace, whether physical or virtual, for producing documents. Often all one needs to know is that the speaker is referring to a *general-desk*, and we don't need to know which of the two subcategories — hard or soft — the speaker has in mind, just as when someone says they've had "a coffee", we get the picture without knowing if it was a *café crème*, a *cappuccino*, or an *espresso*. Likewise, we can perfectly understand a sentence such as "I bought a car today" without needing to know what color the bought car was. Understanding is a mental action that can get along just fine without a great many details. If someone says "my desk is cluttered" or "I spent the whole afternoon organizing my desk", we can understand this perfectly without having any idea if it was a *hard-desk* or a *soft-desk*.

So today, there are three distinct categories lurking in the word "desk" — just as the word "person" can mean a male, a female, or a person of unspecified sex. In this sense, the word "desk" exemplifies a special variety of marking, in which the abstract superordinate category (*general-desk*) shares a name — namely, "desk" — with its two subcategories *hard-desk* and *soft-desk*. Thus once we have two "rival" categories of *desk*, this allows us to put our finger on the essence of the original category by constructing a more abstract concept of *desk*.

Much as acquiring a second language allows one to understand the nature of one's native language more clearly, the emergence of computer-age desks has helped us gain a newer and deeper understanding of our old category *desk*. The advent of home computers changed the venerable old concept of *desk*, making it no longer associated with one indivisible concept. The emergence of three new types of desks — *hard-desk*, *soft-desk*, and *general-desk* — has allowed us to perceive more clearly an essence, hidden up till then, of the original old category. Indeed, the creation of the superordinate category *general-desk* allows us to distinguish between the core property of desks (namely, that they are workspaces) and more superficial properties that, at one time, before the days of home computers, were inseparably linked to material existence and its features, such as how much something weighs, how much stuff is piled up on it, and the shapes and sizes of drawers. In those days, no one had yet imagined that a desk could be a ghostly, immaterial entity. Heaviness, paper-coveredness, and pull-out drawers seemed to be necessary aspects of deskness, but later these aspects were seen to be incidental and not pertinent to a desk's status as a *workspace*. Just as we all eventually transcend the naïve idea that a lump must be the result of someone bumping into something, so the recent development of a variety of notions of *desk* has allowed us to transcend a matter-oriented naïveté about desks, and now we have no trouble imagining desks that weigh essentially nothing, involve no piles of papers, and have no drawers.

For the category *desk*, much as for the categories *lump*, *bottle*, *man*, *animal*, *car*, and so on, the addition of a new level of abstraction showed us some dimensions that were essential about the concept, and some that were dispensable or optional. Without the construction of such an extra level, everyone would have continued to think that the deeper properties were necessarily accompanied by the shallower ones. Moreover, no one would have even thought of trying to make a distinction between shallower and deeper aspects of *deskness*. However, the creation of a fresh new level of abstraction, corresponding to an unmarked (more general) sense of a word, brings out from behind the scenes the difference between a concept's more central and less central aspects — the latter being those aspects that help us to distinguish among subcategories (for example, between lumps caused by banging and lumps having other origins, or between material *hard-desks* and ethereal *soft-desks*).

Pinpointing, in a given context, the incidental or contingent aspects of a concept, as opposed to its deeper, more essential aspects, constitutes an important intellectual step for an individual. On a higher level — on a social and cultural level — the emergence of the concept *desk* in its newer and more abstract sense is analogous to the emergence of the general concept of *lump* within an individual mind.

As we showed in Chapter 2, in the case of proverbs, a concept starts to be impoverished rather than enriched when it is abstracted beyond a certain level — namely, that level at which its essence starts to be lost, even if we don't have any precise criteria telling us where that begins to happen. But despite this risk, abstraction can be enriching: a gradual series of refinements can indeed reveal a unity among a set of situations that at first glance seem entirely different, and, thanks to our faculty of analogy-making, we soon come to see these situations as belonging to a single category, and to feel every bit as comfortable with the new category as we once felt with the old one. This kind of push towards ever-higher levels of abstraction can go remarkably far without adulterating the essence of a category, as we shall now see in short case studies of three familiar categories — namely, *shadow*, *wave*, and *sandwich*.

Of Shadows

Everyone grows up intimately familiar with shadows. In fact, we are constantly shadowed by our shadows — at least when it's light. Early on in life we learn what causes them: shafts of sunlight are blocked by opaque objects from reaching the ground or a wall. Even so, even when we're all grown up, some shadows still strike us as surprising or curious or even strange. For instance, there is something mesmerizing about watching the shadow of the airplane in which one is flying as it comes in for a landing. At first it is just a tiny dark spot far below us on the ground, racing across fields, roads, forests, and rivers, and then it grows and grows and starts to look like an airplane, and in the last few seconds before the landing, it almost feels as if it is rushing up to join its mate; the moment of landing feels like the joyous reunion of a pair of long-separated twins.

There is also the phenomenon of a lunar eclipse. Slowly, slowly, some dark shape moves across the face of the moon. It is eerie and mysterious, and most people probably don't realize exactly what is going on. The truth is, it is just an enormous magnification of what would happen if, in a pitch-dark room, you were to turn on a flashlight and shine it at a suspended ping-pong ball and then were to slowly pass an orange between the source of light and the dangling little ball. Of course the beam of light would be blocked by the orange, and the ping-pong ball would grow much darker; and if, by chance, there were some highly motivated and curious ants poised on the side of the orange that was facing the ping-pong ball, they would in theory be able to stare across the room and observe the darkening of that dangling little ball. Now just blow this picture up by a factor of a hundred million or so. The orange and the ping-pong ball are now floating in the blackness of space, the sun is of course the flashlight, and the ants are human observers. We don't need to spell this out, but what is interesting is that an enlargement on a monumental scale seems to change the nature of the phenomenon entirely. Most people no longer have an intuition for what is going on; it feels cosmic and alien, possibly even filled with prophetic meanings.

Also somewhat disorienting, although more familiar, is the shadow cast by oneself or by other people when the source of light is very low, near the horizon, as at sunset; in such cases, a person's shadow can easily stretch out 100 feet or more across the ground. If at night you are walking down a sidewalk and a shaft of light is streaming from a far-off streetlight behind you, you can detect someone else as they approach you from behind even when they are still a long way off. This, too, is a kind of magnification of the usual phenomenon — an extreme horizontal stretching — and it, too, feels somewhat strange and eerie.

Eclipses and elongated shadows show that an act as simple as a change of size or proportion can stretch the boundaries of a familiar category. Of course the conceptual boundaries can be stretched much further by exploring other dimensions of change. For instance, one's concept of *shadow* grows richer and deeper when one liberates oneself from the idea that what matters is the blockage of the passage of *sun*light. The same phenomenon — exactly the same! — works with moonlight, firelight, lamplight, light emitted by a flashlight, by a television screen, by a cell phone, by a cigarette lighter, or even by a lowly glowworm. Collectively, this provides a significant generalization of the concept of *shadow*, but one can go considerably further, to be sure.

The two photographs on the following page show a lovely old oak tree in two different seasons. They look very similar and yet there is something crucially different. In both of them, the tree appears to be casting a dark shadow, but a careful look reveals that there is more to the story. In the first photo, the shadow is clearly "made of" the absence of sunlight, so to speak, but in the second photo, the day is overcast and there is no sharp source of light above the tree. Therefore, the tree can't be casting a "light shadow". Rather, the white covering on the ground surrounding the tree is *snow*, while the dark patch directly underneath it is the result of a *lack* of snow. What we are seeing is thus a "snow shadow" cast by the same tree, but on a winter's day. The stream of snow that fell vertically from the sky, perhaps a few hours or days earlier, is the analogue to the ceaselessly falling vertical stream of sunlight in the summer photo. But despite all these differences, the two photos are so strikingly similar that it feels as if one is looking at *exactly the same phenomenon.* And indeed, in an important sense, one is doing just that. And that's why "snow shadow" is just the right term for what we see here.

A more established and standard term is "rain shadow". You might at first guess that this term refers to the dry patch on the ground underneath a bridge, an awning, a table, or an umbrella during a heavy downpour — and indeed, why not use the phrase "rain shadow" in that very natural way? — but in geography, it is a technical term with

a rather different meaning. Like a lunar eclipse, a rain shadow is a physical phenomenon on a far larger scale than that of everyday shadows, and it also involves a change of medium, as does the concept of *snow shadow*, and in some ways it also resembles the very long shadows on sidewalks, described above.

To illustrate the idea concretely, let's turn to the state of Oregon. The western third of the state is famous for its rainy weather, due to its proximity to the Pacific Ocean, and of course to the fact that weather patterns generally move eastwards. However, about a third of the way across the state, the Cascade mountain range, running north–south, forms an impenetrable barrier to rain-carrying clouds, and therefore, for the next couple of hundred miles eastwards of the Cascades, there is a vast barren desert — a stark contrast to the lush Willamette Valley just west of the Cascades. The absence of rain for those hundreds of miles is known as a "rain shadow", and just such a shadow is cast by the Cascade Range. The analogue of the downward stream of falling snow or of sunlight is the eastward-flowing stream of rainstorms that suddenly hits the obstacle of the mountains, and whose passage is thereby blocked. Like the greatly elongated late-afternoon or nighttime shadows on a sidewalk, a rain shadow is far longer than the height of the object that is casting it.

What we are exploring is a set of close analogues to the primordial notion of *shadow* that, despite their simplicity, give rise to a much broader category than is usually attached to the word. The core of the *generalized shadow* idea (*shadow*$_2$) is that there is a stream or beam of stuff, for want of a better word (we might call it the "medium") whose straight, linear flow is interrupted by an obstacle, and as a result, beyond that obstacle there is a perceptible absence of the stuff (*i.e.,* of the medium). Irrelevant are: the nature of the medium itself, the speed of the stream, the physical dimensions of the stream and of the obstacle, and the time it takes to create the effect. A sunlight shadow is virtually instantaneous, a snow shadow takes a few minutes or perhaps an hour to appear clearly, depending on the intensity of the storm, and a rain shadow is a phenomenon whose existence has to do with the frequency of rainstorms taking place over a period of years.

The concept of *shadow* can be further extended in numerous fashions. Imagine a large truck driving down a freeway, and going much more slowly than most of the vehicles sharing the road. It is thus being passed continually, both on its left and its right, by faster vehicles. In other words, a constant stream of vehicles (analogous to falling snowflakes) is coming up on the truck from behind. Those in other lanes simply pass by it unimpededly, but those in the truck's own lane cannot do so, as it is "opaque" to them. If they want to pass it, they have to change lanes and go around it, and only after getting somewhat ahead of it can they return to the lane they started in. For that reason, directly *ahead* of the truck, in its own lane, there are no vehicles at all for a couple of truck-lengths or so. Thus in the medium of traffic flow on a freeway, a slow-moving truck always casts a noticeable downroad "vehicular shadow".

In physics experiments in the nineteenth century using long cathode-ray tubes made of glass, a beam of electrons emitted at one end of the tube was powerfully attracted by a strong positive charge at the tube's far end. Physicists didn't really know anything about the nature of this beam, however, and so to find out, they tried interfering with the high-speed stream of unknown entities inside the tube by blocking its flow in different ways, thus creating various kinds of "shadows" at the downstream end of the tube. Of course they knew that they were not dealing with a stream of light but with a stream of mysterious microscopic and invisible entities, and that the notion of *shadow* was thus necessarily being extended analogically, and yet to the physicists who did these experiments, such a conceptual extension did not feel forced or abstract, but natural and simple.

Some shadows are far more shadowy than the rather physical ones we have just discussed. For instance, we can easily understand the sentence "World War II cast a decades-long shadow on the birth rates of many nations." Here we once again have a kind of abstractly moving stream — that of humanity propagating itself down through the years — and the obstacle that this abstract stream hit was war, which inflicted an immense death toll on young men, thus knocking them out of the reproductive stream. The "shadow" consisting of their absence was then felt in many countries. Eventually, after a generation or two, birth rates gradually started to recover, just as in front of the slow truck on the freeway, the downroad shadow that it casts is not infinitely long, but only a couple of truck-lengths or so. Notice how very abstract, in this case, are both the "stream" and the "obstacle" blocking it — and yet, in spite of this high degree of abstraction, the mechanism giving rise to the "shadow" is the same, so much so that it is easy to see the connection of this shadow cast by World War II to the downroad "traffic shadow" cast by the slowly-moving truck.

Consider next the following sentence, ominously foreshadowing World War II, which was taken from an obituary of the famous Hungarian–American physicist Edward Teller: "As both émigré and physicist, Dr. Teller was aware of the Nazis' lengthening shadow." What does this mean? It is reminiscent of the earlier-mentioned very long shadows cast down sidewalks by people walking as the sunset draws near, the people in this case being the Nazis. But the sentence from the obituary does not conjure up in our minds a source of light or an obstacle; rather, we are merely being invited to recall, in a very general way, how an elongated shadow, at the time of a setting sun, can reveal the existence of an ominous threat sneaking up on one from behind, long before it actually materializes. And of course shadows are dark, and as the sun goes down they grow both longer and darker, which increases the sense of somber foreboding.

Then there are the perennial stories of children who have to "grow up in the shadow" of their world-famous father or mother, or of people who suffer calamities and emerge from them "only a shadow of their former self". In these two phrases, as in the case of "the Nazis' lengthening shadow", the physical mechanism that gives rise to everyday shadows is very much in the background, and all that is retained of shadowness is, in the first case, the idea of a "dark zone" where there is such a profound absence of "light" (presumably "underneath" the parent) that it can conceal the presence of even a significant object such as a child, thus causing great suffering, and in the second case, the idea that a shadow is far more ethereal and less substantial than the object that casts it. Depending on one's view of what a shadow "really" is, one may think that these metaphors are finally allowing us to put our finger on the essence of shadowness — or conversely, one may think that these metaphors are merely facile borrowings of some of the more obvious and superficial trappings of shadowness.

Of Waves

Those of us who grew up near the sea have spent countless hours watching waves sweeping in toward land, rising up and turning into whitecaps, then breaking in a loud and violent fashion as they crash onto the beach. Indeed, for many people, the idea of waves in water is synonymous with the deafening crash and dazzling spray of breakers. However, water waves can also roll for hundreds of miles across the open sea without ever breaking at all (these are often called "swells"), and this gives a somewhat different image of the meaning of the word "wave" — and one that, as it turns out, has been far more fertile in its generalizations over the past two millennia.

For those who grew up inland, another kind of wave is familiar: "O beautiful for spacious skies, for amber waves of grain...", and then again, "The wavin' wheat can sure smell sweet, when the wind comes right behind the rain..." Instantly responding to the breeze, tall stalks sway back and forth, with, at times, entire fields undulating nearly in unison. No wonder such waves are celebrated in song.

On a more mundane scale, flags wave in the wind. And we wave to our friends, saying hello or good-bye. That is to say, not unlike flags in the wind, our arms flap back and forth in a roughly periodic fashion for a little while. And if we are teachers, we are inclined at times to succumb to the temptation to give hand-waving explanations of things that are very hard to grasp.

On a far huger scale, America was built by legendary "waves of immigration" washing one after another across her shores. Even if one didn't grow up near the ocean, one can almost see a long series of these mighty swells crossing the Atlantic or the Pacific Ocean, one by one approaching the shoreline, and finally crashing onto *terra firma*, possibly making a huge roar and breaking into spray.

These are the kinds of phenomena that go into the everyday conception of what waves are, and they are also the kinds of concrete imagery upon which early thinkers built as they started to put together a picture of the undulatory nature of certain fundamental and universal phenomena. However, despite their frequent presence in our lives, water waves were not the best sources of inspiration — not by a long shot, as it turns out. There are too many diverse and complex phenomena involved in water waves; thus, there are surface waves (such as the delicate ripples made by water skates), which have to do with surface tension, and there are also tsunamis, which have nothing to do with surface tension, but which instead involve gravitation pulling the water down (think of water sloshing back and forth in a bathtub; when one end goes up, the other is pulled down, and vice versa). And adding to their complexity, water waves of different wavelengths propagate at different speeds, which (as it turns out) is a profoundly complicating factor when one tries to understand waves through mathematics.

From antiquity, the first physicists were inspired by water waves, and as a result they formulated the most basic ideas about them, such as that of *wavelength*, whose very definition betrays its watery origins: "the distance between successive crests" (or equivalently, between successive troughs). Likewise, the *period* of a wave is the time between arrivals of successive crests (or troughs), and the *frequency* of a wave is the reciprocal of the period. Note that these notions don't apply to an isolated breaker, but only to a long series of swells coming in toward a beach. For a physicist, a wave is first and foremost seen as a periodically repeating phenomenon (although, in the end, that requirement, too, can go by the boards...).

Starting out with these basic notions, as well as the extra notion of a wave's *velocity*, which is its wavelength divided by its period, physicists were able, already several centuries ago, to analyze many phenomena involving water waves, such as *reflection* (everyone has seen ripples bounce off the edge of a swimming pool), *refraction* (the slight shift in direction that takes place when a wave crosses over from one medium to another — for instance, ripples moving from one liquid to another, or from a shallow basin to a deep basin), and *interference* (what happens when waves from different sources crisscross). The discovery of these water-wave phenomena, and later of the mathematical laws governing them, was in itself a significant accomplishment, but it was merely a forerunner of far greater achievements in understanding other important natural phenomena.

Already in roughly 240 B.C., the Greek philosopher Chrysippus had speculated that sound was a kind of wave, and some 200 years later, his ideas were developed more fully by the Roman architect Vitruvius, who explicitly likened the spreading of sound waves from a source to the circular spreading of ripples on water. What Vitruvius did is in fact extremely typical of the thinking style of all physicists: taking a familiar, visible, everyday phenomenon and seeing it, in one's mind's eye, as taking place in another medium, sometimes at a vastly different spatial or temporal scale, so that it is inaccessible to one's senses (recall the invisible "shadows" in the cathode-ray tubes). In this case, the familiar phenomenon is ripples, whose wavelength and frequency are extremely apparent, and the new medium is of course air. The wavelengths and frequencies of sound waves are not perceptible and in fact their frequencies are very different from those of ripples or ocean waves. Because of these major differences, it was a very bold act of Vitruvius to apply the same word to two phenomena of which one was very well known and the other was hardly known at all (much like Galileo's daring extension of the word "Moon" to infinitesimal dots that moved, when he observed them through his telescope). It took many centuries more, however, before the theory of sound waves was further advanced, thanks to the work of such insightful scientists as Galileo, Marin Mersenne, Robert Boyle, Isaac Newton, and Leonhard Euler.

Not surprisingly, there were some significant discrepancies between sound waves in air and waves on water. Among the most important is the fact that, unlike water waves, sound waves are *longitudinal*, meaning that they involve motions of air molecules along the direction of propagation of the noise. It's perhaps easiest to explain this by another analogy. When a line of cars moves down a road that has a series of traffic lights, the distance between neighboring cars diminishes each time they must come to a stop, and it increases when they start up again. This is sometimes called a *compression wave*, since the traffic is getting more and then less compressed. Compression waves always are longitudinal, in that the distances grow and shrink along the direction that the wave itself is traveling in. Similarly, the density of air molecules oscillates rapidly as a sound wave passes down a corridor, and the molecules, like the cars on a crowded road, get alternately closer and farther from each other, their relative motion being along the same direction as the sound itself is traveling (once again, that's the meaning behind the term "longitudinal").

Comparing sound waves with water waves seems easy, but there are hidden subtleties. It's obvious to a casual observer that as a ripple passes by, the water at and near the surface moves up and down, and this up-and-down motion is perpendicular to the ripple's direction of travel (which is horizontal). This is called a *transverse* wave. However, it happens that ripples are not that simple. The water actually does another dance at the same time: it also oscillates forwards and backwards, where "forwards" means "in the direction the ripple is going". That constitutes a *longitudinal* oscillation (much as in the case of sound, and also like the cars on the highway). These two motions, transverse and longitudinal, take place simultaneously, at and below the water's surface, and to confuse matters more, the up-and-down motion is not in phase with the back-and-forth motion, but they are, as physicists would say, "90 degrees out of phase". As a result, as a ripple passes by, the dancing water molecules at and below the surface move in perfect *vertical circles*, aligned with the ripples' motion (in the same way as a bicycle wheel is a vertical circle aligned with the bicycle's motion). One last elegant feature of these surface-tension waves is that the circles' radii grow smaller and smaller the further below the surface one descends. As this shows, water waves certainly are not the simplest waves of all.

Yet another major discrepancy between sound waves in air and ripples on water is that whereas water waves travel at different velocities depending on their wavelength (for which they are said to be "dispersive"), sound waves are much simpler: no matter what their wavelength is, all sound waves propagate at the same velocity in a given medium (which entitles them to the label "nondispersive"). This is lucky for us speaking creatures, since otherwise the different waves constituting our voices would all disperse and we couldn't understand a thing anyone said (unless we possessed a far more sophisticated auditory system than the one we have, which works only for nondispersive waves).

In a sense, the leap from visibly undulating water to invisibly undulating air, though humble in a way, was also the greatest leap in the story of the development of the *wave* concept, because it opened up people's minds to the idea of making other daring leaps along similar lines. One success led to another, each new analogical extension making it easier to make the next one. The next big leap — from sound to light — was of course a bold step, but the way had already been paved by the leap from water to sound. To put it differently, the sound-to-light leap was facilitated by a *meta-analogy*, even if it wasn't spelled out explicitly — namely, the idea that one analogical leap (from water to sound) had already worked, and so why shouldn't the *analogous* analogical leap (from sound to light) also work?

Such meta-analogies have permeated the thinking of physicists in the last few centuries: an idea understood well in one domain is tentatively tried out in some new domain, and if it is found to work there, physicists hasten to try to export the old idea once again to even more exotic domains, with each daring new attempt at exportation being analogous to previous exportations. Over the past hundred years or so, making bold analogical extensions in physics has become so standard, so par for the course, that today, the game of doing theoretical physics is largely one of knowing when to jump on the analogy bandwagon, and especially of being able to guess which of many competing analogy bandwagons is the most promising (and this subtle selection is made by making analogies to previous bandwagons, of course!). This highly cerebral game might be called "playing analogy leapfrog". Chapter 8 will consider these ideas in greater detail.

But back to waves. One major difference between trying to prove the existence of sound waves and trying to do so for light waves was that whereas in the seventeenth century, it was relatively simple to devise experiments to determine the wavelengths and frequencies of typical sound waves, no such measurements were feasible for light at that time (the wavelength of visible light is microscopic, and its frequency is enormous — hundreds of trillions of "crests" and "troughs" pass by each second). On the other hand, the *sound* ⇒ *light* leap possessed the happy precedent of the prior *water* ⇒ *sound* leap, which, as we have said, inspired confidence, by analogy.

The first guesses about light as a wave phenomenon were somewhat wrong, as they were based on an overly simplistic analogy with sound; it was assumed that light, just like sound, was a compression wave that propagated in an elastic medium, such as air. Thus light waves were originally conceived of as *longitudinal*, just like sound waves. (In Chapter 7, this kind of assumption will be dubbed a "naïve analogy", and the nature of such assumptions will be scrutinized in detail.) It took careful experiments in the early 1800s by Thomas Young and Augustin-Jean Fresnel to get beyond this naïveté and to reveal that light was not longitudinal but *transverse*, meaning that whatever was oscillating was doing so *perpendicularly* to the direction of motion of the wave, a finding that was very disconcerting, because no one could give a physical explanation for why such a wave would exist. (In a sense, water waves provided a precedent for this finding, since they had an obvious transverse quality, but it was clear that this quality was due to the existence of a special spatial direction defined by gravity, and light moved through space where there was no gravity and hence no special direction, so this removed any promise that the analogy might have seemed to hold out.)

It was only in about 1860 that James Clerk Maxwell came to the astonishing revelation that light waves did not involve the motion of any material substrate at all, but instead were periodic fluctuations, at each point of the three-dimensional space in which we live, of the magnitudes and directions of certain abstract entities called *electric and magnetic fields*. It was as if the medium that conducts light waves consisted of a gigantic collection of immaterial arrows, one located at every point of empty space (actually, two — one magnetic and the other electric), and whose numerical values simply grew and shrank, grew and shrank, periodically oscillating. This was certainly extremely different from the visible, tangible motion of water on a lake's surface or of waving wheat in a field, and some physicists couldn't relate at all to this kind of highly abstract intangibility, but it was too late to go back and undo it. The concept of *wave* was inexorably growing more and more abstract, spreading relentlessly outwards from its original "city center", as is the wont of concepts, always moving out to the suburbs.

It didn't take physicists too long before they started realizing how immensely fertile this concept of *wave* truly was, in the explanation of natural phenomena, ranging from the most ubiquitous, such as sound and light, to all sorts of exotic cases. Any time space was filled with any kind of substance (or with an abstraction that could be likened to a substance), it seemed that local disturbances in that "substance" would naturally propagate to neighboring spots, and so forth, and thus waves would radiate outwards from a source. The disturbance, however, could be very different from ordinary vibration — it could be highly abstract, like the shrinking and growing of invisible abstract arrows. Nonetheless, all the standard old concepts associated with earlier waves could be investigated — *wavelength*, *period*, *speed*, *transverse* or *longitudinal*, *interference*, *reflection*, *refraction*, *diffraction*, and so on, and many of the same equations carried over beautifully from one medium to another.

For instance, *moonlet waves.* That's not the standard term for them, but curiously enough, James Clerk Maxwell's first discovery in physics was the fact that the rings of Saturn are made of billions of tiny "moons", and he proved his theory by showing that if there were compression waves sloshing back and forth inside the rings around the planet, their calculated behavior would perfectly match the data observed by astronomers.

In the early twentieth century, *radio waves* (really just long-wavelength light waves) were used as a host medium for carrying *sound waves.* In other words, sound waves hitch a ride on the much faster medium of electromagnetic waves. Amplitude modulation (AM) is a kind of *transverse* way of letting the hitchhiking sound waves locally distort the medium through which they are traveling, whereas frequency modulation (FM) is essentially a *longitudinal* way of letting the hitchhiking sound waves hop aboard the very fast host waves. FM, in short, is a very abstract sort of compression wave. We can't enter into the details here, but the brilliant though tricky idea of waves riding on waves gradually grew into an ever more common *leitmotiv* in physics.

Later during the twentieth century, *temperature waves* were discovered, in which the value of the temperature of a substance is what oscillates "up and down". In other words, what is moving "up and down" is the number of degrees (Fahrenheit or Celsius) at each point in the substance, like millions of imaginary columns of mercury moving physically up and down in so many imaginary thermometers all through space (and of course the thermometers need not be in phase with each other).

Also discovered were *spin waves*, where the direction of spin of electrons (think of a room filled with millions of tiny spinning tops, some pointing up and some pointing down) can "ripple" across the medium, with spins periodically flipping from up to down and then back to up. And then there are *gravitational waves*, where the amount of gravitational pull at some point in space oscillates periodically in time, as if an invisible ripple were silently shimmering through space and, by its local size, telling thousands of little pebbles floating in space how strongly, and in what direction, they are being pulled by a rapidly shifting, totally invisible force.

Last but not least, among the very most important kinds of waves in all of physics are *quantum-mechanical waves*, sometimes called *matter waves* or *probability waves.* Roughly speaking, at every point in space such a wave has a value that changes over time, and when that value is squared, it tells how likely one is to find a particle in the given spot at the given instant.

We could list dozens of other types of abstract waves, but this modest sampler will suffice. As we have seen, the notion of *wave* in physics has reached an enormous degree of abstraction and sophistication today, and yet all of the latest and most abstract forms of waves are tied by analogy and by heritage to the earliest kinds of extremely concrete, tangible, palpable waves in bodies of water and in fields of amber grain — waves that we can see with our eyes and feel with our bodies.

Of Sandwiches

Oh, the poor fourth Earl of Sandwich! How often is his name abused these days! As is well known (or at least rumored), it was the august Earl who concocted the clever idea, around 1750, of putting some meat (or perhaps some cheese) between two slices of bread. Sandwich's original "bread–meat–bread" pattern spurred an enormous number of copycat variations in the gastronomic world, which we don't need to spell out here.

It might be amusing, however, to mention that one of the authors of this book has been known, to his great gustatory delight, to savor the act of devouring a handful of macadamia nuts placed on a square of American cheese that is then folded back on itself, thus becoming both an upper and a lower layer at once. The analogy is tremendously obvious, and yet there is pleasure in spelling it out: the two layers of surrounding cheese "are" the two layers of bread, and the macadamia nuts "are" the meat. The only question left, then, is whether we also need quotation marks around the word "is" if we assert that this *is* a sandwich. Of course it is *like* a sandwich, but what would the Earl himself say? Or is there any reason to presume that the fourth Earl of Sandwich would be the ultimate authority concerning membership in the category bearing his title's name? Could there in fact be any ultimate authority on the topic? Would it not be a fine thing to create an elite Sandwich Memorial Board that would officially make all such rulings?

Let's move on to sandwiches lying beyond the realm of the edible. Our first variation on the theme — our first inedible extension of this category — is not too far removed from that realm, however. For at least a couple of hundred years, restaurants in large cities have had the practice of hiring hungry souls to serve as walking advertisements, having them stroll the sidewalks wearing wooden or cardboard posters on front and back. Such people are usually called "sandwich men" (although of course a sandwich man need not be a *man*$_1$). Needless to say, there is many a discrepancy between a slice of bread and a piece of wood with slogans painted on it, and likewise there is many a discrepancy between a slab of meat and a living human being; these two facts already cast some doubt on the *sandwichhood* of the described item. But in addition, this kind of "sandwich" is not meant to be *consumed*, except in a very abstract sense — namely, by a visual system. And perhaps more subtly, a human sandwich of this sort challenges the great U.S.A. (that is, the Unspoken Sandwich Axiom), which is the tacit idea that a sandwich must always be *horizontal.* In sum, there are numerous reasons to wonder about the membership of this kind of entity in the category *sandwich.* The question arises as to the limits of the concept, and whether there are any limits at all to it. How far out does sandwichhood stretch, and in what directions? This is a most provocative question.

We are not the only thinkers, we hasten to add, to have posed such far-reaching questions. A bold web site called "The Sandwich Manifesto" addresses head-on the fundamental question "What is a sandwich?" One issue raised there is whether anything that has the form "A–B–A" is a sandwich. For example, do the books on a shelf, sandwiched between two identical bookends roughly a yard apart, form a sandwich? Or is the name "Einstein" a sandwich, given that it is spelled "Ein-st-ein"? Is America a sandwich, with the Atlantic and Pacific Oceans being the slices of bread? The non-identicality of those two great bodies of water, not to mention the non-identicality of "Ein" and "ein" (after all, the first boasts a capital "E" while the second does not), brings up the more general issue of the identicality, or lack thereof, of the two "slices of bread".

On one memorable occasion in Paris, the métro station Odéon was described by a fairly bored métro rider as being "sandwiched" in between two métro stations bearing the names of saints (Saint-Michel and Saint-Germain-des-Prés). This trio of stations might thus, with good reason, be called a "subway sandwich". In a (somewhat) similar fashion, car-borne criminals are sometimes said to be "caught in a sandwich" if two police cars manage to maneuver into a position such that one of them is behind the criminals and the other is ahead of them. Another flesh/flesh/flesh configuration is the sexual position known as "The Sandwich". But leaving the body behind and moving on to the mind, we have rhyme sandwiches: consider a rhyming quatrain whose rhyme scheme is "ABBA", where the "A" rhymes play a *bread* role to the "B" rhymes' *meat.* And surely we would not want to overlook the crucial role of inedible sandwiches in the realm of solid-state physics. Specifically, there are important types of semiconductors called "P" (for "positive") and "N" ("negative"), and from them are formed three-layer structures of the form "PNP" and also of the complementary form "NPN". These are both standardly called "sandwiches" in academic papers by physicists — and such sandwiches are incidentally also members of the category *transistor* (hardly a concept to be sneezed at).

The bold interchange of the *meat* and *bread* roles in solid-state physics raises another fundamental question of sandwichology — namely, whether there are certain categories of things that are more eligible to play the *bread* role in a sandwich, with other categories being more eligible to play the *meat* role. For example, we all know that *peanut butter and jelly* constitutes a fine member of the *meat* category (at least in the context of sandwich-making), but would it ever be able to serve in the role of *bread*? Let us pose this question in a more point-blank fashion. One can envision, without the least difficulty, an "NPN" sandwich in which the "N" stands for a delicious Indian *naan* and the "P" stands for peanut butter and jelly — but what about an inside-out "PNP" sandwich? Or is this simply going too far? Have the ultimate limits of the *sandwich* category been transcended, or could it be that our era is simply not yet ready for such bold new visions?

Arguably the most burning conundrum in sandwichology is under what conditions an entity that has the form "A–B–C" should be counted as a *sandwich.* For instance, if one's Bostonian bosom buddy Bradley ("B", for short) happened to be fast asleep on his comfortable Chesterfield couch ("C", for short), and if Bradley's Abyssinian feline friend Adele ("A", for short) were suddenly to leap atop dormant Bradley, might the resulting A–B–C configuration count as a *sandwich*? And if A had been an *armchair* rather than an *Abyssinian*? Armchairs being presumably a bit more couch-like than Abyssinians, would one not be ever so slightly closer to the canonical A–B–A form?

The last few paragraphs have been rather fanciful, but it is worth noting that the word "sandwich" is routinely used in colloquial speech and even in formal contexts, sometimes as a noun and sometimes as a verb, to denote abstract and definitely non-edible patterns. How many times, for instance, have you casually remarked, "My meeting with the dean this morning was sandwiched right between my dermatologist's appointment and my dentist appointment"? You probably can't even remember! Yes, in an era when using the word "sandwich" to describe all sorts of inedible things has become a routine worldwide phenomenon, we are no longer talking about wild, self-indulgent flights of fancy. We are talking popular culture.

As you can no doubt sense by now, the questions of shadowology, wavology, and sandwichology open up vast conceptual horizons without an abstraction ceiling lurking anywhere in sight.

The Downfall of Proud Capitals

As we have just seen, the extension of concepts by analogy seems limitless. It allows us to see one's homeland as one's *mother*, to see a snow-free area under a tree as a *shadow*, to see a pattern of sequential brakings by drivers along a stretch of freeway as a kind of *wave*, a sequence of three appointments as a *sandwich*, and also to see the story of a person who finds reasons for satisfaction in their failure to purchase tickets for a concert they had dearly hoped to attend as just a differently dressed retelling of the *sour grapes* fable. Sometimes leading to extending the boundaries of a category, such as *bird* or *book*, *moon* or *marriage*, *eat* or *undress*, *much* or *but*, or to the construction of a new and more abstract category, such as *lump* or *desk*, or even a whole range of new concepts, as in the case of *shadow* and *wave*, the process is simply part and parcel of the human condition, and as such is unstoppable. Indeed, almost as if to show off its irresistibility, the process of category extension survives even in extremely austere environments where one would suspect it could not — namely, in the world of proper nouns, a world where everything comes in ones, and thus a world that, by its very definition, would seem to prohibit the extension of categories.

In contrast to common nouns, which are obviously the names of categories, proper nouns, which are singled out by the capital letters with which they start, might seem to be of a completely different nature. Although they are certainly definable through language, proper nouns aren't supposed to need definitions because the set of entities that they refer to seems to be unambiguously defined. Often, they designate one and only one entity: a planet, a continent, a country, a city, a monument, a human being, a work, and so forth — and when they designate more than one entity, as does a first name, a last name, a commercial brand, a nationality, and so forth, the set of entities to which they apply seems nonetheless so sharp and clearly defined that one might have a hard time imagining that there is anything about them that resembles the "halos" that surround the cores of all typical categories, as we have been describing them all through our book so far.

Indeed, one might well go so far as to question the use of the word "category" when there is just one member. What kind of sense does it make to speak of categories such as *Paris*, *Galileo*, *Earth*, and *Moon*, when each of those exists (or at least once existed) in but a single case? But upon analysis, this kind of argument is quickly seen to hold no water. Our irrepressible human tendency to extend categories by the making of analogies applies in the case of proper nouns no less than it does for all other nouns (and other words, for that matter).

Chapter 1 recounted the story of Galileo, the Moon, and the many moons that subsequently came along. That story may have seemed like a very unusual case, but leaps such as that from "Moon" to "moons" (or from "Sun" to "suns") take place all the time around us, far removed from the specialized world of scientific discoveries. In particular, they take place whenever categories are extended by an act of marking, in which a proper noun loosens its belt a bit and in so doing becomes the label of a more general category.

For instance, brand names have often become generic words, thanks to the popularity of the products they name. Thus the old-time General Motors brand of refrigerator called "Frigidaire" became, in the 1920s, an uncapitalized noun in American English (and also in French), just as the brand name "Hoover" for vacuum cleaners became an ordinary uncapitalized noun in British English. Essentially the same story can be told about the brand names "Kleenex", "Coke", "Xerox", "Saran Wrap", "Dustbuster", "Scotch Tape", "Teflon", "Q-Tips", "Jacuzzi", "Frisbee", and so on (and thus we could as easily have decapitalized these words as left them with capital initial letters).

In all these cases, first there is a small category whose members are the products of the specific brand name — and then new products are made that are different enough from the original ones that they seem to call for a new word, yet at the same time, they aren't *fundamentally* different from the original products, since they all share the key characteristics that created the need for the original products. When people want to give a name to these new copycat products that form a halo around the original ones, they will often spontaneously borrow the original brand name but will decapitalize it in order to indicate that this is an extended sense of the original word (much as "Moon" became "moon"). The new members and the old members of the original category now all belong to this new category. Thus the word "kleenex", when decapitalized, stands for all tissue papers, but when capitalized, it stands only for tissue papers of the Kleenex brand.

Now this phenomenon might seem like a desirable thing from the point of view of a popular brand, a demonstration that it is collectively recognized as the most canonical item of its sort. However, although a few companies might welcome such genericizing of their names, more often the shareholders of the genericized companies do not see things this way at all. Indeed, major brand names tend to combat this process very actively, since it tends to dilute the meaning of their name, in the sense that people soon come to hear the word that once was a brand name simply as a bland name without any identity at all.

Brands want to be recognized for their uniqueness, not for their genericity. Who would appreciate it if, within a few years of their naming their very popular dog "Oliver", bandwagon-jumping families in the neighborhood had given every single new dog the name "Oliver"?

The turning of a proper noun into a common noun transforms a once-special term into a commonplace. Who would ever proudly boast of owning "an authentic jacuzzi" or "a genuine frisbee"? In the case of these two brands (and both are indeed brands), the unmarked sense has long since eclipsed the marked sense, and as a result the first letter has been demoted to lowercase status. This type of slide, which entails the loss of legal protection of the brand name, has been dubbed "genericide". This explains why, when the verb "to google" first appeared in the 2006 editions of the *Oxford English Dictionary* and the *Merriam-Webster Collegiate Dictionary*, the Silicon Valley giant instantly launched an intense campaign to restrict the usage of its name, particularly focusing on preventing the proper name "Google" (or rather, the non-proper-name "google") from being used as a verb denoting Web searches regardless of what software is carrying them out.

In the above examples, in which a single word comes to occupy two levels of abstraction, we recognize the telltale signature of the phenomenon of marking. And by coincidence, the noun "mark" is occasionally used in English to mean "label" or "brand name", as in "What mark is your shirt?" Actually, this unexpectedly close relationship between the ordinary word "mark" and the technical term "marking" is not a coincidence, since a commercial brand or logo or mark is a visual identifier, allowing potential customers to distinguish similar-looking products from each other. And the adjective "marked", whose origins have to do with the idea of stamping something with a distinguishing symbol (a "mark"), is used to describe subcategories, much as a commercial mark designates a subcategory of products that all belong to one single overarching category.

Commercial marks (*i.e.,* brand names) such as we've been discussing make up but the tip of the iceberg of the phenomenon of lexical labels that fluidly swivel back and forth between denoting just one single entity and denoting a far vaster category. There are in fact cases where the name of a unique individual, place, or object can, despite its uniqueness, be naturally applied to dozens, hundreds, or thousands of entities.

Sacred Categories

The worldwide unity of the Catholic church is due to the existence of a single spiritual leader, the head of the Vatican: the Pope. Aside from a few historical upheavals that led, at the end of the fourteenth century, to the simultaneous naming of two popes — Urban VI and Clement VII — Catholics have always been able to look to their unique Pope for leadership. He is *the* Pope, and that's all there is to it. However, if the current Pope enjoys the distinction of being the unique terrestrian member of this exalted category (previous members enjoying eternal repose), his title is extremely sought after when it comes to the broader sense of the term, which is to say, the unmarked category.

If one goes to the Web, the papal harvest is rich. Pop Art has its uncontested pope: Andy Warhol. The pope of the personal computer industry is heralded as Bill Gates or Steve Jobs. John Waters is pronounced the pope of bad taste, Robert Parker the pope of wine, Paul Bocuse the pope of gastronomy, etc. Indeed, popes with a lowercase "p" proliferate like flies. Picasso has been called the pope of Cubism, André Breton that of surrealism. Bob Marley has been declared to be Reggae's pope, Ray Charles jazz's pope, and Frankie Ruiz the pope of salsa music. Celtic music, too, finds its pope in Alan Stivell, and contemporary abstract music, not to be left out, has Pierre Boulez for its pope. While the Dalai Lama is anointed the pope of Buddhism, Richard Dawkins is acclaimed the pope of atheism, and lastly — surely to no one's surprise — Pierce's Pitt is proclaimed the pope of pulled pork (where? in Williamsburg, Virginia).

But why quit when you're on a roll? If you persevere in your Web search, you can find popes of positivism, football, Japanese trash cinema, neoconservatism, boxing, free software, haiku, contemporary design, multimedia, management, documentary, TV news, underground cinema, the harpsichord, Scandinavian rock music, manga, rap, tennis, Italian jeans, dog biscuits, coaching, bio-art, broccoli, business calendars, and on and on. Indeed, the list of popes can be extended pretty much without end, and so it makes sense to declare a state of "papal hyper-inflation", meaning that there are so many popes of this, that, and the other thing that at this point the title has lost much of its punch. No matter how narrow some human activity might be, there is always some practitioner of it who is perceived by somebody or other as having sufficient prestige and sway as to merit a nomination to the pantheon of "generalized popes".

And while we're talking about pantheons, they, too, form an interesting case, hovering blurrily somewhere between proper noun and common noun, somewhere between singular uniqueness — "The Pantheon" — and plural ("many pantheons"). Indeed, even when doubly capitalized, as just shown, the name is not unambiguous, there being good reasons to think it designates Il Pantheon (in Rome) and other good reasons to think it designates Le Panthéon (in Paris). In any case, pantheons were originally conceived of as monuments erected to honor a civilization's gods (or perhaps its Gods). The Pantheons in Rome and Paris both represent the highest honor that their respective nations can bestow on individuals of great achievement, serving them as a kind of exalted cemetery. But the category of pantheons is far wider than this, since a pantheon can be a kind of "software temple", or a "temple of the imagination", requiring neither a building nor burials — just a listing of names of a number of important individuals. Accordingly, Albert Einstein is clearly in the pantheon of physicists, and Henri Poincaré, though not buried in the Panthéon in Paris, certainly figures very high in the pantheon of mathematicians.

Various sacred sites and books serve to keep religion on people's minds. Thus the name of Mecca, a destination for millions of pilgrims each year from all around the world, has become, in its decapitalized version, a word that captures the idea of a venerated place — indeed, a cult place — for a particular activity. Below are listed a few dozen meccas that we came across using, shall we say, the pope of search engines. We found meccas of:

> automobile styling, basketball, catamarans, cigars, cinema, cricket, cross-country skiing, entertainment, faded jeans, golf, granite, hang-gliding, hip-hop, hockey, hot-air balloons, 100-kilometer runs, jazz, "made in China", motorcycle racing, mountain biking, 1950's furniture, nudism, obstacle courses, parachuting, petroleum products, piano-playing, psychedelics, rap, rock, rollerblading, rugby, sandwiches, shopping, soccer balls, socialism, sound effects, speed skating, surfing, swing, tea, tennis, terrorism, tourism, the triathlon, videogames, volcanology, voodoo, and wind-surfing.

What is constant in all these meccas — what constitutes the "essence of *mecca*-ness" — is the idea of a *place of supreme importance*, the idea of *uniqueness* (even though for some of the activities two or three would-be meccas vie for the title of *the* mecca), and even the idea of some kind of *sacredness* (even though, for most of these meccas, the activity in question has nothing to do with religion).

The Book of Books — that is, the Bible for some, and the Coran for others — has also been deemed worthy of becoming an abstract category. The category of "Book of Books", representing just *one* book, gets stretched so that it becomes applicable to all sorts of different books in different domains (but presumably just one per domain). Thus there exists a bible of Thai cooking, a bible of ribbon embroidery, and a bible of body-building. And the most reliable book about gardening in an Islamic country might well be called "the coran of gardening", since of course the Bible of Islam is the Coran, and the reverse holds equally well, the Coran of Christianity being the Bible.

To be sure, religion isn't the only field in which this kind of pluralization of proper nouns takes place; the phenomenon occurs in the most mundane of activities no less than in the most otherworldly ones, as we shall see. Indeed, capital letters fall by the wayside left and right on Earth as they do in Heaven.

Pushkins, Chopins, and Galois Galore

A provocative question was posed by mathematicians Frank Swetz and T. I. Kao in the preface of their little book *Was Pythagoras Chinese?*

> Of course, the historical figure of mathematical fame known as Pythagoras and born on the island of Samos in the sixth century B.C., was Greek and not Chinese. But there is another "Pythagoras" equally famous. He is the man who first proved the proposition that "the sum of the squares of the legs of a right triangle is equal to the square of the hypotenuse." For hundreds of years this theorem has borne the name of Pythagoras of Samos, but was he really the first person to demonstrate the universal validity of this theorem?

Indeed, strange though it might seem, it is perfectly possible that the category of *Pythagorases* might have had a member quite a while before the birth of Samos' most famous native.

In everyday conversations, people unwittingly exploit the device of marking to distinguish individuals who stand out from the crowd from more run-of-the-mill individuals. Among the people who excel in a certain field and gain recognition from their peers, a small minority becomes known beyond just a local circle, perhaps for writing a book, or for being written up in newspapers, for acting on stage, or for making a splash in the world of business. A handful of these accomplished individuals then jump over a yet higher hurdle, perhaps by receiving a prestigious prize, or by having their name appear on the marquee of movie theaters, or by rising to hold an office in state government, by becoming mayor of a mid-size city, by hosting a weekly television show, by excelling in a sport, by making a modest fortune in industry or finance, or even by being "famous for being famous".

Among these celebrities, only a minority ever have their name listed in a prestigious catalogue of important people, whether it be *Who's Who* or some kind of encyclopedia. But there is yet another, higher stage of fame, attained only by the cream of the cream of the preceding cream, and this is the stage where one becomes a public category. The names of such individuals, above and beyond designating specific people and their accomplishments, become lexical items that denote abstract categories that can have many members.

A catchy French song called "Le Piano du pauvre" by singer–composer Léo Ferré describes a random Parisian street accordionist as "le Chopin du printemps" — "the Chopin of the springtime". Both the anonymous accordionist and Frédéric Chopin are being honored here — the former for being an "instance" of the latter, and the latter for having been turned into a category that can have instances. We will call this latter honor "canonization". Of course, thousands of people besides Chopin have been canonized. As a matter of fact, with a bit of effort one can turn up scads of colorful expressions based on canonizations, such as the following seventy-odd mind-boggling examples (none of which, believe it or not, was dreamt up by your authors):

> the Bach of the vibraphone, the Beethoven of landscape painting, the Haydn of chess, the Mozart of mushrooms, the Mendelssohn of Hinduism, the Puccini of pop, the Wagner of rock, the Billie Holiday of ballet, the Benny Goodman of duck-calling, the Frank Sinatra of chatterbots, the Elvis Presley of neurology, the Mick Jagger of climate change, the Plato of freemasonry, the Aristotle of the airwaves, the Socrates of snails, the Democritus of modern linguistics, the Euclid of chemistry, the Archimedes of minigolf, the Kepler of etymology, the Copernicus of rodent control, the Galileo of the soccer ball, the Newton of terrorism, the Faraday of window-glass making, the Galois of tobacco science, the Einstein of sex, the Leonardo of ice cream, the Michelangelo of Lego sculptures, the Rembrandt of movie-making, the Picasso of sidewalk art, the Dante of criminal psychology, the Milton of middle-class comedy, the Shakespeare of advertising, the Balzac of the supernatural, the Goethe of Urdu literature, the Byron of the Browning automatic rifle, the Pushkin of feminism, the Tolstoy of 21st-century television, the Proust of the comic book, the Ernest Hemingway of media bloggers, the Thomas Pynchon of internet trolls, the P.T. Barnum of Polynesian pop, the Mae West of tiger taming, the Marilyn Monroe of hip-hop, the Meryl Streep of spitting, the Fellini of photography, the Stanley Kubrick of pornography, the Walt Disney of consumer electronics, the Bill Gates of wastewater, the Rockefeller of video games, the Babe Ruth of bank robbers, the Evel Knievel of oncologists, the Michael Jordan of bagpiping, the Tiger Woods of user-generated video, the Lance Armstrong of tough-guy jokes, the Usain Bolt of cognitive science, the Serena Williams of apathy, the Paul Revere of ecology, the Napoleon of fossil bones, the Rasputin of rockabilly, the Hitler of snuggling, the Franco of fricassee, the Mussolini of mulligatawny, the Mao Tse-Tung of gay soap operas, the Mahatma Gandhi of restaurant criticism, the Che Guevara of tango, the Richard Nixon of superheroes, the Indira Gandhi of astrophysics, the Osama bin Laden of monkeys, the George Bush of Oscar hosts, the Barack Obama of Tamil cinema, the Tarzan of the pole vault, the Sherlock Holmes of Yiddish music...

The creation of a general category through the pluralization of a proper noun, such as a famous person's name, is based on the idea that there is an essence to each very well-known person or thing, be it the Moon, the Mona Lisa, Mecca, or Mozart. This essence can be pinpointed and then distilled from the entity itself; such an act of distillation gives rise to a new abstract category, much as we saw happening in the passage from *hard-desk* to *soft-desk*. It's not difficult to see that the same mechanisms of essence-identification and essence-distillation underlie a different family of expressions, based on the names of famous cultural landmarks, such as the following few: "the Rolls-Royce of dishwashers", "the Concorde of trains", "the Rolex of cameras", "the Leica of sound reproduction", "the Mona Lisa of the British Museum", and "the Taj Mahal of chicken coops", not to mention "the Stradivarius of fly-fishing fishing reels".

One might think great fame is needed for a person or thing to be "canonized" as an abstract public category and thus to be realizable in multiple instances. However, this impression is deceptive: anyone and everyone can be so canonized, albeit at a more local level. One needs merely to be "locally famous" — intimately known to one's family and friends — and that is something that we all are, fortunately, and hence we can all be canonized and pluralized by our kith and kin, and indeed we often are, for we are among their "personal celebrities", so to speak, and they are among ours.

Our Personal Celebrities

Now and then in conversation, people toss off phrases like "Ellen is the Jeff of her family", "I'm the Sam of my circle of friends", "Bill is her David", "She's their George and Priscilla", and so on. The first of these might be used to express the idea that one's friend Ellen has a habit of cracking riotously funny deadpan jokes, especially in the setting of her family, and that this trait of hers is reminiscent of another friend Jeff, who belongs to an unrelated family. In making this kind of analogy, one hopes to cast a fresh perspective on Ellen and perhaps also on Jeff, since taking a fresh point of view via a spontaneous analogy often brings novel insights. As this anecdote suggests, we are all influenced by categories centered on familiar people — our "personal celebrities".

To be sure, Jeff possesses many attributes besides his sense of humor, but the conversation is focused on Ellen and her style of humor, and in that context, only a narrow facet of Jeff is likely to be elicited in the listener. The fact that Ellen is the teen-aged daughter in her family, while Jeff is the middle-aged husband in his, is irrelevant in this context, and is easily ignored. In other words, a context-dependent "essence of Jeff" will be implicitly distilled by the listener with no trouble. This kind of streamlined usage is a highly effective mode of communication, provided the participants in the conversation have the needed background knowledge.

The phrase "the Jeff of her family" amounts to a pluralization of Jeff, suggesting that there could be various Jeffs in various families (or in other groups of people). In other words, it converts *Jeff the unique individual* into *Jeff the founding member* of a category. Of course, since the phrase was just a throwaway remark, this category may not last long in anyone's mind, but it might conceivably plant the seed for a long-lasting and extensible category, so that in later conversations someone might refer to "the Jeff of our family", "the Jeff of that club", "the Jeff of my salsa class", and so forth. But since Jeff has many facets, there could also be other conversations in which another of his facets — say, his perennial pessimism or his frequent griping about his work — would be the "essence of Jeff" that would be implicitly pinpointed and effortlessly understood.

We have no trouble using Jeff as the nucleus of a category. We effortlessly understand remarks like "Sally's no Jeff!", just as we effortlessly understand a remark like "Clint ain't no Mozart!" Or even more explicitly pluralizing him, one could say, "Too bad there aren't a lot more Jeffs in this world!" We pluralize old Jeff just as glibly as we pluralize Mozart, Mother Teresa, Madonna, Steve Jobs, or Joan of Arc.

This type of linguistic playfulness is only one way in which we pluralize our canonized friends. There are other ways we do so, however, which reveal that we see our friends as multi-member categories not just when we consciously decide to do so but also subconsciously, without any prior intention to pluralize. The next section will deal with these kinds of events.

Unintended Slippages from One Person to Another

We have all had the experience of confusing one person with another — and here we are speaking not of their names, which may be very dissimilar, but of their identities. More specifically, we mean the experience of slipping, to one's surprise, from a person to a "similar" person. Errors of this sort reveal unintentional recategorizations: occasions in which person A is momentarily confused with person B, who is very familiar, and the lexical label for person B — that is, B's first name — comes to mind rather than A's first name. This kind of error takes place frequently because categories for people such as our friends can be centered on a single individual and yet blur outwards so as to let in, on occasion, other less central members. Here are some concrete examples:

> Paul just had a violent argument with his wife Catherine. In the heat of the fight, he accidentally called her "Jessica", the first name of his previous wife, with whom he had often had arguments before they divorced. For Paul, during this argument, Catherine became momentarily Jessica, or perhaps "another Jessica".
>
> Richard's daughter is named Marilyn, but dozens of times he has called her "Liz", or come very close to doing so. "Liz" is in fact his younger sister's first name. Marilyn, who is 15, reminds him of how Liz was at that age. Moreover, Marilyn is the second-born, just as Liz was, and her way of acting around her older brother is very much like the way Liz acted around Richard when they were teen-agers. Each summer, Richard takes his family to his old hometown, where his sister still lives, and during those times the tendency is reinforced, and Marilyn becomes even more frequently "a Liz".
>
> Every so often, Phil calls his wife Iris by the name "Betty", which is the name of his long-time assistant at work. This unintended, unconscious confusing of the names of "his two women" makes Phil feel very ill at ease, because whenever he does it, he feels as if the role played by Iris in his life is scarcely any more important than the role of an assistant, which is very troubling to him. This frequent error, repeatedly turning Iris into "a Betty", makes Phil wonder whether boredom isn't creeping into his marriage.

This phenomenon of conflating two people's identities (*i.e.*, seeing one person as an "instance" of another person — that is, as a member of the mental category centered on another individual) can also occur outside of language. When one meets someone new, it sometimes happens that one is reminded, more or less consciously, of someone else, and this déjà-vu sensation can be so strong that we feel we had already met the new person before, and we can anticipate their reactions to what we say, guess accurately their attitudes toward many things, intuit their interests and their sense of humor. (This is what happened when John and Rebecca met Thor, at the end of Chapter 3.) We treat the new person the way we would treat our old friend, and a much greater degree of intimacy is rapidly achieved thanks to this coincidence than would be imaginable with a "genuine stranger".

We are not referring to social stereotypes that are triggered in our mind when we see someone wearing certain articles of clothing, or hear certain words or phrases uttered, or have similar superficial reactions. We're talking about situations where one has the strong feeling of really *knowing* the new person (although one may later discover that this impression was unwarranted), and this intuition is the direct result of a categorization. Once the new acquaintance has been categorized as *a Nancy*, or as any other of our "personal celebrities", then our expectations regarding the new person's inner nature come from this categorization, just as our expectations that a certain animal will bark come from our categorization of it as *a dog*, or that an object will break if it falls come from our categorization of it as *a teacup*, or that an object is edible because of our categorization of it as *an apple.* We might thus expect of *a Nancy* we've just met that she will be gentle, compassionate, and maternal; that when she laughs she will turn her head in a certain way; that she will often lean forward when she walks; that she has a certain style of humor, and so on. On the other hand, we certainly don't expect her name will be "Nancy" — that would just be amusing icing on the cake.

Sometimes unusual circumstances can result in a deep confusion of the identities of two people one knows well, profoundly shaking some of one's cherished categories.

> Dan's wife Ruth had fallen under the spell of another man, and the fear of losing his beloved wife was eating away at Dan day and night. To his surprise, the name "Jeanine" started coming to his mind when he thought about Ruth, which mystified him, since nothing like that had ever happened before, during their six years of marriage. Twenty years earlier, he had been in love with a woman named Jeanine who had returned his interest but then revealed that she was already involved. For a while, Dan and Jeanine had a very intense friendship, and Dan was tormented by feelings of hope and fear, but in the end no romance came about. Gradually Dan's strong feelings for Jeanine faded and after a year, he only thought rarely about her, and by the time he met Ruth, Jeanine was deeply buried in his memory.
>
> And yet all at once, here was this old name bubbling up, reinserting itself into his most intimate thoughts. Even certain aspects of Jeanine's face were contaminating his mental image of his wife when she was not in front of him. Whenever he thought about Ruth, Dan was haunted by the feeling of not knowing who he was dealing with: their intimacy was now tarnished by a sense of duplicity. Long ago, these feelings had been attached to the concept *Jeanine*, and for that reason the name "Jeanine" and even certain features of Jeanine's face were now flooding his thoughts. The unintentional distortion of his wife in his mind revealed to Dan that the same fears of loss he had experienced decades earlier had been reborn. The deep analogical link between Jeanine-back-then and Ruth-right-now was something he had no power to repress.
>
> But eventually, Ruth's infatuation with the other man came to an end and Dan's fears gradually subsided; indeed, one day Dan noticed, to his great relief, that it had been quite a while since Jeanine's name and face had come to mind. For a while, Ruth had been *a Jeanine,* but when Ruth's worrisome behavior came to an end, she ceased being a member of that category.

Categories Based on a Shared First Name

> Seth and Brian were in a museum with their mother. By chance they bumped into Emma, a girl in Brian's class, who was with her parents. After parental pleasantries and pre-teen small talk, the two families parted. All at once Seth blurted out, "It was so *weird* of them to call her *Emma.* She doesn't look *at all* like an Emma." His mother, raising her eyebrows a little, asked, "Oh? So just what do Emmas look like, pray tell?" Seth was momentarily a bit flustered, but he rallied: "I don't know *exactly*, but a *typical* Emma is that girl Emma who lives down the street. *That's* what Emmas look like."

A bit of introspection reveals that the act of pluralizing any of our friends establishes certain expectations, both in terms of looks and personality, based on that person's first name. Thus, if we've "met" somebody solely on the phone or through email, it's nearly impossible not to have formed a kind of advance image of them, no matter how irrational we might know this is. Certain cues — the first name of course being one — exert power over us in creating a subconscious set of expectations concerning this person.

One time J. had an email exchange with the director of an advanced technology research lab, who was named Agnes. It struck him as somewhat odd that an elderly lady would be so highly placed in such a modern, fast-moving laboratory. When one day J. found out by chance that Agnes was considerably younger than himself, he was caught totally off guard, and he realized that, solely on the basis of her first name, he had unconsciously formed an image of his correspondent as being well on in years.

Like Seth, disconcerted by the girl he met who "didn't look *at all* like an Emma", we've all met people whose faces and personalities don't seem to match their name. We find ourselves making frequent errors in naming them; thus we may find ourselves sometimes calling someone "Allen", or at least having that name jump to our mind and nearly get uttered, even though their name is really "Will", simply because the person *looks* to us like "an Allen". How many times have we heard statements like, "The name 'Alex' really didn't fit him. He was dark-haired, slim, tall, distinguished-looking, spoke confidently, and was always well-dressed — the total opposite of my cousin Alex, who's short, blond, brash, plays rugby, and always hangs out in bars."

We all have also had the more common experience of feeling that a friend's name or a new acquaintance's name perfectly matches the person. Thus it strikes us as inconceivable that the people closest to us — parents, children, siblings, close friends — could have had any other names; it seems that their first name fits them like a glove, and that any other one would have been a cosmic miscalculation. And if we take a different first name and imagine sticking it on them, we recoil with shock, finding it absurd, since their true name is such a perfect fit. No other name would have harmonized with their face or their personality. If one discovers that some old friend goes by a name other than the name they were given at birth, one might be tempted to think that the parents really missed the mark in giving their child a name that doesn't fit, but that luckily a different name was found later that fits perfectly.

Such intuitions about the "correctness" of people's first names are simply the result of long-time associations between the name and the person's entire self. Of course when a child is born, parents often waver between two or more names, and later they thank heaven that they made the "right" choice among all those on their short list, since it fits the child so well. But unless one unshakably believes in an amazing power of parental foresight, one has to admit that had any other choice from the list been made, it would, in the long run, have seemed equally predestined, and in that make-believe world, people would chuckle at the thought of any *other* name (including the actual one, in *this* world) having been chosen. And even so, despite all these careful conscious reflections about the arbitrary nature of a first name, one still can't totally squelch a little inner voice from saying, "Emmas look like that girl down the block."

There is a reason for all this: people whom we know well have become stable categories in our minds, and we cannot imagine changing their name any more than we can imagine changing the names that our language gives us for the mundane phenomena that surround us. How absurd it would be if tables were called "chairs", chairs were called "boats", boats were called "cars", and cars were called "tables". How absurd it would be if the labels of salt and pepper, dog and cat, night and day were all swapped, and if laughing were called "flying", flying were called "eating", eating were called "dying", and dying were called "laughing".

We even manage to deal quite handily with what might at first seem a paradoxical situation — namely, the fact that we know several extremely different people, all of whom share the first name "Ian". We like some of them, dislike others of them, and yet it seems to fit them all perfectly. This is very similar to what happens with certain common words such as the noun "story", which designates not only the recounting of an event but also each specific level of a building, or the verb "to file", which means both "to classify" and "to grind down".

Thus a name can act much the same way as other words that designate two or more very different categories. The several Ians we know can coexist in our memory without resembling each other any more than a half-kilogram weight resembles the local animal shelter (even if both are *pounds* of different sorts). This doesn't mean, however, that we lack an overarching category called "Ian", to which are attached all sorts of general traits, such as the most likely age and the most likely cultural background, etc. The existence of such general categories makes us susceptible to thinking that categories based on first names are somehow "universal" or "objective". But this overarching generic *Ian* category coexists with more restricted *Ian* categories that are centered on particular individuals, and these individuals can differ greatly among one another and also from the stereotype of the overarching category, each of them having a first name that seems just right, since after all it is the lexical label that is welded to this specific person.

To sum up the last few sections, we can say that the type of abstraction involved in categorizations resulting in proper nouns leads us to seeing popes, Einsteins, and Jessicas in people who are neither the Pope, nor Einstein, nor Jessica. The same process can also play an essential role in the comprehension of metaphorical statements.

Do Metaphors Necessarily Lie?

If someone says "Andrew is an ass", "Fred is a fox", "Molly is a monkey", "Sid is a snail", or "Sue is a snake", by what right are they assigning these animal traits to humans? How do we know that Sue *isn't* really a slithering reptile? How do we avoid thinking of an actual animal when we hear certain animal characteristics attributed to someone? Psycholinguists Sam Glucksberg and Boaz Keysar have studied the way in which we understand such sentences, which are usually said to be metaphorical. Other related examples would be "Patsy is a pig" or "My job is a prison".

Glucksberg and Keysar suggest that we understand sentences like this by constructing, in real time, "ad-hoc" categories (a notion introduced in the preceding chapter). For example, to understand an utterance like "My job is a prison", we would create on the fly a more abstract category that combines several qualities that constitute the "essence" of the original concept (*prison*), and that shares its lexical label. In this case, thus, the new on-the-fly category is named "prison", sharing the name of the buildings in which convicts serve out their sentences, and the essential property connected with this new category could be verbalized as "an unpleasant place in which one is confined against one's will". A standard prison — a penal institution — is (by construction) a prototypical member of this new category, and the speaker's job is stipulated to be another of its members. The listener thus realizes that the job has the quality of being *an unpleasant place in which the speaker is unwillingly confined.*

This roughly illustrates Glucksberg and Keysar's theory of how we understand sentences of this type — namely, by constructing new ad-hoc abstract categories. Their theory would also explain why we don't drown in a sea of ridiculously literal and confusing interpretations, such as the belief that at the speaker's workplace there are armed guards in uniform, cells with steel bars at the windows, designated exercise hours, and a private little room where you can speak with visiting family members or lawyers. The theory would also explain why we hear "My surgeon is a butcher" and "My butcher is a surgeon" in very different fashions — namely, the abstract ad-hoc categories that we create in these cases are very different from each other. For the first sentence, the new category derives from a fresh and context-dependent essence of *butcher*, while for the second sentence it derives from an on-the-fly essence of *surgeon.*

As for "Patsy is a pig", Glucksberg and Keysar's theory would also explain why no one imagines the poor woman as having a curly tail, wallowing in mud, and oinking — namely, because Patsy is seen as a member of a higher-level category that has been created on the fly by the listener, and which, despite sharing the label "pig", has been stripped of those superficial features of the barnyard animal. We certainly agree that categorization is central here, but we don't see the category to which Patsy has been assigned as fresh, ad-hoc, or creative; we have all heard people called "pigs" hundreds of times. What matters about this category is not its *newness* but its level of *abstractness.*

If we possess an abstract category that shares the same name as a concrete category (*i.e.,* if we are in a situation of marking), then we will refrain from transferring *all* the properties of the more concrete category to instances of the more abstract category.

For instance, we don't need to think hard to realize that it would be folly to imagine *the Mozart of mushrooms* as a wunderkind fungus that composes Viennese music, or *the mecca of wind-surfing* as a sacrosanct spot to which Muslims clad in swimsuits traditionally make pilgrimages atop sailboards, or *the bible of Thai cooking* as central to masses held at ethnic eateries baptized "First Siamese House of Christ", or *the Rolls-Royce of dishwashers* as an ultra-quiet British kitchen appliance parked by valets in fancy garages. We know this because the categories *Mozart*, *Mecca*, *Bible*, and *Rolls-Royce*, having long ago been endowed with a higher level of abstraction by the process of marking, have two types of instances — those that belong to the specific category and those that belong to the more general one. And so we see that the requirement that the category should be *ad-hoc* — that is, freshly minted in real time — is not the crux of the matter. What matters more is that it should possess (at least) two levels of abstraction.

Understanding utterances like "Patsy is a pig" or "Patsy is a powerhouse" is not at all like understanding "Patsy is a pamphlet", "Patsy is a prawn", or "Patsy is a palace". In the latter case, for the sentence to make sense, the category *palace* will have to be creatively extended in real time, without advance warning. Depending on the context, the essential property of the category *palace* might be that it is a rich storehouse of art, or that it is a glittering but nearly unapproachable entity, and so on. If Patsy is seen as a member of this last abstract category, then she will be imagined as being a glamorous or glittery person who exudes an air of aloofness. However, the possible interpretations for "Patsy is a palace" are extremely diverse, and there's no guarantee that we will correctly divine the speaker's intention, nor that all different listeners will arrive at the same interpretation of the remark. By contrast, saying about someone that they are a pig is not a clever new invention, but merely a re-use of a stock phrase that has been part of our language for centuries. In other words, the unmarked sense of "pig" — the word's more abstract sense — has been around for a very long time, and thus using it is neither creative nor ad-hoc.

An ad-hoc category will be created only when the original category — for example, *palace* — doesn't yet have a standard abstract category based on it. The abstract categories involved in the comprehension of conventional metaphors such as "Patsy is a pig" do not have to be constructed on the fly. Understanding such a sentence can be seen as a perfectly standard, run-of-the-mill act of categorization — an act that places Patsy in the abstract category pig_2 as opposed to the concrete category pig_1.

Oddly enough, though, the fact that there exists a concrete category pig_1 makes it seem, at least to some, as if the statement "Patsy is a pig" is not only failing to tell the truth, but even that a surrealistic image is being suggested. This is why such a sentence has traditionally been labeled "metaphorical", as if to suggest that we have no prior abstract category associated with the word "pig" and that we freshly and creatively interpret such sentences each time we hear them. But the fact is simply that the more abstract sense of the word "pig" — namely, "dirty and sloppy" — is an old, familiar category that has the same lexical label as does the curly-tailed animal. Thus, the category *pig* provides a canonical case study of marking. Some people might claim that the sentence "Patsy is a pig" is, in some sense, "less true" than sentences like "Patsy is a partygoer" or "Patsy is a pouter". But if one realizes that for the word "pig" there exists not only a concrete, marked category called "pig" but also an abstract, unmarked category bearing that same name, then a more accurate way of describing the situation is to say that while Patsy is a false member of pig_1, she is a true member of pig_2.

The members of higher-level, unmarked categories are not in any sense "less true" members of their categories than are the members of the lower-level, marked categories from which they have been derived. They are merely members of *other* categories, in other contexts. Thus a tea can perfectly well be a *coffee* at the close of a meal when the server asks, "Who'll have coffee?" Likewise, trucks are perfectly genuine members of the category *car* when little Megan is watching out for *cars* as she walks to school. Boys are genuine *men* when they walk into a men's room. Kittens are genuine *cats*, people are true *animals*, and so on. And thus Patsy is a *genuine* pig — every bit as much a pig as the quadrupeds that supply bacon — but she is simply a pig on a different level of abstraction — in an unmarked, higher-level sense — of the term.

Beyond the marked category *pig* (*e.g.*, that of curly-tailed oinkers) and the unmarked one (*e.g.*, that of dirty and sloppy creatures), it's possible that other subcategories of the abstraction category *pig* could exist. And indeed, if one looks in a dictionary, one will find a subentry for a sense of the word that applies solely to humans. A similar situation holds for many other terms, where an abstract, unmarked category bears the same name as some of its subcategories. And so, when a term is very frequent in a language, its "metaphorical" meanings soon take on a life of their own, becoming autonomous new meanings. Understanding such terms is then much like understanding the word "animal". In some contexts, "animal" will designate an unmarked category that *includes* humans; in other contexts, it will designate a marked category and will *exclude* humans. Neither of these usages, however, can or should be called "metaphorical". Likewise, when we say "Patsy is a pig", it is perfectly reasonable to think (and dictionaries will generally support this) that in addition to the abstract category pig_2 (of very messy, sloppy, animate beings) and the base category pig_1 contained inside it, there is also a different subcategory consisting of *people who are messy.* In fact, as a parallel to what we did for the family of categories sharing the label "desk", it would be more sensible to label the most abstract category "pig_3" rather than "pig_2", and instead to reserve the label "pig_2" for human beings who are messy and sloppy, with the label "pig_1" applying to the omnivorous farm animal.

A Quick Panorama of Metaphorical Usages

In summary, then, we can identify three different types of understanding of "metaphorical" sentences.

The first type involves situations where the metaphorical expression is completely standardized, and in order to understand the sentence one does not need to invoke an abstract category (that is, an unmarked term). The unmarked, abstract category can certainly exist, but it is simply not needed for the understanding process, and hence is bypassed. This can be illustrated in the case of fluent English speakers who hear the sentence "Jack's house is a dump", dealing with it in much the same way as they would deal with "Patsy is a pig". We assume that the abstract category *dump* — the unmarked sense — existed in their minds long before the sentence was heard, as did the more limited subcategory of *dump* that applies just to messy dwellings (hotels, houses, rooms, etc.). The understanding of "Jack's house is a dump" is thus straightforward, simply taking advantage of this concrete sense of the word "dump", which applies specifically to messy dwellings. Moreover, much as the farm-dwelling type of pig and the messy human type of pig share the property of being sloppy and making loud noises, garbage dumps and messy dwellings share the trait of being unpleasant to see and to be in. This is because there is a more abstract category — the *unmarked* sense of "dump" — that includes them both and binds them together. In the mind of a fluent English speaker who hears "Jack's house is a dump", this abstract category is present but only as a kind of silent backdrop, and it is not needed in order to understand the sentence, because the more specific subcategory was long ago constructed and this guarantees that the sentence will be easily understood. In other words, much as with *desk* and *pig*, we are here dealing with three distinct categories: $dump_3$, an abstract category that includes dirty, unpleasant places of all sorts, and two subcategories that are more specific — $dump_1$ for garbage dumps, and $dump_2$ for messy dwellings.

In a closely related type of situation (but still in the realm of the first kind of metaphorical understanding), the abstract category may be lost in the fog of etymology and may no longer exist in the collective mind of native speakers. In such cases, the abstract category belongs to the history of concepts but not to their psychology. These cases include familiar expressions that are perfectly understood by native speakers but whose origin is mysterious and for which one feels at a loss to connect the two different meanings of the same term (such as the two senses of the verb "to bore", which, although they probably sprang from a common origin, are not heard by English speakers as being in any way related). For instance, consider the sentences "Walter is a wolf", "Tim is a turkey", "Denise is a dodo", and "Belle is a bitch". In each case, one can get involved in endless guessing games as to why this particular kind of animal is the source of this particular metaphor in English, because there is no obvious superordinate category in one's memory that includes the two in a natural way. But the key to understanding such sentences is that there already exists in the listener's mind a non-animal category that shares the same lexical label as the animal category.

In the second type of metaphorical understanding, there is likewise no need to construct a new abstract category on the spot because it is already in one's mental lexicon, but that category has to be made use of in order for the sentence to be understood. Thus, suppose someone were to exclaim, "This damn word processor is a pig!" A suitable scenario to evoke such an outcry would be if one's word processor frequently made wrong hyphenations, ugly interword spacings, and random insertions of characters. In this case, one would have to utilize the most general unmarked category of pig_3 in order to understand the utterance. This would be a completely natural act of categorization, not requiring the construction of a new category, because one can simply take advantage of the abstract category pig_3. To understand "This damn word processor is a pig", the unmarked sense of the word "pig" is activated, just as unmarked abstract senses of words are activated in all sorts of everyday situations — as, for instance, when somebody refers to Liz's pickup truck as "Liz's car", or refers to an atheist friend as "the pope of popsicles", or announces, "I need to buy some kleenex", even if the Kleenex brand isn't sold in the store.

We come now to the third type of metaphorical understanding. Here the abstract category does not exist *a priori*, so the listener is forced to construct it in order to understand the statement. This would be the case for utterances such as "Bill is a bridge", "Steve is a stone", "Florence is a firefly", "Patsy is a prawn", and so forth. In such cases, one has to distill on the fly a new essence from an already-existing category in order to see Bill, Steve, Florence, and Patsy as members of new ad-hoc categories. The challenge is similar to that for a French speaker who is told by an English speaker, "My new car is a lemon." Since this metaphor doesn't exist in French, the French speaker would have to concoct a new ad-hoc category in order to understand the statement, and there is no guarantee of success, even if to English speakers it's obvious that the key property of lemons in this case is their sourness, which theoretically would allow anyone to make a new abstract category that includes both citrus fruits and cars.

And how would we English speakers understand a French person who told us, "The movie last night was a turnip"? French speakers understand immediately that the movie was mediocre, because "être un navet" ("to be a turnip") is a stock phrase applying to films, and so to understand the sentence they simply exploit this familiar pre-existing category. They do not need to jump to a yet higher level of abstraction, that of *turnip*$_3$; the level of *turnip*$_2$ suffices. By contrast, for people who are not familiar with this expression, it will be necessary to concoct a new abstract ad-hoc category based on some new-found essence of the concept *turnip*, and of which certain films will be members. Thus some people might imagine that what's crucial for this kind of abstract *turnip*-ness is *being purple*, or *growing underground*, or *being used in salads*. Then again, to someone more in tune with the speaker's tone, it might occur that *insipidity* or *blandness* is the key quality.

In all of these cases, whether the category is abstract or concrete, pre-existing or invented on the fly, the understanding of metaphorical statements depends on applying a category to a situation. Our take-home lesson is thus:

That pizzeria is not a greasy spoon, and yet... that pizzeria is a greasy spoon.
A human being is not an animal, and yet... a human being is an animal.
Richard Dawkins is not a pope, and yet... Richard Dawkins is a pope.
Karen's work is not a prison, and yet... Karen's work is a prison.
Joan of Arc wasn't a man, and yet... Joan of Arc was a man.
A mint tea isn't a coffee, and yet... a mint tea is a coffee.
Your car isn't a lemon, and yet... your car is a lemon.
My truck isn't a car, and yet... my truck is a car.
Patsy is not a pig, and yet... Patsy is a pig.
This book doesn't weigh a ton, and yet... this book weighs a ton.

Mathematics is Not Always Cut and Dry

If there's any domain that people think of as having precise and unambiguous concepts, it would have to be mathematics. Here, where contradictions and blurriness should play absolutely no role, one would naturally suppose that the subjective, context-dependent phenomenon of marking, which by definition conflates two categories by assigning them the same label, would surely be nonexistent. And yet this is not the case. Even in mathematics, our human style of fluently jumping between categories, relying on context to make things clear, trumps the desire for pure logicality, as we'll now see.

Tom is in seventh grade. He's just finished a geometry class in which his teacher gave him some homework problems on the topic of quadrilaterals. The first exercise said, "Write 'S' in each square, 'R' in each rectangle, 'Rh' in each rhombus, and 'P' in each parallelogram." Tom diligently carried out his assignment, and the figure below shows what he did.

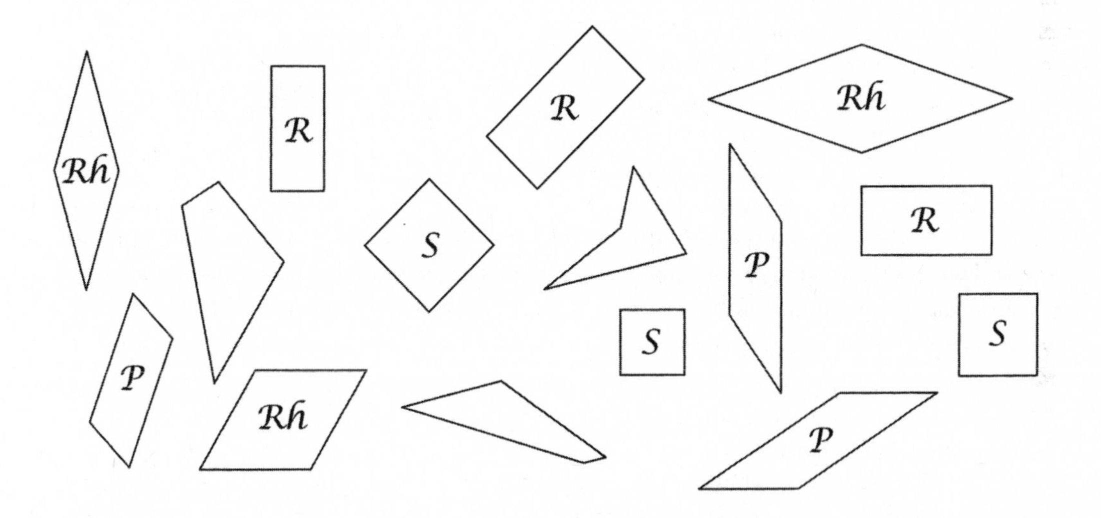

Tom was very careful not to fall for any trick questions. For example, he wasn't fooled into thinking that the square balanced on one of its corners was a rhombus. And he also correctly identified rectangles that were tilted, and even rhombi that were tilted at strange angles.

The assignment that came next was called "definitions", and it featured the following questions:

1. How do you recognize a square?
2. How do you recognize a rectangle?
3. How do you recognize a rhombus?
4. How do you recognize a parallelogram?

Tom had learned all of this material very well, and he replied as follows:

1. A square has four right angles and four equal sides, parallel in pairs.
2. A rectangle has four right angles and four sides, parallel in pairs.
3. A rhombus has four equal sides, parallel in pairs.
4. A parallelogram has four sides, parallel in pairs.

Tom's teacher graded the homework and when Tom got home, he proudly told his parents that he'd gotten a perfect grade on each of the exercises. Is there anything worrisome in this situation? Well, yes, something is wrong; in fact, there are interesting contradictions here. Indeed, what would have happened if Tom's acts of writing letters inside shapes had been guided by the written answers that he gave?

> *A square has four right angles and four equal sides, parallel in pairs.* So far so good. This works for all the squares in which Tom wrote 'S', and no other figure is described by the phrase.
>
> *A rectangle has four right angles and four sides, parallel in pairs.* Here things are a little trickier. Tom looked for figures having four right angles and sides parallel in pairs. And yes, the rectangles in which he wrote 'R' all satisfy this criterion, even the tilted ones that almost fooled him — but the squares with 'S' in them are *also* described by this phrase. So why didn't he write both 'R' and 'S' in all the squares?
>
> *A rhombus has four equal sides, parallel in pairs.* This category also is tricky. That is, all of Tom's rhombi with 'Rh' in them are indeed described by this definition, but the squares, once again, are also described by it. So why didn't Tom write 'Rh' in each square (as well as 'R' and 'S')?
>
> *A parallelogram has four sides, parallel in pairs.* The plot thickens... Of course, every figure in which Tom wrote 'P' satisfies this criterion, but nearly all of the other figures do too: all the rhombi satisfy it, as do all the rectangles and squares. And so, if Tom's placement of letters had been consistent with his written answers, he would have had to put a 'P' in twelve of the figures, indicating that all but three were parallelograms.

Why, then, did Tom's teacher give him a perfect grade, when in fact his answers to her two assignments have just been shown to be inconsistent?

Are Squares Rectangles?

Indeed: are squares rectangles? The answer comes down to a classic case of marking. From a mathematician's point of view, a square is certainly a rectangle, because it satisfies the criteria that define rectangles. In that sense, the question is unambiguous, and the answer is simply "yes".

If one looks at the definitions of the various types of quadrilaterals, one can easily draw a diagram showing their relationships:

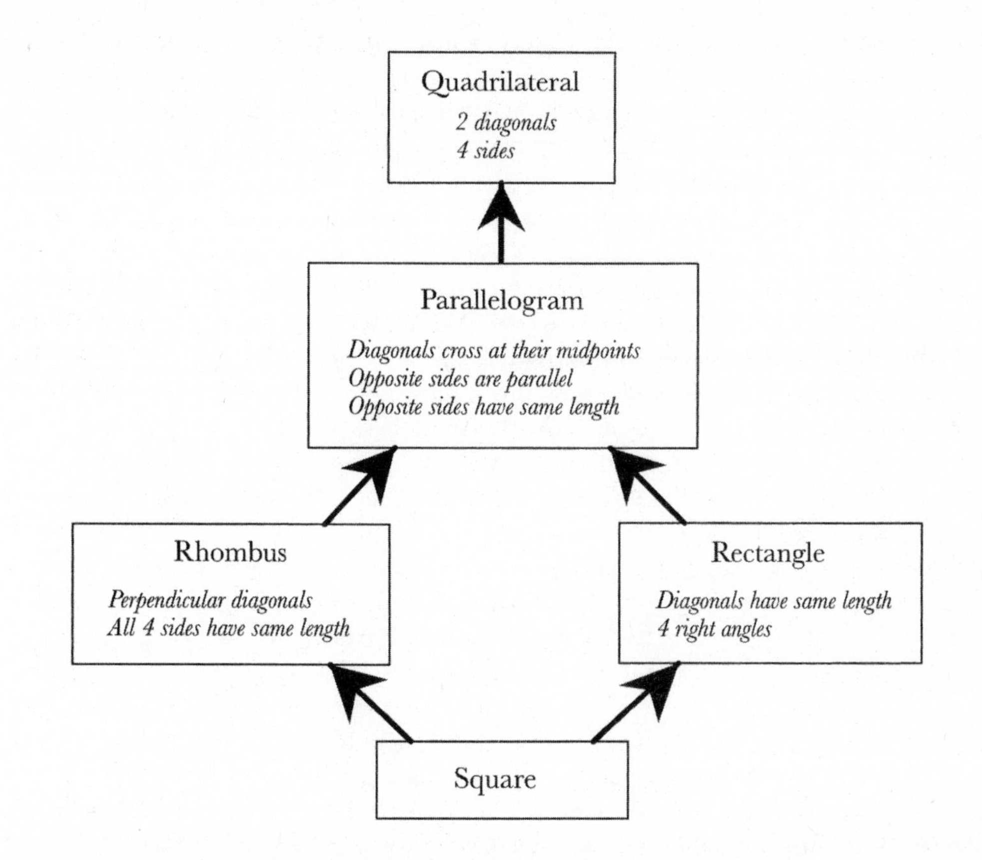

The different categories shown in the figure illustrate the different points of view that one can adopt for any of the quadrilaterals drawn. According to the diagram above, a square belongs to five different categories: *square*, *rectangle*, *rhombus*, *parallelogram*, and *quadrilateral*. If it's seen as a quadrilateral, a square has two diagonals, but there's nothing special to say about them. If it's seen as a parallelogram, a square's two diagonals slice through each other exactly in their midpoints. If it's seen as a rhombus, its diagonals are perpendicular; and if it's seen as a rectangle, they have the same length as each other.

And yet this chart of categories in the world of quadrilaterals isn't sophisticated enough for us to anticipate how Tom's geometry teacher would react if one of her pupils were to draw nothing but a single square as the answer to the following exercise: "Draw a square, a rectangle, a rhombus, and a parallelogram." Strictly speaking, she should be very respectful of the precise understanding of the definitions implicit in such an answer, and she should give the young showoff the highest possible grade. So let's suppose that she was a good sport, and that she did so. Even so, she would have to be taken by surprise by this playful answer, and depending on her personality, she might be either charmed or exasperated by the highly unusual interpretation of her question. But why would she be so surprised to see such an answer, given that it's perfectly correct and elegantly economical, to boot?

Our chart of quadrilateral types, given above, doesn't allow us to predict how Tom's teacher would react if Tom, on his first assignment, were to write four different labels in each square (indicating that it is simultaneously a square, a rectangle, a rhombus, and a parallelogram), and if he were to write two labels in each non-square rectangle (indicating that it is both a rectangle and a parallelogram), and if he were also to write two labels in each non-square rhombus (indicating it is both a rhombus and a parallelogram). Such labels would come directly from a careful application by Tom of his own definitions (in his second answer) — definitions that were so warmly applauded by his teacher and that were given top grades. Would she have given him full credit for answering exercise #1 in such an unorthodox fashion, when we know that in fact she gave him full credit for a totally different, much more conventional set of answers? One would hope that she would be delighted to find in her class a student who is so insightful, and that such thoughtful answers would be rewarded with full credit, and without reservation — but that would depend on whether she appreciated or disdained unconventional viewpoints.

Indeed, categories similar to *non-square rectangle* — for example, *non-square rhombus* and *parallelogram that is neither a rectangle nor a diamond* (and thus not a square either) — are needed to bridge the gap between rigorous mathematical definitions and informal human concepts. To be more precise, we tend unconsciously to presume "non-square" when we think "rhombus" — and yet square rhombi exist, of course. However, they are such a special case that we have to distinguish them from "normal" rhombi, which are not square. Once again, we find ourselves square in the territory of marking. For example, we could posit two new categories, $rhombus_1$ and $rhombus_2$, analogous to car_1 and car_2. Just as the marked category car_1 excludes trucks while the unmarked category car_2 includes them, $rhombus_1$ would exclude squares while $rhombus_2$ would include them.

Every square is thus a member of $rectangle_2$ but not necessarily of $rectangle_1$, and every rectangle (whether of type 1 or 2) is a member of $parallelogram_2$ but not necessarily of $parallelogram_1$. And so we see that the various types of quadrilaterals fall into a more complex diagram than the one we saw earlier. It looks as follows:

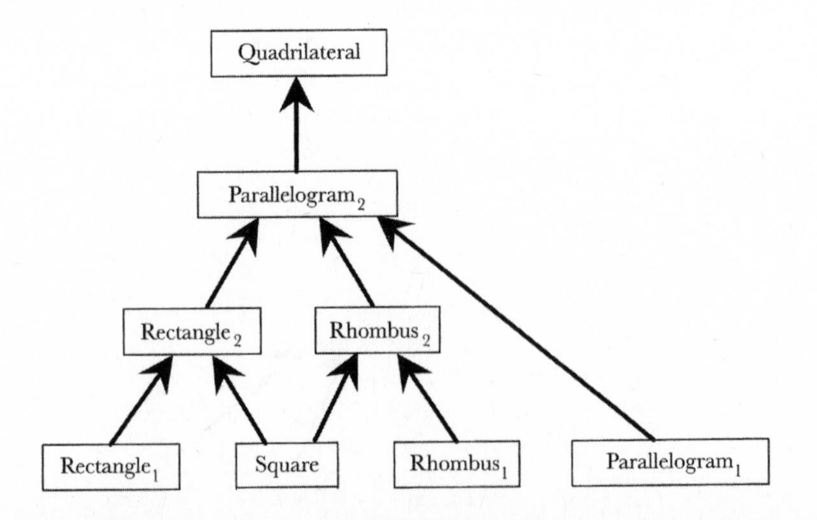

There are now eight categories, organized on four levels of abstraction; all eight are needed to capture what we might call "expert knowledge", which comes from two sources: first, an understanding of rigorous and formal mathematical definitions, and second, a sense for how these words are actually used in various contexts by someone who has fully mastered this small but slippery mini-domain of math. It's not hard to see why this can give schoolchildren headaches.

We aren't saying that this kind of hierarchical diagram is an explicit structure (like the sequence of letters of the alphabet) that gets consciously committed to memory by all people who understand quadrilaterals well. To the contrary, no memorization is involved at all. If one has a mathematical bent, each link in the diagram comes directly from a clear understanding of how two particular concepts are related, and the consequences flow out almost trivially. And yet ironically, this kind of crystal-clear understanding rests tacitly on an ambiguous way of using words: sometimes a square is seen *in contrast to* a rectangle, while other times it is seen as *a kind of* rectangle.

All this goes to show that even in mathematics, where utter precision is expected, it's commonplace to have terms that stand for two or more concepts, hence are ambiguous. In this particular case, to be an expert in the domain, one needs to have various categories such as *non-square rectangles* (that is to say, *rectangle*$_1$), and *rectangles whether square or not* (that is, *rectangle*$_2$). Altogether, the number of such terms comes to eight, which exceeds the number of lexical items by three. And it is by no means easy for people to acquire the proper organization of the categories associated with the words "square", "rectangle", "rhombus", "parallelogram", and "quadrilateral".

Indeed, a study we undertook in France showed that the majority of university and middle-school students are unaware of how some of these categories include others, and many simply refuse to accept the idea that squares are rectangles or rhombi. They may go so far as to invent new properties, if one asks them for definitions. For example, they may insist, "A rectangle has to be wider than it is high" or "A rectangle has two pairs of equal-length sides, but not all four sides have the same length", or then again, "A rhombus has four equal-length sides, but it has no right angles".

This reveals that the most general interpretation of these words — that is, the unmarked sense — is usually not recognized, and that students' improvised definitions mostly describe *marked* categories (such as *non-square rectangle*, *non-square rhombus*, etc.). The most frequently observed idea in the minds of these students, all of whom had supposedly mastered all of these notions (indeed, these matters were very fresh in the minds of the middle-school participants, and were assumed to be part of the background knowledge of the undergraduate participants), is just a two-tier hierarchy with *quadrilateral* on top, and *parallelogram*, *rectangle*, *rhombus*, and *square* below.

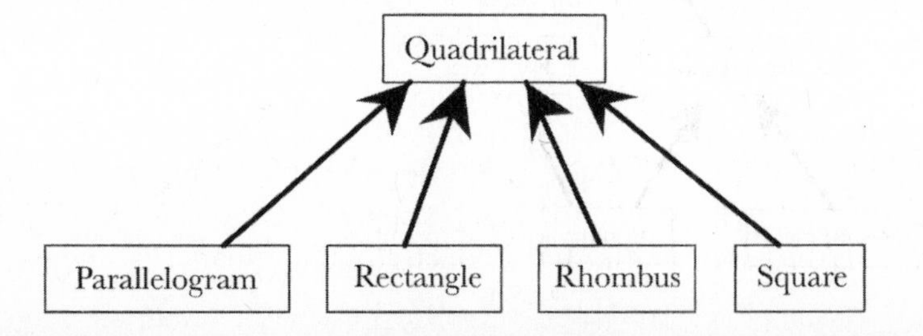

Students who gave slightly more sophisticated answers tended to add just one middle level — *parallelogram* — which they interpolated between the top level (where *quadrilateral* was found) and the bottom (on which resided *rectangle*, *rhombus*, and *square*):

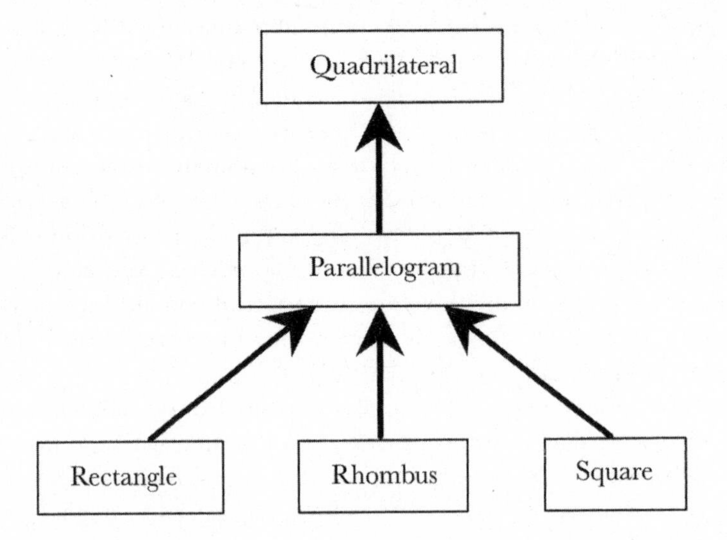

Our investigations showed that an expert-like hierarchy having *four* levels (as in the figure two pages back) was present in only a very small percentage of the participants.

As we have just seen, part and parcel of becoming an expert in a domain involves, over and above possessing many categories, organizing them efficiently. Thus, to understand quadrilaterals well, one has to know a good deal more than just the fact that squares, rectangles, rhombi, and parallelograms are all special types of quadrilaterals; one also has to know the relationships that interconnect them. This theme — that an efficient organization of categories is part and parcel of expertise — is important, because expertise is far from being limited to narrow, technical domains. Expertise is what anyone acquires who has deep knowledge of *any* domain; everyone is an expert in their everyday environment, as well as in their profession and in their various hobbies. To be sure, expertise does not always require high degrees of creativity or insight, although of course it doesn't exclude those.

The Verticality of Expertise

You might aspire to become a past master in dog breeding, or perhaps in the etymology of English words, or in the varieties of tea, or in the world of banking, or in Spanish literature, or opera or history or anatomy... If you succeed, some will call you a "walking encyclopedia" of your specialty. Mr. Martin takes great pleasure in spending afternoons reading about dogs. Aside from learning the names of many breeds, including "poodle", "bulldog", "German shepherd", "basset", "chihuahua", "golden retriever", "grayhound", "pit bull", "borzoi", "Saint Bernard", "fox terrier", "Dalmatian", "doberman", "labrador", "dachshund", "spaniel", and so on, he has learned to recognize them all in photos. He has also learned various little factoids about each breed, such as that golden retrievers have a very gentle character, that German shepherds have almond-shaped eyes and ears that stand up, that greyhounds are very attached to their masters, that bulldogs may have originated in Tibet, that Dalmatians are the mascots of firehouses, that chihuahuas are the smallest breed, that fox terriers owe their existence to fox-hunting, that Saint Bernards are named after a hospice in the Swiss Alps, that labradors are excellent guide dogs, that poodles are probably descended from water spaniels, and so forth.

This is typical of anyone who is tackling a new domain. One familiarizes oneself with the most salient entities belonging to the domain and one learns to tell them apart, since one surely can't claim know a domain well with recognizing its principal denizens. Thus categorization involves being able to make *distinctions.* But since the categories constituting any domain are also linked by many types of relationships, acquiring a new category means connecting it mentally with one or more prior categories; thus categorization also involves making *associations.*

The distinction between *genus* and *species*, borrowed from biology, is another way of describing this process. Any new entity belongs to a certain *species* (a narrow category), which in turn belongs to a certain *genus* (a wider category). Thus, the first time you run into a certain animal you might simply call it a *dog*, and later you might learn that it is a *dachshund.* Possessing or acquiring such knowledge gives one the sense of having reached a decent level of mastery, even if one is not yet an expert. The simple act of categorizing an entity as a *dog* allows one to tap into prior stocks of knowledge one has, such as *barks*; *has a stomach*; *may bite*; *may have and transmit rabies*; *is prone to drooling*; *has a life expectancy of between ten and fifteen years*; *may be dangerous to children and even adults*; and so on. To these general pieces of knowledge one can add specific facts about golden retrievers, such as *has golden fur*; *has a gentle disposition*; *is rather large*; *weighs between 60 and 90 pounds*; *is playful*; and so on. Just as Mr. Martin proceeds in this manner as he learns about dogs, so do we all as we learn about any domain.

Is this what becoming knowledgeable in a specific area is all about? Does the acquisition of knowledge consist in learning about more and more species that belong to a given domain? Is a beginning enthusiast someone who has acquired ten or fifteen categories, all subsumed under one umbrella category, and who knows a little bit about each particular one? Does an expert differ from this in having hundreds of examples all of which are richly mentally annotated?

Yes, in part; but expertise goes considerably beyond that. Aside from certain very technical and esoteric domains, being an expert does not mean having memorized a list, even if the list is enriched with specific tidbits about each of its items. If Mr. Martin wished to become an expert on his town's telephone directory, then memorization would suffice; he would be an expert simply by virtue of having learned many names, addresses, and phone numbers by heart. But in more conceptual domains, expertise doesn't come from amassing knowledge of ever more species belonging to a single genus; to be sure, experts will generally have more species under their belt than will novices, but that is by no means the full story.

Changing Category to Change Viewpoint

Thus Mr. Martin, in order to slake his thirst for knowledge about dogs, will construct new categories at various levels of abstraction, and will create links among them. His way of understanding dogs will go beyond just knowing many breeds; it will include different ways of categorizing them, thus allowing him to glide back and forth from one viewpoint to another.

Thus, for instance, there is a standard distinction among four main subcategories of *dog*: the *molossoids*, which have massive heads, short muzzles, small floppy ears, and heavy bones (*e.g.,* bulldogs); the *lupoids*, which have triangular heads and pricked-up ears (*e.g.,* German shepherds); the *braccoids*, which have wide muzzles and floppy ears (*e.g.,* Dalmatians); and the *graioids*, which have wide fine heads, thin limbs, and small ears facing backwards (*e.g.,* greyhounds). Yet further refinements are possible, including such subcategories as *pointers*, *hunting dogs*, *retrievers*, *water dogs*, *sheep dogs*, *terriers*, *dachshunds*, *sled dogs*, and so on. The process of refinement can also take place within a single breed; thus the *smooth-haired Catalan sheepdog* and the *Czechoslovakian wolfdog* are two types of sheepdog among perhaps fifty, and in certain elite circles, it would be a terrible *faux pas* not to know the difference between Afghan, Scottish, and Hungarian greyhounds.

The gradual enrichment of one's personal repertoire of categories will necessarily involve the creation of a number of different levels of abstraction, since categories are always interrelated in many ways. Thus a *Bohemian wirehaired* is a kind of *griffon*; a *griffon* is a kind of *pointer*; a *pointer* is a kind of *hunting dog*; a *hunting dog* is a kind of *braccoid*; a *braccoid* is a kind of *dog*.

One can of course pursue links very far upwards, leading towards abstraction heaven. How many levels, for instance, would you suppose there are between *dog* and *animal*? Two or three, perhaps, such as *carnivores*, *mammals*, and *vertebrates*? It's true that these are three of the rungs on the ladder, but they are by no means the only ones. For a Latin-oriented zoologist, a dog is a *canis lupus familiaris*. This is part of the broader category of *canis lupus* (which means $wolf_2$ — note that a dog is a $wolf_2$ but not a $wolf_1$ — we're back to marking once again), of which the *gray wolf* is a different example. Then a *canis lupus* is part of the family of *canidæ*, like its cousins the fox and the jackal (at this level, claws are not retractable). And the *canidæ* belong, in turn, to the *caniformia*, a level of classification that they share with (among others) otters, walruses, bears, skunks, and raccoons. Above that there are the *fissipedia*, and at this level dog meets cat, or more technically put, *caniformia* and *feliformia* find themselves united. These are in turn subsumed by the great order of *carnivora*. From there we gradually ascend to the *placentalia* (where we humans join dogs, cats, and skunks, our common link being that our embryos grow inside the mother's body and are fed by a placenta), then the *theria* (which reach out to include kangaroos), then the *mammalia* (a familiar landmark in a strange landscape), then the *gnathostomata*, the *vertebrata* (having a skeleton and a spinal column), the *chordata* (having the nervous system above the alimentary canal), the *deuterostomes* (characterized by certain processes that take place during embryogenesis), the *eumetazoa* (many-celled animals), and finally, at long last, one level up, we hit the *animalia* (which is to say, *animals*!). Although some of these levels are debatable today (taxonomy being a field in constant flux and filled with controversy), let's say that there are at least a dozen vertical steps, if not more, between *dog* and *animal.*

This ascent in the world of biological categories, so sheer that it might make one dizzy, wasn't meant to imply that expertise always involves so many different levels of abstraction. Usually abstraction hierarchies are far more modest than this. However, the fact that one can find a case where so many levels are stacked one on top of the other shows that abstraction hierarchies are not just a far-fetched fantasy; to the contrary, whether they have many or few levels, hierarchies of this sort are inevitable ingredients in the human process of acquiring knowledge.

Consider the world of typefaces, for instance. There are book faces and advertising faces, there are sans-serif faces and faces with serifs, there are "old style", "transitional", and "modern" faces, and many other ways of classifying typefaces exist. For any standard typeface, there are roman and italic varieties, and each one comes in many different point sizes and weights, and specialists can add shadings of various types.

In most domains, familiarity with even three or four levels can be a quite telling sign of knowledge. Novices often remain stuck at only two levels — a genus and some species within it — and the construction of just one more level may be a crucial conceptual leap ahead. As we will see later in this chapter, and even more clearly in the final chapter, the leap upwards to a new general category — a vertical slippage, so to speak — can open up important perspectives, whether in the simplest activities of everyday life or in the most exalted of scientific discoveries.

It would be wrong to suppose that one particular way of cutting up a domain into categories is "the right one". Any nontrivial domain is going to admit of rival category systems that allow it to be sliced along different sets of axes. Take the world of dogs, for instance. Many other ways of categorizing dogs exist than those that are considered "official", and each different breakdown comes from a particular way of dealing with dogs. Hunters have one, veterinarians have another, dog-show organizers yet another, not to mention pet-store sales clerks or directors of zoos — and each of these people will have built up an expert-style organization of categories in their head, made by creating new categories and by inserting new levels where useful. It's almost certain that they will have different systems of categories and will organize them differently, and none of the systems will coincide with the official zoological set of categories.

To take just one more among innumerable other domains, for an aficionado of rock music, *classical music* might well be just one single tiny little category, blurring together a few dimly familiar but esoteric names such as Beethoven and Mozart, maybe even Bach, but *classical music* will be nearly invisible in the rock-music fan's richly interlinked network of subcategories of *music.* Of course the reverse could easily hold for a lover of classical music, who has a giant and complex network of subcategories of *classical music* (including rich subnetworks for the works of many specific composers), but who, in symmetric fashion, casually lumps a huge variety of types of modern popular music under the single bland and monolithic-sounding label "rock music", thus betraying an ignorance that would mortally offend the rock-music lover.

In sum, we all build up our knowledge by constructing categories, linking them together, and structuring them by abstraction. In general, neither novices nor experts do this consciously, but if one examines any domain at all carefully, one finds that it is filled to the brim with categories and interconnections among them, and that such links form such a complex pattern that it would astonish an outsider, who would have been tempted to see the domain in the simplest possible way: as just one main overarching genus with a few species one level below it.

Nice Work if You Can Get it!

Any domain, no matter how limited it might seem from afar, can be refined forever not only horizontally (the number of categories) but also vertically (the levels of abstraction). The quest to mirror reality perfectly would require an infinitely fine-grained set of categories, but each of us is capable of only so much refinement. How far to go in this quest? The answer is not theoretical but practical, depending on what goal one has set for oneself with respect to the chosen domain.

Take, for example, the domain of professions, seen from the viewpoint of people who specialize in thinking about them. Everyone is familiar with such distinctions as that between *employees* and *free-lance workers*, or between *blue-collar* and *white-collar workers*, or between *management* and *labor*, or between jobs in the *industrial sector* and in the *service sector*. But this set of coarse-grained distinctions is as nothing compared to the variety that exists in the charts of professions created by the aforementioned specialists in types of jobs. For example, in the construction sector, there are *waterproofing and insulation workers* as well as *pipelaying fitters* and *wood-fastener installers*, not to mention *plywood-panel assemblers, disk flange operators*, and *metal-fabrication duplicator-punch operators*. Among textile workers, one finds such jobs as *fur garments patternmaker*.

Hierarchical listings of jobs include such general categories as *manufacturing and processing*, below which one finds, for example, *central control and process operators*, and one level further down there is the category of *petroleum, gas, and chemical process operators,* and in that category a sample job is *pipeline compressor-station operator*. Following down other branches of this same tree whose highest-level abstraction is *manufacturing and processing*, one runs across such professions as *crayon-making machine tender* and *paper-bag-making machine operator*, and even that of *tennis-ball-maker operator*. Last but not least, if one starts at the top level of *logging and forestry workers* and descends to *chainsaw and skidder operators*, one encounters such exotic-sounding professions as *grapple skidder operator*.

In Web sites specializing in job searches there are typically on the order of twenty high-level nodes, such as *fabrication and construction industries*, below which one will find hundreds of more detailed nodes, such as *metal-forming, -shaping, and -erecting occupations*. Then below that level, one may find all sorts of more specialized nodes, such as *structural metal and platework fabricators and fitters*, and below that a seemingly endless list of specific kinds of professions. Such complexity may seem surprising, but this is merely an overview, and within each sub-area there are yet further specializations. No matter where one turns, be it the medical profession, the publishing profession, the fashion industry, or whatever, there are several layers in the classification and hundreds if not thousands of job types. One gets the sense that one could zoom in almost forever.

Let's look at a profession usually listed at the lowest, most specific level of such a hierarchy — namely, that of *university teachers and researchers.* For people who work in that world, however, the phrase "university teachers and researchers" isn't specific at all; it stands instead for a huge high-level rag-bag. Their world of jobs is richly structured, distinguishing among *full professor*, *associate professor*, and *assistant professor*, as well as *instructor* and *lecturer*. It includes *visiting professors*, *visiting researchers*, and *visiting scholars*. And let's not forget *teaching assistants* and *research assistants* (both *graduate* and *undergraduate*) — and so far we haven't even breathed a word about specific disciplines here!

In sum, if one wanted to make a "complete" chart of professions, it would involve a diagram with hundreds of thousands of categories, if not more. To be sure, no one has such a diagram in their head, so making such a diagram is not crucial for somebody whose goal is expertise, but being a specialist in some domain inevitably means that one has internalized some local portion of this complexly structured knowledge.

Buon appetito!

We now shift to a very different domain from that of professions — a most concrete and everyday domain, which at first glance would seem far removed from subtle considerations of categories. We are speaking about food, and in particular about the world of pasta, which offhand seems far simpler than the world of jobs. It might seem unlikely that there could be too much diversity when it comes to products made solely from semolina and water, perhaps eggs. And to be sure, in many countries, "pasta" means little more than "spaghetti" and perhaps "macaroni"; however, if one goes to Italy one will find an impressive variety of products in the pasta section of any grocery store. In fact, there are at least 80 types of pasta — 200, if one allows synonyms — including such little-known ones as *creste di gallo* ("rooster combs"), which are small-sized noodles (if one dares use such a crass-sounding term for such a delicacy) that are especially suited for soups and salads, *strangolapreti* ("priest-stranglers"), which, as their name suggests, are rather heavy, and are often made with spinach, and *torchi* ("torches"), especially suited for thick sauces.

But the world of pasta can be broken down in many ways, not just in the stereotypical fashion featuring many species underneath a single top-level genus. There are varieties that one can easily make at home, and ones that are better left to the professionals. There is fresh pasta and there is dry pasta. And most of all, there are various natural subclasses determined by shape, which profoundly affect the uses of various types of pasta in cooking, and which are, in fact, the primary reason for all the diversity that exists. Thus the smaller noodles, such as *anchellini* ("mini-hips") and *primaverine* ("springtime noodles"), are intended mostly for soups, and then there are types with rather fanciful shapes, such as *cocciolette* ("potlets"), *dischi* ("disks"), and *fusilli* ("little spindles"), intended for use with salads and vegetables. The long ribbons and string-like shapes, such as *linguine* ("tonguelets"), *tagliatelle* ("little slices") and *spaghetti* ("stringlets"), harmonize well with sauces. In particular, *caserecce* ("homemade noodles") and *conchigliette* ("little shells") go best with meat sauces, while others, such as *cappelletti* ("hatlets") and *ravioli*, are intended to be stuffed with tasty fillings. Every cook who delights in using pasta will have a well-developed mental network of pastas and connections linking them, revealing an unconscious structuring of mental categories.

À votre santé !

A good plate of pasta merits a good wine, and in the world of wines, the range of categories seems to be inexhaustible. Rank beginners will make only the *red/white* distinction, while those who are slightly more in the know will refer to the type of vine (*cabernet sauvignon* or *merlot*, for example) and to the source country (France, Italy, Chile, Australia, the United States) or region (Bordeaux, Bourgogne, Napa Valley); however, expertise in wines involves going far beyond these kinds of facts. It can lead to distinguishing a particular vintage of a particular year of a particular grower of a particular *sous-appellation* of a particular *appellation* of a particular country — for example, a Château Beaucastel Hommage à Jacques Perrin 1998 belongs to the *sous-appellation* called *Châteauneuf du Pape* within the *appellation* known as *Côtes du Rhône*, and it comes from the prestigious vineyard *Domaine Beaucastel.*

Does *Château Beaucastel Hommage à Jacques Perrin 1998* constitute a typical category for a true wine connoisseur? Absolutely. It has many characteristics that people will use to describe it, such as its color, its bouquet, its taste, how it ages, its market value, the dishes it goes best with, the ideal temperature it should be served at, places where one can obtain it, ratings it has received from professional tasters, and so forth. The few thousand bottles actually produced of this exquisite elixir constitute the *extension* of this category — its concrete members. Of course, despite the seeming precision of this description, there are all sorts of hidden blurs at the fringes, as there are for any category: a counterfeit bottle, an unlabeled bottle, a half-consumed bottle, a poorly conserved bottle, a bottle that has been boiled or dried out, an empty bottle, a few drops spattered on a tablecloth after a glass was poured, and so forth. And no less than any other proper noun, this category lends itself to pluralization, as is illustrated by the grower of Château Beaucastel who proudly declares, "You should taste the new *Jacques Perrin 1998* that I produced here in my vineyard in 2005."

To give an idea of the richness of categories in this arena, we point out some of the categories that necessarily inhabit the mind of a wine lover who states, "In 1996, Dominique Laurent did a great job with *Gevrey-Chambertin Premier Cru Les Cazetiers*", and which are unlikely to exist in the mind of a more modest wine lover who says, "I prefer French wines to Spanish wines." In the former case, not only are the high-level categories *wine* and *French wine* implicitly referred to, but moreover it is taken for granted that the listener is familiar with several yet narrower categories — namely, *bourgogne, Gevrey-Chambertin, Gevrey-Chambertin Premier Cru* (since most members of the category *Gevrey-Chambertin* do not come from the first harvest), and even *Gevrey-Chambertin Premier Cru Les Cazetiers* (since there are quite a few other types of *Gevrey-*

Chambertin Premier Cru aside from those called "*Les Cazetiers*"). To pin the wine down fully, our hypothetical commentator can be both precise and concise, saying, "*Gevrey-Chambertin Premier Cru Les Cazetiers Dominique Laurent* from 1996", thereby distinguishing it from other growers who have produced related wines and from other years in which close cousins were produced. Such a fine-grained call is a clear sign of high expertise.

Why Abstraction is Central for Expertise

In any normal domain of expertise, even the most knowledgeable of specialists can't come close to knowing everything about the domain. The breadth of knowledge that one would need to assimilate in order to be an expert in every nook and cranny is so vast that setting oneself such a goal makes no sense. And of course the idea of being an expert "in every nook and cranny" is in itself problematical, since when one uses a magnifying glass, every domain shatters into yet further subdomains. And yet we don't expect experts to throw up their hands and say "I don't know" whenever they are asked a tricky question in their domain. This is because true experts have knowledge not just about many specific cases in their domain, but also, through analogical links, a set of expectations about cases that are far less familiar to them. Thus in the domain of wine, we might imagine asking an expert, "What you do you think of the *Château Vinus 1995?*" Our expert might start right up and give all sorts of details learned either first-hand, through tasting, or second-hand, through conversations or reading. In this case the game is very simple, since the category relied on is precisely *Château Vinus 1995.* On the other hand, our expert may know nothing specific about *Château Vinus 1995*, and might therefore reply, "I don't know", which would be a perfectly truthful answer, and yet hardly useful at all, and also hardly reflective of what we think expertise is.

A more reasonable answer from a true expert would be grounded in knowledge of more abstract categories than just the particular wines themselves. For example, our hypothetical expert, though not knowing the wine itself, may have knowledge about the Château Vinus vineyard and its winemaking operation, whose reputation is top-flight, thanks to the excellence of the soil and the skill of the head of the outfit. Furthermore, the year 1995 may be well-known to our expert for its fine *libournais* wines (those coming from the right bank of the Dordogne River, where the Château Vinus is located). Thus our expert can compensate for specific lacks of knowledge by exploiting more general facts about related wines and wineries. Perhaps the Château Vinus winery is totally unknown to our expert, but the label on the bottle says that this wine's appellation is *Fronsac*, which has a reputation for reliability and for making high-quality and long-lasting wines. And thus, this knowledge about the *appellation*, enhanced by further knowledge about the year 1995, allows our expert to give an educated guess concerning the probable level of quality of *Château Vinus 1995*, although less certain than if it were directly known.

If our expert comes from Australia and is not particularly fond of bordelais wines and their *appellations*, then there are other avenues that may well furnish predictions concerning the Château Vinus 1995, such as the fact that 1995 was generally speaking a good year for wines in France as a whole, and the rule of thumb that bordeaux wines from good years tend to age well (the year and the fact that it is a bordeaux are written right on the label). In other words, without knowing anything specific about the French wine in question, a wine connoisseur from halfway around the globe can still have clear opinions on it, which can surprise and impress people who are less knowledgeable.

Thus we see that in times of need, one category can stand in for another. If you have no knowledge of A, then use your knowledge of B, where B is a "cousin" to A, close to it either horizontally or vertically. Experts never have access to all categories, but a genuine expert has a dense enough mesh of categories that specific gaps at various levels can be gracefully sidestepped by the process of analogy-making, and this helps to fill in missing knowledge in any specific area of the domain.

Whenever one changes one's categorization of some aspect of a situation, one changes one's perspective on the situation. Experts have so many potential perspectives that even in an unfamiliar situation, they can very often find a highly pertinent one. Specific, concrete categories are precious to experts because, as they are all genuinely different from one another, they furnish the most precise insights that have been gained over a lifetime. On the other hand, general, abstract categories are also useful to experts because they summarize many cases at once, and also because they are closer to the "essence", the "conceptual skeleton", of concrete situations. In sum, then, down-to-earth categories allow one to be precise, while highly abstract categories allow one to be deep — and precision and depth are the two most crucial keys to expertise.

Variations on the Theme of Random Killing

> On September 29, 1982, a twelve-year-old girl died in Chicago, launching a news story that would reverberate in the news media for some time. In the next couple of days, six more people died of the same cause. Careful cross-checking by police allowed them to discover the common cause: containers of Tylenol had been opened, capsules had been coated with cyanide, and the bottles had then been placed on store shelves. Announcements were made on radio and television, and police cars roamed Chicago neighborhoods, warning people through loudspeakers not to consume any Tylenol. Aside from the five bottles that caused the seven deaths, three more containing poisoned capsules turned up in the search on store shelves. As the bottles came from different factories, the hypothesis of a worker doing the tampering at the factory was discarded. The remaining possibility was that someone had tampered with the pills after they had already reached the stores. The murderer, whose identity was never discovered, must have bought bottles in several different stores, tampered with them, and then reshelved them.

Unfortunately, there are always people inspired by such macabre deeds. Let us try to enter into the mindset of a copycat killer who wanted to follow in the footsteps of this psychopath. What would such a person do? Make a beeline for the nearest grocery store and tamper with the Tylenol bottles on its shelves? It's a possible thought; after all, the original killer barely scratched the surface of the national supply. However, to stick with the category of *Tylenol* would be rather naïve, since the manufacturer instantly ceased production of Tylenol capsules, made a recall of all such products on store shelves, launched a media campaign to tell people to throw away any capsules in their possession, and made an offer to exchange Tylenol capsules for Tylenol tablets for free. Making the most literal-minded copycat move is thus out of the question.

Where might such a person turn instead? To another over-the-counter medicine? Once again, making such an upwards leap in abstraction is tempting; indeed, it's what anyone would think of at first. And indeed, some genuine would-be copycat killers actually had this precise idea. Some even tried to commit the "perfect murder" by distributing tampered bottles of medicine to stores and simultaneously planting one such bottle "innocently" in their own medicine cabinet, thus murdering their spouse and hoping it would be seen instead as a random Tylenol-style murder.

The Food and Drug Administration decided that there was no reason to suspect that would-be killers imitating the unknown Tylenol murderer would limit themselves to Tylenol itself. In fact, to the FDA officials, it seemed that any over-the-counter drug had a fair chance of being targeted for tampering by psychopaths. The Tylenol episode thus spurred the creation of a national requirement that made sealed packaging obligatory. This would make any tampering evident before the product was opened. Today, all over-the-counter medications feature such protection.

Now a would-be copycat killer, seeing that these most obvious doors had been barred by the FDA, might be discouraged, but there remain other doors to check out. A small upwards leap in abstraction will suffice: one need but shift from the category *medications* to the more abstract category of *edible products.* At this point, there is an embarrassment of riches. The job can be done by using a syringe to inject a tomato, a melon, a grapefruit, or an orange with a small dose of a lethal liquid. Any fruit or vegetable will do. And who would notice a tiny hole in the cardboard cover of a yogurt or sherbet container? Then there are drinks. It would be simple to stick a needle into a bottle of milk or juice without causing a leak. And then there are cabbages, onions, broccoli, carrots, olives, and other items in the produce racks, needing merely to be sprinkled with some kind of poisonous substance. With such thoughts, we have clearly long since left behind the down-to-earth category *Tylenol*, and have passed through two more abstract categories: *medicine* and *edible product.* Why not jump yet further upwards?

One might be able to spread a thin layer of a noxious substance onto *consumer goods* such as silverware, glasses, pots, pans, and plates; this could do in random buyers who don't rinse things before using them. For that matter, toothpastes and rubbing creams could be tampered with in advance. One could go yet further, dreaming up ways to tamper with computer batteries so that they will catch fire after a while, thus burning the computer, any nearby papers, the room, the house, perhaps neighboring houses…

But let's jump upwards from *consumer goods*, further broadening our vision of the possibilities of random murder. One could stand on a lonely hill with a rifle and fire on random people in the streets below. One could strap explosives around one's waist and board a crowded bus or walk into a nightclub or a busy marketplace.

From Tylenol murders to various types of terrorist acts, we have moved upwards in abstraction, maintaining only the conceptual skeleton of *random murder.* Each leap upwards in abstraction corresponds to creating a wider ring of possibilities around a common core, moving gradually outwards in semantic distance from the original event. This is the quintessential human way of imagining novelties — starting with small variants, then working one's way outwards to more and more radical ones, the discovery of which is much less likely. As is always the case when the answer to a riddle is given, abstraction seems very simple if one displays its fruits in front of an observer's eyes, but that impression is deceptive. If the pathway meandering through the space of categories seems like child's play, that is an illusion. When you have to make the leaps yourself, each one presents an obstacle, and few people will get far in the endeavor. The best proof of the difficulty of this kind of progressive abstraction is its rarity.

In the end, how far did the FDA go? It did not impose any requirements on how fruits and vegetables were sold, nor on other types of buyable goods, and so today, one still finds such items on the shelves of stores exactly as they were back in the pre-Tylenol-murder days — without protection. Why did the FDA stop at the level of medicines, and not go any further? Why did it arbitrarily draw the line here?

The reason is that the type of rising-abstraction semantic pathway that we sketched out above cannot be followed at will, and cannot be taught in schools as a sure-fire creativity technique. The number of ways a given object, action, or situation can be categorized is gigantic (recall Mr. Martin's wineglass, tadpole home, and spider holder), and the vast majority of potential categories are simply not available to the conscious mind, even to a mind actively looking for new ways to categorize something. However, there are circumstances in which a useful discovery can be the fruit of a simple-seeming upwards leap. In such cases, the seeming ease of an act of mental agility not only is esthetically pleasing but also lets one escape from mental ruts and see things in a genuinely novel fashion.

Getting Off the Beaten Track

"Why does opium make one grow sleepy?" "Oh, it's all due to its dormitive virtue," proudly claims the candidate physician in Molière's play Le Malade imaginaire, convinced that he has given the perfect answer to this question.

"Thinking outside the box" — who has not heard this mantra chanted dozens of times? It purports to reveal the key mechanism of creative thought. It suggests that whenever we have a mental block, whenever we get stuck in a rut, whenever we find ourselves trapped in a box canyon, whenever we have painted ourselves into a corner, whenever we are getting nowhere fast, whenever our wheels are spinning, whenever we are "pedaling in sauerkraut", as they say in French, why then, if we merely know this appealing sound bite, it will come to our aid like a loyal St. Bernard dog, lifting us up out of our foggy, unproductive mindset. Like a genie in a lamp, the magic recipe of *thinking outside the box* will liberate us from our cul-de-sac thinking and will open up fantastic new vistas. There is even an electricians' union whose motto is "Trained to Think Outside the Box!", which sounds to us like a contradiction in terms (unless by "the box" they mean "the fuse box"). Or then again, as another advertising slogan put it a while back, whenever things looks hopeless, just "Think Different!" This notion that creative thinking involves avoiding the beaten path will surely not catch anyone off guard, but unfortunately it is no more insightful than stating that what makes opium induce sleep is its "dormitive virtue". Yet it is amazing how humble are the conceptual slippages that have sparked intellectual revolutions, sometimes small, sometimes large, sometimes on a tiny level, sometimes on a global level. Indeed, creative insight depends surprisingly often on very small but subtle conceptual slippages.

The Bottleneck

> Jillian and a few friends set off for a Sunday picnic. They spread their tablecloth out in a lovely glade in the middle of a forest. Everyone is enthusiastic, the weather is perfect, and a light breeze brings a refreshing coolness. The table is soon covered with delicious things to eat. But then, as the bottle of carefully selected wine is fished out of the basket, Jillian realizes she left her corkscrew at home. There's no one in view to ask for one. All at once, it occurs to Jillian that instead of pulling the cork *out*, she can simply push it *into* the bottle! It takes a bit of work, but in the end it works perfectly.

This solution is hardly the stuff of genius, but it is nonetheless a clever idea whose origin is worth looking at. It consists in the conceptual leap from *pulling* the cork to *pushing* the cork — actions that are essentially diametric opposites. And yet, what allowed Jillian to make this directional slippage is an act of recategorization — namely, she glided from the concrete, physical idea of *pulling the cork out* to the more general idea of *accessing the wine*, or to put it more abstractly, she glided from the narrow, physical idea of *pulling out what is blocking access* to the broader but more goal-focused idea of *gaining access to what is desired.* This is clearly an act of abstraction. In this situation, *pulling out* is a member of the category *gaining access*, and the recategorization allowed Jillian to imagine other ways in which she might gain access to the beverage. Jillian's mental move is reminiscent of that strange ambiguity, mentioned earlier, of the English verb "to grow", which can be used to describe not only changes where something increases but also changes where something decreases. And thus Jillian escaped from her box canyon by imagining the idea of *reverse pulling*, which amounts to pushing.

Such a mental slippage resulted from a new encoding of the situation, seeing it in a more abstract fashion than at the outset. Jillian realized that her primary goal was not to pull the cork out of the bottle but to get some wine; with this goal clearly in mind, she could reperceive the obstacle that was between her and the wine, and so she vaguely mused to herself, "If I can somehow *slide* this cork, whether inwards or outwards, it will slip out of the bottleneck, and that will give me the access I seek." The seemingly paradoxical aspect of Jillian's picnic dilemma is that the act of abstraction, which drastically reduced the attention she was paying to certain concrete aspects of her mental representation of the situation and which was thus a *cognitive impoverishment*, turned out to be the key step in finding a solution.

This phenomenon is connected to the fact that one is never able, in a single moment, to think of all the properties of an object (or of an action or situation). Rather, one is usually trapped by the more specific, concrete, and salient aspects of the entity in question, for those are what most obviously distinguish one category from others. For example, seeing the challenge of opening a wine bottle as a *pulling* of its cork tends to make less salient the idea that the cork is *pushable.* However, by the act of simplifying and thus perceptually *impoverishing* the situation, replacing the initial category by a more abstract one, the representation of the situation is paradoxically *enriched* by revealing characteristics that had previously been hidden. That, in a word, is the paradox.

The apparent simplicity of this creative strategy of *enriching one's perception by engaging in an impoverishing act of abstraction* does not preclude it from being a powerful mechanism. It allows one to transcend an irrelevant point of view that has led to a dead end, which is the result of an unfruitful initial categorization of the desired goal.

Boxes that Hold Us Up, and Voluminous Bodies

The just-described phenomenon of recategorization can be seen in a psychological experiment known as the "candle problem", invented in the 1940s by the German gestalt psychologist Karl Duncker, in which subjects are given the challenge to attach a candle to the wall. Aside from the candle, they have at their disposition a full matchbook and a full box of thumbtacks. The insight that yields the solution involves emptying the thumbtack box and then using some thumbtacks to affix it to the wall, after which the candle can be placed on top of it. Since the box is filled at the start, the challenge is subtle and not too many subjects find the solution. However, it is rendered easier, though by no means trivial, if at the start the box is empty and the thumbtacks are simply lying around on a table.

The main challenge in this problem is that of categorizing the thumbtack box not merely as a *container* but also as a potential *stand.* When the thumbtacks are contained in it, the box is spontaneously and naturally perceived as a *thumbtack box*, and that is seen as its sole identity. At a more abstract level, one can perceive it anew, as a *box*, whose central function remains that of containing things. To reach the solution, a still higher level of abstraction is necessary: that of perceiving this same object as a *stand.* Of course, at this level of abstraction, any number of objects can be so perceived, including tables, chairs, shelves, drawers, and so on. In any case, the empty thumbtack box thus recategorized can be used in a new manner.

Discoveries of great import can result from just this kind of mental slippage. For example, if the old "Eureka!" legend can be trusted, Archimedes' discovery of the "principle of Archimedes" in hydrostatics came from just such a shift in categorization. The historical accuracy of the tale may be questionable, but in any case it epitomizes both the subtlety and the richness of adopting a more abstract viewpoint concerning an object one is looking at, even if it involves ignoring nearly everything that one normally associates with it.

> King Heron asked Archimedes to determine whether his new crown was made of pure gold or whether, as he suspected, the goldsmith to whom he had given the gold was a crook and had adulterated it by mixing silver into it. It was simple for the King to check that the crown weighed exactly the same as a standard gold ingot. But how to know if the metalsmith had replaced some of the gold by exactly the same mass of silver? If only he could determine the crown's *volume*! In case of adulteration, the crown's volume would have to be greater than the ingot's, since silver is less dense than gold. Unfortunately, though, the crown was far too irregular to allow Archimedes to determine its volume by using geometry. What then to do? According to legend, Archimedes hit upon the solution while taking a bath. He noticed that as his body slid into the water, it made the water level rise exactly in proportion to the volume that was submerged; it occurred to him that the same would hold for the crown and a gold ingot. And so, by submerging them and observing the changes in water level, he discovered that the crown's volume was greater than the ingot's, and so the metalsmith was indeed a crook.

Archimedes made an analogy between the crown and his body, noting that either one, when fully submerged, would displace a volume of water equal to its volume. Now if one names various categories to which a crown belongs, one comes up with such things as *symbol of royalty*, *made of precious metal*, *worn on head*, and so on. These have little if anything in common with categories that come to mind in thinking of a human body. Only when one rises to a highly rarefied level of abstraction — all the way to the category *object endowed with volume* — does the analogical connection appear explicitly. One has to ignore many facts, such as that a human body is alive, in order to map it onto a crown, which is not alive. Here, then, is a case where an analogy depends on a categorization far more abstract than anything that springs to mind spontaneously. The creative act in this story thus resides in the ability to relate entities that are extremely different in every way except one, which allows the discovery to be made.

The preceding discussion might give the impression that we are trying to portray creativity as the natural outcome of meticulous searching for commonalities among situations that, on first glance, would seem to have little or nothing in common. Unfortunately, though, even if we are exploring the mechanisms underlying creativity, we cannot pass along a recipe for genius, because a *deliberate* process of search for features linking two situations plays little or no role in the matter. The process of abstraction here described does not emerge from systematic, conscious search. Indeed, it often surprises the discoverer as much as it would surprise an observer.

When leaps are made at a very high level of abstraction, small though they may be, they can give rise to highly important discoveries, as the Archimedes story shows. This is certainly part of the explanation of the mystery of creative discovery, but it would be misleading to see it as a recipe, for in general, it is only *after the fact* that two seemingly distant situations strike an average person as being analogous, and the common traits that give rise to an abstract analogy linking them may have been hidden by unconscious presuppositions, tending to make the sought-after generalization very hard to find.

Of Mice and Men

There is a well-known recent invention that came out of one simple abstraction and resulted in a major revolution. In the early 1980s, communicating with computers was not easy. Aside from a tiny coterie of people who had worked at or visited a handful of advanced and little-known research centers in the preceding decade, no one had ever seen or used a mouse. Everything was done by typing symbols on the keyboard. One had to explain to the machine, using arcane strings of symbols expressing commands, what one wanted it to do, and it would then execute those instructions. All human–machine interaction was conducted via such a "command language". A human would order the machine to do something, and the computer would obediently carry out the order, as long as the command was syntactically well-formed.

Over the last few decades, things have changed enormously. Today, just about the only use for keyboards is to write text, as in an email or a book; users almost never write out commands to be obeyed. When we want some action to be taken, we use a mouse (or the successor to a mouse on laptops — a touchpad). People who have grown up with this natural-seeming technology have a hard time imagining how things could ever have been otherwise. That this forward step was dependent on technological advances is self-evident, but it also depended on a small-seeming but nontrivial intellectual leap that we will now discuss.

A mouse is a kind of electronic counterpart to a biological limb; it is a prosthesis with magical powers. While our biological limbs allow us to act on the three-dimensional physical world all around us, the mouse constitutes our interface with the two-dimensional virtual world on the screen. Like Alice, who followed the white rabbit into its hole and thus discovered a world of wonders, we gain access, via the mouse, to an immaterial world. The mouse is our rabbit.

Indeed, choosing a file to work on, opening it, closing it, shifting it from one folder to another, copying it, printing it, deleting it, and so forth — all these are ways of interacting with immaterial objects. To carry out such actions in the material world, our biological limbs serve us well. To do such things in the virtual world, we rely on a mouse. We sidestep the writing of commands; instead, all we do is act and see what happens, exactly as we do in the material world.

To say that the situation has changed since the command-language era would be a vast understatement. It's as if we had once had to do everything without using our limbs at all, always being forced to describe what we wanted in some technical jargon, or as if we had always been separated from the computational world by a thick sheet of glass, our sole channel of communication being slips of paper. After writing down our desires, we would slide our slip of paper through a narrow slot in the glass made for this purpose alone (corresponding to the "Enter" key), and the computer would then execute our written command. Such a *modus operandi* would obviously be extremely inefficient.

Why do we claim that the invention of the mouse was more than a technological advance — indeed, that it constituted a marvelous conceptual breakthrough?

> Is it possible to see the invisible? No, since the invisible is precisely that which one cannot see.
>
> Is it possible to touch the intangible? No, since the intangible is precisely that which one cannot touch.
>
> Is it possible to move an object without touching it, to open or close a folder without touching it, to throw an object into the wastebasket without touching it? Moving, opening, closing, and getting rid of intangible objects — these are things that we do continually with our mouse.

Acting on the immaterial world is no longer a self-contradiction. The mouse plays the role of a biological limb, but acts on intangible entities. Thus with the help of a mouse, one can flip through a virtual book, adjust the volume of a loudspeaker, place items into a "shopping cart" in a virtual store, and so forth. Immaterial entities can be acted upon no less than physical objects.

The mouse is the fruit of a substantial mental leap. From time immemorial there has been an unbridgeable gap between the categories of *material* and *immaterial.* Thanks to this gap there has always been a clear-cut distinction between the world we touch and perceive and the world inside our minds, between what is real and what is imagined, between the concrete and the abstract, between matter and pattern, between tangible and intangible. If the idea that a physical action can have an effect on a distant entity is strange, the idea that it can have an effect on an *immaterial* entity is even stranger; indeed, it verges on the paradoxical. Thus the creation of the mouse involved generalizing a property that till then had been limited to the world of material objects — that is, the property of being actable-upon. A new category — that of *actable-upon objects*, far broader than the category of *material objects* — came into being when the mouse was conceived of. This new category toppled our prior assumptions about how we relate to the world around us, dramatically altering our ontological categories — that is, the set of basic categories with whose aid we carve up the world.

What Makes *Homo Sapiens Sapiens* Sapiens?

An auto mechanic in the small town of Martinsville, Indiana had the curious idea of hoisting an old VW bug onto a sturdy stump in front of his garage, in order to attract the attention of people driving by. After all, it's quite striking to see a car perched on a tree stump, several feet above the ground. And come Christmas season, the ex-bug was often duly decked out with Christmas lights of various colors, and as a result, no passing driver could fail to see it as a member of the category *Christmas tree.*

And when one checks into any hotel these days, chances are good that one will be handed a plastic card and pleasantly told that it is the key to one's room. Well, if this card belongs to the category *key*, then of course the slot near the door handle in which one "swipes" it will belong to the category *lock*. And the traveling couple who emerge from their motel room right next door to the VW repair place in Martinsville with their keys in their pockets and who spot the nearby Christmas tree aren't exactly in the prototypical situation that the words "key" and "Christmas tree" would tend to evoke in one's mind...

This small example, just one among myriads that we could have chosen, is the result of the special capacity that distinguishes humans from (other) animals. Every normal person, from the most primitive to the most sophisticated, has this capacity, and it comes in various degrees, much as does any human quality, and it drives our mental development from the earliest age. Of course the capacity we are referring to is that of endlessly extending categories through analogy. We humans construct categories, extend them, multiply them, link them together, and glide gracefully among them. If we think in a subtle fashion, it is due to the richness of our repertoire of concepts and to our flexibility in seeing freshly-encountered situations in terms of our concepts. We categorize in ways that go well beyond the most common meaning of the term "categorize", for not only do we categorize objects, actions, and situations in pre-existing categories, but at the drop of a hat we transform any already-perceived entity into a category, a process exemplified by proper nouns that are run through the categorization mill and emerge as full-fledged plurals — ranging from one's sister to the corner druggist, from Lincolns to Leonardos, from Picassos to Puccinis, from Mozarts to Mataharis. To put it in a nutshell, we humans are compulsive category-extenders, insatiable tourists who cannot get enough of the world of categories, old hands at semantic slippages, similarity-spotters par excellence, and last but not least, inveterate abstractors.

The phenomenon of marking illustrates perfectly our irrepressible tendency to extend categories. Even as we keep a lexical label fixed, we relentlessly widen the set of objects or situations to which it applies as our knowledge gradually expands, and meanwhile, behind the scenes, there is always a strong and stable set of analogical links — a conceptual skeleton — uniting its many members. Think, for instance, of the *lumps* that, for a child, are inevitably the result of bumping into something, or of the *bottles* that the child imagines as made for holding liquids; for an adult, these are far richer categories all of whose members share a deep essence. Semantic enrichment via analogy is what justifies the use of a single word or phrase as the label of more than one category, allowing us to glide smoothly up and down a spectrum of abstraction, choosing the appropriate point along the way as a function of the context in which we find ourselves. Thus Megan, who would never make the egregious error of calling even a single truck in her toy chest a "car", will instinctively and naturally avoid trucks (and motorcycles, bikes, scooters, etc.) when her father tells her to "Watch out for cars!" just before she heads off for school. And thus Roy, who would never dream of ordering a coffee, can sip his tea while "having coffee" with some friends.

The process of marking plays an important role in clarifying what the essence of a concept is. From our first concrete examples of any concept, we proceed to more abstract ones (thus from tangible paper documents found on our wooden desks to intangible electronic documents found on our screens, or from postal addresses to email addresses). Once there are two levels of abstraction (as in these examples), then the earlier, lower level is the marked category, while the later, higher level is the unmarked one. The two levels share an abstract essence that was not visible to those who knew the marked category alone. Thus only when files and folders became familiar to people in their electronic guises did the heavy solidity of one's traditional desk come to be seen as a dispensable attribute of *deskness.* The same holds for the traditional desk's properties of *occupying considerable space in a room* and *possessing drawers.* Thanks to technological advances, the essence of the concept *desk* was revealed as being the fact that desks are *workplaces for documents*, a description that applies equally well to old-fashioned wooden desks and new-fangled electronic ones.

Such a process can carry one far afield from the first members of a category, for the pathways that slowly reveal the hidden essence are many, and finding them takes time. They may involve several levels of abstraction, as we saw with the notion of *waves.* Such extensions of categories are very important for our thinking, since they help us see what is central and what is peripheral in a given concept, and thus they show us how to generalize appropriately.

We are all able to overlook certain superficial qualities and to focus on more important ones in order to see what two situations have deeply in common, and this ability to abstract, which we depend on constantly, allows us to use our previous knowledge efficiently as, over the course of our lives, we confront one new situation after another. Abstraction is thus the key tool that allows human beings to move gracefully from one category to others, and to perceive the world efficiently and to interact with it profitably.

It is in their way of doing this in a given domain that novices and experts differ. Novices depend on a very coarse-grained categorization of what is out there — a single genus and a number of species beneath it. This is a perfectly good way of starting out in any domain, since it serves the basic purpose of categorization: to make distinctions based on recognizable features. For example, a *square* (one species) is a *quadrilateral* (the genus) that has four equal sides that come in parallel pairs, and it also has four right angles. As one's familiarity with a domain grows, class inclusions become increasingly common, through the addition of categories at intermediate levels (sub-genera). This often takes place by using a concrete term as the name of a more general category — that is, through marking. Thus, for a zoologist, who is by definition an expert, a *dromedary* is a kind of *camel*, while for most people the two categories are mutually exclusive; in like manner, a *square* is a kind of *rectangle* for mathematics-savvy people, while for most people the two categories are mutually exclusive.

The proliferation of levels of abstraction grants us the freedom to shift viewpoint whenever one of the categories that we are currently focusing on seems to be leading us into a box canyon. But the recognition of *which act of categorization* is responsible for the lack of progress is extremely difficult, and much of the subtlety of creativity lies here. For example, who would have been able to say, *a priori*, that it was an overly hasty categorization of the box of thumbtacks that caused the impass in Duncker's classic candle problem? Recognizing this fact — that is, being able to put one's finger on the categorization that was at fault — is very difficult, but that skill can open up new horizons, because a more abstract categorization, by bringing into focus certain features that had been unseen till that moment, allows one to circumvent the dead end. From the picnic in the woods to the principle of Archimedes and the invention of the mouse, the key move is making the appropriate upward leap in the space of categories.

We humans all think thanks to categories, and as we collectively expand our shared repertoire of categories, we all continually expand own own personal repertoire. We have very abstract categories on which we unconsciously base our understanding of the world around us, such as the category of *what is actable-upon* and the category of *objects with volume*, and we also have very concrete categories, which allow us to distinguish between a *Yorkshire terrier* and a *Staffordshire terrier*, to hold forth on the qualities of a *Gevrey-Chambertin Premier Cru Les Cazetiers Dominique Laurent 1996*, or to select between *anchellini* and *primaverine* as an ingredient for a minestrone. It would seem that the flexibility of our category systems is the key to what distinguishes *Homo sapiens sapiens* from a Yorkshire terrier.

❧ ❧ ❧

CHAPTER 5

How Analogies Manipulate Us

❧ ❧ ❧

At the Mercy of Uninvited Guests

Is it possible that analogies have the power to manipulate us, to twist us around their little fingers? Certainly; in fact, they do so in two senses of the term "manipulate". First, analogies often arise in our minds without our even being aware of them: they *invade* us surreptitiously and seize center stage. Second, analogies *coerce* us: they force our thoughts to flow along certain channels. And thus, far from being just appealing and colorful pedagogical or rhetorical devices, analogies are wily creatures with a will of their own. They shape our interpretations of situations and determine the conclusions of arguments. Put otherwise, an analogy will not be content with merely crashing the party; having shown up, it then dictates the rest of the evening.

The word "manipulation" tends to exude negative connotations, but analogies are not always manipulative in a harmful way. Being nudged, pushed, or even forced by an analogy can of course have bad consequences, but that is the flip side of a phenomenon that is largely positive. The fact is, the interpretation of a situation is inseparable from the analogies (or categories) it evokes. Our categories are thus organs of perception; they extend our physiological senses, allowing us to "touch" the external world in a more abstract fashion. They are our means of applying the richness of our past experience to the present; without them, we would flail about helplessly in the world.

Where We are Headed in This Chapter

Although this chapter concentrates on how analogies manipulate us, it is a kind of microcosm of the whole book, in that it runs through analogies on all scales, ranging from unnoticed throwaway quips to monumental decisions that changed the course of history; between those extremes, it passes through categories that are awakened in everyday situations and that help us to deal with the commonplace events of our lives.

Perhaps the most inconspicuous way in which we are manipulated by our analogies is revealed by the speech errors that people commit when, to put it crudely, they get their wires crossed, as a result coming out with streams of words that are distorted and lack sense. We will show how the relentless pressure to categorize in real time is the culprit here, and we will also see that the phenomenon is not limited to language, since our actions, too, often betray the conceptual wire-crossing taking place in our heads.

Leaving the world of errors, we will then move slightly upwards in size and visibility, to consider analogies that people make despite themselves, and for no obvious reason. Such analogies arise without warning, usually lead nowhere, and nearly always are rapidly forgotten. And yet their very insignificance is, for our purposes, highly significant, as it reveals the ubiquity of analogy-making — how it is a perpetual activity of our minds even when there is no purpose behind it.

The next kind of manipulative analogy that we will look at has greater import, because it shows that a simple analogy can frame how we see situations and can shape our perceptions, our reasoning, and our decisions. The light shed on a situation by the analogical evocation of an inappropriate category can be so blinding that we are led unawares into cognitive box canyons, which can induce great confusion. To convey the importance of this phenomenon, we will show it in a number of striking situations.

When one is deeply engrossed in an activity or is powerfully struck by a very unusual event, the intense level of interest may cause a swarm of analogies to invade one's mind uninvited, analogies that would never occur otherwise. This phenomenon reveals an Achilles' heel of analogy-making, which is that it can lead us to absurd interpretations of situations, and thus to very poor decisions. And yet our susceptibility to making unusual analogies under the influence of our obsessions is also a source of great creative potential. To be sure, most of the similarities that we notice when under the influence of an obsession bring no great insight, but every once in a while such a connection can give rise to one of those miracles of the human mind that we call "strokes of genius". Indeed, someone who is taken over by an obsession will see analogies to their obsession popping up everywhere, on every street corner, everywhere they turn, day and night, and each little thing that happens is potentially mappable onto some aspect of the obsession. And sometimes, though not very often, something exceptional will come out of this unstoppable drive.

Perhaps the most fascinating way in which we are manipulated by analogies, this time at a rather high level of cognition, involves analogies that make decisions for us, not only behind our backs but even, at times, over our dead bodies. This happens when we find ourselves in a situation where an analogy to a past situation is so blatant and irresistible that it simply forces itself on us, and we feel compelled to obey it. In such a circumstance, we find ourselves convinced, without rational explanation, that whatever happened *in the past* will inevitably also happen *once again* in the new situation, even if it conflicts with our desires or leads us to an irrational conclusion. When logic vies with a strong analogy, the "analogic" wins hands down.

And now we shall raise the curtain on Act I, which is devoted to micro-analogies that manipulate us on the smallest of scales, insidiously and incessantly.

Speech Errors as a Rich Window onto the Mind

In the midst of the most fluent native speech, intelligent, articulate speakers often come out with words or phrases that, if framed in print, would look very odd, and occasionally they utter sequences of words or pseudo-words that, if they were slowed down and highlighted, would make even a schoolchild snicker — and yet, since it all happens in real time and goes whizzing by far too fast for most listeners to pay attention to both its form and its content, seldom does anyone (speaker or listeners) hear anything in the least strange. But if one is deeply fascinated by language and thought, one's ear is likely to be more open and more sensitive, and sooner or later one may begin noticing small oddities that for most people are merely parts of the bland background.

Thus it was for us, especially this book's senior author, who for more than forty years has been passionately and meticulously collecting speech errors, and whose collection merged, some twenty-five years ago, with that of his friend David Moser. This book's junior author joined the team more recently, and his careful observations add yet another dimension. The combined collection includes many thousands of errors of countless types, and a lengthy book could easily be devoted to it alone. In what follows, we shall have to settle for dipping into just a very small subset of it.

Theorizing about speech errors has come a long way since Sigmund Freud, but Freud's ideas are still deeply entrenched in the popular mind. The Freudian view of speech errors is that they are *a window onto the subconscious*, revealing dark secrets hidden in the psyche of the person who blurts them out. For this reason, errors that manage to slip out despite the conscious mind's attempt to censor them constitute a deeper, truer message than the overt, intended message. In short, a speaker's innermost soul is inadvertently revealed by their "Freudian slips". Today, this idea is still widespread but is not taken as seriously as it once was, and yet, if one interprets the phrase "a window onto the subconscious" in a less emotionally charged manner, it is still perfectly valid. Indeed, speech errors constitute audible traces (or visible ones, in the case of written errors) of mental processes far below the conscious level. Like animal tracks in a forest, they can be "read" by a careful observer and much can be gleaned from them.

After Freud, the field of error research came into its own in the early 1970s, with the work of the linguist Victoria Fromkin, who assembled a large corpus of errors and analyzed them as manifestations of unconscious processes. Significant extensions to Fromkin's work were made by, among others, the psychologists Gary Dell, Donald Norman, and David Rumelhart, who developed influential computational models of error-making. The French psychologists Mario Rossi and Évelyne Peter-Defare, as well as Pierre Arnaud, have worked in the area as well.

The goal of our discussion of error-making is to show not just the diversity of errors we all make, but also, more importantly, how our errors reveal certain facets of the ceaselessly churning mental activity of categorization through analogy-making. The first few sections of our discussion will deal with what we'll call *lexical blending*, an unconscious process that seamlessly but incoherently mixes together two or more stock lexical items, ranging all the way from long idiomatic phrases to monosyllabic words.

Rivals Mesh to Make a Mishmosh

I *watched at* him for a long time / Is that *Buckminster Palace* over there? / She grew up in a *working-collar town* / His research, now mostly forgotten, was an important *stepping block* along the way / I always carry my notes with me as a *safety blanket* / I'm *thick in the middle* of a major project / It was just a *last-of-the-minute* impulse / She always does things *on the last minute* / It seems she's *having second doubts* about it now / America has let its railroads *fall to pot* / He's had a rough time this season, *right out of the get-go* / Please *keep that into account* / She seems very much *up your wavelength* / Everyone is just *itching at the chance* to see him / I should *count my lucky stars* / His campaign speech was such *a pack of cards* / She went completely *off the rocker* / Well, I can't say that he's *shaking the world on fire* / I have to *share my shoulder of the blame* / When I saw how many people had come, it just *blew me over* / Our old car finally *bit the bucket* / Let's not *mince hairs*, please! / Okay, I'll *give a stab* at answering your question / That's the real *meat and butter* of their business / I slept till noon and *woke up like a baby* / I don't want to *be in any touch* with that guy any more / He's *as loony as a tune* / I *didn't know diddly-word* about their plans / My son *complains like a broken chimney* / I was unaware of what went on *under the scenes* / Little elves darted *hither and skither* / I might be able to *lay my fingers on it* / As a teen-ager, I was hugely *insecure of myself* / I made a great speech error *yesterday night*! / Doug is *crazed with* salsa / She was hoping to *follow in his shoes* / I *try like the plague* to avoid people like him / We're hoping to *piece apart* the details of the process / In the meanwhile, we'll *keep our eyes out for it* / But *for the meantime*, that'll be enough / His nutty ideas *aren't worth the time of day* / They'll probably *snub their nose* at you / The French *turn down their noses* at rosé wine / I wouldn't *fall for that trap*! / I woke her up when *she was just getting asleep* / Hey, you did a really *nice piece of job* there! / In their eagerness to slander him, they *stooped at nothing* / That'll surely lead you down *a tricky slope* / *What's up, Dolly?* / *Whoa, Betsy!* / Well, *for heavenly days*! / *Goodness grief!* / Wow, that's *pretty something*! / It's going to be smooth sailing *from now on out*! / Give me a call *when you have a time* / And then I *piped in*, "Go for it!" / *Oh, I kay* — you're probably right / His three teen-age daughters really *run him nuts* / I didn't realize I was *treading on anyone's sacred cows* / I *bought the bullet* on that one / It's not gonna *make a big deal*, right? / I hope you can *steer your way clear* to making it to my party / My parents are *breathing down my back* / It's *no skin off my teeth* / By then, the campaign was *in high swing* / You actually need *magnifying glasses* to read it / Some of my colleagues *from even further aflung* will be coming / They're always *touting Princeton's horn* / When she tried to say it in Italian, she *stumbled all over her face* / I have no problem with *taking second seat* / I was thinking how *difficult on her* this must be / The magnitude of the challenge *caught me by surprise* / They had to pull him out of there *kickin' and draggin'* / Watch what you eat, or you're going to have a *heart-a-choke* / My cell phone isn't within my *hearsight* / My mother is a bit *hard of seeing* / This new place is really *my oyster bed* / None of the profs *would hesitate twice* about consulting Wikipedia / I would *walk through water* to have my kids get accepted at that school / I tell you, that guy is *one cool potato*! / He was *en his route* to Singapore via Bangkok / I wonder what Bach would sound like to someone with a *tin-deaf ear* / When I saw his name in the paper, *two and two just clicked*...

All these blends are utterances that we heard and transcribed while listening to ourselves, our families, our friends, our colleagues, and also the radio and other media. Readers who are fluent speakers of American English will probably enjoy trying to pinpoint the contributing phrases (keeping in mind that the number often exceeds two).

Many of the above "mushy meshes" are humorous, and thanks to having been set in a frame and highlighted in italics, they may seem blatantly obvious after the fact. And yet in the absence of frames or italic highlighting, only a very careful listener will consciously notice clunkers of this sort, even when the two (or more) "notes" are played loudly. Indeed, most lexical blends are picked up on by absolutely no one — neither by the speaker nor by listeners. Strange though it may seem, fluent speakers generally "hear" only the idea that was *intended*, rather than the words that hit their eardrums.

Although lexical blending is seldom noticed, it is rampant, and with some training and some careful attention, any interested observer of language and thought can start noticing the phenomenon and writing down examples of it. Moreover, recorded speech can convince skeptics that everyday speech is in fact riddled with lexical blends that escape the ear.

Our error collection was launched decades before the Web existed, and of course we never dreamed that such a powerful and public linguistic treasurehouse would one day arise. By doing simple Web searches, we have come to realize that quite a few of the errors that we once presumed were unique events in the history of the world have been independently recommitted by dozens or even hundreds of people as they type up their Web sites. It's quite amazing. Interested readers can do searches for various blends and find how often they appear. A few of the multifarious blends that we've collected are still one-of-a-kind events, but many are a dime a dozen.

Once one has gotten sensitized to noticing lexical blends, blend-collecting can become quite addictive. The great appeal of blends is the rich window that, taken collectively, they offer onto the mind. They reveal that behind the scenes, as people struggle to convert their thoughts into words under the real-time pressures of everyday communication, the analogies they make in the intense effort to simplify and compress complex situations down to their essences are in constant unconscious competition with each other. Although the frenetic competition more often than not results in clear-cut hands-down winners, at times it will happen that no one category is a clear winner, and as a result, two or more lexical items — the standard verbal labels of the competing categories — will vie with one another to get themselves uttered at the same time. In such cases, snippets of the rival items will get interwoven in the speech chain and thus will emerge strange, unpredictable hybrids like the many phrases exhibited above. We will now examine a handful of examples more carefully.

A Zigzagging Cognitive Pathway inside a Dizzy Dean's Brain

One day while talking on the telephone, a doggedly determined dean came out with the following remark about a famous and admired researcher who was the target of an intense recruitment campaign at the dean's university:

We'll pull no stops unturned to get him to come here.

Smiling inaudibly, the listener at the other end whipped out his little notebook from his back pocket and transcribed these words instantly, lest he forget them, and ever since that day, this short remark has been one of our favorite examples of the phenomenon of lexical blending. Despite the many years that have gone by since then, this remarkable blend still affords us today a glimpse of the meshing and churning of the metaphorical gears going on inside the dean's brain way back then.

Why did the ardent dean not correct this strange phrase that had popped out of his own mouth? Surely he didn't think that "to pull no stops unturned" is a standard English idiom — so why didn't he pause, backtrack, and self-correct? Well, exactly the same question can be asked about all blends and all speakers, for we all make similar errors that simply go unnoticed. Most readers will probably chuckle at many of the defective phrases that we have blatantly framed in this discussion, but the fact is that every day, every reader of this book, just like both of its authors, comes out with similarly flawed constructions, but since no one explicitly puts frames around them, they are not very likely to be noticed.

Two rival categories in the dean's mind were the *pull-out-all-the-stops* category and the *leave-no-stone-unturned* category. The lexical label of the first category is a phrase that originated in the playing of pipe organs, and it means that the music will be blasted out into the church as loudly as possible. (A stop is a device that blocks a particular organ pipe, and thus the pulling-out of all stops means that all the organ's pipes will sound.) Since first being uttered, the phrase has of course spread by analogy and taken on the meaning of going all-out in trying to achieve a goal, not holding anything back.

As for the second category, its lexical label alludes to a desperate search process in which something has been lost and one wishes to check in every possible place — under every metaphorical stone — no matter how unlikely it might be. Once again, the idea is that of going all-out in the pursuit of a cherished goal.

The two rival analogies that sprang to life in the dean's mind involved seeing the faculty-recruitment situation as one in which a crucial and important goal was being pursued by every possible means, as urgently and intensely as possible. We can exhibit the roles played by the two component phrases as follows:

1 3

"We'll *pull* out all the *stops* to get him to come here."

"We'll leave *no st*ones *unturned* to get him to come here."

2 4

The numerals "1" through "4" show the zigzagging order in which fragments belonging to the two rival lexical items were picked out. Note that fragment 2 consisted not just of the word "no" but also of the initial consonant cluster "st" of "stones", which coincides with the initial consonant cluster of fragment 3, which is the word "stops". In all likelihood, this fortuitous phonetic overlap of the words "stones" and "stops" was a contributing element that made it easier to follow this zigzagging pathway, rather than some other pathway, weaving (or rather, interleaving) the two phrases into a single seamless-sounding output stream.

It's not terribly surprising that both of these stock phrases were simultaneously activated in the dean's mind; in fact, it seems so reasonable and even inevitable that both got activated that a reflective person would have to wonder, "If the dean got tripped up by two parallel analogies being built simultaneously in his brain and bringing the two rival lexical items to mind at once, then how come I myself have often managed to use just *one* of these phrases without getting tripped up by interference from the *other* one?" When you stop to think about it, it's a little bit like asking, "How come skilled pianists so seldom hit two adjacent notes on the keyboard simultaneously?"

Striking Two Notes at Once on the Keyboard of Concepts

As a matter of fact, beginning pianists *do* make such errors all the time, to their frustration, but gradually they figure out techniques that allow their fingers to plunk themselves down on *just the one note that is needed*, despite the extreme nearness of other notes. This is a kind of small motoric miracle that we will not attempt to explain, but it serves to make readers aware that, quite analogously, there is a small *categoric* miracle going on every time that we come out with an uncorrupted word or phrase. We usually come out with *that word or phrase alone*, pure and uncontaminated, not blended with any of its semantic near-neighbors. But how is it that our "mental fingers" almost never strike two neighboring "notes" at once?

In truth, our mental fingers often *do* strike two notes at once, activating two or more rival categories. As a result, human speech is peppered with all sorts of tiny defects that are lingering traces of the silent battle raging behind the scenes. If one pays very close attention to any native speaker of any language, one will hear slightly deformed vowels, slightly prolonged consonants, voiced consonants that should actually be unvoiced (or vice versa), slight pauses between words, and many other subtle phonetic distortions, all of which are surface manifestations of the seething activity below, rife with interlexical competitions of which the speaker is nearly always totally unaware.

How Many Contributing Phrases?

Some observers of the phenomenon of lexical blending have asserted that every blend is necessarily a splicing of *exactly two* contributing lexical items, but as we briefly said earlier, this is an untenable hypothesis. In any large corpus of blends, there will inevitably be cases where three (and at times even four or more) items were involved.

As a matter of fact, in the blend made by our delightfully dizzy dean there is a third stock phrase in English that was most likely involved. That phrase is "We'll pull no punches", which, just like "We'll pull out all the stops" and "We'll leave no stone unturned", means that one will tackle the challenge not half-heartedly but with as much ardor as one can muster. Moreover, "We'll pull no punches" and "We'll pull no stops unturned" start out with exactly the same three words, so that "We'll pull no punches" seems quite likely to have played a role in the process, although on the other hand it's far more common to say "We won't pull any punches", so there is room for doubt.

We turn our attention now to some more blatant cases of triple or even higher-degree blends. In a radio interview, an author was describing San Francisco's famous City Lights Bookstore during the beatnik era. He said:

> Destitute poets were always browsing its well-stocked shelves, and on occasion someone would walk out with a volume tucked under their arm. The clerk would just turn the other eye.

That last phrase glides by remarkably smoothly; no one would have any trouble understanding it. The interviewer didn't snicker and say, "*Turn the other eye*, eh?" But no matter how smooth it sounds, "turn the other eye" is not an English idiom; in fact it makes no sense at all. There are, however, several English idioms that are, in various senses, close to it. One is "look the other way", another is "turn a blind eye", and a third is "turn the other cheek". There is also the shorter idiom "turn away". We won't speculate as to how these four idioms might have contributed to the utterance, nor will we claim that we are sure that *exactly* these four lexical items and *no others* contributed.

Indeed, one of the problems with the retrospective analysis of lexical blends (and until there are incredibly sophisticated real-time brain-scanning mechanisms, and until we understand the brain infinitely better, there can only be retrospective analyses based on plausible guesses) is that although sometimes it seems obvious what the contributing phrases had to be, at other times it is highly debatable. We error-makers do not have privileged access to our unconscious mechanisms (they are "under the scenes"!), and we are not necessarily more reliable analysts of our own utterances than outsiders are, so the phrases that we ourselves suggest as being the "culprits" are not necessarily the right ones. In any case, it would be naïve to think that all contributing phrases contribute *to an equal degree.* So what we are faced with is, first, a list of *plausible* contributing phrases, which could include many more than two, and then the thorny (in fact, unanswerable) question as to *how much* each phrase contributed to the final blend (not in the sense of *how many words* it contributed, but in the sense of *how much influence* it wielded).

Here is a lexical blend with an unclear number of ingredients:

> What is it the hell he wants, anyway?

Two contributing phrases are clearly "What is it that he wants?" and "What the hell does he want?" But there is also a potential third contributor — "What in the hell does he want?" — and even possibly a fourth — "What in hell does he want?" Although we contend that there is (or at least there was, at the time of the utterance) an actual, scientific fact of the matter as to which of these phrases *did* or *did not* contribute to this error in the brain of the particular speaker, we acknowledge that such questions are completely unapproachable, given today's level of understanding of the human brain.

Here's a striking case of multiple influences:

> My dad really hit the stack when I got home so late.

What the speaker meant is that her father grew very angry very fast, which is to say, he both *hit the ceiling* and *blew his stack*. Any native speaker of American English will immediately realize that these two expressions were involved behind the scenes, and perhaps will think that that is the whole story. However, that's very unlikely. Each of these phrases contains roughly half of the final blend, but there is another very common American idiom — namely, "hit the sack" — which *as a whole* sounds almost identical to the expression that was uttered. Even though the meaning of "hit the sack" — "go to bed" — is utterly unrelated to sudden bursts of anger, it is hard to believe that the "phonetic pull" or "sonic attraction" of that standard phrase played no role in this blend. It was like a huge planet gravitationally pulling the speaker towards it.

In this lexical blend and in all others, the phrases that contribute are summoned up out of their dormant state either because of some kind of *conceptual proximity* to an appropriate concept, or because of *phonetic proximity* to some lexical item (frequently the nascent blend itself). And either kind of proximity means that an unconscious analogy has been made. To crudely summarize this particular unconscious mental process in just a few words, once the erroneous hybrid phrase had been assembled (by splicing together the fruits of two rival *semantic* analogies), it "sounded right" because of its phonetic similarity to a very familiar phrase (*i.e.*, thanks to a *phonetic* analogy), hence it became more alluring, and this gave it a big boost towards getting uttered.

The following amusing phrase escaped from the lips of a distinguished professor of cognitive science as he was paying gracious tribute to an administrative assistant who was stepping down after many years of service:

> She'll be hard shoes to fill!

To the perplexed listener who quickly captured it for posterity in his little notebook, this sentence sounded so extremely strange (indeed, on the verge of incomprehensibility for a moment or two) that he was astounded to see that no one in the assembled crowd of cognitive scientists was smiling, chuckling, or writing it down. How could such a weird, garbled phrase go unnoticed by scores of people whose profession is the study of the mysteries of language and cognition? And yet it did.

The contributing phrases in this case clearly include the two idioms "She'll be a hard act to follow" and "It'll be hard to fill her shoes", but it's also likely that the simpler phrase "She'll be hard to replace" played a role here, since, like the blend itself, it uses the words "She'll be hard", without the indefinite article "a".

We conclude this section by looking at a blend that strikes us as having even a larger number of plausible contributors than the blends we have discussed so far:

> Things are looking glimmer and glimmer these days.

It's almost certain that at least two of the adjectives "glummer", "gloomier", and "grimmer" were involved in the behind-the-scenes story of this very pessimistic blend. Each of them bears a unique and strong phonetic resemblance to the nonexistent adjective "glimmer". It's also possible that one or both of the adjectives "slimmer" and "dimmer" might have played a role. And finally and most intriguingly, there is the standard phrase "a glimmer of hope", containing the very word "glimmer" that appears inside the blend. Curiously, though, inside "a glimmer of hope", the word "glimmer" is a *noun*, whereas inside the blend it is an invented *adjective.* Although the meaning of "a glimmer of hope" is cautiously optimistic, it's a phrase that tends to be uttered only when a very worrisome situation is being described, so it's a close cousin of the overall pessimistic feeling that lay behind the blend.

We hasten to add that not all six of these plausible contributors were immediately obvious to us. It took us much rumination to come up with them, but once we had done so, every one of them struck us as quite plausible, four of them even being very strong candidates. So here we seem to have six different potential ingredients, and yet it's impossible to know which of them were actually involved, and how deeply, and in what ways. And of course there could be yet other words or phrases that played a contributing role but that we didn't think of. All in all, then, the question of what *really* took place behind the scenes in a particular lexical blend is, as this example shows, fantastically slippery. All we can do is make educated guesses.

Single-word Lexical Blends

Most of the blends that we've looked at so far involve phrases made up of several words, but blending can also take place on a smaller scale, using exclusively short words. The resulting blends pass by extremely fast, however, which makes them even less likely to be heard. For instance:

Our book is maistly about analogy-making.

This oddity, featuring an obvious blend of the two rival adverbs "mainly" and "mostly", has been heard many times since it was first added to our collection. When seen in print, it stands out like a sore thumb, but in the flow of very rapid speech, it often sails by completely unheard.

Another simple but subtle one-word blend is the following:

I don't want to dwelve into it.

In this case, the participating phrases were "to dwell on it" and "to delve into it". Both "dwell" and "delve" are fairly unusual words. Daring to use a word at the fringes of one's vocabulary is a bit like making a long, risky leap to hit a note near either end of the keyboard; obviously the chances of hitting two neighboring notes at once go up greatly when one lunges at a target that is far from one's most familiar territory.

And yet sometimes very high-frequency words are blended as well, as the following case shows. The phone rang and Danny's father picked it up. "Is Danny there?" asked a young voice. The father heard himself reply:

I'm not sure — I'll go seck.

Being an ardent error-collector, he instantly jotted down his own error with glee, at first thinking it was just a cute amalgam of two one-syllable words — namely, "see" and "check". It occurred to him to wonder why he hadn't combined them in the reverse fashion ("I'll go chee"), but then he realized that there was a third component playing a key role behind the scenes — namely, the very common phrase "Just a sec!", which had certainly been one of his thoughts as he put the phone down, even if he didn't utter it. If there had been any competition between "I'll go chee" and "I'll go seck", this extra hidden phonetic factor would have easily tipped the balance in favor of "I'll go seck."

Here's another blend of single words, one that we've heard many a time, and once again, it tends to go by so fast and to sound so natural that it's easy to miss:

Why, shurtainly!

A fight between "Sure!" and "Certainly!" is shurtainly what lurks behind the scenes.

Herewith follows a small pot-pourri of lexical blends at the single-word level:

Monosyllabic blends: That's a *neece* idea! / She's pretty *darmn* young for him, if you ask me / A couple of unwarranted assumptions have *slept* into the text / Will you please *stut* that out? / I decided to *skwitch* to a later chapter / We'll board as soon as the *plight* has unloaded / He's a big *wheeze* in the NSF / I'll come over pretty *shoon* / He vanished into the *maws* of the volcano / She's scared of little beasties that *prounce* all about...

Polysyllabic blends: He's such an *easy-go-lucky* guy / It's a pretty tall building, fifteen *flories* high / An outburst in which she *bystepped* all logic / She'll probably arrive somewhere *arout* ten o'clock / I did it as *quiftly* as I could / Yeah, I see what you're *griving* at / He's got lots of *oddbeat* ideas / Wow, you're quick on the *updraw*! / I'm going *outstairs* to get the paper / So far they just have two *kildren* / She's out in California, visiting her *farents* / Thousands of people jammed the square to show their concern for the *frailing* pope / You've just gotta keep on *truggin'* / And then he slowly pulled on his *slousers* / His oversight caused a *humendous* wave of guilt / They were very *viligent* about including bike paths in the plan / Don't be such a *slugabout*! / She was the *spearleader* of the project / She's such a self-centered, *inconsisterate* person / I agree *full-heartedly* / That woman who lives across the hall is such a *biddy-boddy* / You should probably add a short *appendum* to your article / What in the world motivates that kind of *zealousy*? / It was a real *annoyment* / For his birthday they're preparing a big *shing-ding* / He was bored to death by all that *grudgery* / With a catchy title like that for your talk, you're bound to get a good *showout* / Her sense of humor could do with some *retunement* / Of the two parents, I'm the *pushycat* / Why does the phone always ring at some *un-god-awful* hour?

Blurting out the Exact Opposite of What One Means

Lexical blending sometimes results in one's saying the exact opposite of the idea that one intends to convey. For instance, when the fan belt in a certain woman's brand-new car suddenly snapped, she frustratedly commented, "I don't think those things just break for no reason at all!" Her husband, who concurred and wanted to lend her moral support, replied:

I don't agree!

inadvertently blending "I agree" and "I don't think so either", so that his remark came out 180 degrees opposite to his intention.

A professor had been invited to submit an article to an anthology and was glad to accept but he had no time to write anything new. Luckily, he remembered a piece he had written some years earlier but never published. However, on rereading his old piece, he saw that it was not very much on the topic, so he reluctantly had to abandon that avenue. Then a colleague suggested to him that he use a different unpublished manuscript, and he was very enthusiastic about that idea. But when he looked up this second old manuscript, he discovered, to his shock, that it, too, was very far from the topic — in fact, considerably more so than the article he'd already given up on. In frustration, he commented, strongly stressing the two words in italics:

Unfortunately, *this* article's even *less* irrelevant than the first one was !

Unwittingly, he had combined "even more irrelevant" with "even less relevant" and in so doing he came out with a statement diametrically opposed to what he was thinking.

Biplans

A certain kind of blend involves two (or more) different *thoughts* that occur to the speaker at the same time. We will refer to such a blend as a "biplan", since it seems to blend two different plans for the utterance. A biplan is different from the lexical blends described in earlier sections in that the analogical link between the two rival items is more abstract, and for that reason the blended ideas seem further apart.

A typical example of this phenomenon took place when Francisco exclaimed:

It sure is thirsty!

He was simultaneously thinking the thought "It sure is hot!" and also the thought "I sure am thirsty!", and he telescoped the two thoughts into one short sentence, which, though compact, made no sense. Being hot and being thirsty are two quite distinct physical discomforts, and *it's hot* situations and *I'm thirsty* situations have less overlap than do most of the blended situations that we've covered in the past few pages.

Nonetheless, both types of situation involve a core essence of *physical discomfort wanting to be remedied*, and it was to an analogy residing at this rather high level of abstraction that Francisco's blend was due.

Among the biplans in our collection are the following ones:

> *I like these ones nicer* than those / I hope that she gets a hold of him today, but *I wouldn't cross your fingers* / She's *finishing the touches* on her new cookbook / When I heard the news, I *couldn't help not to* think of you / It all *depends to* what you do now! / Sorry, but I just *can't make it won't work out* / I thought I looked everywhere, but *I hadn't occurred to me* to look in the drawer / *Be carefully* if you're driving in this weather / *Get well better!* / *Where* in the dickens did I *do with* that stupid knife? / *I'm gonna 'bout to* take my shower / Oh, sorry — I *completely remembered* that I'd promised to send it to you! / There's really no need to thank me; it was *my problem*! / Boy, that package *weighs a lot of money*!

In several of these it's apparent that there are rival grammatical plans that are duking it out behind the scenes. In the final one, the two rival plans involve a competition between the thoughts "it weighs a lot" and "it's worth a lot of money".

Here is a biplan that took place in Italian but that is simple to understand whatever one's native language is. Alberto was speaking on the phone with a friend, and the conversation was winding down. Trying to finish up in a courteous fashion, he said:

Grao!

When he heard this nonsense word come out of his mouth, he was quite embarrassed by it. It was the awkward result of an attempt to say "grazie" ("thanks") and "ciao" at the same time, so he rapidly tried to save face by trying once more, intending to say just one of them. Unfortunately, though, he hadn't quite made up his mind which of the two rival thoughts he really wanted to express, and so the second time what came out of his mouth was again wrong, though somewhat less wrong. It was this:

Giao!

which was nearly "ciao" except that the initial consonant sounded like an English "j" instead of English's "ch" sound. What had happened was that a tiny residue of the initial voiced consonant cluster "gr" of "grazie" remained inside the new word's initial consonant, altering the intended unvoiced "ch" sound by voicing it. This is a good example of how two components in a blend can contribute to it to different degrees, "ciao" clearly having had the upper hand, but "grazie" not letting itself be shut out entirely. Here again we see a blend brought about through an abstract analogy — namely, the analogy between *grazie* situations and *ciao* situations, which are both centered on the speaker's intention to conclude a conversation politely.

Our final example of a biplan is taken not from speech but from an email message that one of us received from a friend:

I hope my package got there in one shape.

Here the contributing phrases, based on common idioms, involved distinct though related thoughts: "got there *in one piece*" and "got there *in good shape*". The resulting blend is rather humorous.

As this example and many others show, there is no hard-and-fast line between blends that involve genuinely distinct thoughts (*i.e.*, biplans), and blends that involve very closely related or synonymous thoughts (*i.e.*, more standard lexical blends). It's a matter of degree and thus a subjective judgment call, but to our minds it's a useful distinction, although of course, as is true for any category, its edges are far from sharp.

Conceptual-proximity Slippage Errors

Autumn had arrived, and Sandra, looking out at a yard completely covered with fallen leaves, commented:

We've got to peel the lawn.

What she meant was: "We've got to rake the lawn."

Sandra used a word that is typically used in connection with potatoes, cucumbers, apples, or other vegetables or fruits, as if the lawn had acquired a skin that some appropriate mechanical device could just strip off. Had she been writing a poem about the advent of autumn, her word choice might be considered a novel metaphor for getting rid of the leaves; it would have seemed creative and insightful. But this was simply an everyday utterance, and as such it is reminiscent of the children's sentences featuring surprising verb choices, such as "undress the banana", "eat some water", "nurse the truck", and others, which we discussed in Chapter 1.

There is, however, a noteworthy difference between this case and those. The children had not yet acquired the fine-grained semantic mesh that would have allowed them to find the appropriate verbs. They simply lacked the crucial concepts and the associated lexical items, and so their utterances were cases of people doing an optimal job, given the limited set of concepts at their disposal; Sandra, by contrast, simply retrieved the wrong concept, given the much broader conceptual repertoire that she possessed (and she did not retrieve and apply the concept *deliberately*, as a poet might).

What mechanisms could lie behind this anomalous choice of concept and of word? The same ones, essentially, as are responsible for phrase blends, as discussed above. In the case of those blends, though, the battles were never won outright by a single word or phrase; instead, two, three, or more winners split the spoils in the utterance that was produced. The difference between those cases and Sandra's "peel the lawn" comment is that in Sandra's case, the battle *was* won outright, but it was won by an impostor. This can be compared to a pianist who, under intense real-time pressures, aims at a broad region of the keyboard and who, rather than hitting two keys at once, hits just one, as intended, but unfortunately it is a clunker rather than the right note.

To describe such an event, we can say that a quite broad zone of a person's conceptual space was activated, containing a number of concepts with distinct lexical labels, and one of the less appropriate concepts in that zone somehow wound up winning the battle for utterance. In the case of Sandra's utterance, in the broad swath of conceptual space that was activated in her brain by the sight of the leaf-strewn lawn, aside from the concept *to peel* there was almost certainly a set of closely clustered concepts, some of which would seem fairly close to the activity of gardening, such as the concept *to rake*, and very likely a set of other concepts, including *to sweep*. Indeed, had she said "We've got to sweep the lawn", it would still have sounded like an error (or a mildly creative metaphor), but it would have sounded less wrong (or less creative), since *to sweep* is clearly much closer to the concept of *to rake* than is *to peel*.

Any slip of this sort reveals a potential analogy that is stored in a particular speaker's brain by virtue of the conceptual proximity of two concepts. Given that each concept in our mind is surrounded by a "semantic halo" consisting of other concepts that are semantically close to it, this means that in everyone's brain there are millions of such latent analogies, ready and waiting to reveal themselves through conceptual-proximity slips, though of course most of them will, alas, never be granted the chance to show their eager faces, since such slips, although fairly frequent, are not frequent enough to reveal the entire semantic halo surrounding each particular concept.

We'll begin our brief tour of such errors with a few examples in which a broad zone in conceptual space seems to have been activated and one of the concepts inside it got chosen more or less randomly, reminiscent of a pianist whose hand rises so high above the keyboard that when it comes down, it doesn't land on the right note but on a random note in its general vicinity.

> At a dinner party, a guest remarked that he had recently seen "a marvelous Finnish movie about a woman transporting a turkey across the United States in an airplane". His wife gently corrected him, saying that it in fact it was an *Icelandic* movie about a woman transporting a *goose*.
>
> For non-Scandinavians, it would be relatively easy to mix up any two Scandinavian countries, and it also seems extremely plausible for a person to confuse two varieties of large fowl standardly consumed on festive occasions. But it would have been very unlikely for the guest to have described the film as, for example, "a Peruvian movie about a woman transporting a hummingbird" or "a Vietnamese movie about a woman transporting an ostrich". Such slippages strikes us as highly implausible because we intuit that the interconceptual distances in anyone's mind are much greater, making the analogies much more improbable.

> L. was cleaning out her closet of some childhood toys. Picking up a stuffed whale, she said, "I'm going to toss out this little lion", and then felt embarrassed.
>
> Both whales and lions play "regal" roles in their respective habitats, which explains why they would lie fairly close to each other in L.'s memory, and which also explains why it would have been unlikely for L. to refer to the toy she was holding in her hand as "this little mosquito", "this little snail", "this little zebra", or "this little octopus".

A woman was explaining to a friend why he hadn't been notified of her party. "I put a bunch of PS's in my email, and I thought I'd included you."

The woman didn't mean "PS's" but "cc's". She simply confused one common two-letter abbreviation having to do with mail (and email) with another such item.

Something in the kitchen smelled nice, and when A. went in, he found his daughter fixing a cake. But something troubled him, and he said, "Move your cookbook further from the sink!" When she looked at him strangely, he repeated what he had said. She again looked at him with confusion and said, "But that's not the sink!" He tried to patch up what he'd said, but what came out was, "Your book should be far away from the kitchen!" It took him a few more seconds before he could articulate what he meant, namely "Your book should be far away from the stove!"

What we see in this example is that the distance in conceptual space in A.'s brain between *sink* and *stove* — and also between *kitchen* and *stove*, for that matter — was smaller than one might suspect. Or to be more precise, it was relatively small under the specific constellation of cognitive pressures created by that moment and that context.

K. said, "I saw a beautiful bird in my yard but I couldn't identify it, so I called up a bird-watching friend and gave her as clear an explanation as I could." Then K., hearing her own error, said, "Uh, I mean 'description'."

The concepts of *explanation* and *description* are far less tangible than such concepts as *peel*, *goose*, *lion*, *PS*, and *stove*, but intangibility does not reduce the likelihood of an interconceptual slippage taking place as a result of semantic proximity.

P. was talking about various ingredients one could put into a vegetable soup, and he wound up by saying, "And of course pasta is temporary — or rather, it's *optional*."

This example, centered on a semantic slippage between adjectives instead of nouns, serves to remind us that what goes for nouns goes for other parts of speech, and for longer phrases as well.

Professor H. said, "I have a wonderfully free position at my university: instead of having to teach standard courses, I get to dream up egocentric seminars on all sorts of topics."

Was this a Freudian slip, revealing the deep dark secret of Professor H.'s unbounded egocentricity, or was it a more innocent event? We would opt for the latter, suggesting that it was merely a vanilla conceptual-proximity slippage error, in which the long adjective "idiosyncratic" was replaced by the slightly shorter adjective "egocentric". Both adjectives have meanings that could be expressed approximately as "determined by my own personal desires", with "egocentric" leaning more towards the idea of "doing exactly what I want" and "idiosyncratic" leaning more towards the idea of "indulging my personal quirks", and so it's not in the least surprising that both of them would have been simultaneously activated in this particular context. Moreover, the two words are phonetically very similar — in fact, so similar that it's very likely that exactly this same semantic slippage has been committed hundreds if not thousands of times before, the world around.

Some semantic slippages seem so weird that one might think that whoever made them must be deranged or must suffer from a highly deprived conceptual and lexical repertoire, and yet the people who made the following strangely childlike errors were all perfectly normal adults.

> W. and his wife were watching a DVD on television at home. The opening scene was so dark that it was hard to make anything out, so W. got up and walked over to the window, saying, "I'm going to turn off the shutters."
>
> L. asked, "When do the stores finish this evening?"
>
> When F. was asked, "How old is your house?", he replied, "When we were considering buying it, we were told that it was born in the late 1930s."

The above examples are reminiscent of the utterances "They turned off the rain" and "Turn your eyes on, Mommy", which were made by children and were cited in Chapter 1, but since the people who came out with the above oddities were all adults and were thus at a far more advanced stage of cognitive development than young children, their utterances have to be considered errors. Such slippages show that despite the refinement of a mature and sophisticated person's conceptual repertoire, semantic connections that were forged during their childhood remain latently present for the entirety of their life.

> A. woke up extremely groggy and said to her husband, "Too much sunlight — could you please pull the eyelids shut?" In fact, she meant "the curtains".
>
> When one is in such a woozy state, one can blurt out very peculiar things. This particular slip reveals a potential analogy in A.'s brain, by virtue of conceptual proximity, between the concepts of *curtain* and *eyelid.* This seems rather odd, but how else can one account for her strange remark?
>
> S. worked in a large company and had misplaced a document. "Just a second — let me go check in my bedroom", she commented (instead of saying "in my office"), thus unwittingly revealing a certain degree of mental confusion between her personal and professional life.

Many conceptual-proximity errors are characterized by involuntary slippages along the dimension of time, showing that a category such as *yesterday* may extend much further out, by means of unconscious analogy-making, than one would *a priori* expect. The following examples illustrate this kind of slippage:

> A very common error — one made by virtually every teacher from time to time — is to start a class by saying, "Yesterday we were talking about..." usually meaning two days earlier, but sometimes a week earlier. A variation on this theme took place when a teacher opened class after two weeks of vacation by saying, "Last week, we were talking about..."

Once again in this same general ballpark, a colleague was discussing the courses he had taught during the previous semester, and he came out with "The courses I've so far taught today" when he in fact meant "this year", thus revealing a latent analogy stored in his brain between short-term and long-term varieties of *now*-ness.

A mother was explaining that her son would soon return in a couple of days from Chicago and said, "He won't get a lot of sleep, since his grandparents will be picking him up bright and early on Saturday, and he only gets back from Chicago very late yesterday." Of course she didn't mean "yesterday" but "the previous day".

This mother had intensely projected herself into her son's point of view, imagining how he himself would feel when his alarm clock went off on Saturday morning, and she described Friday night from *that* vantage point, as having already gone by, rather than from her *actual* vantage point, in which Friday night was still a couple of days in the future. In other words, for this mother, the intensity of identification with her son briefly reduced the semantic distance between the concepts of *the previous day* and *yesterday* so much that the one was able to slip into the other.

While all the previous slippage errors were based on simple and obvious analogies, the errors in the next set are based on subtler analogies, which involve some intangible function (the abstract action performed by an entity, the abstract role played by a person, and so forth).

He and she were discussing an event due to occur on October 20 and she wanted to know what day of the week that would be. He impatiently said, "Go look at the map!"

What he meant was "Look at the calendar!" The slip is easily explained, a calendar being an evident temporal analogue of a spatial map, but nonetheless, in this context, no one would intentionally say "map" instead of "calendar".

A student was relating the story of the tragically early death (at age 20) of the great French mathematician Évariste Galois, and he said, "And so, the night before the fateful debate, Galois stayed up all night and in a frenzy wrote down all his ideas…"

The student knew very well that Galois had died as a result of a duel, not a debate, but the concepts *debate* and *duel* were semantically close to begin with in his mind (as they are in ours as well). Also, the presidential campaign was in full swing at the time (not "in high swing", as we originally wrote here!), and televised debates had just taken place, making it much more likely for these particular wires to be crossed.

The *debate/duel* analogy is just one of myriads of potential analogies that are hidden in each human mind but whose existence one wouldn't suspect *a priori.* The student's error, however, reveals that this analogy was indeed lurking in his mind and simply needed the right opportunity to show its face. The intensely political atmosphere of the period primed the concept of *debate*, which had the consequence of reducing its distance from the concept of *duel.* Moreover, the likelihood of such a slippage was enhanced by the (essentially irrelevant) fact that the image of writing one's ideas down on paper is much more easily connected with the concept of *debate* than with that of *duel.*

D. wanted to make a comment about the fall on skis taken by California governor Arnold Schwarzenegger, but the first word that came out of his mouth was "George", from "George Bush".

What could have led to D.'s confusion between these two people? In the first place, both were very salient right-wing politicians of roughly the same age — but there was more than just that linking them in D.'s mind. Whereas Governor Schwarzenegger had recently broken his leg in a tumble taken while he was standing stock still on a ski slope, President Bush had famously nearly choked to death one time while eating a pretzel. The analogy linking these two mishaps of celebrity politicians is subtle. The shared conceptual skeleton involves a serious threat to a person's physical well-being that comes about unexpectedly during a seemingly trivial and innocuous activity. It's hard to imagine that there would be an *a priori* link in anyone's brain between the ideas of *eating a pretzel* and *standing on a ski slope*, or between the ideas of *breaking a leg* and *choking on something in one's throat.* On the other hand, the abstract idea that *no matter how famous one is, one can get hurt in a trivial-seeming accident* may well have been a key part of D.'s unconscious encoding of each of the two mishaps, and if so, that common element would have catalyzed his retrieval of the first name "George" in this context.

A science writer who had just published a biography of the physicist and Exploratorium founder Frank Oppenheimer was having tea with a friend, and she stated that Frank Oppenheimer and his more famous brother Robert had both been infatuated with communism in the 1930's but had rejected it after a short while, and in that connection, she casually added that in those early days of the Soviet Union, Benjamin Franklin, too, had traveled there but he had come back very disillusioned with it. "Benjamin Franklin?" exclaimed the friend, amused. For her part, the science writer was shocked that she had come out with that clearly wrong name, and yet she couldn't recall who she had in fact meant. A few days later she finally did recall who it was — Thomas Edison.

This confusion was due in part to the fact that in the science writer's mind, both Edison and Franklin were strongly associated with electricity, but there is far more to the explanation than that. Firstly, both individuals were saliently American (otherwise, why not Ampère, Maxwell, Faraday, or many other famous scientists?). Furthermore, both were self-taught inventors, and both are commonly associated with folksy wisdom (think of Franklin's *Poor Richard's Almanac* and all the sayings associated with him, and Edison's famous quotes, such as "Genius is 1 percent inspiration and 99 percent perspiration").

Though we cannot be sure which of these factors played a role, let alone how large a role, we see that there are a number of analogies linking Benjamin Franklin to Thomas Edison, any of which could have given rise to this slippage. And there are further factors that could have tipped the balance in favor of the name of Benjamin Franklin. For instance, it might have been relevant that his last name coincides with the first name of Franklin Delano Roosevelt, a contemporary of the Oppenheimer brothers and a figure indelibly associated with the relations between the United States and the Soviet Union during that period.

A woman was talking about her nephew, who at age 38 had finally gotten married, and she said, "His mother's thrilled. Now that Peter's pregnant, she's hoping to be a grandmother any day now."

But no one was pregnant — surely not her nephew, but not his bride either. The aunt merely meant "now that Peter's married". However, since the conceptual distance between *married* and *pregnant* is fairly small, and in this highly emotional context even more so, it was very easy for her to slip from one to the other. Of course it wasn't the aunt herself who was ardently dreaming of having a grandchild, but she had projected herself very effectively into the mentality of her sister.

Among the most curious of conceptual-proximity substitution errors are ones based on slippages between a concept and its opposite concept. Here are a few examples:

P. has noticed that on countless occasions over the years he has used the verb "to read" when he meant "to write", and vice versa. He has also observed that many of his friends have the same tendency, and moreover, he has seen that this is a reliable tendency across languages. Moreover, when speaking French, P. has often caught himself on the verge of blurting out the word "mort" ("dead" or "died") when meaning to say "né" ("born"), stopping a hair short of making this most embarrassing slip-up.

David and his aging father Jim were driving by a cemetery. Jim commented, "This is where all four of your grandkids were born." David, who had a daughter of five but no grandchildren, was bewildered by this absurd-sounding remark. But after a moment's thought he realized that there was a great deal of sense to it — indeed, it was completely true — if he simply replaced two key concepts by their opposite concepts: "This is where all four of your grandparents are buried."

Jim's remark was in a certain sense more incorrect than if he had said either "This is where all four of your grandparents were born" or "This is where all four of your grandchildren are buried", since those sentences contain only *one* error apiece. And yet, although Jim's utterance contained *two* errors, it was much more self-consistent; it is as if one of the two conceptual slippages had brought the other one along on its coattails, as a result of an unconscious desire on Jim's part to be internally coherent.

Struggling to recall the name of an acquaintance, a woman said, "Unfortunately, my brother's not home, because I can't ask him."

In this error, one frequent conjunction ("so") was replaced by another one ("because") having the opposite meaning, thus inverting cause and effect.

To be considered opposites, two concepts must share a great deal. For example, *big* and *small* are opposite *sizes*; likewise, *light* and *dark* are opposite degrees of *brightness*. The fact of inhabiting opposite ends of a spectrum is what makes these pairs of concepts be located very near each other, and it gives rise to the possibility of slippage between them. Ironic though it may seem, oppositeness, which naïvely makes one think of a maximal distance, is actually a type of conceptual nearness; it simply resides at a more abstract level than one usually associates with categories (for example, *brightness* is more abstract than *light* and *dark*). Thus the two extremities of any life are birth and death, and grandchildren and grandparents are both linked to a given person by being two steps away in the sequence of generations, either upwards or downwards. Likewise, reading and writing are both activities connected with printed text, one involving the "decoding" of text and the other carrying out "encoding"; and finally, "because" and "so" both express causality but see it from opposite points of view.

Many conceptual-proximity substitution errors are triggered by situations where a concrete analogy (visual, auditory, tactile, gustatory, etc.) and a more abstract one (functional or role-based) reinforce each other, each contributing some pressure towards the slippage, and where the joint pressure due to the simultaneous analogies becomes irresistible. Herewith follow some examples, with the mutually reinforcing analogies spelled out.

A couple emerged from a pizza place with a hot pizza in a cardboard box. The man pointed to the rear of his bike and said, "I'll take the pizza home in my trunk" (meaning in the basket located above the rear fender of his bike).

The *functional* analogy here is between parts of a wheeled vehicle that are designed to carry items of any sort. The *visual* analogy is that both a car's trunk and this particular bicycle basket were located behind the "driver's seat", and moreover this particular basket happened to be a rather large one. Had the basket been located above the front wheel or had it been very small, then a slippage from "in my basket" to "in my trunk" would have been less likely to occur.

W. called the doorknob of the bathroom "the faucet".

The *functional* analogy involves a small object that controls or regulates a much larger object. The *visual* analogy is that both fit comfortably in the palm of one's hand and both work by twisting. There is also an aspect of priming involved, since it was not a *random* room's doorknob that was called "faucet", but a doorknob in a room that featured faucets aplenty.

Two friends were at the edge of a lake. One of them saw a hang-glider with a very thin white line diagonally descending from it towards a motorboat. Pointing skywards, she exclaimed, "Look at that glider being pulled by a boat!" The other replied, "Oh yes, I can see the string!" Several times more she called it a "string", until her friend smiled and said, "Don't you mean 'wire'? You're probably thinking of a kite."

The *functional* aspect here is that wires and strings are both used to pull things in all sorts of contexts. The *visual* aspect is that kite strings often go extremely high into the sky and they link high-flying devices to people who are down on the ground, and moreover, kite strings are usually white, just as this tethering line seemed to be. Moreover, a hang-glider seen from far away looks very much like a kite. Even this visual analogy, however, has abstract aspects to it, since the role of the person flying the kite was being played by the boat, the role of the ground was being played by the water, and of course the role of the kite was being played by the hang-glider.

S. referred to the subtitles in a film she was watching as "the footnotes".

The *functional* analogy is that both subtitles and footnotes are, generally speaking, short written aids to comprehension, while the *visual* analogy is clearly that both occur at or near the bottom of the visual field.

A brother and sister were emptying out their late parents' house of clutter that had accumulated over some fifty-plus years. One of them referred to the basement of the house as "the attic".

The *functional* analogy is that both areas of the house had been used for storing the same kinds of old, musty items over a period of decades. The *spatial* analogy is that both the basement and the attic were very large areas of the house, and both could be accessed only by taking stairs.

A. once complained, "I have a wart on my foot and it hurts when I walk." He meant "blister".

The *functional* analogy is that warts and blisters are unpleasant growths on one's skin. The *visual* analogy is that they have roughly the same size and look somewhat alike.

It's convenient to try to distinguish between perceptual and functional analogies, but the distinction is far from being black-and-white. One of them is based on what is sensorially obvious and the other is based on what is inferred indirectly, even if the inference is so rapid that one doesn't feel one is drawing any intellectual conclusions at all. For this reason, what is perceptual and what is functional are nearly inseparable. For example, *roundness* is a visual attribute and *rolling* is a functional quality, but saying "Round things roll" sounds like a vacuous utterance. We "see" that the back of a chair is intended to brace someone's torso and that the seat is intended to support their posterior, and these functional qualities seem to be direct perceptions, just as, in our list of semantic slippages, the fact that a car's *trunk* is a *container* seemed a trivially obvious fact, and much the same for the fact that a *wire* and a *string* serve as *connectors.* When the link between perceptual and functional analogies is very strong, then we can speak of strong "affordances", to use the term devised by the American psychologist James Gibson in the late 1970s to describe the way that an object can implicitly suggest the actions that one can carry out on it.

By contrast, there are cases where it seems clear that the pieces of knowledge activated during perception go beyond what the senses perceive directly. For example, the facts that a dog can protect or can threaten a human being do not directly follow from perception, but the act of perceiving a dog allows these facts to be activated and recalled; thus visual input facilitates access to functional knowledge. The fact that footnotes and subtitles are both pieces of writing aimed at facilitating comprehension, and the fact that blisters and warts are both unpleasant-looking skin growths that one wishes to rid oneself of, are functional pieces of knowledge that any adult has in dormant storage and that can be activated by visual perception. (The same could be said for the relationship between the concepts *round* and *roll*, but such a basic connection is so deeply ingrained that it is hard to think of it as having been learned.)

Actions Meet Words

Are the foregoing types of errors limited to the domain of language? Although the term "slip" is generally thought of as referring to speech errors (as in "slip of the tongue"), the same kinds of phenomena show up outside language, in the world of physical actions. Indeed, speech errors are just one manifestation of a broad phenomenon that concerns categories in general, and which reveals itself in various contexts, including physical actions. Action errors, too, have their own eloquent fashion of showing how conceptual wires can get crossed thanks to analogies.

Sometimes we activate appropriate categories but apply them to inappropriate targets. A typical example is when, after pouring ourself a cup of tea with a bit of milk, we pick up the teapot to put it "back" into the refrigerator, when of course our intention was to put the milk back. "The same thing" happens when, after finishing our bath, we reach out and turn on one of the water faucets, when in fact what we intended to do was to open the drain to let the water out. Our intention (to induce some water to flow) was eminently reasonable, but our action brought about a flow in the wrong place and the wrong direction. Or else we can't find our computer's charger because earlier, in a moment of distraction, we placed it in the proper pouch but of the wrong container — our backpack's little pouch instead of our computer bag's little pouch — thus getting the category *pouch* correct but missing on the specifics.

> C. often placed her cell phone next to her brand-new mouseless computer, and she often reached for it as if it were a mouse. However, it never worked as a mouse, even though it was just about the size of a mouse and was right where a mouse ought to be.
>
> Having pulled into a gas station, D. noticed that the gas pumps were on the wrong side of his car. Instead of making a U-turn or moving across to the parallel set of pumps, D. executed *both* of these maneuvers, and lo and behold, he found himself with the new set of pumps on the wrong side of his car, exactly as before. This was an *action* biplan, similar to *speech* biplans in which a person winds up saying just the opposite of what they intended to say ("My bracelet came unloose").
>
> One time E. wanted to add a bit of sugar to his coffee and simultaneously to tap his ashes into the ashtray; he wound up pouring sugar into the ashtray while tapping his ashes into his coffee cup, obtaining an undrinkable coffee but a sweetened ashtray.
>
> Then there was the sleepy teen-ager F., who poured his milk right into the cereal box instead of into the bowl that he'd just filled with cereal, and his younger brother G., who, on waking up in the middle of the night, tried to locate his cell phone by shining light around the room from… his cell phone.

As we see, these kinds of action slips are directly comparable to the slips of the tongue discussed above. It makes no difference that some involve words and others do not; the underlying cognitive mechanisms are the same. Our next example is a striking case in which words and actions are intimately mixed.

The Error that Boggled the Mind of the Error Connoisseur

It was just about time for the German lesson of Doug's son Danny, and Christoph, Danny's tutor as well as a family friend, arrived a bit ahead of time. Father and tutor chatted for a moment in the kitchen, and Doug asked Christoph if he would like something to drink — a Coke, some fruit juice, anything he felt like. Christoph replied, "A glass of water would be perfect, thanks." But such asceticism didn't jibe with the Christoph that Doug knew, so to entice his guest he opened the refrigerator and said, "Cranberry juice, apple juice, orange juice, Coke, milk, coffee, tea?" But Christoph simply said, with a gentle smile, "Thanks, but really I'd just like a glass of water." A bit puzzled but seeing no reason to push any further, Doug turned toward the cupboard where the glasses were, and at the same time he pulled his wallet out of his rear pocket, extracting a one-dollar bill from it. Then, in a friendly fashion, he proferred the bill to Christoph. Only when the latter looked at him in a nonplussed manner did Doug realize that something had gone very much awry. Staring bewilderedly at the piece of paper in his hand, he exclaimed, "What on earth am I doing?"

A moment's thought, however, shed light on what almost surely lay behind this mysterious act. "Do you know what just happened?" he said to Christoph. "Your request really threw me, in its minimality. Who would go for flavorless water over a delicious drink? And yet I was fully intending to give you your ascetic choice, but at the same time I was also aware that I owed you $20 for the lesson, so I pulled out my wallet. As I tried to carry out the two unrelated actions simultaneously, my wires got badly crossed. My intention to give you some water got blurred in with my intention to pay you: I unconsciously blended the two goals. Offering you one dollar showed that I'd conflated the simplest possible *drink* with the simplest possible *bill*!"

Indeed, lying at the very core of Doug's action error was an excellent analogy, since the abstract idea of *the minimal version of something desirable*, originally triggered by Christoph's insistence on a mere glass of water, had been deftly carried over by Doug from the domain of *drinks found in my kitchen* to the domain of *bills found in my wallet.* In a manner reminiscent of the lexical blends that we've discussed at considerable length, Doug had changed horses in midstream, nimbly hopping from *plainest beverage* to *plainest bill.* This slippage had doubtless been caused by his bafflement at Christoph's ascetic choice of drink, but the analogy had remained totally unconscious, for otherwise Doug wouldn't have been perplexed by his own action.

Curiously enough, there is a classic adjective in English that applies perfectly to both sides of the bridge in Doug's mind — namely, "watered-down". In a literal sense, a glass of water is (quite obviously) the most watered-down drink possible, while in a more abstract sense, the most "watered-down" bill is (also obviously) the one-dollar bill.

Playfully flipping this analogy around, we can imagine the following exchange between Doug and Christoph. "While we're waiting, can I offer you a little bit of cash? Take a gander at all these lovely bills in my wallet! I've got a five, a ten, even a twenty — whatever you like!" "Oh, thanks a lot, but a one-dollar bill will do just fine…" Thereupon, Doug goes to the sink, pours a glass of water, and proffers it to Christoph.

Though Many are Called, Just One is Selected

As we've seen, speech errors and other anomalous actions are the visible traces of a ceaseless unconscious competition between categories, under various pressures. Most of the time, just one of the competitors handily wins out, and in such cases, no auditory or written trace is left of the hidden contest. In that sense, listening (especially without a trained ear) to a smooth, fluent-sounding conversation is a bit like browsing through a photo book of the Olympic Games in which only gold-medal winners are ever shown; one would never suspect that behind each winner's beaming smile, there was a long and arduous series of merciless competitions over a period of years, beginning with local competitions inside each country, then wider ones, until finally the championship event took place. For every winner, there are countless unseen and unsung losers. But there are occasional ties, and such special circumstances are reminders, although only in a small way, that a fierce competition guided the process from start to finish.

Everything depends on the ear that is listening. After all, the category *error*, just like any other category, is not a precise box with razor-sharp membership criteria, but has shaded degrees of membership. If one's ear is attuned to this kind of thing, one will recognize in the short pauses, in the slight tremblings, in the "uh"'s and the "ah"'s, in the aberrant intonations, and in various other minute sonic distortions that punctuate most people's speech, the telltale signs of that seething competition. Rather than witnessing just gold-medal winners proudly strutting their stuff, one gets at least a fleeting glimpse of the silver and bronze medalists, and this serves as a hint about the existence of the rest of the hidden iceberg of the selection process.

Thus speech errors, whether blatant or subtle, are cues that remind us of the frenetic hidden mental processes taking place in parallel and collectively contributing to the utterance of each sentence we speak. These cues are seldom noticed by anyone, but the fierce inter-category competition of which speech errors are perceptible traces pervades every moment of our lives as thinking beings.

Although speech errors don't provide any useful service for the speaker, they do so for observers of thought and language, thanks to what they reveal about cognitive processes. And this is a quality that they share with a certain odd sort of analogy that we are now going to talk about: although these analogies are eager to attend the party, once they've arrived, they wind up doing nothing.

Analogies that Serve No Purpose (Other than Telling Us about Thought)

> H. is heading off to celebrate a dear friend's sixtieth birthday. He gets into his car, sets out for the freeway, drives to the airport, and flies across the country. Friends meet him when he lands and take him to their house. The evening is a smashing success; candles forming a "6" and a "0" are blown out while many a knowing wink and nudge are exchanged about the consequences of turning 60. The next morning, H.'s friends take him back to the airport and he flies home. When he lands, he locates his car in the parking garage, heads for the freeway, and drives home.

> Nothing has changed about the road, which H. has taken a thousand times, yet the drive home is oddly different from the previous day; this time he keeps noticing the numeral "60" whenever it's displayed on his car's digital speedometer, and each time it reminds him, if only for a split second, of his old friend and the party he just attended.

The symmetry of the "out" and "back" trips provides a marked contrast to the change in H.'s brain. For all the years he'd owned a car with a digital speedometer, seeing the digits "60" on it had never had any effect on him, though of course it had shown that reading thousands of times. Yet this time it was different, since the concept *sixty* was highly activated in his mind, thanks primarily to the party, of which it had been the theme. This idea of *sixty*, activated in H.'s head, lurking silently in the depths of his brain, was avidly seeking analogues. Secretly and calmly it was waiting to pounce — and when the speedometer showed "60", it made its move! Of what cognitive interest was the newly-found link between the "sixtiness" of his friend's age and that of the speedometer's display? None. This "discovery" served no purpose in any way. All that happened was that a fleeting thought sped through H.'s mind for a second or two, bringing no new insight, not even the tiniest hint of a new idea, only to vanish moments later into thin air. So why are we focusing on an event so feckless and ephemeral?

It is tempting to think that analogy-making must be a subtle and deep intellectual process that always serves a grand cognitive goal and is always utilized in a serious and efficient manner. Yet all sorts of analogies that we humans come up with lead nowhere at all and lend no support to any prior mental project we might have had on any scale whatsoever. Quite to the contrary: since analogies emerge as an automatic by-product of our cognitive mechanisms, which are at all times searching for familiar-looking landmarks in unfamiliar landscapes before us, they frequently have no depth or insight. In a word, many of the analogies we make are utterly pointless and lead nowhere at all.

And yet, the very uselessness of the two "60"'s tells us something important. Yes, analogy is the linchpin of such crucial and deeply human activities as understanding, reasoning, decision-making, problem-solving, learning, and discovery — and this fact is already proof of the centrality of analogy in human cognition. But the fatuity, gratuity, and vacuity of the analogy linking the two "60"'s shows *even more clearly* how fully analogy pervades human thought. It shows that the search for samenesses linking situations is so profoundly inherent to thought that trivial, meaningless analogies are tirelessly produced by our subconscious minds, for no reason at all, out of the blue. Analogies pop up and momentarily seize our attention wherever the mind roams.

The analogy between the two "60"'s is not in the least anomalous. If we had a neurological technique for capturing fleeting thoughts on the fly, we would harvest myriads of pointless analogies just like it. But it's their very uselessness that condemns them to being forgotten mere instants after they are born; fleeting analogies of this sort have such a short life expectancy that they disappear only fractions of a second after being created. It thus takes both practice and interest in the mind's workings to notice such evanescent analogies at all. To shed more light on this kind of analogy, we'll take a look at a few fleeting analogies that were caught on the wing by their creators.

A Visit to the Land of Pointless Analogies

D. owns two cars that look very similar. Both are rather old blue Ford station wagons, and each of them was purchased close to the birth of one of his children. For this reason he often calls them "Dannycar" and "Monicar", which E. finds amusing. One day, as E. and D. are jogging together in a city park, E. asks D. why he bought two cars that look nearly identical, and just as D. has finished answering, they chance upon a woman on their pathway who is walking two black poodles that look very much alike. E. is instantly reminded of D.'s two blue cars. Later that afternoon, E. notices two identical backup disk drives that are sitting near each other on D.'s desk, and they remind him not only of the two station wagons but also of the two poodles. He jokingly suggests to D. that they should be called "Dannydisk" and "Monidisk".

It's a Saturday evening, and a Parisian family is enjoying an evening at home. As the youngest son gazes out the window, he makes an odd observation: directly across the street, in the facing apartment, the neighbors have hung up exactly the same chain of lights as is hanging in their own house, a type of decoration that is found in only one store in all of Paris. This greatly surprises everyone, because it's very disorienting to see a rare object of one's own in someone else's house.

The next day the family goes out for brunch together. It's a cool morning, and the mother has donned a very unusual colored scarf that she bought several years ago and is particularly fond of. As they are eating their meal, she notices a woman at a nearby table who is wearing exactly the same scarf. Without a moment's hesitation, she is reminded of the two identical chains of lights.

Hopefully, we can agree on at least one thing: that none of these analogies has any lasting intellectual value; indeed, they are all completely useless! After all, what earthly value could it have for you to be reminded of two rather old blue Ford station wagons when you run by two black poodles, or to be reminded of two black poodles when you glance at two identical backup disks? Of what use could it be to you to be reminded of two identical chains of lights when you happen to spot two identical scarves? None of these analogies brings new information or suggests fresh perspectives to the people involved, allowing them to deal more effectively with life. To be sure, all these analogies are perfectly sensible and valid — they even leap out at the eye — but none of them casts any bright light of insight on this complex world, nor do they help us resolve any pressing dilemmas.

Analogies Made Without Rhyme or Reason

It's an appealing notion that analogies are inevitably produced in the service of some goal, in support of reasoning, in acts of problem-solving, and so forth. But the universe of human thought is also teeming with intimate, private analogies that have no trace of purposiveness, that are utterly unrelated to logic, and that contribute in no way to the solving of any problems that one might be facing.

The mechanism that gives rise to such analogies creates, or brings to mind, a shared category, but not the kind of category that we wish to preserve for a long time. Such categories thus tend to be instantly forgotten unless there's a special reason not to do so. In the case of the two blue station wagons, the two black poodles, and the two disk drives, the shared category is that of *two very similar-looking items that unexpectedly belong to the same person.* As for the light chains and scarves, the category is *two indistinguishable items owned by different people and observed at the same time.* And finally, the more abstract category *two very similar-looking items whose simultaneous presence is surprising* allows the two sets of analogies to be united in one higher-level category. But no matter how one might choose to group them, these analogies are useless. Our point here, though, is not to focus on their uselessness but to see what they can tell us about the human mind.

The fruitless analogies we've been discussing were all quickly discarded. But as in any game of chance, such as a lottery, not only the winner took a chance, but so did every single loser. The point is, we're constantly noticing odd, random resemblances around us because our brains are always on the lookout for insights into reality, always using the past to try to make sense of the present, always making spontaneous connections, always throwing bottle after bottle overboard in the faint hope that one of them might reach land. Most, of course, sink to the bottom of memory's ocean and are lost forever. These wasted efforts are the price that we pay for a rare but great success — a ship in trouble saved, a kidnapped passenger rescued, whatever metaphor you prefer. And speaking of ships in trouble, we now present another extended analogy that originated on the high seas and that shows that on occasion, even the most useless and most quickly discarded of analogies can reach certain heights of abstraction.

A Fleeting Analogy Caught Just Before it Vanished Forever

Thomas has just boarded a plane en route home after a vacation. He pulls out the book his sister has given him, *Hunting Mister Heartbreak* by Jonathan Raban, and starts to read the first chapter. In it, the author tells of taking a freighter, the *Atlantic Conveyor*, from Liverpool to Halifax when he was a young man. One passage describes what the ship's captain did when he received a report that a hurricane named Helene had been spotted near Bermuda and was heading northwards in the Atlantic on a collision course with the ship's route in a day or so. He instantly decided to turn the ship southwards so as to skirt the southern edge of hurricane Helene instead of crashing head-on into it. Such a maneuver would cost them a day, but it would keep them out of danger. It worked like a charm and they arrived in Halifax "in one shape" (as someone once said).

Thomas finishes reading Chapter 1 just before his flight lands. Once he's off the plane, he heads down the wide concourse toward the baggage claim. It's not a busy time of day and he's proceeding at a brisk pace when all at once, to his left, a woman wheeling a suitcase appears out of the blue and walks straight in front of him, cutting across the corridor without at all looking where she is going. In a trice, Thomas makes a deft little swerve leftwards and then swerves back, just barely avoiding a collision with the oblivious woman and her suitcase.

> This general sort of avoidance maneuver is a dime a dozen when one is walking down city streets — we adults are all past masters at the everyday sport of "sidewalk squeeze-bys"— and one doesn't ever give such things the least thought; yet in this case, a shadowy notion flits briefly through Thomas's mind. *He's* the freighter *Atlantic Conveyor* and the broad concourse is the Atlantic Ocean. Of course, left is south, right is north, and one second is roughly one day. The sudden appearance of the woman "to the south" is the warning of a hurricane spotted near Bermuda, her blithe crosswise march is the hurricane's northwards path, and last but not least, Thomas's deft little dodging maneuver is the captain's maneuver in dodging Helene. But none of this does he verbalize as such; it's far from the center of his attention. Indeed, despite its complexity, his analogy was so inconsequential and meaningless that it was about to disappear without a trace, but by chance Thomas noticed it fluttering at the fringes of his consciousness and caught it on the wing just before it sailed out of view forever.

Why would Thomas's unconscious mind have come up with such an ephemeral and pointless analogy? Not one iota more useful than the analogies of the previous few sections, this momentary likening of an oblivious woman to a threatening hurricane did not help Thomas solve any problem, nor did it help him fathom the motivations of hurried travelers in airport concourses, nor did it yield the least insight into the capricious fury of hurricanes or other violent atmospheric phenomena. At best a cute piece of cognitive fluff, Thomas' fleeting analogy was entirely purposeless.

After the fact, one might think that this analogy was handed to Thomas on a silver platter. And yet, as obvious as it seems, its emergence was not inevitable but somewhat dicy. One can easily imagine someone else who, having just read the same chapter, steps off the plane and makes a little swerve to dodge someone crossing the concourse, but doesn't notice any link between their footwork and the just-read passage. And if Thomas hadn't just read the chapter, he too would merely have lumped his mundane avoidance-maneuver in with all the thousands of mundane avoidance-maneuvers that he routinely makes each month, and the rest of his day's cognitive activities would not have been in the least affected. In short, Thomas's analogy, although it had some abstract and subtle qualities, was neither compelling nor significant, and its creation was not an inevitable cognitive event.

If we compare this analogy with the ones discussed in the past few pages, we can say that the analogy between the two "60"'s is the hollowest of all. No insight results from spotting the identity of a friend's age (in years) and the speed (in miles per hour) of a vehicle. As for the analogies involving cars, poodles, disk drives, chains of lights, and scarves, all these depended on categories such as *two very similar-looking items whose simultaneous presence is surprising.* There is a slight bit of abstraction here but not much complexity. By contrast, in the case of the hurricane and the traveler, we have a scenario involving agents acting in time and space: two entities simultaneously moving in perpendicular directions in a confined region, a sense of calm interrupted by a sudden threat of danger, a quick calculation, a deft sideways maneuver, and a successful avoidance of a collision. This category is complex enough to allow instantiations of extremely different sorts. But even so, it is just as useless as its predecessors, and Thomas's analogy nearly poofed out of existence without ever being noticed.

Although these just-discussed fleeting analogies managed to get themselves noticed, myriads of other small and superficial mental comparisons that we routinely make each day as by-products of our ceaseless, frenetic search to make sense of our lives do not ever get noticed as such. No sooner do we come up with them than we unconsciously dismiss them as dull and irrelevant, and they are nipped in the bud: squelched before they are noticed.

A Revealing Blindness

Whereas the analogies just presented are of interest to us precisely because of their near-total lack of interest, it's certainly not the case that all analogies are totally useless. Some analogies are so pervasive that they totally shape our perception of the situations we encounter. This happens when our most basic sensory tools for interacting with the surrounding world become our main gateway, possibly even our unique gateway, to the encoding of situations in memory. Concreteness then merges tightly with abstractness.

To make this idea vivid, let us imagine a small tribe living in a remote region, and let us suppose that in their language there is a word for "stick", and that this word means not only a piece of a tree but also the abstract concept of *punishment.* Let us further imagine that the word meaning "vine" not only means a dangling plant but also stands for the concept of *friendship*, that the word meaning "water" represents not only the liquid but also the concept of *life*, and so on. Thus the word for *forest* denotes not just a wooded area but also all of humanity, the word for *rain* has a second meaning of *inevitability*, and the word for *mountain* does double duty, sometimes meaning *eternity.*

Would you find this tribe's confusions between abstract and concrete concepts to be primitive? Would you be startled to find one and the same word used to designate a tangible object and also an intangible notion? Would you be inclined to conclude that this tribe's members are incapable of distinguishing between the concrete world (that is, what the senses perceive) and the abstract world (that is, the world of pure ideas)? Would you find the tribe's blendings as strange as, for instance, someone who mixed up visual perception with intellectual understanding?

But then consider a Western observer who declares, "The proposed distinction between concrete visual perception and abstract intellectual understanding is crystal-clear, enlightening, and furnishes a brilliant, luminous perspective. However, it blindly leaves certain facets of the issue in the shadows. Indeed, upon looking at matters more closely, one feels pushed to shift viewpoint concerning the boundary line separating sight from insight, for it appears blurrier than one might suppose, at first glance."

If you don't *see* the irony in our hypothetical *observer*'s remarks, then let's quickly take another look at the relation between vision and understanding. Although the quotation above sounds like a diatribe ranting against blurring the notions of vision and understanding, it actually blurs the two notions considerably. Thus the rant itself, in the very act of lambasting people who fail to distinguish between visual perception and intellectual understanding, employs a hailstorm of phrases that *clearly show* that we do indeed *see* understanding in terms of vision. Many of the words in the quoted diatribe — "crystal-clear", "enlightening", "brilliant", "luminous", "perspective", "blindly", "shadow", "look at", "viewpoint", "insight", "appear", "blurry", and "glance" — come straight out of the experience of vision. Moreover, if we hadn't *highlighted* these terms, they wouldn't strike a casual reader as being particularly *colorful* or as representing a specific *angle* on the matter. In the end, then, one has no choice but to admit that we all blithely *blur* the concrete and the abstract, and that a deep familiarity with vision is indispensable if one wishes to make headway in the understanding of understanding.

Thus, when we speak of ideas that are clear, murky, limpid, or opaque, we are far from using poetic metaphors. We are simply using the only way we have of speaking of ideas, which is to use analogies, and in this case, analogies based on vision. Vision is probably our most sophisticated sense, but in any case it is the sense that contributes most richly to our vocabulary, so it's not too surprising that so many vision-based phrases are used to describe things that are far from being visual in a strict sense. But vision is certainly not the only one of our five major senses that is recruited by our language to describe other kinds of phenomena. In fact, all five senses help us to build bridges between the sensory world and that of emotions and ideas. For instance:

> Without anyone's lifting a finger, one can be *touched* by a kind *gesture*, *struck* by a beautiful scene, or *hurt* by a *jabbing* remark.
>
> Without one's taste buds being stimulated in the least, one can *taste* the joy of victory, find a movie to be *tasteless*, be in a *sour* mood, or make a *bitter* remark.
>
> Without perking up one's ears whatsoever, one can *hear* from a friend, declare that an idea *sounds* crazy, find a shirt to be too *loud* or a guest to be a *crashing* bore.
>
> Without opening one's eyes for a split second, one can *see* things in one's own way, *watch* the trends, and think that things are *looking* ominous.
>
> Without any air passing through anyone's nostrils, one can *smell* a rat, think that a plan *stinks*, and resent someone else's *sniffing* around one's mate.

The words "sensitive", "sensation", and "sensitivity" are all equally applicable to concrete and abstract situations. Thus one can be sensitive to the cold and to an insult; one can have a sensation of burning and of déjà vu; one can have developed great sensitivity to penicillin and to beauty. Indeed, the word "sense" is itself perhaps the most striking illustration of the way in which we constantly intermix our physiological and our psychological senses. In a word, our physiological senses are embodiments of our psychological senses.

This last metaphor, linking two senses of the word "sense", both describes and constitutes the notion of "embodiment" — the theory that all of our concepts are intimately dependent on our bodies. It could be taken as the motto of the movement called "embodied cognition", which has taken wing over the past couple of decades and is now very influential, involving neuroscience, psycholinguistics, philosophy, cognitive psychology, developmental psychology, robotics, and artificial intelligence. Affective science — the scientific study of emotions, trying to establish connections between our emotions and our intellects — is connected as well to this movement.

What all these highly diverse fields share is the belief that one's interaction with one's body and one's environment constitutes the heart and soul of human thought. The concepts that one creates, as well as one's way of reasoning, are seen as emerging from such interactions. Such a vision has no room for disembodied symbolic thinking, guided solely by the rules of logic. In other words, people do not mentally juggle with patterns of unanchored, meaningless symbols. Rather, thought is anchored in two fashions (that is, the meanings of our concepts come from two sources). Firsly, thought is anchored in the past through analogies, and secondly it is anchored in the concrete world through the body, which has participated in so many experiences.

Embodiment and Abstraction

The embodied approach to understanding cognition shares a fundamental thesis with our approach — namely, that all human thought is rooted in the life experience of an individual, as well as in the shared experiences of social groups, linguistic communities, and entire cultures. In *our* formulation, the thesis asserts that people think through the medium of their concepts, which are built up and retrieved through nonstop analogy-making, which is carried out in response to the exigencies of living in a physical body embedded in the physical world. The embodied approach to cognition, however, places little stress on the notion of abstraction, as if to imply that raw experience suffices for thinking, with abstraction being merely a luxury add-on.

Let's recall, however, that abstraction is the result of recognizing and isolating what different concrete concepts have in common. The fact that new situations constantly remind us of other ones that we (or our friends, family members, etc.) have encountered before, and the fact that complex situations constantly bring vivid, down-to-earth phrases to our lips, together show unmistakably that our minds are powerful engines built to seek hidden, deep commonalities linking one concrete thing to another one. This amounts to a drive to discover or create abstractions.

Take the very concrete-sounding stock phrase "heading down the home stretch", for example. This phrase is often used to describe a major undertaking that is about to be completed but that will still require some significant effort, and whose completion will be a source of great joy and relief. An appropriate use of this idiom amounts to an act of categorization of a certain type of situation that involves seeking a long-term goal. The categorization of such situations is mediated by a spatial analogy in which (a) the goal looming up ahead corresponds to the finish line of a footrace, (b) the various stages of the project that have already been passed through correspond to the distance that has been already covered in the race, (c) the current stage of the project corresponds to the spot where one is right now in the race, and (d) the last few challenging subgoals one still has to accomplish correspond to the final straight 100-meter stretch of an oval-shaped quarter-mile track, which separates a runner from the finish line.

Is this an *embodied* metaphor? That is, does it depend on living in the physical world and having a body? The answer would seem to be yes, in the sense that it has to do with having run footraces on tracks, or at least having watched them. But it is also very abstract. Anyone who has run or watched races can easily parlay their concrete experiences into an understanding of all sorts of far more abstract challenges featuring a distantly beckoning goal that will take considerable time and effort to reach. Moreover, there are quite a number of common English idioms that exude something of the same flavor, as is shown by these sentences: "Jones got off to a slow start in his campaign but he sprinted to the finish", "The architects are in their final spurt now", "It would be a real shame to give up on my novel now that the finish line is in plain sight", "A few months ago I was discouraged about my thesis but now I've gotten my second wind", "This project will take a lot of stamina", "We took a short break from work to catch our breath", "At last the city council is moving at a good pace", "She had to jump over a lot of hurdles to get that job", "If reading this book seems like a marathon, writing it was an ultramarathon", and so on.

Although the above sentences all sound quite concrete, they are also all highly abstract, because nearing the completion of a long-term project and finishing an arduous footrace are similar only on an abstract plane. Indeed, if the expression "heading down the home stretch" denotes from the get-go an *abstract* category having to do with the achievement of long-term goals, then we see that the human mind depends on abstract categories no less than on concrete ones. On the other hand, if it denotes a *concrete* category applicable to footraces alone, then we see that the human mind, in linking that concrete category to all sorts of other superficially distant concrete phenomena, reveals a great capacity for abstraction. In either case, abstraction is key, and to leave it out of one's theory of thinking is to miss the boat by a wide margin.

Cleanliness is Next to Godliness

Western culture is pervaded by the notion that *to be moral is to be clean.* This analogy manifests itself in many ways. For instance, the role of baptism in Christianity is to wash oneself of sin (the Bible exhorts: "Arise and be baptized, and wash away your sins"). Washing oneself plays a similar role in Islam, and Hinduism also sees the purity of the body as being of vital importance. Our everyday language is filled with implicit references to the connection between morality and cleanliness. Thus one can *do dirty deeds, have unclean thoughts, use filthy language, be a pig, get dragged through the mud, besmirch someone's reputation, get one's foul mouth washed out*, and so forth. On the other hand, one can perfectly well *have a clean conscience* or *a spotless past* or *an unsullied reputation*, or one can *be unstained and untainted by any allegations.*

In 2006, the psychologists Chen-Bo Zhong and Katie Liljenquist undertook an experimental study to test how deeply this analogy is incorporated in people's psyches at an unconscious level. They showed that it is very deeply rooted, and they dubbed their finding the "Macbeth effect", after Lady Macbeth, who hoped that she could wash her hands of the murder of Duncan, King of Scotland.

In an experiment, the psychologists asked subjects to recall some action that the latter had carried out in the past. Half the subjects were told to think about a bad deed they'd done, and the other half were not required to do so. After this part of the experiment, as a separate and presumably unrelated task, all the subjects were asked to fill in the blanks in the following letter-strings, so as to make English words:

W _ _ H SH _ _ ER S _ _ P

The "bad-deed" subjects often suggested words that were linked to cleanliness, such as "wash", "shower", and "soap", while the remainder of the subjects tended to suggest words having a more neutral emotional character, such as "wish", "shaker", and "step". It thus appears that thinking about a bad deed activates concepts that are linked to the concept of *cleaning oneself* (an immediate consequence of the naïve analogy, as an immoral act makes one dirty).

In another experiment, all subjects were asked to recall some bad deed they had done, and then (through an innocent-seeming maneuver by the experimenter) half of them were given an antiseptic tissue with which to wash their hands. The other half were given no tissue, and thus didn't wash their hands. Lastly, each subject was asked if they would be willing to give some help to a student who was having trouble. It turned out that those who had not washed their hands were significantly more willing to aid the unknown student than those who had washed their hands. It would thus appear that washing one's hands makes people feel cleansed of their past sins, diminishing their general sense of responsibility towards others, while people who have freshly recalled some piece of bad behavior but have not been given the chance to "expurgate" it are more inclined to be good samaritans.

Like many analogies, the association between morality and cleanliness insinuates itself unconsciously, and people tend to think of the tight link between morality and cleanliness as a *fact* rather than as just a possible way of looking at things. The analogy implies not only that doing a bad deed makes one dirty, but also that bad deeds can be compensated for by washing oneself (and in fact, that good deeds are a kind of soap that will clean one up after one has perpetrated a bad deed). The fact that people usually do not question this point of view prevents them from seeing their personal morals as a set of values that are determined in part by the culture they grew up in, rather than as absolute and eternal values.

Categories as Blinders

Calling categories "blinders" is an admission that our categorization mechanisms, which help us navigate fluidly through many a lofty realm of abstraction, can on occasion mislead us, via insufficiently flexible insights. Sometimes such blinders impose a viewpoint on us that just doesn't dovetail accurately with the situation. Locked in, then, by our overly simplistic perception, we tend to miss things that are totally obvious to people who have categorized the situation in a more fitting manner.

Such an event befell E., an American philosopher interested in the ancient philosophical doctrine of essentialism, and who was spending some time at a French university. Not long after her arrival, she saw a poster announcing a lecture entitled "Aux sources de l'essence" to be given in two weeks by a Greek geologist. E. was a bit confused by the fact that the lecturer was a geologist rather than a philosopher or a historian, and also by the photo on the poster, which showed several oil rigs pumping away in a desolate desert landscape, but she decided that this was an elegant metaphor for how the fertile human mind, in its perpetual search for essence, probes hidden subterranean zones. In the end it was that photo, plus the speaker's connection with ancient Greece, that persuaded her to go. When she turned up at the talk, she was delighted that the speaker started out using the oil-rig metaphor front and center, but when, after five minutes had passed, he was still holding forth about oil and oil rigs in the Arabian deserts, as if his ingenious metaphor were more important than the true essence of his talk, E. was confused. But then, all at once, she recalled that the French word "essence", although it sometimes means what *we* mean by "essence", is also the standard term for "gasoline", and it hit her that the talk's title must have meant "The Sources of Gasoline" rather than "The Sources of Essence". Feeling quite sheepish, E. realized that she had read deep metaphorical meaning into the photo where none had been intended.

Why did it take E. several minutes of the lecture to overcome her original impression? Firstly, the activation of the English word "essence" had given her a mistaken idea of the lecture's topic, and secondly, the manipulative power of analogy had reinforced this *essence* idea for the preceding couple of weeks. In fact, E.'s initial interpretation was repeatedly reinforced by the rest of her thought processes. Thus instead of taking seriously her initial doubt, triggered by the photo of oil rigs, and instead of starting to think, "Something smells fishy here", she embraced the photo as a charming metaphor, letting it buttress and entrench her initial categorization.

Kind Hearts and Cor(o)nets

In much the same vein, our friend David Moser, not only an inveterate error-collector and keen observer of cognition but also a sophisticated jazz musician, told us of a phone call he had many years ago, after placing a classified ad in the local paper in order to sell his old cornet. A would-be buyer read the ad, liked what he imagined, and called David up. From David's point of view, their conversation went like this:

Caller: Hi — I'm calling about your used cornet.

David: Yes?

Caller: It's only $500?

David: Right.

Caller: Well, what's wrong with it?

David: Nothing. It's in good shape.

Caller: Well, how many miles does it have on it?

David: Uh, it's about ten years old.

Caller: Oh. What color is it?

David: The usual gold finish. It's got a few dents in it, but it's not too tarnished.

Caller: Is this just a standard cornet or what? Can you describe it?

David: Well, fairly light, standard bell. It has three spit valves.

Caller: Can I come over and take it for a spin?

David: Sure, just call me before you come over.

This droll double-edged dialogue will most likely remind readers of implausible mistaken-identity scenes in so many plays, movies, and operas, painstakingly concocted by clever script-writers so as to flow seamlessly and make one fall off one's seat with mirth. In this case, though, the interchange really did take place, with each person hearing the other one's words as making reasonable sense for quite a while before any red flag was raised.

For the caller, what was under discussion was, without any doubt, a used Dodge Coronet, and with this "fact" solidly in mind, he was able to fluently integrate David's comments alluding to the car's "usual gold finish", as well as to the odd feature of its "standard bell" (whatever that might be), and even to its "three spit valves"! Perhaps the caller took the "spit valve" allusion as a knowing wink on David's part — one car expert talking to another — alluding to some technical problem with the valves in a couple of the car's cylinders. Or perhaps he unconsciously transformed David's words "three spit valves" into "three *split* valves" in order to make the unexpected phrase more plausible, as if talk of "split valves" in a car would make perfect sense to any self-respecting American male.

As for David, he just as clearly had in mind a small trumpet, and wasn't in the least prepared to think that the caller might not share that image. And so, when the caller asked about the brass instrument's mileage and color, even inquiring if he could "take it for a spin", David wasn't tipped off that something was wrong. To him, these queries, though oddly put, were just fresh, cute metaphors — idiosyncratic ways on the part of the would-be buyer of angling in on the key matter of the condition of David's cornet.

To put this in the context of our overarching theme, let us point out that we began with an eager used-car buyer — so eager that he was able to read the printed word "cornet" (with a lowercase "c") in the newspaper ad as the name of a car — and in his mind this evoked the image of a Dodge Coronet. This was the initial triggering of a category. Once the category was activated, it took charge of the interpretation process, filtering everything in the environment, distorting all the cues and giving rise to reactions that, to someone not biased by the category, would have grown more and more implausible. As it happened, everything collapsed at the end of this phone conversation, just as happened in the case of philosopher E. attending the lecture on "sources of essence". Slowly, the evidence opposing the entrenched category mounted, and finally it was strong enough that the stubborn initial category was overthrown.

Unwitting Default Assumptions: A Pitfall of Categorization

For a quite different example, consider the story of the father who dies in a serious traffic accident, and whose son, a passenger in the same vehicle, is taken by ambulance to the hospital in critical condition. An emergency operation is needed to save his life. The surgeon on duty comes quickly into the operating room, suddenly goes white as a sheet, and exclaims, "I can't operate on this boy — he's my son!"

It's extremely common for people to read and reread this story many times, always bumping into the irreconcilable problem that the boy cannot possibly have two fathers. What is going on? Has the surgeon misperceived the boy's identity? No. Or was there perhaps not just a biological father but also an adoptive father? No. Did the father somehow get resuscitated and miraculously make it to the hospital before his wounded son? No. In fact, all these questions and answers, despite giving correct information, serve only to further entrench the problem, which is caused by categorical blinders. There is in fact a perfectly reasonable explanation of this situation, which we are confident that every reader, after sufficient reflection, will find.

The manifold ways in which we are continually blindsided by our categories cannot be overemphasized. Take, for instance, the story that you just read about the traffic accident. Did it at any point occur to you that the vehicle involved was a bus? Virtually no one envisions it that way, even though nowhere in the short paragraph recounting the story does it say "car". It simply never crosses our mind that our initial categorization could have been wrong. The default assumption that comes along with the concept of *traffic accident* is that it involves a *car*, as opposed to, say, a bus, a truck, a motorcycle, a bicycle, a mobile home, and so forth. Such default assumptions are at times very deeply entrenched, and in some cases they are nearly impossible to detect and overthrow. That is why it is so very hard for many readers — even today, when women have made so many strides along the pathway to social equality with men — to figure out how the surgeon's remark makes perfect sense.

Flitting from One Essence to Another

Categorizations that lead to a mistaken understanding of a situation can have major consequences for someone trying to solve a problem. Indeed, a poor categorization can make a simple problem difficult, if not unsolvable. An external observer might take someone's being stumped by a simple problem as a symptom of incompetence, when actually the would-be solver is merely trying to solve a problem other than the one that was posed. Just as E.'s biases led her to misconstrue the word "essence" in the lecture title, so the solver has misconstrued the problem because of biased categorization.

Various studies in cognitive psychology have brought out the nature of such mistaken categorizations and are helping to overturn the stereotype that says that if someone fails to solve a problem, it's only because they didn't find a successful strategy. That would indeed be the case if the person had perfectly understood the problem statement but didn't know how to proceed from there. But research has shown that one of the most frequent sources of trouble in problem-solving is a misunderstanding of the problem statement, meaning that the wrong categories are mobilized. Once the proper categorization is found, finding the solution is often quick and easy.

An example will help make this clear. The study in question involves the famous Towers of Hanoi problem, studied by many psychologists. In a very simple version, there are three disks — one small, one medium-sized, and one large — with holes in their centers, which have been stacked in a pile on one of three posts — post A — and which must be moved to post C while respecting three constraints: (*i*) only one disk can be moved at a time; (*ii*) only the uppermost disk in a pile can be moved; and (*iii*) a larger disk can never be placed on a smaller one.

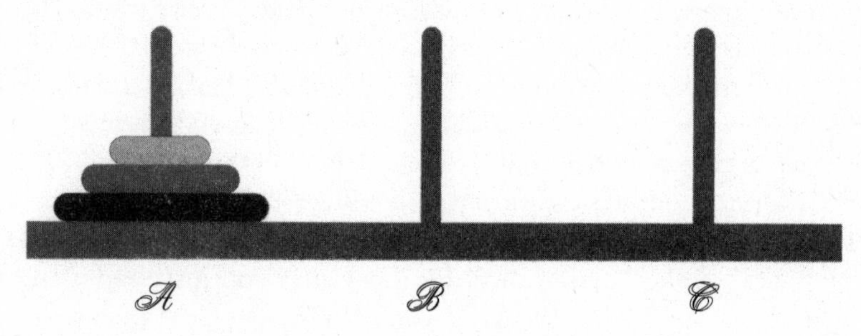

Many experiments have been done on the topic of how people tackle this challenge and other "isomorphic" challenges (so called because they are in fact just superficial disguises of the same problem, and exactly the same technique will solve them all — for example, the disks can be replaced by acrobats, and the third constraint then says that no acrobat can ever stand on the shoulders of a smaller acrobat). We'll now shine a spotlight on some interesting results. Cognitive psychologist Jean-François Richard has shown that a significant percentage of elementary-school children, at around age eight, solve this problem with a very large number of moves — around thirty or so — whereas only seven moves will suffice if one knows exactly what one is doing. However, the strategy adopted by these children is anything but random. It seems that they throw in one extra rule that isn't part of the official statement of the problem. This rule, which they tacitly assume without realizing it, states that one cannot move a disk directly from post A to post C (or vice versa). In other words, these children don't allow themselves to jump from A to C or the reverse, requiring instead that any disk moved from either A or C must always have B as its landing-spot. If you tackle this alternate challenge yourself, you will soon discover that the addition of this rule slows the solution process up considerably: the minimal solution jumps from 7 moves all the way to 26.

Why would so many children impose this extra rule on themselves, essentially tying one hand behind their back? It all comes down to the concept of *motion*, which seems to be perceived somewhat differently by adults and children. Children often implicitly suppose that moving from one post to another means passing through each intermediate position, whereas adults have a more sophisticated conception of motion, which doesn't depend on the way one gets from point A to point C. When someone says, "Next Friday I'm going to Arkadelphia", it could be by car, by bus, by train, or by plane, and the details of the route make no difference. Adults have no problem distinguishing between a trip as a *state change* ("Today I'm in Bloomington and on Friday I'll be in Arkadelphia") and a trip as a *process* ("My Bloomington–Arkadelphia jaunt will carry me through Indianapolis, Little Rock, and several other towns").

The tendency to see a motion as a *process* rather than as a *state change* is closely related to the idea discussed in Chapter 4, according to which various stages of abstraction of a concept involve stripping away one after another of its less crucial aspects (such as removing from the concept of *desk* the notions of *weight*, *volume*, and *substance*, in the end leaving just the core notion of *workspace*, or removing from the concept of *key* the notions of *metallic*, *long*, *thin*, and *irregular in shape*, in the end leaving just the core notion of *transportable entity that performs unlocking*, such as the magnetic cards that one gets in hotels). It appears from the aforementioned experiments that a good number of eight-year-olds have trouble separating the abstract notion of a *move* from the more concrete notion of the *route* by which the move is accomplished. The former doesn't care about how it took place and thus allows direct jumps from A to C or back, while the latter insists on passing through all the salient intermediate points. Just as Mica, in the previous chapter, could not help hearing in the word "hump" an earlier collision that had caused the hump, so many young children cannot help hearing in the word "move" the idea of *passing through all intermediate stations.*

Precisely the same phenomenon can be seen in adults when one "dresses" the Towers of Hanoi problem in other "clothing", thereby yielding one of its isomorphs. Jean-François Richard and his colleague Évelyne Clément studied an isomorph of the Towers of Hanoi in which, rather than moving the disks, one can change their size. In that situation, adults tend to behave exactly as do the children who move the disks. That is, a good fraction of them unconsciously add the extra rule that one can't change a small disk into a large one (or vice versa) — one has to pass through the intermediate size. In order to jump from size A to size C, they feel they need to take *two* "steps", stopping momentarily at size B. The cause of this unexpected mental rigidity, which markedly slows these adults up in their solution of the problem, is the same as for the children: they use an overly concrete (and thus naïve) notion of the concept *size change*, not separating the final size from the process or "route" taken in getting to it. For most people, the prototype for the concept *size change* is that of biological growth: one grows from babyhood to adulthood in passing through childhood and adolescence. This model of the concept of *size change* is naïve, though, and adults who get stuck in this trap, unconsciously borrowing the naïve model in tackling the Towers of Hanoi isomorph, are fated to take much longer than those who do not.

It thus appears that one single mental phenomenon is at work both in the "Sources of Essence" trap and in the Towers of Hanoi trap — namely, memories of analogous situations impose themselves without consulting the person who matters — the person in whose head all this is happening. The standard excuse that one gives is along the lines of "I didn't quite catch what they told me", or "They sure didn't make it clear what it was all about", or "I didn't pay close attention" — but in fact a more accurate explanation is that one unconsciously succumbed to an inappropriate categorization.

When the entire process is unconscious, as in the cases we've been discussing, only through very meticulous research will the hidden analogies be revealed. Sometimes the existence of categories shared among people in a culture or subculture furnishes a particularly clear demonstration of the way categories influence the perception of a situation, since people who do not belong to this community would be amazed by a point of view that strikes them as extremely odd.

The following example of this phenomenon will speak directly to readers who live in large metropolitan areas, and may intrigue those who live in less traffic-congested locales. It shows why the loveliest spot in a city isn't necessarily what one might think.

What is San Francisco's Loveliest Spot?

Union Square? Chinatown? Twin Peaks? The Great Highway? The Cliff House? Pacific Heights? The Golden Gate Bridge? Fisherman's Wharf? Golden Gate Park? The Presidio? The Marina? The Palace of the Legion of Honor? The Top of the Mark? Coit Tower? The Ferry Building? Lake Merced? West Portal? Russian Hill?

Surely, for a non-resident of the City, one of the above would fill the bill, but a true San Franciscan sees things differently. Finding a place to park one's car in the City without worrying about getting an astronomical fine or having to go pick it up at the pound can verge on the miraculous, especially in certain areas and at certain times of day. Thus it is not infrequent that one finds oneself driving up and down steep hills, back and forth on broad avenues, crisscrossing one's prior pathway umpteen times in desperate search of a parking spot, knowing full well that the likelihood of finding one is microscopic. This plight is so common and so upsetting that San Francisco drivers tend to be powerfully drawn to vacant parking places, finding high esthetic value in simple concrete rectangles as long as it is legal to park in them. The rarity of such a spot turns it into a precious entity.

It's not uncommon to hear one local say to another, as they walk down the street, "Just look at that beautiful spot! Wow!" The intimate relationship that San Francisco drivers have with untaken parking spaces thus engenders the category of *lovely spots.* As anyone who's driven in the City can testify, spotting a lovely spot when one doesn't need to park in it always evokes a tempting counterfactual scenario of chancing upon exactly that empty spot just when one needs it desperately, and thus a sense of "if only" or "too bad" is triggered.

The abstract category of *lovely spots* deeply affects how these people perceive the physical spot. This example shows how powerfully categories impose a view of the world. If Kazimir Malevich's famous *White Square on White Background* is a member of the category *works of art*, then surely Market Street's *Gray Rectangle on Gray Background* is a member of the category *lovely spots.* The intense feeling of longing that the gray rectangle inspires in so many people shows how irresistible is the psychological force that pushes for categorizing it in that fashion. And yet such a categorization, for all its emotional intensity, doesn't drive people into paroxysms of irrationality. So far as we know, no one has yet succumbed to the siren song of a lovely spot by suddenly throwing their Saturday-evening plans out the window, screeching to a halt, and parking their car then and there, fatally seduced by the lovely spot that was winking at them.

Certain categorizations, however, have very powerful influences on our thought and behavior. For example, people's perception of the October 11th crash was very different from what it would have been had the event occurred a few years earlier.

The Irresistible Strength of Analogies: The 10/11 Crash

In the middle of the afternoon on October 11, 2006, an airplane crashed into a tall building in New York City. On first hearing this breaking news, no one, unless they had spent the last few years on a remote island, could fail to think of the destruction of the World Trade Center on September 11, 2001. One would automatically assume it was a terrorist attack, and one would retain a lingering feeling that this must be the case even after learning that it was only a four-seater private propeller plane, not a huge jet, that the only deaths were of the pilot and copilot, that the fire in the building was quickly extinguished, and that the building was never in danger of collapsing.

The analogy with the events of September 11th sweeps in immediately, profoundly coloring one's perception of this event. It would thus be hard for anyone to imagine, on first hearing about this event, that it was simply a random accident of the sort that takes place frequently and in many different ways all over the world — an unfortunate incident, to be sure, but with limited consequences, and without any link to religious fundamentalism or terrorism. That's a most unlikely first thought! As a matter of fact, the Dow Jones average declined for a short while after the collision was announced.

The September 11th events cannot fail to be evoked front and center when one hears about the October 11th crash. Because of September 11th, the broad category of *terrorist attack* is instantly activated, and so is the more specific category of *September 11 attack*, perhaps even more strongly so. Thus one tends to think, "Oh, no — September 11th all over again!" or "Is this another September 11th?" Together, the broader and the narrower category provide the inevitable framework for understanding an event that would have been perceived totally differently had September 11th never occurred.

Someone might object, saying "Hold on, now — 'September 11th' isn't the name of a *category* but of an *event*!" We would counter this claim by reminding the objector of categories such as *popes*, *bibles*, *meccas*, *Bachs*, *Einsteins*, *Picassos*, and *Rolls-Royces*, discussed in Chapter 4, which show that apparently single-member categories, no matter how unique or *sui generis* they might seem to be (even the super-specific *Château Beaucastel Hommage à Jacques Perrin 1998*), will be quite effortlessly pluralized by the similarity-driven human mind, when the proper situation shows up.

Other clear signs of the plural nature of the category *September 11th* (more often called "9/11") are the standard use of such phrases as "India's 9/11" (the bombings of many buildings in Mumbai, including two luxury hotels, in late November of 2008), "Russia's 9/11" (which has been exploited by various political groups to denote various massacres of innocent people), "Pakistan's 9/11", "Spain's 9/11", and so forth. In particular, the bomb attacks that took place in Madrid on March 11, 2004, claiming the lives of over 200 people, soon acquired the label "Spain's 9/11", and Spaniards very understandably quickly took to using the analogy-drenched nicknames "11-S" (for the attacks on New York and Washington) and "11-M" (for those on Madrid).

On the Web, one can also find such phrases as "the 9/11 of the seventeenth century", "the 9/11 of New Orleans", "the 9/11 of World War II", "the 9/11 of the Scriptures", "the 9/11 of rock", "the 9/11 of hormone replacement therapy", "the 9/11 of commercial shipping", and on and on. Moreover, one easily finds scads of Web sites that refer, using explicit plurals, to such categories as "the 9/11's of the future", "the 9/11's of the 1960's", "the 9/11's of history", and so forth.

In brief, the concept of *September 11th* is now a common category having many members of different strengths (Pearl Harbor being a fairly strong member — and, amusingly enough, in a sad way, 9/11 itself being a fairly strong member of the category *Pearl Harbors*). Just think how many birthdays and wedding anniversaries were sabotaged in the years following 2001 simply because, many years earlier, someone happened to be born on the 11th of September, or because a couple had perfectly reasonably picked that date for their nuptial celebration. And for quite a while after 2001, all sorts of receptions, openings of shows, and other public events were carefully scheduled so as to avoid being stigmatized by the irrepressible association with the "radioactive" date September 11th, which poisoned everything that came close to it.

One Thing Changes and Everything Changes

As we have seen, analogies constrain one's perception of situations. Whether we're talking about members of the category of *lovely spots* or that of *9/11's*, analogies constitute filters through which the world is seen. This statement may seem surprising, since it's easy to forget that whoever one is, one sees the world through filters that powerfully control the flow of one's thoughts. But the lessons learned from the preceding categories apply throughout everyday life, and they apply to any idea, however small or large, that occupies center stage for a while.

> One day, Y. saw a brand-new ceramic elephant in a friend's apartment, and they exchanged a few words about the art object. An hour later, Y. was walking down the sidewalk with his wife, and all at once he was stunned to see exactly the same object in a store window. At the moment he spotted it, he was doing many things at once — talking, walking, listening, pondering, avoiding obstacles of all sorts — and wasn't in the least thinking about the elephant in his friend's apartment. And the fact is that in the preceding weeks, Y. had walked by this same store window dozens of times and never once had noticed it, and yet the elephant had been in it every single time, as was attested by the thick dust on it, and as the store owner confirmed when Y. went in and asked. In other words, Y. hadn't *seen* it even once in the preceding months, although it had been in his visual field dozens of times, and yet on this day, it had jumped right out at him as if it were ten times brighter than anything else in the store window.

This little episode provides a useful illustration of the constantly ongoing process of *filtering*. Because certain concepts had just gotten activated, the previously unnoticed elephant became, in this new context, cognitively salient; it thus moved above the threshold of Y.'s attentional filtering system and became visible.

As this anecdote shows, our perception is profoundly biased, but this is fortunate rather than problematic, for our biases are generally very useful and efficient. If our brains tried to pay equal amounts of attention to all things around us, we would drown in confusion. Thus our categories act as filters, and as such, they are crucial elements of our mental life, allowing us to deal with the flood of stimuli constantly bombarding us. Since our categories are our organs of perception of the world, whatever affects our system of categories affects our perceptual organ. The rather haphazard course of our thoughts as we drift through life deeply colors our way of seeing the world.

Thus learning a new fact or having a new experience can profoundly alter our perception of our environment. A pregnant woman sees pregnant women everywhere around her, and after giving birth she runs into newborn babies everywhere she goes. Someone who decides to embark on psychotherapy soon finds out that everyone they know has done the same thing. If one starts going down the pathway of divorce, all at once divorce stories start cropping up whoever one talks to. If one indulges oneself in a new car, one is shocked to see exactly the same model turning up on every street corner. If one starts noticing a tiny gesture or microscopic verbal tic in a friend, all of a sudden it becomes the dominant feature of the friend's face or speech, even though one had never noticed it before. If at a party one makes the acquaintance of the mother of a child who attends the same school as one's own child does, and who regularly picks up her child there, thereafter one notices her every single day at school despite never having seen her before for years.

The Power of Obsessions

Categories that are activated in one's mind are always on the lookout for instances of themselves in one's life. The more highly activated they are, the fewer cases they miss, and the more fluid and creative they are in spotting instantiations of themselves in all sorts of guises.

The preceding examples have shown how this holds for relatively concrete and familiar categories (*ceramic elephants*, *pregnant women*, *divorce stories*, and so forth), but we are often involved in blurrier situations whose boundaries are extremely ill-defined. In such cases, our mind's unconscious scouring of its surroundings for resonances with its active categories still goes on just as feverishly as in the simpler situations, but the search takes place at a more abstract level of perception. This abstract filtering of the world can give rise to connections that would seem very strange to an observer who didn't share the obsession, whether fleeting or long-term, of the analogy-spotter.

What we perceive is the result of a compromise among our environment's offerings, our repertoire of categories, and our current concerns. If a concern verges over into an obsession, it seizes control and everything in sight winds up being perceived over and over again in terms of this obsession. Thus after the death of a loved one, the themes of death and sadness are bound to pervade a person's perceptions. Virtually every object and situation is tinged with the loss. A tilted parasol evokes tears, reminding one of a tree about to fall, of an imminent ending, of universal mortality. A stopped watch represents the fact that time has ceased to exist for the deceased one. A cloudy sky is seen as death hovering above, sending down a pall of gloom. A spoiled peach, a withered rose, a chipped cup, a broken toy, a tipped-over garbage can, a closed shutter, a lowered awning, a dented car, a hunk of ham — each of these is death once again. And then on the other hand, merrily laughing strangers, tenderly gazing lovers, a kissing couple, a happy family strolling by — all these symbols of joy in the world suddenly become cruel attacks reminding us of people's blithe indifference to the sufferings of others, and highlighting the bleak solitude of every suffering soul.

Obsession thus tries to exploit every cue coming from the environment. Otherwise put, the source of an obsession will give rise to a cornucopia of analogies applying to situations of every possible sort, and at the same time it will tend to drown out all competing analogies. The dark thoughts of the previous paragraph, for instance, all come from analogies to death, and each one is perfectly justifiable. But they clearly are just a few among a myriad possible ways of perceiving the situations described. Although such driven analogies make a certain degree of sense, nonetheless, when one stands back, they sometimes seem quite forced — so much so that they aren't really convincing. And yet, the boldness and intensity with which an obsession invariably scans the world in search of fresh instances can occasionally give rise to new and insightful thoughts, even if this is the exception rather than the rule. Indeed, an avid quest for analogues of a given phenomenon increases one's chances of coming across important new perspectives that, without the obsession's driving force, would never see the light of day. If you try the same key in a thousand different locks, perhaps one time it will work.

It's almost impossible to imagine someone coming up with a revolutionary new insight without being steeped in the domain in an obsessive or near-obsessive manner. But since we are encroaching on Chapter 8's discussion of scientific discovery, suffice it to say for the moment that great physicists, great mathematicians, and great scientists of any stripe are invariably involved with great passion in their discipline. Louis Pasteur once famously observed that "Chance favors the prepared mind", and obsessed minds are nothing if not prepared! Were their owners not passionately obsessed, they would never be able to spot connections that for a long time had escaped the eyes of all their colleagues. This brings us back to the idea, considered in the previous chapter, that creativity cannot be turned on and off with a simple switch: in order to come up with creative analogies, one has to be possessed by an idea.

Let us momentarily recall Archimedes' "Eureka" moment in his bathtub. To understand how this discovery was made, one has to take pressure into account — and here we don't mean the pressure exerted by the liquid on all objects immersed in it, but the pressure exerted by the monarch on his faithful servant Archimedes. It's not hard to imagine that in those days, it was not looked upon kindly if a royal request was not met, and thus one can easily imagine poor Archimedes pacing up and down, racking his brains for any possible way to get a handle on the volume of the damned crown. He would start seeing volumes everywhere, volumes where other people see nothing of the sort. Thus starting with the idea that a *crown* has a volume, he might soon slide to the idea that a *door* has a volume, then that a *chair* has a volume, that an *animal* has a volume, that a *person* has a volume, that *I myself* have a volume, that *parts of my body* have various volumes, that *the water that I displace in my tub* has a volume… Aha! That's it!

Of Hammers and Nails

Obsessions bring out curious connections that no one would dream of otherwise. This is reminiscent of a maxim originated by the psychologist Abraham Maslow: "If the only tool you have is a hammer, it is tempting to treat everything as a nail."

As a high-school student, J. had such a "hammer", for he was obsessed by pinball machines, playing on them several hours each day. Getting higher and higher scores assumed increasing importance in his life, and suddenly one day he saw every human life as the trajectory of a ball in a pinball machine. His analogy rested on the vision of life unfolding in the same random and unpredictable way as the ball rolls. Birth, mapped onto the ball's launch, was followed by a tumultuous life full of swerves and traps, corresponding to the ball's bounces. At all moments there was a risk of perishing, and death was inevitable, whether the playing had been brilliant or mediocre.

K. was a devoted rider of horses; she lived among them and adored them. Her understanding of the human world was rooted in her understanding of the equine world. Her profound "horse sense" constituted the key allowing her to unlock all the complexity of human relations, even convincing her that her insights into humanity were deeper than those of other people around her.

L. was a dog-fancier from earliest youth, and he based his social relations, including quite successful business connections, on rapid intuitive links that he made between each person he met and a selected breed of dog. Into each new acquaintance he read character traits that came along with their "breed", and on that basis he made all his decisions about how to treat friends and colleagues. L. thought of himself as a Saint Bernard, and he had friends and colleagues whom he saw as poodles, bulldogs, German shepherds, fox terriers, and so on.

M., a physics graduate student, was so deeply plunged into the world of particles that he built his understanding of social relations on how particles interacted. Even very recondite quantum phenomena (such as quasi-particles, superconductivity, virtual exchanges, and renormalization) had their counterparts in human relations.

N., when a teen-ager, fell in love with golf. In her parents' yard, she made an eighteen-hole course and spent many hours playing on it each week. Her days and nights were profoundly impregnated by the vocabulary and imagery of golf: irons, woods, putters, balls, greens, par, birdies, eagles, bogeys, sand traps, fairways, and so forth. During this period of her life, wherever she went, N. would see, in every lawn or grassy knoll or meadow that she passed, a potential hole or putting green. When her family moved to Switzerland for a year, every car trip in the surrounding countryside was the occasion for her to fantasize golf courses wherever she went.

A few years later, N.'s flame for golf was replaced by no less ardent a flame for photography. Barely weaned from her golf mania, she turned into a photo maniac. Every landscape, every scene, every gesture or expression of everyone she met was caught in stop-motion in her mind and framed inside various possible rectangles.

Pinball machines were for J. what horses were for K., what dogs were for L., and what particles were for M. Each of their domains provided a rich wellspring of analogies, and each individual grounded their personal model of humanity in these analogies. All of these individuals were obsessed, and each one benefited, in some fashion, from their obsession. And just as we can move from one city to another, so we can move from one obsession to another, as the case of N. shows.

For outsiders who hear about such extensive, systematic, and long-term families of analogies for the first time, they may sound unnatural and somewhat weird. How can human beings, in all their richness, be understood by invoking images of mysterious invisible particles, balls in pinball machines, or horses or dog breeds? It sounds like imagination given free rein and running wild. It even suggests that an analogy can be drawn between virtually anything and anything else, provided that some kind of obsession lurks behind it all, acting as the driving force.

And indeed, there are resemblances to be exploited wherever one's gaze falls. As the logician and philosopher Nelson Goodman observed, any two situations have an arbitrarily large number of properties in common. For instance, a crown and a human body have in common the characteristic of not being located exactly one mile away from the center of the sun, nor at 1.1 miles, nor at 1.2 miles, and so forth. Although this remark undoubtedly has philosophical relevance, its psychological relevance is minimal, since it's obvious that humans do not look at all possible properties of all things that they look at — just a tiny fraction of them. For example, no one cares about the fact that Queen Elizabeth's crown is not located at exactly π miles from the center of the sun.

Nonetheless, the incredible fluidity of the notion of *similarity* allows people to come up with connections between entities that *a priori* would seem utterly unrelated, simply because obsession-driven hunts for resemblance always wind up finding results. The fruits of such avid searches, however, are not random or arbitrary. A passion for horses or dogs does not instantly turn these animals into sources for analogies that yield insights into triangle geometry, quilt design, fly-fishing, or who knows what else. On the other hand, a *double* obsession could surely give rise to such analogies. That is, a simultaneous fanatic for, say, Euclidean geometry *and* for fly-fishing would doubtless find plenty of phenomena in these two domains on which to found analogies, for in this case the search would be intense on both sides.

Real-life Homilies Unearthed by a Pac-Maniac

Whenever someone's passion for something increases, their personal likelihood of analogy-making (or analogy-finding) goes up. The following story shows how a young man immersed himself to excess in the world of a particular video game. From this intense experience he drew a significant harvest of analogies, showing how far a profound obsession can carry someone.

T. bet a friend that he would be able to beat him in Pac-Man, the famous video game of the 1980s, in which the player is represented on the screen by a yellow circle with a wide-open mouth. This quaint old icon seduced T. and several of his friends, and as they grew increasingly sucked into the Pac-Man world, they spent ever vaster amounts of their time playing. For hours every day and even more hours every night, T. would sit at his computer and play and play at this game.

To do well in Pac-Man, one has to avoid getting eaten by four enemies called "ghosts", and of course the goal is to acquire as many points as possible. There are two types of pill that the player can swallow, and one of them makes ghosts edible for a short time. If possible, one wants to eat all four ghosts, since the second ghost is worth twice as much as the first one, the third one four times as much, and the fourth one ten times as much. Sometimes a fruit icon will appear at random on the screen, and swallowing it will confer special advantages such as extra points, higher speed, a temporary invulnerability, or the doubling of all one's points until the next stage starts.

T.'s days and nights were so steeped in Pac-Mania that he came to see all of life through the filter of this game. It may seem odd, but he formulated what amounted to a whole philosophy of life thanks to a long series of analogies he made while playing. Herewith follow fifteen of T.'s Pac-Man–based maxims:

(1) In Pac-Man, one endeavors to eat one's prey and to avoid being eaten by predators. T. observed that in real life as well, some people are stronger than you and some are weaker. Thus *life is a pecking order in which everyone eats smaller beings and is eaten by larger ones.*

(2) In Pac-Man, one continually jumps back and forth between being a pursuer and a pursuee; indeed, whenever one swallows a certain type of pill, all of one's pursuers instantly turn into prey. So T.'s second motto was: *to beat an enemy, first you have to weaken it.*

(3) Whenever a ghost is a predator, it can pursue you without stopping, but whenever it is edible, it flees the moment you approach it. T. thus learned that people adjust their behavior to the exigencies of the moment and that *even big hoodlums dash for cover whenever they meet their match.*

(4) The more skilled and better-trained you are, the less likely you are to get eaten, and the more likely you are to be able to eat others. A novice, however, no matter how lucky, will never get very far. T. learned from Pac-Man that *in order to excel, you have to work hard.*

(5) T. noticed that in some games he never got any fruit, while in others it was abundant. He thus enjoyed happy occasions on which fortune smiled on him and bitter ones on which he constantly had to fend for himself. He philosophized thus: *in life, some people have all the luck while others have none.*

(6) The most flagrant type of unfairness in Pac-Man is the fact that eating a cherry doubles all one's future points. Thus anyone who has eaten a cherry can get a pretty decent score without being a very strong player, whereas without that boost, you have to take big risks and fight very intensely. T. generalized this observation to the world of people: *some people are born with a silver spoon in their mouth, while others have to sweat their whole lives through.*

(7) Taking risks is indispensable, because otherwise one misses all one's chances to raise one's score. Playing in a risk-free way assures you of a mediocre outcome. So T. rediscovered for himself the old proverb *nothing ventured, nothing gained.*

(8) Risks don't always pay off. Anyone who obtains a very high score has taken risks, but players who take risks and who are every bit as skilled as their rivals often die. They pay dearly for their boldness. And so T. discovered how merciless the world is, which is to say that *bravery and death go hand in hand.*

(9) Sometimes it looks as if eating a ghost will pay off big, but at the last moment the ghost catches you off guard and turns into a predator, devouring its former pursuer. Everything flips in a split second. T. learned that *taking too big a risk can turn your world upside down.*

(10) T. had experiences in Pac-Man in which hesitating just a tenth of a second cost him everything. It's not enough to make good decisions; you've got to be able to make them on a dime. T. concluded that *life doesn't give you second chances.*

(11) T. retained a bitter memory of games that were extremely promising but where just a moment of distraction made him slip into the clutches of defeat in just a few seconds. He saw that *one moment of carelessness, and all your hard work can go up in smoke.*

(12) Sometimes a delicious fruit would appear when the ghosts were edible as well, but when T. tried to eat them all, he usually wound up with nothing instead of a grand feast. Thus he was led to the credo that *you shouldn't try for too much.*

(13) Sometimes T. took extreme risks and perished when, had he been more restrained, he would have won fewer points but would have survived. In this manner, he learned that *a bird in the hand is worth two in the bush.*

(14) Whether madly pursuing prey or rapidly fleeing from a predator, T. chose his direction by paying close attention to the distances involved. But sometimes much better results came when he headed off in precisely the wrong direction. In this way, Pac-Man taught T. that *the shortest path is not necessarily the best path.*

(15) There were times when the ghosts' movements were such that no matter what he did, T. simply was done for. He learned that *although one is alive now, death may be only moments away.*

These fifteen "life principles" came out of a video game lacking all metaphysical pretension, a game whose sole *raison d'être* is to entertain. This shows how analogy pervades one's thinking when one is driven by an obsession. The maxims to which T. was led by his Pac-Man addiction would not have been formulated by someone who was a relative newcomer to the domain; people with just a few minutes or hours of experience with Pac-Man would say, "The only purpose in this game is to get as high a score you can while killing the ghosts — that's obvious!" It's most unlikely that their reflections would go any further than that.

T.'s maxims may seem, like the other obsession-based analogies cited above, to be exaggerated and artificial, in the sense that they seem unrelated to Pac-Man. But on the other hand, his analogies are as legitimate as can be. The connections between the maxims and specific phenomena in the game are always clear as a bell; the analogies are spot-on and also they teach lessons — lessons that someone could learn from real-life experiences. The difference is that in Pac-Man these lessons are learned in a highly compressed time frame and are intensified, since even the strangest circumstances will reappear fairly regularly when one is a Pac-Man addict. During his period of Pac-Man addiction, T. lived tens of thousands of different "lives".

It is striking that a simple video game contains enough richness to lead a player to insights about life that are essentially philosophical, all based on analogies that link the trivial "lives" of a Pac-Man entity, involving little more than eating or getting eaten, with the full complexity of human life. In this respect, the Pac-Man microworld is similar to the Copycat microworld, as we will see in the next chapter.

Irresistible Analogies: Are They Meaningless or Meaningful?

We now direct our attention towards another frequent type of analogy — specifically, analogies suggested by extremely salient resemblances, which perhaps are superficial and perhaps are not. That's the key question. Although such analogies are irresistible, do they deserve to be given the time of day or not? We tackle this issue with the help of an example.

Shortly after giving a lecture, Professor F. was contacted by a reporter who worked for his university's public relations bureau. She wanted to interview him for a university publication. They met, and as the interview progressed, F., who was not married, found her more and more attractive. He also discovered, to his chagrin, that she was married, but against his better judgment, he invited her to lunch, and she, against her better judgment, accepted. For the next two months, they met frequently for intimate conversations at lunch, always completely platonic but always sparkling. At last F. decided to see if there was any chance of passing from platonism to hedonism, but his hopes were turned down in an unambiguous though compassionate manner.

Five years later, F., still unmarried, accepted an offer from another university. No sooner had he arrived on campus than a reporter who worked for a magazine published by the university contacted him in the hopes of conducting an interview by telephone. Rather than replying to her, "Great idea — we can both save some precious time that way!", F. suggested that they meet in person instead. As is obvious, the analogy between the two situations, although it was based on resemblances that were quite superficial, exerted a nearly irresistible pressure on F.

We are all constantly prone to listen to inner voices that whisper to us, "*This* situation is extremely reminiscent of a situation that you were in before and that you remember clearly, and so you're very lucky! Just make your decision on the basis of your prior experience!" Let's consider another case of this sort.

G. once wrote a column for a well-known magazine. When he was hired, he was unaware that every issue was eventually published, in translation, in many countries around the world. One day, out of the blue, G. received a package containing various international editions of the magazine, featuring his columns in many different languages. At first he was delighted to see his ideas being spread around the world, but on carefully examining some of the European translations (he had studied a few European languages and was fascinated by them), he was deeply disappointed, as he saw that his very careful writing style had not been faithfully reproduced at all and had been replaced by a style that he found boring and flat. He felt betrayed.

Some thirty columns later, the magazine's publisher contacted G. by phone and told him that he had just learned that the directors of the Korean counterpart of the magazine were hoping to publish a volume specially devoted to all of G.'s columns. What an honor for G.! So would G. give the go-ahead to publish this anthology? G., however, just hemmed and hawed. The publisher pressed on: "What's the problem? This is a great opportunity!" G. then admitted that he hadn't been pleased with the European translations of his columns, and that although he didn't speak a word of Korean, he was worried that the same thing might... The publisher cut him off and said, "Now, now, old boy, not to worry! The Korean edition of our magazine uses only the very best translators that can be found — super-experts, extremely sophisticated. That's a guarantee. So it's a deal?"

What would the reader do in G.'s shoes? On the one hand, there was strong pressure coming from the magazine's upper echelons to publish a book in Korea, which would have been good for G.'s reputation. On the other hand, there was an analogy whose relevance was not in the least clear. What possible link could the disappointing performance of a few European translators have with the performance of unknown translators in a country that was so distant geographically, linguistically, and culturally?

In the end, G. turned his publisher down. The latter had assured him that the Korean translators were as good as they come, but what did *he* know about it? It's inconceivable that he would have said to G., "I know how painful it is for an author to be mangled by a translator. Since you didn't like any of the European translations, your doubts about the quality of translation into Korean are highly merited."

It's far easier to imagine the publisher saying something like this: "Few things in the world are more different than Korean and European translators! To be sure, some European translations of our magazine have now and then left a bit to be desired, but that problem is completely localized. Korea is another ball of wax. A completely different mentality reigns over there — a far more serious one. Of course I understand your temptation to make an analogy, but believe me, there's simply no connection. It's as if you were telling me you would never drive a Korean car after having test-driven only cars built in Eastern Europe. Trust in us, old boy; if we assure you that these are faithful and reliable translators, you can count on us and on them." But G. didn't buy this line — after all, "Once burned, twice shy"! And a week later, he received a letter telling him that his column would be terminated. His analogy had cost him dearly.

What, then, was the bedrock set of beliefs on which G. thought it reasonable to ground his analogy? Was it sensible to leap from the quality of a German translator to the quality of a translator in an extremely distant language? And what about the fact that it wasn't a question of just *one* low-quality European translation, but *four*? And how to take into account the differing levels of G.'s knowledge of those four tongues?

At what point is it sensible to extrapolate to an unknown world an event that took place in a known world, and on the basis of what kinds of knowledge and what kinds of prior experiences? If a translation into a given language is of low quality, why should a translation into another language, done by another person, also be of low quality? Does one earn the right to make such an extrapolation if there have already been *twenty* abject translation failures in twenty different languages spread out over all the different continents? Or does one earn the right to make such a leap long before that moment? What does it all depend on, in the end? On the distance between the countries involved? On their cultural distance? On the relatedness of the languages? Does one need to have a deep knowledge of all the languages concerned? Or is it enough to have seen problematic translations into just one or two languages, whatever they might be?

Bagels Belonging to a Single Batch

In each of the two foregoing anecdotes, someone presumes that what once took place in one situation has a good chance of taking place again in a new situation that brought the older one to mind. Thus the professor who, five years earlier, fell head over heels for a university public-relations writer who called him up for an interview cannot suppress the tingling feeling, when he's contacted by a public-relations writer at his brand-new university, that there is a pretty decent chance that he'll find *her* to be just his type as well. And the magazine columnist who was disappointed by the translations of his columns into a few languages that he can read can't help but suspect that the translations into languages of which he can't read a word will be just as flawed.

These anecdotes are cases of *induction*, a form of thought in which one extrapolates to new situations certain observations that one made in one or more prior situations. Such an extrapolation can extend to a sizable category (for example, presuming that all Dutch people are smart after having met a few who are smart, or guessing that Paul is always late on the basis of a couple of late arrivals), or it can be limited to just one other case, or a handful of other cases (for example, presuming that the almond flan, soon to be wheeled in, will of course be delicious — after all, the chiles rellenos were terrifically tasty! — or leaping to the conclusion that Jack and Susan must be just as blond as is their strikingly blond brother). Induction has no claim to logical validity, in the sense that no rigorous rule of reasoning allows a person to arrive at an inductive conclusion with absolute certainty. No known law compels all female university public-relations writers to be heart-throbs for a certain male professor, nor is there any universal principle entailing that all the translators employed by a certain magazine must share an identical level of mediocrity. To be sure, there are powerful, compelling analogies that link the situations in these scenarios, but that doesn't make the conclusions that these analogies suggest anything close to certain.

Of course, the fact that a given conclusion is not derivable through watertight logical reasoning doesn't mean that the conclusion has to be false; far from it! It's perfectly possible that the Korean translation will be miserable, and that the professor will be smitten beyond belief when he meets the public-relations writer in his new university. In short, the *likelihood* of a conclusion should not be confused with its logical *validity*. And indeed, what usually matters in everyday life is how likely something is rather than how logically deducible it is. The professor isn't looking for an ironclad *proof* that he will fall in love with a woman he hasn't yet met; he merely wants to know if the chances are decent that she will appeal to him. And what matters for the columnist is the high probability that the Korean translations will leave much to be desired. Neither of these people cares a hoot about whether their conclusions can be reached by following strict logical reasoning. What matters overwhelmingly to them is simply the question of likelihoods. Moreover, the conclusions reached by these two individuals have very different levels of probability. Whereas the romantic hopes of the starry-eyed professor can strike one as pretty unlikely, the fears of the skeptical columnist may seem somewhat more realistic. And perhaps each of these individuals, in making an "illogical" analogical extrapolation, accurately intuited its likelihood of being right.

To live one's life in this world, one has to trust one's own judgments about what is and what isn't likely, far more than worrying about fine points of logical validity. If, on some grade-A gray day, you suddenly decided to reject every inductive conclusion you reached simply because it wasn't the result of an ironclad form of reasoning, then your thinking would have to grind to a halt, because every thought that anyone has, no matter how tiny it is, no matter how spontaneous or mindless it might seem, is an outcome of this kind of mental activity that has no logical validity.

If A. were to ask Z. a question as innocuous as "How are your fries?", then Z., in order to be strictly logical, would have to reply, "Well, the six I've partaken of so far were most savory, but since I haven't tasted any of the others on my plate, I have no basis for commenting on how they are." In some theoretical sense, Z.'s answer may be defensible, but anyone sane would find it grotesquely pedantic. A sensible person would reply, without even thinking twice, "They're delicious!", because such a person would unhesitatingly extrapolate outwards — from the titillating twinges of their taste buds triggered by their initial nibblings all the way to the whole plateful. The analogy between the already-savored fries and their yet-to-be-ingested cousins would seem too compelling to allow any other answer even the slightest chance of coming to mind.

Numerous psychology experiments have identified various factors that contribute to the apparent validity of such generalizations, which is to say, to the feeling that such analogy-based extrapolations are *credible.* For example, *the greater the number of previously observed cases* is, the more secure people tend to feel in concluding that the same outcome will take place in a fresh new case. Thus, the more fries one has tasted from one's plate, the more secure one will feel in guessing that the entire plateful of fries is good.

The presumed *diversity of members of the category* is another factor: that is, the more diverse one believes the category to be, the more cautious one will be in accepting the analogical extrapolation. Thus, if the first few people one sees in a certain town are all greatly overweight, one is less likely to jump to the conclusion that everyone who lives there is obese than if one knows that one is dealing with a trait that is generally very uniform among samples, such as electrical conductivity. If the first few samples one tests of a new material all conduct electricity well, then one is very likely to conclude that anything made of that material will conduct electricity well.

The greater the *diversity of observed cases* is, the greater will be one's confidence in an analogical extrapolation to the whole population. Thus, the more different dishes one has ordered in a restaurant, the more justified one will feel in making a claim about the overall quality of that restaurant's food, whereas if one has always limited one's orders to just the vegetarian dishes, no matter how many there are and no matter how many times one has ordered each one, one will be less likely to jump to a broad conclusion about the restaurant's overall quality.

Another factor that is likely to influence one's tendency to believe in one's analogical extrapolation to a large population is the degree to which one thinks one has observed *typical members of the category.* In other words, one is more likely to generalize outwards from a case that one takes to be a prototypical member of the category than from a case that one takes to be peripheral. Thus, in estimating the quality of a restaurant, one is going to place more trust in one's judgment of its baked salmon than in one's judgment of its olive bread or of its mocha latte.

A modest metaphor that we call "bagels from the same batch" will help us to unify the preceding considerations. All of the bagels that are cooked in a single batch are presumably interchangeable in most ways, in the sense that they should all be equally salty, equally tasty, equally warm, equally soft, etc. The general question of whether one can extrapolate from a given situation to a different situation then becomes the question of how much these situations belong, so to speak, to "the same batch". Do all the French fries on one's plate count as "bagels from the same batch"? In all likelihood, yes. And what about all the various translators who are hired by a given magazine? When it comes to their competence in translation, can they all be seen as "bagels from the same batch"? And what about all the public-relations writers hired by different universities — are they all "bagels from the same batch"? When things are cast in this light, we see the key question "Is it analogy-making or is it categorization?" coming back and grabbing center stage, because how one answers the analogical question "Are these two items essentially bagels from a single batch?" will depend on the degree to which one perceives the items as belonging to a single category.

Sometimes the answer is so obvious that one would cringe at someone even posing the question explicitly. For example, all the copies of a novel printed at the same time by the same printing plant are clearly bagels from the same batch, and it makes no sense to imagine someone wondering whether *one* of those copies of the novel would be a good read if they have already read and enjoyed *another* of those copies. On the other hand, recognizing *sour grapes* situations as such amounts to seeing bagels from the same batch, but on the basis of traits that are far less immediately obvious — and when a scientist makes a great discovery by jumping by analogy from one phenomenon to another within a single domain or even across domains, it's because that scientist saw bagels from the same batch where all of their colleagues merely saw a pile of highly diverse breakfast edibles. A different way of asking the question "Do these situations share a single essence?" is thus to ask oneself, "Are these essentially bagels from a single batch?" The analogies that invade our minds and channel our thoughts, most often doing so unbeknownst to us, and sometimes doing so helpfully and sometimes misleadingly, are those that strike us intuitively as being bridges built between bagels belonging to a single batch.

The Tyranny of Analogies

Analogy can play an even more coercive role. Sometimes analogies not only arise naturally out of situations, as we've just seen, but they can then shut out all other viewpoints. In such cases, we are dealing with, so to speak, "the tyranny of analogies".

> K.'s grandfather adored redwoods. When he was old and very ill and everyone knew the end was near, his son decided to take him for one last time to see Northern California's beautiful Avenue of the Giants, a 31-mile stretch of road that features some of the most spectacular redwoods in the world. This was a wonderful moment for the grandfather, allowing him to spend some special moments of tender remembrance with his son, among these trees that he had always so much cherished. Not long after this trip, the old man passed away, peacefully and serenely.
>
> Forty years later, K's father was himself an old man and for several years had been in declining health. He too had a lifelong love of redwood trees, and one day it occurred to K. to do for her father what he had done for his own father — to take him on a trip to the Avenue of the Giants, so that for one last time he could be close to the magnificent trees he had always adored. She dearly wanted her father once again to experience this rare grandeur, and she hoped to share it with him.
>
> But alas, it was not to be, and the blame lies entirely with the human faculty of analogy-making. Had K. even dared to hint that she wanted to take her father to see the redwoods, the analogy with *his* trip with *his* father would have instantly leapt to the mind of everyone in the family, and most of all to K.'s father himself, of course. K.'s suggestion would have meant, pretty much explicitly: "Dad, soon you're going to die, and I dearly want to take this last trip with you before you do." In other words, taking this trip would have been an unmistakable message from K. to her father, telling him that his family saw him at death's door. "Better not go there!", K. said to herself, inadvertently making a double entendre.

This intergenerational analogy has great power, because the fact that it would instantly spring to everyone's mind totally blocked K. from doing something that her father would have loved to do and that she herself thought would be a beautiful gesture. And yet there is no causal power linking one trip to the other. There is no reason for a sane person to believe that taking one's father on a trip to see huge redwood trees will bring about the latter's death in short order. "The Curse of the Avenue of the Giants" might be a good title for a murder mystery, but it should nonetheless not have inspired great fear in K.'s father or in any other member of the family.

Though not grounded in logic, this analogy constitutes a greater pressure than could emerge from any kind of formal deductive reasoning. Everyone knows that if *A implies B* and *B implies C*, then *A implies C*, and yet such reasoning patterns don't exert great psychological force on us. We know that there are often hidden traps in what appear to be valid modes of reasoning. The way that the redwood-tree analogy, which has no logical power, takes over our minds willy-nilly forms a stark contrast with the way that we react to formal reasoning.

If one day we were to read that the proof of some famous theorem — say that of Fermat's Last Theorem, discovered toward the end of the twentieth century — had been shown to have flaws in it, we would not keep on believing in its validity. We have all heard arguments that sound logical and rational but that lead to conclusions that are blatantly false, and so we have learned that sometimes it pays to be prudent when faced with scientific-sounding reasoning, even when it sounds ironclad. In contrast, in the case of K.'s analogy, we are overwhelmed by the stark power of the resemblance; it is so strong that the conclusion is irresistible. In short, it is easier to be suspicious of a logical argument than of an argument by analogy.

The analogy between the two trips to the redwoods, though it lacks any basis in logic, and though its conclusion may well be totally false, is so blatant that it cannot be shrugged off. Without in the least believing in any kind of "paranormal analogical force" that will kill her father, K. feels completely trapped, because she, like any other human being, is incapable of suppressing the heavy pall that the analogy would cast over such a trip. Even if K. and her whole family had a long talk about it, and everyone, including her father, were in total agreement that it was just an analogy with no meaning, there would still remain in everyone's mind the salient vision of the sword of Damocles that the analogy would have brought into existence, which would be the thought, "Suggesting this trip to Dad inevitably says to Dad that we all think his days are numbered." Indeed, a long family conversation about it all would only strengthen and entrench the analogy's grip on everyone's mind.

The two trips, so far apart in time and yet so similar in the minds of all involved, cannot psychologically be pulled apart. The mapping is so salient that the human mind concludes that the two stories have to end identically. It's a classic case of the proverb "One does not speak of rope in the house of someone who was hanged." No one would believe that taking the trip would actually *bring about* the father's death; it is simply that the idea of taking the trip would inevitably be contaminated by the image of *that other trip* one generation earlier, and the analogy would impose itself heavily and sadly.

Human thought simply is this way. Certainties do not come from following rigid deductive laws; indeed, conclusions reached through such reasoning strike many people as suspect for that very reason. By contrast, categorizations brought about by analogy-making impose their conclusions in a manner that is hard to resist. If it's a dog, then it ought to bark. If's it's a chair, then one ought to be able to sit on it. If it's night, it's hard to see. If I suggest the trip, then Dad will think that we all think he's at death's door. This particular case has shown that analogies can make up our minds for us in a very firm fashion. Just as we cannot help thinking "four legs" when we think "table", or "feathers" when we think "bird", or "genius" when we think "Einstein", so K. and her family wouldn't have been able to resist thinking "imminent death of Dad" if they were to think "a trip for Dad so that he can savor, one last time, the Avenue of the Giants".

A Double-edged Analogy

At this juncture in our book, another analogy in the same vein imposes itself on us. It begins when Chilean physicist Francisco Claro, accompanied by his wife Isabel and their three children, set out from their native land for Indiana University in Bloomington, where they were going to spend Francisco's first sabbatical year ever. This couple was among the best friends of the American couple, Doug and Carol.

Many similarities linked these two couples. Doug and Francisco were just a couple of years apart in age, and both had gotten their doctorates in physics working under the same professor at the same university; both adored Bach and Chopin and played piano frequently for each other. As for Carol and Isabel, they were good friends, both had gotten degrees in librarianship and had worked as librarians, and each of them had a kind of soft Latin beauty.

Several months after Francisco's arrival at Indiana University, Isabel started having a series of agonizing headaches. After being taken to Bloomington Hospital, she was diagnosed as having a brain tumor, and was instantly transferred to a much more sophisticated hospital in Indianapolis, about sixty miles to the north, for further tests and procedures. The tumor was found to be very large, and indeed, before operating, the surgeon described it to Francisco and Isabel as "the size of a lemon". Never in their lives had they known such a terrible fear as at that moment. And yet, in the course of the operation on Isabel, which was necessary to forestall her death, it was discovered that the tumor was benign, and not only that: it was encapsulated and was thus able to be removed very naturally. Isabel recovered quickly from the surgery, and there were no lasting consequences.

Seventeen years after this event, Doug left his native land with his wife Carol and their two children to spend their first-ever sabbatical year in the university town of Trento, Italy. After a few marvelous months there, Carol started having a series of excruciating headaches. Doug took her to the hospital in Trento, where they did tests and discovered a brain tumor. She was instantly transferred by ambulance to the much more fully equipped hospital of Verona, some sixty miles to the south, for more extensive testing and procedures. The tumor was found to be very large, and indeed, before operating, the surgeon described it to Doug and Carol as "the size of a lemon". Never in their lives had they known such a terrible fear as at that moment.

The analogy between these two situations is incredibly strong, so strong that readers might think it is all just an invention, but it is all true down to the finest detail. In each case, we are in the presence of a young family during their first sabbatical year ever, undertaken in a foreign country; these families are analogous for the reasons given above, and the strength of their friendship lends great strength to the analogy. In both situations, a tumor is detected in the brain of the wife after a series of terrible headaches; in both cases, the patient is instantly transferred from the local hospital to a much better equipped hospital in the nearest large city, roughly sixty miles away; in both cases, the tumor, once measured, is described as being "lemon-sized".

Now given the strength of this analogy so far, no one could hear it and not be at least tempted to conclude that "history was inevitably going to repeat itself" in the case of the two couples — that is to say, that Carol's tumor, like Isabel's, would be found to be benign and encapsulated and would be removed perfectly, with no consequences, and that everything would be fine afterwards. Indeed, for Doug and Carol, this belief was more than just tempting — it was enormously powerful, and it allowed them to face this horrifying situation and even to feel optimistic, as if this whole story was simply one part of the parallel unfolding of the two couples' lives, and thus was inevitably bound to have a similar ending. And although all this turned out to be wrong, Doug and Carol were sustained until nearly the very end by this compelling analogy.

The Prison of the Known

We are constantly confronted with the new and the unfamiliar, and we deal with it through the help of myriads of analogies. Those same analogies, however, manipulate us, turning us into prisoners of the familiar. The Indian thinker Jiddu Krishnamurti wrote a great deal about how the memory of our past shackles us as we grapple with the present. In his writings, he vaunts the idea of acquiring new perspectives that are *not* shackled by our memory, for in his view the chains of memory do not allow a pure, true, genuine, and deep perception of oneself, of others, of one's environment, or of situations that one encounters ("That memory is knowledge, that knowledge is going to interfere… — obviously."). "Freedom from the Known", the title of one of his most famous works, clearly expresses this viewpoint.

As we have just seen, there are nearly invisible analogies that crop up in the tiniest acts of cognition and there are large analogies that, by staring us in the face, force us to take decisive forks in our lives; moreover, our categories, selectively activated by our momentary concerns and our momentary obsessions, filter our perception of our surroundings and control our thoughts. In fact, it is the known that manipulates us at all times and in all ways. We depend intimately on the known, on both very small and very large scales. And thus there can be no doubt that looking at the world in terms of one's past experiences is an undeniable fact of human existence. Yes, as Krishnamurti says, we all are shackled by the blinders of our categories; indeed, they follow us like shadows, acting as indispensable collaborators of our sensory organs and as inseparable partners in our perceptions. In this sense, analogies manipulate us and control us shamelessly, boldly inserting themselves left and right between us and what surrounds us, and even between ourselves and our selves.

Analogy pervades our thoughts from top to bottom, controlling every aspect of our interactions with the world. The fact that it controls us so intimately leads to the inexorable conclusion that we can think only in terms of what we know in some fashion or other. We are like blind people who have always lived among other blind people, unable to imagine the existence of senses beyond touch, smell, hearing, and taste. Even the most daring ideas of science-fiction authors and the wildest visions of surrealistic painters come from combining commonplace concepts from our everyday world; thus, such creators dream up such notions as *a flying three-headed lion*, or *an intelligent lake*, or *a machine that can reverse the direction of time's flow*, or *an invisible person*, or *a cross between a human and a spider*, or *a person who can see into the future*, or *a new kind of force between particles that creates fabulous amounts of energy*, and so on. The book *Codex Seraphinianus* is an illustrated encyclopedia several hundred pages long that portrays a fictitious world in enormous detail, a world that in nearly every way diverges from our own, but which, at the same time, is constructed entirely from conceptual building blocks that are completely mundane. Even when people reach such high peaks of creativity, they do so totally through their conceptual repertoire that comes from their mundane existence.

Each of us is continually creating extensions of or variations on what we already know, and at the base of this huge edifice lie our most primitive needs. And our constant quest to meet these primitive needs leads us to undertake activities having seemingly unlimited levels of sophistication. The need for food gave rise to *haute cuisine*; the need for warmth gave rise to high fashion; the need for shelter gave rise to architecture; the need to move about gave rise to vehicles of innumerable sorts; the need to mate gave rise to erotic art and innumerable love songs and poems; the need to reproduce gave rise to families and their interactions; the need to exchange goods gave rise to huge networks of interdependent economies; the need to cooperate gave rise to governments; the need to understand the world gave rise to science; the need to communicate gave rise to a thousand constantly-evolving technologies... We humans have created an unlimited cornucopia of elaborate variations on the themes of what we know, but we are incapable of going beyond that.

What, then, is this goal of "freeing oneself from the known"? The known has two closely linked facets: it is a *constraint*, in that it biases our perceptions through the filters it imposes, but it is also a *guide*, in that it offers us the possibility of constantly changing points of view. Like railroad tracks, which give a train the ability to move great distances but also force it to follow the linear path they define, our categories allow us to say and predict a vast variety of things, but since in each case we adopt only one particular point of view, all other points of view are temporarily suppressed.

Stripped of all past experiences, a human being would be incapable of seeing, distinguishing, or understanding anything at all. Seeing the known as an obstacle to human thinking is like seeing tracks as an obstacle to a train's motion. While this is true in one sense, since tracks keep trains from wandering all over creation, it is quite absurd in another sense, since trackless trains would go nowhere at all. Likewise, seeing analogy as a manipulative force is correct in one sense, because we are all relentlessly pushed around by our analogies, but it's absurd in another sense, because without analogies no thinking would be possible at all.

Of course, a seven-year-old girl who is delighted with her brand-new revelation that "shaving cream is like toothpaste" is a prisoner of her prior knowledge: her category of *toothpaste* necessarily biases her perception of her father's shaving cream. But this "prison" has to be contrasted with the "freedom" that she would enjoy if she had no knowledge of the concept *toothpaste* — or of the concept *white*, or of the concept *cream*, or of the concept *substance*, and so forth.

No one can deny that our knowledge of the world constitutes an extremely tight set of constraints, but it is precisely this set of constraints that imbues our thoughts with their marvelous novelty and freshness. This brings to mind the final four lines of a graceful little ode written by James Falen, a gifted translator of Alexander Pushkin, the great Russian poet whose poetry features constraints of all sorts:

There are magic links and chains
Forged to loose our rigid brains.
Structures, strictures, though they bind,
Strangely liberate the mind.

Hoping to get rid of one's categories acquired through experience would be like wishing to jump right into the most advanced stages of Alzheimer's disease. We humans have the great luxury of being able to look at our world through all sorts of filters — all manner of categories — but a virgin perception, untainted by any prior concepts, is a chimera. The known, as it is an intrinsic extension to our physiological senses, is part and parcel of our perception. To summarize matters pithily, if, in making one's way in the world, one were offered the choice between having a backlog of known things to depend on or having nothing known to depend on, it would not be tantamount to a choice between living in chains and living free as a bird, but rather, to a choice between living in a complex maze pervaded by patterns or living in perpetual blindness.

Yes, analogies manipulate us, and yes, we are enchained by them. This is a fact that we simply must recognize. Not only are we prisoners of the known and the familiar, but we are serving a life term. But luckily for us, we have the power to enlarge our prison over and over again, indeed without any limits. Only the known can free us from the known.

CHAPTER 6

How We Manipulate Analogies

❧ ❧ ❧

Sticks for Stirring Become Javelins for Rowing

Emmanuel and Doug have just gone out for their habitual afternoon coffee break. "Two *crèmes*," says Doug to the server. When the drinks arrive, Doug pours some sugar in his cup, takes the spoon from the saucer, and while stirring, says, "Here in Europe, nobody is surprised to see coffee served in real cups, with real saucers and real metal spoons. It's par for the course."

"Sure!" says Emmanuel. "But why would that surprise you?"

"It was like that in the U.S. when I was a grad student," says Doug nostalgically, "but nowadays, wherever you go, coffee is always served in big tall cardboard or styrofoam cups, and instead of a spoon all you get is a super-thin little wooden or plastic stick. Why doesn't anyone ever complain? It's as if, uhh… It's as if some tourists came to a lake and wanted to rent a rowboat and then, instead of being given oars, they were given a pair of *javelins*. Can you imagine that? And on top of that, imagine that nobody uttered a peep in protest."

Caricature Analogies: A Creative Communication Tool

What Doug just came out with is a *caricature analogy* — a very common sort of cognitive act consisting in the dreaming-up of a new situation that differs greatly from the original one, at least on the surface, but which, at a deeper level, is "exactly the same thing", and which has aspects that cannot help nudging the listener towards the conclusion desired by the speaker. Such a process is generally triggered when one is desirous of sharing a strong personal reaction, such as indignation, to a situation. Often one fears that a direct and straightforward recounting of the situation itself will be too bland to get anyone else to feel one's intense sense of indignation. And so, dipping judiciously into one's vast system of categories, one tries to concoct a fictitious situation in a distant domain yet very much like the situation at hand, sharing a "conceptual skeleton" with it, and hopefully vivid enough to pull listeners in and get them to see the original situation through one's own eyes. We now illustrate this phenomenon through a series of examples drawn from daily, often quite mundane, interactions.

A scientist seeking a job abroad wrote to a colleague: "I love my country, but doing science here is like playing soccer with a bowling ball."

Doug said to Carol, "The German word for 'tortoise' is 'Schildkröte', which literally means 'shield-toad'." Quite tickled, Carol replied, "Shield-toads, eh? And I suppose that over there they don't have *eagles* but *feather-cows*?"

Carol asks Doug, "You don't have a pen on you, by any chance, do you?" Doug, who makes a point of never being without a pen, replies to his wife's question just as he has replied to it a hundred times before: "Is the Pope Catholic?"

At a party an adult asks a teen-ager how old she is, and she says, "I'm seventeen." Her father reminds her that she's still just sixteen. "Oh, come on, Dad!" she retorts. "My birthday's just two weeks away!" He counters, "Sure, sweetie. And today's September 29th. So October has already started, I suppose?"

A journalist asked Paul Newman why he remained faithful to his wife, actor Joanne Woodward. (Their marriage lasted fifty years, until Newman's death.) He answered, "Why go out for a hamburger when you have steak at home?" And when he was asked why he missed the ceremony in which, in his sixties, he was at long last being awarded an honorary Oscar, Newman explained, "It's like chasing a beautiful woman for eighty years. Finally she relents, and you say, 'I'm terribly sorry — I'm tired.' "

S. says to her father, "This morning I went to the store with Jack [her brother] and this girl named Jill." Perplexed, her father replies, "What's with calling her 'this girl named Jill', eh? She's been over here at least a dozen times, and I always had to drive her home, and each time I spent ten or fifteen minutes talking to her! Don't you remember all those times? You might as well have said to me, 'This morning I went to the store with a guy named Jack…' "

A feminist slogan of the eighties said, "A woman needs a man like a fish needs a bicycle."

A common bumper sticker on American cars during the Vietnam war said, "Fighting for peace is like fucking for virginity."

"Love without jealousy is like a Polish man without a mustache." — Polish proverb.

A famous physics professor gets an email from a post-doc he's never heard of. She starts out with "Dear Dan", and then says that she's about to submit a "revolutionary" grant proposal and hopes that "Dan" will "come on board" and co-sign it. He is shocked by the note's presumptuousness, and says, "So if I were the Queen of England, would she have called me 'Lizzie'?"

A sports announcer was taking contemptuous note of the fact that some people were pushing for chess to become an official Olympic sport. He scoffed, "And so what's next on the list — Monopoly? Clue? Tic-Tac-Toe?"

A writer about cancer who wanted to make fun of alternative medicine wrote: "A doctor who treats stomach cancer by advocating a diet of special plants is like the captain of the Titanic spending his time rearranging the deck chairs instead of issuing an SOS."

A physics professor said, "Imagining relativity before the equation $E = mc^2$ was discovered is like imagining Pisa before the Tower of Pisa had been built."

Fred: "So did you receive my *in*vite in the mail?" Jim: "Yeah, not only did I get your *in*vite, I already sent you my *ac*cept."

A teen-ager: "Pay ten dollars to get to see the previews for *Harry Potter*?! No way! I'd rather have my tongue stapled to the wall!"

"So religious fanaticism is going to disappear during this century? Wonderful! And you know what? I have a great bridge I can sell you for just ten dollars!"

M., who's always loved puns, is annoyed that many bright people ridicule them. She says, "Why do so many thoughtful people hold the art of wordplay in contempt? It's as if some sport were a lightning rod for vicious mockery by sports lovers. Imagine that every time any pole vaulter gracefully cleared the bar and landed safely in the pit, the announcer were to groan, 'Grotesque! What a lousy jump! It stank!', and the crowd were to hiss loudly. Why do so many people do just that every time that they hear a pun, no matter how clever it is?"

A couple introduce their cat Snoopy to a friend, who scratches his head and asks, "Why did you call a female cat 'Snoopy'? Everyone knows Snoopy is a male beagle!" The wife answers, "It's because she's constantly snooping. That's her signature. Her name has nothing to do with *Peanuts.* The connection never came to our mind!" The friend replies, "Oh, come on! It's as if you were telling me, 'When we named our son "Adolf Hitler", it never crossed our mind that anyone would think of the Nazi Führer!"

A young woman reveals to her brother how badly her fiancé has verbally abused her for years. She tearfully adds, "I'm just so used to it; I've never known anything else. Sometimes I really do think I deserve all his insults." Her brother replies, "That's crazy. You're like a whipped dog who accepts cruelty as normal, while all the other neighborhood dogs have kind masters who pet them all the time, and who couldn't imagine hurting their dog."

Two friends on a walk notice a café called "The Corkscrew". Chuckling, one says, "Do you think the same people own a cocktail lounge called 'The Coffee Grinder'?"

A dad plays blindfold chess with his daughter, beating her in 40 moves. He mutters to a friend, "Oh, I played so badly!" The friend replies, "You remind me of a talking dog who complains that he can't get rid of all the split infinitives in his speech."

At a dinner party, P. is sitting next to a doctor who states his view that midwives should all be women, because only women know what it is to give birth. P. reacts, "By that logic, breast-cancer specialists would all have to have had breast cancer, and only someone handicapped could sell wheelchairs. And of course, a bald person couldn't be a hairdresser."

Investment guru Warren Buffett commented that the huge profit-making opportunities opened up by the global financial crisis made him feel "like a hungry mosquito at a nudist camp."

A trouble-shooting site on the Web tries to explain why its Webcam seems so slow. "Why does our video camera run so slowly? Well, the amount of information it has to transmit is very large, and standard telephone lines and an old modem struggle to process all of the data. It's like trying to route the entire Mississippi River through the plumbing in your house. It just doesn't fit, but we do our best!"

Such a list could be extended forever. It shows that caricature analogies jump to the lips of anyone and everyone, most frequently provoked by an intense and sudden reaction to a situation, such as indignation or surprise. The analogy can take on many external forms, such as "Thinking X is as dumb as thinking Y", or "Doing X wouldn't be any less absurd than doing Y", or "Now that they've accepted X, are they going to accept Y and Z as well?", or "You might as well believe in Y if you believe in X", or "If X is true, then Y is true too", and on and on. Caricature analogies are often based on extremely salient entities in their respective domains, such as Albert Einstein, the Titanic, the Mississippi River, Mount Everest, or McDonald's, or on commonplace, hackneyed facts, such as that the earth is round, that a week contains seven days, that the Pope is Catholic, and so on.

Unconvincing caricature analogies can of course be dreamt up, just as far-fetched categorizations can be made. After all, the human mind often widely misses the mark in its attempts to zero in on the gist of situations it faces! For example, when an overzealous computer executive was announcing a modest incremental advance in the technology of chips, he grandiosely declared, "Compared to our latest new chips, the old generation of chips is like a rusty can opener next to a brand-new Ferrari!" The mere fact that this is a cute caricature analogy doesn't suffice to make it convincing.

A Caricature Analogy in Slow Motion

When in lectures we explain to audiences what a caricature analogy is, people are often stimulated by the idea and some launch right into the deliberate construction of caricature analogies themselves. When such attempts work well, some people think they know all there is to know about caricature analogies. This, however, is a pipe dream. Just as being able to drive a car doesn't make one an expert mechanic, so being able to come up with caricature analogies doesn't make one an expert concerning the underlying psychological mechanisms.

What kinds of mental process give rise to this phenomenon, which ranges from the mundane to the highly creative? Among the countless new situations that we face each day, what is it about a few special ones that launches us on a quest for an analogy based on a crystal-clear but totally imaginary situation? How does one put one's finger on the conceptual skeleton of the situation inspiring the search for a caricature? How does one choose a suitable alternative domain, and then export this same gist into it?

In order to cast a little light on these matters, let's delve into the first example — the scenario of javelins used as oars. Doug wants to get across to Emmanuel his annoyance at the way flimsy wooden or plastic sticks are offered in America as coffee-stirrers. He has, however, a suspicion that his friend, who has not witnessed the gradual slide in American customs, will need some help in order to be brought to the point where he sees things more or less as Doug does.

Doug could say, "It's as if they were giving us some *needles* to stir our coffee with." And indeed, at first he feels tempted by that, but refrains; it would be too extreme and too crude. Mentally replacing thin wood sticks by needles would amount to twisting a knob to turn up the situation's degree of absurdity. Turning such a knob, though easy, is merely indulging in exaggeration, and Doug knows that exaggeration usually reduces the credibility of what one is saying. So he wonders how he can quickly convey the crux of what bothers him without going into a long, heavy-handed explanation.

The crux of Doug's annoyance is the craziness of giving a stick rather than a spoon in order to stir a liquid — not just coffee, but any liquid. So he wants to put his finger on this gist by caricaturing it in a *different* domain, where its absurdity will stand out like a sore thumb. But for his caricature to be effective, the new domain has to be as familiar as possible. The challenge is thus to find a conceptual slippage where coffee is replaced by a generic liquid and where the use of skimpy sticks is blatantly ridiculous.

The mental pressure is clearly pushing for a *coffee* ⇒ *water* slippage, since water is the liquid we all know best. But what familiar activities are there in which one churns up water with some object, and in which it crucially matters *how much purchase* the stirrer has on the water? (The paltry purchase afforded by the stirring tool is, after all, the conceptual skeleton's backbone.) A few possibilities come to mind: the propulsion of a boat by a propeller or a paddle wheel, the act of swimming, and that of rowing or paddling. But to maintain the conversation's momentum, it's crucial for Doug to choose one of these in the blink of an eye.

For Doug, underwater propellers and paddle wheels are not terribly familiar objects and so he skips over them, moving on to the domain of swimming. Here it's the *arms* that have purchase in the water, but replacing a swimmer's arms by some kind of thin objects is not easily visualized; one would have to imagine a bizarre surgical operation, which would make the caricature feel very forced, not graceful and natural. Another possibility would involve the swimmer's *hands*, which do the brunt of the propulsion work. One might imagine a swimming coach who says, "Turn your hands so they present the *least possible* surface area; make them slip *maximally easily* through the water! Minimize your hands' purchase!" But for a coach to come out with such nonsense is so implausible that the caricature wouldn't have a snowball's chance in hell of success.

The remaining domain is that of rowboats. Luckily a rowboat's oars look a bit like oversized spoons, and since they are relatively familiar, human-sized objects, it's very easy to replace them, in one's mind's eye, by other entities. No need to imagine a surgical operation, a nutty swimming coach, or technological savvy! All Doug needs to do is find some good substitutes — that is, some familiar objects that are roughly oar-sized but that would have far less purchase on the water, and that would also recall the trendy slender sticks for stirring coffee (after all, the goal is to create a clear parody of the offensive recent convention). Since the concept *needle* was already activated in his mind, he imagines a giant needle — a needle as big as an oar — and all at once the image of *javelins* jumps to mind. Yes, javelins are very slender and smooth, and yes, they are just the right size, and of course they would have no purchase whatsoever in the water. This amusing conceptual slippage strikes him as a pretty good choice, and so, smiling internally, he takes the plunge and blurts it out.

To concoct a convincing caricature analogy is a challenging cognitive activity in which one hopes to bring someone else around to one's own viewpoint. Sometimes one is indignant or outraged, and that's the feeling one wishes to induce in others; other times, one wants to convey a sense for why one is confused about some topic. This is the case in our next example, and we'll again explore the hidden search mechanisms.

The Highest Peak in a Carefully-selected Mountain Chain

A., an American, receives an email telling him that the pantomime artist Marcel Marceau just died. He mentions it to a French friend, who says she's saddened to hear the news but wonders why anyone would have bothered to send A. an email about it. Perplexed by this reaction, A. says, "What!? If the Eiffel Tower had collapsed this morning, wouldn't that have been email-worthy?" Let's look at what pushed A. to make a caricature analogy and how it came into being.

The trigger was his French friend's casting doubt on the importance of this highly French event, and so A. wants to express, and in a vivid manner, his astonishment at her attitude. To this end, he focuses on Marceau's world renown rather than on his artistry. Since A. had always considered Marceau to be a very major icon, and since the concept of *France* is highly activated in his mind, it's no surprise that the Eiffel Tower would pop up as a quintessential member of the category *icons of French culture.*

But why did A. choose an *inanimate* French icon rather than a famous French person — say, Descartes, Napoléon, or Louis XIV? Well, there were various mental pressures here — that is, blurry constraints — pushing in specific conceptual directions. First of all, it makes no sense to receive an email announcing a death that took place a very long time ago. Secondly, jumping to a radically contrasting domain is a more effective rhetorical strategy (recall this is why Doug shifted from the domain of stirring coffee with very thin rods to the domain of rowing in a lake with very thin rods). And lastly, why didn't A. choose Mont Blanc, the city of Paris, or even France as a whole? Because the disappearance of any of those three would be a nearly unimaginable catastrophe; the Eiffel Tower's collapse seems far more real.

It's clear that many diverse caricature analogies, not just one, are applicable to this (or to any) situation. Thus A. might well have said, "Wouldn't you have sent me an email when the Twin Towers were destroyed?", or perhaps "And if a nuclear explosion had obliterated Paris, wouldn't that have been email-worthy?", or then again, "When John Lennon was shot, I certainly would have appreciated a phone call letting me know about it." Each of these examples conflicts in one way or another with the mental pressures we hypothesized above, which guided A. in the creation (or selection) of his "Eiffel Tower" caricature analogy, but on the other hand, each of them has its own brand of logic, making it at least a plausible candidate for utterance.

A Quick Cascade of Caricatures

We now present a cascade of three caricature analogies, spewed out one after another in ten seconds at most, and all looking extremely simple. But the mechanisms giving rise to them were far from mechanical. Moreover, this episode shows that a whole "bouquet" of caricature analogies can be triggered by a single situation.

> M., sixteen years old, barefoot and in shorts, pulls out the ironing board in the kitchen, places a skirt on it, and turns on the iron. Her father says to her, "Please put on some shoes — that looks dangerous to me!" But M. merely snaps back, "How come each time I *cook* you don't tell me to put my shoes on? And why didn't you tell me to cover my *legs*, also? And aren't you going to insist that I put on *gloves* whenever I iron?"

What led M. to replace the activity of ironing by that of cooking? Her peeved tone, mocking her father's cautious attitude, implies that he might also hallucinate danger lurking whenever anyone cooks, analogous to the danger of an iron falling off the board. Perhaps she imagined something very hot falling off the stove onto her feet (boiling oil, a hot noodle, a frying pan...). But in order to imagine such a mini-scenario, she would have to draw on years of memories of experiences in kitchens. We thus see that the use of an imaginary mini-scenario was crucial in the creation of the first caricature analogy.

And what led M. to transform her father's request that she cover her *feet* into the rather silly request that she cover her *legs*? Is it merely because the feet of any earthbound creature are attached to its legs, thus making *foot* and *leg* close cousins in her conceptual network? Perhaps, but she also had to take into account how she herself was dressed, because the slippage *feet* ⇒ *legs* would be unjustified if she hadn't been wearing shorts (there would have been no danger to her legs if they were covered). Just as in the previous case, a mini-scenario (the falling iron grazing her bare leg) was needed to give rise to M.'s second caricature analogy.

And how did she transform the idea of covering her *feet* with *shoes* into that of covering her *hands* with *gloves*? Did she simply exploit the proportional analogy "*foot* is to *shoe* as *hand* is to *glove*"? If so, why wouldn't she just as likely have thought of the formal analogy "*foot* is to *shoe* as *head* is to *hat*", or "*foot* is to *shoe* as *neck* is to *scarf*"? Why didn't she say (and why would she never have thought of saying), "Aren't you going to insist that I put on a *hat* on when I iron?" M. didn't entertain the scenario of an iron jumping up to hit her head any more than she entertained many other improbable mini-scenarios (*e.g.*, the iron jumping into her mouth, or suddenly falling apart, and so forth). All of this means that there was no reason for the slippage *feet* ⇒ *head* to occur to her. On the other hand, the slippage *feet* ⇒ *hands* makes perfect sense in this context, because an iron, even without falling, can still burn the hand of the person using it. And thus this mini-scenario played an indispensable role in the genesis of the third caricature analogy.

In summary, this series of casually tossed-off retorts shows that the mechanisms giving rise to caricature analogies can take all sorts of aspects of the situation into account, and that the convincingness of a caricature depends crucially on certain mini-scenarios that unfold, lightning-fast, in the mind of the caricature generator. Such mini-scenarios are stereotypical members of previously known, very familiar categories of situations, on which the generator draws in order to devise a strong caricature analogy. For all this to happen in a flash is most impressive.

Explanatory Caricature Analogies

The caricature analogies presented so far have all had a mocking nature, aiming at conveying a sense of outrage or confusion. But not all caricature analogies are mocking. They can also be efficacious explanatory tools. Consider S., who for many years has been unreliable with everyone he knows. One day, out of the blue, he announces to his best friend T., "I've been rethinking everything in my life and I've completely changed! I'm never going to be flaky again." T. replies, "Congratulations, but you know, an ocean liner can't turn around on a dime."

T.'s caricature analogy has no trace of mockery. The image evoked in his mind by S.'s well-intentioned declaration activates the word "turnaround", which denotes both the abstract concept of *major alteration in life* and the more concrete concept of *U-turn*. The latter brings to mind the image of a very quick U-turn made by a vehicle traveling at high velocity. Actually, what makes it hard for a speeding object to flip around very quickly is its momentum, which involves both its speed and its mass, and so T. tries to think of a familiar highly massive entity moving along at a substantial clip. Among the most salient candidates is a train (but trains, being confined to motion along preset tracks, don't make U-turns, so they're ruled out); other candidates are a very massive ship, a missile, or perhaps even a planet in orbit. Why didn't T. say to S., "You know, a missile can't spin around in a tenth of a second", or "Unfortunately, a planet can't just jump out of its orbit"? Perhaps because these objects are less familiar than ships, and perhaps also because missiles are so light that they might seem agile, in contrast to an ocean liner, which doesn't seem maneuverable in the least. In addition, boats move in the viscous medium of water, preventing them from turning easily. In short, the naïveté of thinking that an ocean liner could carry out a very rapid about-face is easy for anyone to see, and so the image T. chose seems very apt for the situation.

It would nonetheless have been possible for T. to use the image of a missile or of a planet, had he been in a slightly different frame of mind. Likewise, had T. isolated a different conceptual skeleton in S.'s claim, he might have cast his caricature analogy in a different direction. Thus he might have quoted a proverb such as "A leopard can't change its spots" or "A zebra can't change its stripes", suggesting that it is nearly impossible for a person's character ever to change profoundly. Or thinking about how hard it is to bring about bodily changes, he might instead have said, "A sprinter can't become a long-distance runner overnight" or "You can't lose 50 pounds in a week."

Quite clearly, the production of caricature analogies is no more a deterministic process than is the choice of words, phrases, proverbs, and so on. Thus even if certain analogies are particularly likely to be chosen in a given context, there are plenty of others that are easily imaginable, even if they are less likely to be produced.

This type of creative cognitive act is an outcome of several facts about the mind. First is our irrepressible human tendency to abstract situations that we encounter in order to pinpoint their most central core; second is the fact that we possess a vast repertoire of mental categories at many levels of abstraction; and third is our ability to select a fresh new category that is vivid and familiar and that very closely resembles the original situation.

Caricature Analogies that Tickle Our Fancy

The device of caricature analogy is sometimes used simply to express something ordinary in a colorful fashion. Such rhetorical flourishes, not just humorous but seemingly offhand, are frequent in sports announcing. Here are two such remarks made by baseball announcers on American television:

> "Trying to throw a fastball by Henry Aaron is like trying to sneak a sunrise past a rooster." — Curt Simmons, pitcher who became an announcer.

> "Trying to get a hit off of Sandy Koufax is like trying to drink coffee with a fork." — Willie Stargell, member of the Baseball Hall of Fame.

Many readers will doubtless pick up on the fact that the second of these quips is reminiscent of the caricature analogy in which oars for stirring are replaced by javelins for rowing. And yet someone might think that the image of failing to drink coffee by using a fork to scoop it up was chosen pretty much at random by Willie Stargell as a humorous example of an impossible challenge, and that he might just as easily have come out with the phrase "Trying to get a hit off of Sandy Koufax is like trying to construct a skyscraper with one's bare hands" or "is like trying to draw a square circle". But substituting either of those metaphors into Stargell's quip results in an utter *non sequitur.* There was a good reason for Stargell's choice. The image he evoked of coffee easily dripping between the long, straight tines of a fork has much in common with that of a ball easily slipping past a long, straight bat. Stargell's choice of metaphor was thus very well-tuned, and it would have been far less effective had he alluded to a randomly chosen impossible task, even if, we repeat, he was probably unaware of the cognitive mechanisms pushing him towards his witty choice.

It's tempting to ask what would happen if batter Henry Aaron found himself up against pitcher Sandy Koufax. Given the very striking descriptions that we've just quoted of these two individuals' abilities, one has the feeling that it would be a case of the famous category *an irresistible force meeting an immovable object*, and one can't help but wonder: what would give?

Caricature Analogies Help Us Explain Things to Others

Spontaneously invented caricature analogies can be excellent tools for explanation, as the following example shows. A boy of age ten meets a vegetarian for the first time in his life and tries to understand what lies behind her choice. He asks her, "Why won't you eat a hamburger, since the cow is already dead?" The woman explains to him that eating a hamburger is merely the last link in a complex economic chain that stretches all the way back to feedlots and cattle ranches, but the boy sticks to his guns, insisting that the cow was already dead and that that's the only thing that matters. At this point, the woman changes tack and says, "Think about those large-size bottles of soft drinks that stand upright on supermarket shelves. They're all lined up on a slope, one behind the other. When a customer removes the closest one, all the ones behind it slide down, so the second one winds up at the front, the third slides into second place, and so on. At the very back, the last bottle slides down, leaving an empty space that a store worker will have to fill with another bottle, or else the store's soft-drink supply will run out soon enough. Well, when you eat a hamburger, it's like removing the closest bottle. The hole that it leaves at the far end of the chain will have to be filled, or else the meat supply will run out pretty quickly. Sure, the specific animal whose meat you are eating had already been killed, but your seemingly innocent act will help bring about the killing of *another* animal, far away and unseen by you."

This caricature analogy draws on a drastically simplified picture, since the full situation is clearly too unfamiliar and remote for the boy to grasp. The long chain that slowly moves upstream towards the source — from the hamburger itself to the grocery-store display case, then to the butchers in the back of the store, then to a large meat-transport truck, then to the slaughterhouse, then further back yet, to an animal-transport truck, then to a feedlot, and finally all the way back to the pastures of some far-away farm — that whole long chain, invisible and spread out over a vast territory, has been deftly replaced by a short straight line containing just a handful of bottles that can slide on a gentle slope. And the fact that a piece of meat is but a tiny part of a large animal has been left out of the picture, since it is not crucial to the key idea, which involves only the notions of *supply* and *demand.*

We thus see that caricature analogies are not limited to situations where a speaker wishes to communicate a sense of outrage or perplexity; they can also be the natural outcome of a desire to transmit subtle ideas in a vivid, clear manner.

Caricature Analogies Help Us Explain Things to Ourselves

We can use caricature analogies to help us understand a situation we find ourselves in, a thought we are having, a sudden feeling triggered by an event, and so forth. In such cases, the caricature may reveal or set in sharp relief a conceptual skeleton of which we had been only fleetingly aware. Such analogies, rather than being created to show somebody else something, are created for ourselves alone. A familiar example will demonstrate how the explanation of some phenomenon — in this case, an explanation that a driver comes up with — can come about thanks to analogies that one manipulates.

As any San Franciscan knows only too well, parking in Baghdad by the Bay is always hellish. One night, though, in the wee hours of the morning, as E. was driving home through downtown San Francisco, he suddenly had a surreal sensation. On Market Street, where it's normally hopeless, he saw empty spots left and right. All around Union Square, perhaps the city's most nightmarish zone, were dozens of parking places crying out to be taken. And as he arrived in his own neighborhood, always jam-packed with cars, his gaze was met with vacant spots galore. E. felt mounting outrage as he stared at these priceless gems for which any driver would have given their eyeteeth and yet for which he had no use at all. Should he simply stop and park in one of them, to savor this astonishing godsend? No, that would be too silly... E. realized he had no choice but to resign himself to letting these amazing rarities (which in this case were a dime a dozen) sail by untouched, one treasure after another. It gave him little consolation to think that in just a few hours, all these streets would once again be clogged to the gills, obliterating this improbable sight.

E. tried to understand more deeply just why he felt so frustrated to see all these "lovely spots". A few analogies came to mind and helped him out. He said to himself that these spots were like a mountain of gold in front of someone who had no tools to extract even the tiniest nugget from it, or like the discovery in an attic of a huge fortune in banknotes no longer accepted as legal tender. Next, he imagined himself utterly stuffed after wolfing down a mediocre meal and then having to decline a whole series of delicious gourmet courses brought to the table.

As he continued ruminating, E. decided that what he was seeing was not so precious after all. He started coming up with happier analogues: "This is like being admitted for free into the city's trendiest nightclub moments before it closes, when no one wants to be there any more, and the last few souls there all drunk as pigs and looking sickly in the bright glow of the overhead lights that have been turned on for the cleaning service that's about to arrive." Then again, "This is like being given free access to the Fillmore Auditorium the morning after the Rolling Stones perform there, when the place is totally empty." And also, "This is like being on a secluded tropical island, normally the playground of billionaires, but in monsoon season." In this fashion, a new vision emerged in E.'s mind: rather than being hugely frustrated by vast riches before him that he couldn't touch, he was now simply in a terrific place but at a terrible time to be there.

We see that caricature analogies can help us strip a new situation down to its core, shedding fresh light of all sorts on it, bringing out in the open one or more conceptual skeletons that were vaguely lurking in the shadows when the situation first arose. Such caricatures spring spontaneously from the "halo" of rapid generalizations that any new scenario of interest tends to evoke in the mind of a curious person, and they help reveal the conceptual skeleton at the root of such generalizations.

In this case, two different types of caricature analogy allowed E. to take two possible perspectives on the situation he was facing. The first type, mapping the empty spots onto treasures that seem ripe for the taking but that are inaccessible due to an unfortunate combination of circumstances, helped him to understand just why, as a San Francisco driver, he was so frustrated. Tending to reinforce his initial impression, these first caricatures had little pedagogical use, but at least they allowed him to see a bit more clearly why he was feeling so unlucky. In contrast, the second wave of caricature analogies, as their common underlying conceptual skeleton started emerging into the light, helped him find a happier way of looking at things. His plight was now mapped onto various scenarios in which timing is everything: all these lovely spots, totally available because it was in the middle of the night, were like concert tickets after a great concert is over. Different essences can thus be brought out by different caricatures.

The Best Ones are Always Snatched up First

A caricature analogy can sometimes clarify subtle phenomena that at first came to mind only as intuitions without any clear logic to them. When one runs through a number of caricatures, the hidden conceptual skeleton that one had only vaguely sensed at the outset starts to come out into the open much more clearly.

To see this, let's consider a paradox that users of big-city bicycle rental systems (such as are found in Washington, Miami, Montreal, and a few other North American cities, as well as in various other cities around the world) have almost surely all noticed. Rentable bicycles are locked to long racks in many parts of the city; with a credit card or membership card, you can unlock one, ride it here or there, and then leave it, locked once again, in another rack, where other people will be able to use it in turn. But alas, one soon notices that there is scant cause for joy when one walks up to a rack where there are only a few bikes left. Indeed, chances are high that they're all defective, so the rack is as good as empty. Contrariwise, when a rack is full, one generally has a wide choice among many bikes in great shape. At first, this fact about nearly-empty racks might seem surprising, for it involves a "double penalty": not only are there fewer bikes to choose among, but adding insult to injury, the few that are there are almost sure to be unusable. Thus a rack's bike-sparseness is doubly rough on any hopeful rider who arrives, since sparseness is also a sign of defectiveness. You may already have figured out why this phenomenon, rather than being surprising, is in fact to be expected; if so, more power to you. But if you don't see it yet, then you might appreciate some help in identifying a much more general phenomenon of which this bike-renting paradox is just one particular example.

A set of analogies — explanations more than caricatures — can help clarify the root cause of the bike-renting paradox. Imagine you are in a grocery store and you head for the produce department because you know there's a sale on cherries, and when you get there you find that there are only a few boxes of cherries left on the shelf. Well, you can be pretty sure that *those* boxes are full of damaged and rotting fruit — they are the dregs that no one wanted to take. By contrast, if the shelf is bursting with boxes of cherries, then damaged or rotting fruits are only a very small proportion of the whole, and even if customers shun them as they choose their boxes, the poor ones won't dominate, so your chance of picking up a box of barely edible fruit will be fairly low. As another example, consider the perennial lament of single people who are no longer so young: "At my age, anyone who's still available is bound to have some fatal hidden flaw!" And one can also think of those straggling few items of clothing, usually very unappealing, that remain unpurchased on the shelf at the tail end of a big sale — or then again one can imagine a factory visitor who is astonished to observe that, among the very rare items that are rejected by the quality-control checkers, not a one is up to par. But no surprise here — indeed, that's the whole point of having quality control! In the case of rentable bicycles, each cyclist who participates in the program cannot help but play the role of a quality-control checker. A bike that has visible problems will be scanned quickly and passed over. This simple analogy shows why there is no reason to be surprised that bikes on sparsely populated racks are nearly always defective.

Collectively, all these analogous situations reveal an interesting fact. Without any abstract explanation having been provided, a new category has nonetheless come into focus — that of *items whose rarity is a symptom of defectiveness, because a limited supply guarantees that a process of selection will take place in which the best items are systematically taken first, and the items that remain to the end will have been rejected by many potential consumers.* This is why, whenever one spots a rack with just a handful of bikes in it, then those bikes, rather than being akin to gemstones, are more akin to ugly ducklings.

The lesson here goes beyond the utility of arguments based on concrete examples. To devise a good explanatory analogy, it won't suffice to make it down-to-earth; in fact, adding lots of concrete details can easily cloud up an explanation, making the gist get lost in the fog. What's crucial in devising a good explanatory analogy is how clearly the invented scenario embodies an abstraction. That is, a successful explanatory analogy is one in which an abstract conceptual skeleton has been very naturally realized in a concrete situation, so the two go together like hand and glove. The conceptual skeleton is then likely to jump out effortlessly to anyone who hears the argument.

For example, a situation in which customers are free to choose the items they are buying (cherries, or items on sale) brings out as a salient fact that when there are few items remaining, they're the ones that didn't pass muster in the eyes of many customers, and so they are *of course* very likely to be defective. This is a clear and helpful image. The metaphor of quality control, where the only items available to purchase are ones that have failed an official test, is equally clear and helpful, because it effortlessly brings out the idea that *stragglers tend to be of low quality.* These explanatory analogies effortlessly lead one to the conceptual skeleton that they all share. They make understanding far easier and more direct than would an isolated verbalization of the conceptual skeleton, such as the legalistic phrase we wrote out a couple of paragraphs up: "items whose rarity is a symptom of defectiveness, because a limited supply guarantees that a process of selection will take place in which the best items are systematically taken first and the items that remain to the end will have been rejected by many potential consumers".

Humans automatically try *on their own* to extract a conceptual skeleton from any concrete situation — that's how they understand — whereas being handed a highly abstract phrase of legalese (such as our long phrase above), no matter how accurately it might capture the conceptual skeleton's essence, is likely to lead only to puzzlement. This idea has important repercussions in education, to be discussed in the next chapter.

Analogies Underlie All Our Major Decisions

In their book *Mental Leaps*, psychologists Keith Holyoak and Paul Thagard tell the droll story of a fictitious researcher in decision-making who has the luck of being offered a job in a prestigious rival institution, which throws him into a major dilemma. If he accepts the job, his salary and professional prestige will both take leaps, but on the other hand, the move would be a huge emotional upheaval for his entire family. A colleague whom he privately asks for advice reacts, "Hey, what's with you? You're one of the world's experts on how decisions are made. Why are you coming to see *me*? *You're* the one who invented super-sophisticated statistical models for making optimal decisions. Apply your own work to your dilemma; that'll tell you what to do!" His friend looks at him straight in the eye and says, "Come off it, would you? This is *serious*!"

The fact is that when we are faced with serious decisions, although we can certainly draw up a list of all sorts of outcomes, assigning them numerical weights that reflect their likelihoods of happening as well as the amount of pleasure they would bring us, on the basis of which we can then calculate the "optimal" choice, this is hardly the way that people who are in the throes of major decision-making generally proceed. What we all do in such situations is to draw on one or more analogues in our memory. Here are a few types of familiar and important situations in life — categories we all know — in which people automatically behave in this way.

- accepting or rejecting a job offer;
- making or not making a marriage proposal;
- staying where one lives or moving;
- divorcing or not;
- going back to school to get another degree or not;
- getting a pet or not;
- having children or not;
- making an addition to one's house or not;
- buying a new or a used car.

Let's take a look at the first item. When one has received a job offer, one thinks it over, comparing the potential job with jobs one has already had, and particularly with jobs of the same sort. What could be more self-evident? And if one has never had a job that would afford a useful perspective, then one quickly jumps, without even thinking about it, to the experiences of people one knows. This means, for instance, that one recalls a job that one's sister once had, and one imports it to one's own life, carrying over all the things she described — the good points, the bad ones, the complications, and so forth. To be sure, one can't import *everything* she said about her job, since most of it is irrelevant to the potential job; one has to know how to arrive at a key nugget of knowledge, abstracted from her tales, which seems relevant. And then this nugget is imported into one's own life much as if one were trying on an item of clothing before purchasing it. When the stakes are high, as they are when a job is involved, one doesn't stop there; one looks for other perspectives to shed further light. After a certain number of tryings-on of this sort, one generally homes in on just one of the explored perspectives and makes a decision after careful consideration of it alone. "Always being caught between two bosses who never agree on anything? My sister once had a job like that and it gave her so much grief — so no thanks."

The idea is simple: the only way we have of making decisions, whether they are small or large, is through analogy — that is, by making analogies with a spectrum of previous experiences (whether personal or vicarious) that have been brought to mind by the pressing decision.

Even when a decision comes down to something so basic as a single figure — the salary of a new job, for instance, or the square footage of an addition to a house, or the number of days of rain per year in a new city — one ponders this number by making comparisons with prior cases that are more familiar: someone whom one knows well and who gets roughly the same salary; a room that is roughly the same size as the proposed addition; another city where one has lived that gets roughly the same amount of annual rainfall. Such comparisons are inevitable, because the past that we have lived through is all we have for thinking about the future.

Thus comparison, whether it is conscious or unconscious, voluntary or forced, is inevitable. Sometimes the analogies themselves are in the driver's seat; other times, we analogy-makers are in the driver's seat — and all intermediate shades between those extremes exist. The process is passive at times, and other times it is very active, but whether it manipulates us or we manipulate it, whether we are the leader or the follower in the dance, analogy is our perennial dancing partner.

Analogy Wars

Decisions at much higher levels than those just discussed are also made by earthbound human beings — global decisions, decisions that may profoundly affect thousands or millions of human lives for the better or for the worse. These decisions are always based on analogies to historical precedents. We are speaking of political decisions, of course, and specifically of wartime decisions.

In global situations, the influence of such analogies is so far-reaching that it squelches all other potential sources of decisions, since politicians and governments cannot afford to appear internally inconsistent. A country, in its quest for internal consistency, can even be forced by an analogy that is very blatant into taking a stance that would seem to run against its best interests. Such a situation arose for Greece during the Falkland Islands War between Argentina and Great Britain in 1982; because of a screamingly obvious analogy, Greece took sides in a curious fashion.

The Falklands lie very close to Argentina, but despite this, they are a British territory. One would have expected, for numerous reasons, that Greece would side with Argentina in this conflict. For instance, thirty years earlier, Greek Cypriots had revolted against the British rule of Cyprus, and this led to Cyprus being given to Greece. This was one clear reason for Greece to side against Britain. In addition, Argentina was, like Greece, a developing country with a fundamentally Mediterranean population (an obvious analogical force, to be sure). And finally, the geographical proximity of the Falklands to Argentina, as well as the international political climate against colonialist aims of all sorts, would seem to have made it an open and shut case.

And yet Greece didn't do what these various analogies would have led one to expect; it sided instead with Britain. What secret lobby swung the Greek government in this direction? Well, there was simply a more compelling analogy, which involved Greece's bitter decades-long dispute with Turkey over the ownership of Cyprus. The analogy mapping Turkey to Argentina and Greece to the United Kingdom leaps instantly to the eye, as both cases involve islands ruled by countries that are much farther away than the countries that seek to possess them. For Greece to side with Argentina in this dispute would therefore have been "the same thing" as siding with Turkey in the dispute over Cyprus; indeed, Cyprus could well be called "the Falkland Islands of Greece"! A choice by Greece to align itself with Argentina would thus have played straight into the hands of its rival Turkey, and would have seriously undermined Greece's position vis-à-vis the ownership of Cyprus. Analogy forced Greece's hand.

Analogies as Critical Tools for Fighting Wars

In his fascinating and carefully documented book *Analogies at War: Korea, Munich, Dien Bien Phu, and the Vietnam Decisions of 1965,* the political theorist Yuen Foong Khong tackles the thorny question of the reasons that underlay the choices made by the American leaders, both military and civilian, during the Vietnam War, particularly in its earlier phases in the 1960s. Khong analyzes a range of historical precedents that were available to be drawn on, including the Korean War, conducted by the United States some fifteen years earlier, the defeat of the French in Indochina, the Munich conference in 1938, in which Sudetenland was conceded to Hitler, the Berlin crisis of 1948, which gave rise to an airlift of food and other supplies, the long insurrection in Malaysia, lasting from 1948 through 1960, and the economic crisis in Greece in 1947. Another potential analogue that Khong suggests, and that we will look into shortly, is simply called "the thirties".

Khong explores the role played by analogies that were publicly mentioned by the decision-makers during the Vietnam War, asking whether the American leaders made their decisions solely by "pure reasoning" (that is, bypassing all analogies) and later gave official justifications of their decisions by pointing to historical analogues in order to win over the masses, who, rather like children, are always looking for simple explanations. Through the analysis of many transcribed conversations among the American leaders, Khong concludes that, quite to the contrary, in every major Vietnam War decision, one analogy or another was the key factor. In other words, analogies were not merely cynically exploited as façades to hide the real reasons from the naïve public. Moreover, in each case, for one historical analogue to emerge victorious in the war among rival analogies, it took intense fighting over a long period — sometimes several years — among the main participants in the decision process.

Can People Reason Without Using Analogies?

The fact that analogies were always central to the decision-making process hardly comes as a surprise to us, but what about the alleged alternative — namely, decisions supposedly made through a "pure reasoning" process, supposedly in the total absence of analogies?

Khong tries to explain this distinction by proposing a few types of non-analogical arguments that the American leaders could have used during the Vietnam War. Among these are arguments based on *containment* and on *domestic politics.* Let's consider the former. The word "containment" evokes a very concrete image. The military idea of *containing an enemy* (in this case, communism) with its strongly physical verb "to contain", is based in part on human-size phenomena, such as a cage holding a dangerous animal or a cell holding a prisoner, and also on larger-scale phenomena, such as a surging crowd held back by a rope, a metal chain, or a police barricade, or a river threatening to flood a city and held back by sandbags, a dike, or a dam. The military idea of containment is also based on situations in team sports such as basketball or soccer, where one team has managed to surround the other team and to keep it far from any possibility of scoring, not to mention chess situations in which some of the opponent's key pieces are holed up in a corner of the board and are stuck there, perhaps completely immobilized or able to move only in ineffectual fashions. Clearly all of these are indeed analogies (albeit with phenomena from everyday life), so Khong's desired distinction is already on shaky grounds. But then he tries to draw a distinction between analogical thinking rooted in everyday, non-military experience, such as these scenes, and analogical thinking rooted in grand historical precedents. Does this new distinction stand up to scrutiny?

To respond to this question, let us consider one of the most frequent concepts of the Vietnam era: the so-called "domino theory". The idea is based on the image of a chain of dominos that will all fall in a chain reaction if the first one of them topples. The standard analogy was between this tabletop situation and the countries of Southeast Asia, with a toppling domino being, of course, a country falling under the domination of communism. This seems to be an everyday metaphor rather than a grand historical analogy. But there is more to the story. Let us go back and look at the historical precedent called "the thirties", to which we briefly alluded above. Khong defines this historical precedent in the following manner:

> The 1930s is a composite analogy composed [in the mind of the analogy user] of one or more of the following events: Japan's invasion of Manchuria, Mussolini's annexation of Ethiopia, Hitler's reoccupation of the Rhineland, the Munich conference, and Hitler's invasion of Czechoslovakia. From the perspective of Dean Rusk [who was the American secretary of state at the time] and Lyndon Johnson, the two major users of the 1930s analogy, the prototypical event of this period was Munich [which took place in 1938], which they interpreted as Western appeasement of Hitler, an act that made World War II inevitable. I will use "Munich" and "the 1930s" interchangeably in this book.

We thus see that this is a mixture of several different analogues — a mental superposition of a number of different historical precedents. We will return shortly to this interesting idea, but for the moment what is of interest to us is to compare this definition with a remark that Khong makes about domino theory. He writes:

> The question of Munich is primarily one of stakes. The Munich analogy magnified the stakes of Vietnam for the United States because it envisioned a 1930s syndrome in Southeast Asia. In this sense, the Munich analogy was the intellectual basis of the domino theory. American policymakers from Eisenhower to Nixon remembered the crumbling European dominoes of the 1930s only too well; they were convinced that the spread of communism — the fascism of the 1960s — would lead to a similar catastrophe. Failure to stop the Asian dominoes from falling — with South Vietnam as the Czechoslovakia of the 1960s — would require the United States to fight communism later and under worse conditions; it would also probably cause World War III.

This observation shows us that the image of a chain of dominos falling in a few seconds, "victims" of earth's gravitational pull, was tightly linked, in the minds of the American decision-makers, with the historical image of a "chain" of European countries that "fell" to fascism during the 1930s, and was also linked with the frightening image of a "chain" of Asian countries that would also "fall" to communism during the 1960s. In other words, "dominos" and "thirties" are also synonyms.

In short, the non-military, non-historic image of dominos falling on a tabletop was profoundly mixed, in the minds of the leaders, with the political, military, and historical image of a set of countries that would fall, one after another, to invading forces (whether in Europe or in Southeast Asia). Given this, can one sharply distinguish between grand historical analogies and humble, mundane analogies that come from everyday life? Clearly not, since, as we've just seen, sometimes dominos on a table are seen as countries during a *recent* war, other times as countries during a *current* war, and yet other times as just plain old wooden dominos. (As Sigmund Freud might have said, "Sometimes a domino is just a domino.") Anyone who knows the domino metaphor and the historical facts cannot help mixing all these images in an intimate fashion. (Later in this chapter we will carefully consider this kind of mental blending of situations. We use the term "frame blends", while pioneering frame-blend researchers Gilles Fauconnier and Mark Turner opt for "conceptual integration networks".)

As we have already observed, most clearly in Chapter 3 and Chapter 4, any specific event — perhaps the falling of some dominos on a table, perhaps a famous historical defeat — can be encoded at various levels of abstraction, which means it is perfectly possible for us to see as one and the same phenomenon the succumbing of a series of neighboring countries to an evil empire's domination and the successive toppling-over of many dominos neatly lined up in a chain. This universal fact of human high-level perception allows us to see far beyond the concrete details of situations and to connect events that superficially are enormously different from each other.

We now come back to the notion that Khong proposed of a "non-analogical argument" based solely on abstract concepts. The problem is that whatever abstract concept is under discussion (*dominos*, *containment*, etc.) in a military or political context, it will necessarily evoke, simply because of the words it involves, familiar everyday images. Furthermore, in the mind of anyone who has a decent knowledge of history, an abstract concept of this sort will also evoke a wide range of historical analogues, at differing levels of awareness (for example, activation of a concept of this sort could evoke a range of historical precedents, such as a moat around a medieval castle, the walls of medieval cities, the Great Wall of China, the Maginot Line, and so forth). The fact that a word such as "containment" seems abstract and somewhat bland does not mean that it is devoid of a metaphorical substrate and that it evokes no historical precedent. To the contrary, the power of abstract words comes from the fact that they evoke a set of concrete images, all derived from experiences that one has had, either directly or vicariously, over the course of one's life. This rich halo of familiar and historical analogues constitutes the imagery that is evoked in the mind of a decision-maker, and which in turn gives rise to life-and-death decisions.

Pluralization and Schemas

Khong's book abounds in vivid expressions such as "No more Munichs!", "another Hitler", "a series of Koreas", "the next Chamberlain", and "replays of 1917". As we saw in Chapter 1, when we discussed words like "Mother" and "Moon", a one-member category is, from the moment that it comes to exist in a mind, capable of being pluralized and used just like any highly abstract notion. (Chapter 4, too, in the section dealing with Platos, popes, Mozarts, meccas, Bachs, and bibles, gave many examples of this sort.) In the above expressions (and others) in Khong's book, a term like "Hitler" or "Munich", seeming on its surface to denote *a particular historical entity* (a person, a place, a defeat, a war, a strategy) is used as the label of *a general category*, of which there could exist dozens or perhaps hundreds of instances, sometimes actual, and sometimes imaginary. As this makes clear, it is not even remotely possible to draw an ironclad distinction between "concrete" and "abstract" usages of a concept in one's memory.

Nonetheless, Khong tries to draw a clear line between *analogues* and so-called *schemas.* A *schema* is defined as the mental superposition of several historical precedents, whereas just one historical precedent all by itself would be an *analogue* (one side of an analogical bridge, to recall the image from Chapter 3). It's tempting to imagine that this is a sharp and unambiguous distinction, but as we just saw, the concept of *thirties*, although Khong treats it as if it were *one* specific event, is not one event at all; he himself defines it by blurring together at least *five* different events, and so, by his own definition, it is a quintessential schema! And yet this plurality does not keep him from speaking about "the analogy to the thirties", as if such a thing involved making a mapping to *one single* historical precedent, every bit as concrete, local, and specific as the Munich agreements or Mussolini's annexation of Ethiopia. Thus we see that sometimes Khong says that a schema is an abstraction and that seeing an event as covered by a schema is *not* the making of an analogy, while other times he treats a schema and an analogue as completely indistinguishable (both serving as ends of analogical bridges).

In Chapter 3, we saw that when someone reminds us of another person, what takes place inside our head is the building of an analogical bridge between two mental representations, and that when we categorize an object that we see, such as a cup, the same mental process is involved. In the case of the reminding, the mental entity that gets freshly activated is our memory of a *specific* person, whereas in the case of the categorization, the freshly activated mental entity is based on a lifelong series of perceptions of *different* objects (or situations), and we have little or no recollection of the "founding members" of the category. Despite this difference, the two processes are cut from the same cloth, since analogy-making is at their core; that is, in both cases, what is going on is an act of analogical mapping that builds a link between a fresh new mental representation and an older mental representation stored in our brain.

Khong's notion of a schema is not significantly different from a many-membered category such as *cup*, while his notion of a particular historical precedent is essentially the same as the mental representation we have of an old friend. One might think that mental processes involving schemas versus those involving historical precedents are very different, but all that is involved in either case is the building of analogical bridges between mental representations. Moreover, as we saw in the case of the slowly accreting *Twain–Grieg–Einstein* concept in Chapter 3, a schema can slide gradually from very concrete to very abstract, which means that it makes no sense to try to draw a sharp dividing line between making an analogy and using a schema.

Let's take a concrete example. If you are arriving for the first time in a lawyer's office, you might have this anticipatory thought: "I'm probably going to have to wait for a long time in this waiting room, just as I did the last time I went to get a physical at Dr. Blahblah's office." Or you might instead think: "I'm probably going to have to wait for a long time in this waiting room, just like when I have doctor's appointments." Then again, you might well have this thought: "People who have private practices and fancy offices always make you wait a long time." In the first case, which would seem comparable to one of Khong's historical precedents, the source situation (the checkup with Dr. Blahblah) is equally concrete as the the target situation (the meeting with the lawyer), and you superimpose your recent experience of a long wait in your doctor's office onto this new situation. In the other two cases, which would correspond to Khong's schemas, the source situations are more abstract: a generic visit to your doctor, or an even more generic visit to any kind of professional. But in all three cases, you are using a familiar source analogue in order to make educated guesses about a brand-new situation. As soon as you scratch a little, you find that, under the skin, the putative distinction between using an analogue and using a schema vanishes into thin air.

Domestic Politics and the Associated Brain Mechanisms

In a section toward the end of his book, instead of pressing on with his major thesis that analogies are omnipresent in political thinking, Khong promotes the curious idea, pushed by certain historians, that analogical thinking plays *no role at all* in decisions about *domestic* politics. One can't help wondering how the mental mechanisms involved in thinking about domestic politics could differ fundamentally from those involved in thinking about foreign events. Indeed, such a distinction is most implausible.

Imagine a historian of physics suggesting that physicists, when tackling questions in thermodynamics, always depend on analogies, but that when tackling questions in electrodynamics, they never use them at all. Now why would one area of physics be *resistant* to a certain set of thought mechanisms while another area of physics *required* those exact mechanisms at all times? The idea is as silly as suggesting that Capricorns always use analogies but Geminis never use them. The fact is, as Chapter 8 will show, that all areas of physics depend on analogies. And what holds for thinking in physics should (by analogy!) hold just as much for political thinking. Indeed, there are good reasons for believing that the mechanisms underlying human thought are universal.

Are We Humans Really So Superficial?

If psychologists working on analogy were asked to name the most solidly established experimental finding in their discipline, surely the winner in such a poll would be the notion that *surface-level features are the key to memory retrieval.* In his book, Khong puts great stress on this empirical finding (and we could quote dozens of similar claims in articles describing experiments on analogy-making): "One of the most interesting findings of researchers working on analogical problem-solving is that people pick analogies on the basis of superficial similarities between the prospective analogue and the situation it is supposed to illuminate."

Experimental studies have indeed demonstrated that subjects who are shown a source situation and who are then given a target situation are usually unable to see any connection between the two unless they share surface-level traits. Furthermore, in such experiments, when two situations have a superficial resemblance, then the second one invariably brings the first one to mind, no matter whether it is appropriate or not (that is, irrespective of whether there are deeper reasons to connect the two cases). For instance, if subjects first tackle an arithmetic problem concerning items bought in a store, then any other problem concerning purchases will instantly remind them of the initial problem. But if the theme of the first problem is experimentally manipulated — say it becomes a visit to a doctor's office instead of a store — then the participants will almost surely see no link between the two stories, even if the solution method for the first problem applies perfectly to the second problem.

If such a broad claim were true — if in cognition the superficial always wins out — it would have profound consequences concerning both the quality of human thought and the utility of analogy as the basis thereof. If this claim were true, one would have to face the sad fact that we humans are simple-minded creatures able to react solely to the most obvious aspects of what we encounter, unable to see beyond façades constantly leading us astray, incapable of spotting deeper essences lying behind the scenes. In a word, it would mean that our brains would be like bulls constantly charging madly at beguiling red flags everywhere. The lure of the superficial would lead us to believe that everything that glitters is indeed gold, that seeing a swallow inevitably means that spring has sprung, and of course that books must always be judged by their covers. And analogy-making would come out looking pretty sorry, thanks to this law of the surface's winning appeal, for it would be revealed as a primitive mode of thinking that relies on facile and misleading resemblances between things. This would really be grist for the mill for analogy's numerous detractors. And from there it would be but a short step to conclude that we humans should promptly seek ways around this crude and unreliable method of thinking in favor of a more rigorous, more deductive method based on the mental manipulation of abstract symbols and guided by the precise time-tested laws of logic. Indeed, this would be the logical conclusion to draw!

The cognitive psychologist Dedre Gentner, well known for her research on analogy, has proposed, with some colleagues, an evolutionary interpretation of the lure of the superficial (without necessarily subscribing to this interpretation):

> These findings may leave us feeling schizophrenic. How can the human mind, at times so elegant and rigorous, be limited to this primitive retrieval mechanism? An intriguing possibility is that in the evolution of cognition, retrieval from memory is an older process than inferential reasoning over symbolic structures. We could thus think of our surface bias in retrieval as a vestige of our evolutionary past, perhaps even a mistake in design that we have never lived down.

If the ideas in this passage are valid, then the bias towards superficial analogy-making is a ball and chain that has shackled humanity since time immemorial and that continues to plague us to this day. Is our bias towards superficiality merely an unfortunate legacy bequeathed to us by a *Homo sapiens* but a *Homo* less *sapiens* than we are? If so, then hopefully all we need to do is wait a few hundred thousand years for natural selection to purge this bias from our systems.

The domination of surface-level features in experiments on memory retrieval is a very robust phenomenon, confirmed by a great number of studies. There can be no doubt about the correctness of the finding. And yet to understand it properly — especially to understand why the evolutionary interpretation cannot be taken seriously — one has to go into the psychologists' labs and get one's hands dirty. The dirt one has to deal with, in this case, is the experimental paradigm that has guided nearly all experimental investigations of analogy-making, and which we'll call the "source–target paradigm". In this paradigm, subjects first study a source situation, which typically is a problem whose solution is given to them; then, at a later time, a new and unsolved target problem is presented to them and they try to solve it.

What makes this paradigm so attractive to psychologists is how easily, following it, one can design experiments that can be performed quickly on a large number of participants. Furthermore, it is a powerful technique since, by varying the source of the analogy, one can compare the behavior of groups of participants each of which is exposed to a different source, or possibly to none at all. As a result, the source–target paradigm has totally dominated the world of experiments on analogy-making, because (alas!) there are some good reasons for using it, as we mentioned above. We say "alas" because this paradigm, much like a medication that is effective but that has serious side effects, has unfortunately helped to propagate misleading ideas about analogy-making.

The Achilles' heel of this paradigm — indeed, its fatal weakness — is that the analogies studied in experiments based on it have little to do with analogies made outside the laboratory, in "real life". Both the humble analogies that we all make on a daily basis, as necessary to our survival as the air we breathe, and the flashes of genius that every so often light up the scientific landscape and give rise to a revolutionary new theory, are cut from a radically different cloth from the analogies that are generally studied in the laboratories. The community of researchers who investigate analogy-making experimentally has mistakenly extrapolated the results of their experiments following the limited source–target paradigm to the entirety of analogy-making. There is great irony here, since the researchers who assume this paradigm to be representative of all of analogy-making have themselves fallen for a misleading analogy.

Unfortunately, the source–target paradigm has a serious defect that undermines the generality of the conclusions that experiments based upon it produce. This defect stems from the fact that the knowledge acquired about the source situation during the twenty minutes or so of a typical experiment is perforce very limited — often consisting merely in the application of a completely unfamiliar formula to a word problem. By contrast, when in real life we are faced with a new situation and have to decide what to do, the source situations we retrieve spontaneously and effortlessly from our memories are, in general, extremely familiar. We all depend implicitly on knowledge deeply rooted in our experiences over a lifetime, and this knowledge, which has been confirmed and reconfirmed over and over again, has also been generalized over time, allowing it to be carried over fluidly to all sorts of new situations. It is very rare that, in real life, we rely on an analogy to a situation with which we are barely familiar at all. To put it more colorfully, when it comes to understanding novel situations, we reach out to our family and our friends rather than to the first random passerby. But in the source–target paradigm, experimental subjects are *required* to reach out to a random passerby — namely, the one that was imposed on them as a source situation by the experimenter.

And so, what do the results obtained in the framework of this paradigm really demonstrate? What they show is that when people learn something superficially, they wind up making superficial analogies to it. It would hardly be an earth-shaking revelation that people who have been given a single five-minute juggling lesson turn out to be lousy jugglers. Or suppose that subjects were taught the rules of chess in two minutes and then were made to play a few games. Could one validly draw the conclusion "human beings employ very primitive strategies when they play chess"? Of course not. And yet this is the character of the "scientific conclusion" that superficiality trumps depth whenever people make analogies. Are we poor human beings really so constantly gulled by surface appearances?

We Go as Deep as We Can Go

No, we are not constantly gulled by surface appearances — virtually never, in fact. But defending this viewpoint requires some explanation. We have just seen that the troubling and counterintuitive "fact" of superficiality trumping depth has been experimentally demonstrated only in the limited case of subjects' domains of incompetence, and this already greatly diminishes the finding's impact. But we might further ask why it is that novices are so often seduced by surface-level features. Are we humans really so shallow that we are interested only in what glitters? That's most doubtful. In order to understand these findings, then, we will need to go a bit more deeply into the nature of what is commonly said to be "superficial". Indeed, so far we have used the term in a rather casual fashion, relying on people's usual views about its meaning. But now we have to be more specific about what we mean.

A superficial feature is an aspect of a situation that can be modified without touching the core of that situation. Thus the color of my gearshift — originally black, it was recently repainted yellow — has no effect on how it works as a gearshift. Color is quite obviously a surface-level feature for the category *gearshift*. Likewise, when it comes to problems on a test, superficial features are those that can be modified without affecting the problem's goal or the pathways allowing it to be solved. For example, if a new problem has to do with shopping in a mom-and-pop grocery store as opposed to shopping in a supermarket, or with going to the dentist's office rather than going to the lawyer's office, its core is unlikely to be affected. On the other hand, when features are crucial to a category — when their modification changes the category itself — then one speaks of *structural features*. Thus a car whose motor has been removed, or worse, a car that has been compressed into one cubic foot at the wrecking yard, loses its *car*-ness. In the case of problems to be solved, structural features are those whose alteration would change the goal of the problem or the pathways to solving the problem. They are those features that one needs to pay attention to in order to find the solution, whereas superficial features are those that one can ignore.

Even if these definitions seem to be totally reasonable, they give rise to a couple of paradoxes concerning the phenomenon of superficiality's dominance over depth. The first one comes from the fact that, by definition, a novice in a particular domain cannot tell what the essence of a concept in the domain is. In other words, the distinction between *surface-level* features and *deep* features doesn't apply to novices, because to them any trait that they perceive could equally plausibly be shallow or deep. (This brings to mind the case of desks, discussed in Chapter 4, where attributes that once seemed essential to *deskness*, such as being bulky and covered with papers and having pull-out drawers, were revealed to be superficial only when virtual desks came on the scene.) Given that novices have no clue as to what's deep and what's shallow, how could their remindings always involve shallow features? How could it be the case that a person whose cognitive system can't even distinguish between shallow and deep features would invariably "smell" and be fatally lured by the shallow ones when reminding is involved? This seems perverse. Experiments have shown that surface-level features guide the retrieval of analogies, but the reason behind this cannot be that they are surface-level as opposed to deep features, because this distinction doesn't exist for the novice. And so it falls to us to explain the findings in some other manner, which we will do below.

That's the first paradox associated with the thesis of the dominance of the superficial in memory retrieval. The second is that the thesis amounts to a bizarre principle of cognitive anti-economy. Normally, principles of economy, which describe the efficiency of our behavior under the various and sundry constraints imposed on us by life, are taken to be inviolable. More specifically, if we agree that superficial features are defined by their irrelevance to a situation or to the goal of a problem, then how could a principle of cognitive economy explain the fact that the sources in analogical memory retrieval are selected precisely because of the *irrelevance* of their features to the given situation? Why would memories be triggered by *irrelevant* rather than *relevant* similarities? Again, this seems like a perverse if not fatal strategy for supposedly thinking beings to be guided by.

Fortunately, the resolution of these two paradoxes is quite simple. Novices have not built up the deeper categories of the domain, hence they don't perceive them. In other words, it's not the case that novices perversely and paradoxically favor shallow aspects over deeper ones, systematically snubbing what really counts. Rather, novices try as best they can to recognize what is important and relevant in a new situation, but lacking crucial knowledge, they most often cannot do so, and therefore they have to settle for shallow and most likely irrelevant features. As a result, the analogies they draw to prior situations tend to be shallow rather than deep, but this is only because the deeper analogies are not available to them, given the current state of their conceptual repertoires.

Rather than positing that memory retrieval — a process as ceaseless and as indispensable for cognition as the beating of one's heart — is archaic and useless (which is essentially what the above-cited evolutionary interpretation implies), we might more reasonably posit that the mechanism of memory retrieval is, in fact, perfectly efficient, as long as one takes into account certain constraints that we will now clarify.

The fact that novices are incapable of distinguishing surface-level features from deep ones, that novices have no little bird whispering to them, "This is a relevant feature, but that one is not!", leads us to adopt a different viewpoint about novices' memory retrieval. What, then, do novices do? What do we all do? Among all the diverse memories to which we are led by our knowledge (or lack of knowledge) of a given situation, we rely on whichever one is the most *salient.*

In short, what we use to guide our retrieval of memories when we are in an unfamiliar situation is not what is most *superficial*, but what is most *salient* to us. That is to say, the features that guide our retrieval of a specific memory are chosen not because they reside on the surface, but because — quite to the contrary — among all the potential retrieval cues to which we have access, they are the *deepest* ones.

For different people, the salience of a given feature depends on their expertise in the given domain. The research of psychologist Michelene Chi long ago confirmed that categories constructed by experts in a given domain are very different from those constructed by novices, and that experts don't rely on the same cues as novices do in categorizing new situations. Novices pay attention mostly to superficial aspects, simply because to them, those are the most salient qualities. For example, novices in physics will tend to group together all problems involving pulleys, or all problems involving springs, whereas people who have more experience in physics will tend to group together problems that involve the same physical principles, such as the conservation of energy. Thus an expert, instead of labeling a particular problem "a pulley problem", as would a typical novice, will call it "an energy-conservation problem". Why is this? It is because in their attempt to find connections between problems, novices, unaware of the key features to look for, rely on features that, to an expert, seem irrelevant; experts, by contrast, have already built up categories that critically depend on the key features.

Only as an individual's knowledge of a particular domain increases can there be a gradual evolution from categorization based on obvious features to categorization based on more abstract ones. When people eventually attain a high enough level of abstract knowledge of the important types of situations that frequently arise in a domain, then they are able to reliably apply this abstract knowledge by analogy to new situations whose façades are very different from the situations that they encountered during their learning phase.

This explains why a chess novice can easily see surface-level facts in a chess situation ("My queen can take this pawn!"), but will not be able to see deeper things ("This fork is threatened by my opponent's knight"), for such subtleties, transcending the novice's perceptual and conceptual repertoire, escape the novice's view. In order to acquire the ability to see such things, one has to gain expertise. Novices can easily be enticed by the surface-level traits that they perceive in situations, since (at least to their untrained eyes) there is nothing deeper around to lure them away from those traits. The distinction between *surface-level* and *deep* does not apply to them, even though it may well exist objectively. This is why we are casting doubt on the psychological relevance of such a distinction for novices. The reason novices are drawn to objectively superficial features is simply that those are the *only* features they are able to perceive.

Suppose that a researcher were to give non-speakers of Bengali a hundred poems in Bengali and then asked them to sort the poems into a few sensible categories. The piles that would result would of course be determined by surface-level attributes, such as a poem's overall length, a poem's average line-length, whether a poem looks typographically uniform or highly irregular, whether a poem uses indented lines or not, and perhaps a few others, but that's about as far as it goes, because those sorts of categories — visual or geometric, and unrelated to meaning — are the only ones that are available to the non-Bengali speaking "novice". By contrast, if the subjects of the experiment were university-educated native speakers of Bengali, we would see extremely different categories emerge. These would have to do with what a poem talks about, the era that it exudes, its use of formal or informal language, its tone and register, its classical illusions or lack thereof, its grammatical complexity or simplicity, its use of colorful language, and on and on. All of these categories are of course utterly invisible to the "novice".

There is thus no paradox about the novices' categorization behavior, because it's not the case that the novices are perversely *opting for shallow categorizations* over deep ones; they are simply making the *deepest possible* new categories they are able to make, given their inability to read a single word of Bengali. Meanwhile the experts, who in theory could see and pay attention to all the same shallow features as the novices are using, are ignoring those features, almost as if they didn't see them, just the way seasoned drivers ignore the color of the gearshift they are using; moreover, the deeper semantic and stylistic features that guide the native speakers of Bengali jump to their eyes every bit as rapidly and effortlessly as the shallower features jump to the eyes of the non-speakers of Bengali. For an expert in the domain, the *deep* features are not elusive or hidden; rather, they are the *most salient* features!

Let's also note that this phenomenon does not just depend on knowing the Bengali language. It is plausible that poetry specialists who knew no Bengali whatsoever could, on looking at a sampler of Bengali poems, discover some insightful categories that even an average native speaker of Bengali might never think of. Of course these would have to involve form rather than content, but they could still be of interest to Bengali-speaking poetry lovers. We are thus brought back once again to the idea that the features people tend to notice in any domain are as deep as their perception allows, which is a function of the set of categories they have evolved over their lifetime.

It's for this reason that *the deepest clues available* are what guides a person who is searching for analogies between fresh situations and ones in memory. If a novice is not guided by deeper features, it's because novice-level knowledge doesn't afford glimpses of such features. It's not what is *superficial* but what is *salient* that catches one's attention, and this applies equally to novices and to experts. The difference, however, is that as one gradually acquires greater expertise in any domain, the identifying features of deeper categories gradually grow more salient. This is a genuine principle of cognitive *economy*, for it states that no matter what one's level of expertise is, when searching in memory for an analogous case, one goes as deep as one's expertise allows. And we should point out that when we speak of expertise, we do not have in mind just the narrow and technical kind of expertise possessed by a handful of specialists in some arcane domain, but also the expertise that each of us has picked up concerning our everyday environment. We are all pretty much experts in what surrounds us, and that's a lucky thing, since it's exactly the kind of expertise that comes in handiest in life.

Thus if we (presumably experienced drivers) get into an unfamiliar car to drive it somewhere, we are immediately able to use its gearshift quite efficiently by making analogies to the relevant aspects of other gearshifts we have used before, and we are not deflected from its main functions by such superficial features as its color, its size, its shape, its degree of dirtiness, or where it is located in the car. All those qualities may be seen or felt, but they are instantly abstracted away, leaving as its *most salient features* the number of gears it has, and where they are located in space relative to each other. Those are the things that matter for driving, and it's no coincidence that they are also the most salient features to seasoned drivers. For someone to be a seasoned driver means precisely that what matters for driving pops right out to them effortlessly.

Essences Are Revealed by Surfaces

Superficial and *deep* seem to be diametrical opposites. A quick trip to the dictionary will tell you that the words "surface" and "depth" are antonyms, and the concrete imagery associated with these words reinforces this sense of opposition. Specifically, an object's surface is apparent and accessible; it is where the object meets the rest of the world. Think, for instance, of the surface of the earth, or the surface of a house, or of the ocean, of a stone, of a ball, of a mattress, of a wall, of a body, of a plant. Conversely, the depths of an object are internal, hidden, far from the object's "skin". Think of the depths of the earth, of a forest, of a lake, of a house, of a wound, and so forth. If one interprets these words at a more abstract level, their opposition remains every bit as strong; this is clear from their standard usages. Thus, "staying on the surface" means limiting oneself to appearances, to what is immediately obvious; this contrasts starkly with "going into depth", which means being profound, transcending appearances, getting at the meat of things, seeing beyond the surface, not being distracted by what first meets the eye.

The adjective "superficial", which we have often used as a synonym for "surface-level", reveals that "surface" has a second meaning. Not only does "superficial" refer to what lies on the surface and is thus immediately accessible, but it also frequently expresses a value judgment — indeed, a negative one. To call a book, a film, a piece of music, or a scientific idea "superficial" is clearly to imply that it is just "fluff", neither deep nor serious. This idea that "superficial" and "deep" are opposites gives surfaces a bad reputation.

Words, however, are often misleading, and we would suggest that the apparent contrast between surfaces and depths is, in fact, merely a surface-level contrast. Although dictionaries tell us that they are antonyms and our intuition tends to uphold this prejudice, the fact is that this opposition is only apparent. The royal road to the depths of a thing, to its core, to its essence, is precisely what lies at its surface. The surface gives us clues — deep clues — as to what is hidden inside, affording glimpses of the depths at the core. Surface and depth are thus related as are glove and hand, the former constituting the outermost layer of the latter.

The idea is not difficult to grasp. Psychologist Eleanor Rosch has described how our categories respect the *correlational structure* of the world. That is, not all combinations of properties are equally likely; rather, certain properties tend to co-occur in our environment. For example, the property of *flying* is correlated with the properties *laying eggs* and *having a beak*. In other words, when we perceive surface-level features, that activates in our minds other features that are correlated with those first features. These secondarily activated features are ones that our experience tells us tend to be present when the first ones are, but in themselves they are not instantly perceptible. In this way, what we see on something's surface leads us to its hidden depths, and thus allows us to draw meaningful, insight-lending analogies with what we have known before.

Our physiological senses react to surfaces, and our brain uses this input to activate certain categories, which give us clues to the gist, the crux, the core, the essence of what we are dealing with. Psychologist Myriam Bassok has carefully studied this notion of "induced structure", focusing on how it applies to the way that students learn in schools. Furthermore, the relevance of her findings goes well beyond the educational system; indeed, they apply to the way that we relate to the world around us. Thus a feather is likely to be light; a peach is likely to have a pit; something round stands a fair chance of being able to roll; and kicking something that looks like an anvil is not, in general, highly advisable.

Psychologist James Gibson proposed the notion of an object's *affordances*, meaning the possible actions naturally suggested to a person who perceives the object. Thus a doorknob might tempt us to turn it, a button would cry out to be pushed, a bag would invite us to place things in it or remove things from it, a switch would suggest "Flip me!", and so on. When an artefact's perceptual affordances are consistent with the purposes for which it was designed, then the artefact will be easy for people to use. This idea is well known to designers, who use the term "transparent" to designate an artefact whose mode of use is apparent from its design. Thus an important design principle would be that of making an artefact's instantly available surface reveal its deeper nature. The basic idea is that surface and essence should, ideally, be closely related.

To be sure, the world is full of traps that lead people to insist that appearances are deceiving and to be very wary of assuming that wherever there's smoke there's fire. There are all sorts of ways of being deceived by appearances. Indeed, this theme lies at the core of the notion of *fauxthenticity*, which we discussed in Chapter 3, and it also gives rise to con games, and even to the false belief, held by teachers who are tired of grading, that they can evaluate their students' understanding by very superficial rapid-reading techniques. But these are rare cases; in most cases the connection between surface and depth is not misleading. It is a minority of cases that are misleading, although of course the consequences of such cases can be huge. In most situations, the surface-level cues that we pick up quickly furnish a reliable guide to the situation's essence. This is why we survive and even manage quite well in the world.

Stereotypes, although they have a bad reputation, are in fact crucial to our survival. They tend to simplify things greatly, but they can nonetheless be extremely helpful. In the terms we have been using, a stereotype is a category that, thanks to easily perceived surface-level features, gives us access to a "shallow" kind of depth that has a decent chance of being correct, although the frequency of exceptions is large enough to warrant further refinement of the category. The surface, or more precisely how the surface is perceived, evolves as expertise evolves; as a result, properties that once did not seem to lie at the surface become easily visible, making new kinds of depths available. Experts have categories that evolve over time, allowing them to make observations that are doubly opaque to novices. Firstly, experts are able to see features that elude novices, because what is salient to experts is not salient to novices; secondly, experts associate hidden traits with these subtle surface-level traits, whereas novices almost certainly are totally unaware of those hidden traits.

The reason analogy is so extremely efficient is that appearances are indeed great indicators of essences. This is why reliance on surfaces is not a poor strategy in life. It's just that in selecting *which* of a situation's innumerable surface-level features to rely on as clues, one has to do one's best at separating the wheat from the chaff. This is what the development of expertise does for us. Experts see things that are hidden to novices. They perceive cues that novices either do not see at all or else take for irrelevant, and these surface-level cues give them access to deep perceptions. And thus surfaces become more and more imbued with depth.

Miniature Me-too Stories in a Miniature Domain

There can be no doubt that one of the key questions about human thinking is how we encode situations that we encounter so that later in life they can be spontaneously retrieved from memory. We have just described the life-and-death consequences of how episodes in the global realm of world politics get encoded (and later retrieved) in the minds of decision-makers, and in Chapter 3 we spoke of the same issue but in humbler contexts, such as when a father is watching his one-year-old son engrossed in play with ants and leaves when at the side of the Grand Canyon. We are now going to take another look at the same vital issue, but in a much humbler context — an artificial domain that, some three decades ago, was carved out in order to lay bare many of the central issues of cognition in the clearest possible way.

In Chapter 3, we spoke of the *me-too* phenomenon, typified by cases where you tell a story and then a friend spontaneously reacts, "Exactly the same thing happened to me!" Ironically, these words are a clear tip-off that it was a quite *different* thing that happened to your friend, since what you hear your friend tell is a story involving a different place, a different time, different people, different events, and different words — and yet despite all these differences, you know perfectly well why your friend said, "Exactly the same thing happened to me!" Although on one level their story was totally unlike yours, on another level, a more abstract one, the two are indeed the same. One and the same conceptual skeleton can describe two very different events.

We will now look once again at the me-too phenomenon — this time in the austere microdomain of Copycat. The domain's innocent-sounding name comes from the fact that when children play at being copycats, their imitations of each other often wind up being surprisingly flexible and creative. It might seem on first glance that being a copycat is nearly mechanical, requiring no imagination or fantasy, but that turns out to be far from the case. For example, if five-year-old Cora waggles her ponytail with her hand, what should six-year-old Xavier, who has no ponytail, do? Well, he could pull a lock of hair just above his forehead, or he might even tweak his nose. If Cora fiddles with one of the buttons on her blouse, Xavier might slide the zipper on the front of his sweater up and down. If Cora removes a barrette from her hair, maybe Xavier, who has no barrettes in his hair, will take off his pair of glasses — and so on. In short, there is plenty of room for verve and playfulness in the act of being a "mere" copycat.

As for the Copycat domain, it is focused on short sequences of letters of the alphabet, dubbed "strings", as well as on tiny "events" that can happen to these strings — that is, changes that might take place to denizens of that mini-world. Let's plunge right in with a sample me-too in the Copycat world. To make it come alive a little bit, we'll pretend that the letter strings can talk. We tune in just as the string *abc* is telling some friends (various letter strings) about the time it got changed to *abd.* Among the listeners is *pqrs*, who pipes up, "Hmm, that's funny — exactly the same thing happened to me the other day." Then when *pqrs* tells its story, it turns out that it got changed to *pqrt.* And so… was it *exactly the same thing*, or was it a quite different thing?

If someone wished to see the two micro-events as different, all they would need to do is point out that *abc* contains three letters while *pqrs* contains four, that the two strings have no letters in common, and lastly that putting a *d* in the third position has nothing to do with putting a *t* in the fourth position. Therefore, these are totally dissimilar events! Surely that would burst the balloon of the "exactly the same" claim.

And yet it's just as easy to flip perspective, to ignore nearly all details, and to declare, "What happened to *abc* was that its rightmost letter got replaced by its alphabetical successor, and that's also what happened to *pqrs.*" From that perspective, thanks to a little abstraction, the two strings, though of different lengths and having no letters in common, underwent exactly the same change, and so, yes, their two stories were "exactly the same".

What Gets Encoded When an Event Takes Place?

As we have just seen, "same" or "different" is in the eye of the beholder, and it all depends on what you attach importance to, how much importance you attach to it, and what you consider to be irrelevant. In the real world, we can't possibly take everything into account all the way down to its most microscopic details, and so we necessarily must ignore almost everything about every situation that we encounter, and that means we unconsciously make a highly selective encoding of it when we store it in memory. We have to strip everything that we experience down to a caricature of itself. The same idea holds in the Copycat microworld.

To be sure, the changing of *pqrs* into *pqrt* is less complicated than the plot of *War and Peace* or than an embarrassing case of mistaken identity at the grocery store, but even in the microworld, the drastic-simplification principle just uttered applies. If you were *pqrs*, you would probably have encoded the event that happened to you in a more abstract fashion than simply storing in your memory the "raw film", which would merely record the concrete facts in their most boring details — "My entirety was replaced by *pqrt*" — without making any effort to see the *essence* of what took place, which is that most of *pqrs* was left completely alone and only a small part of it changed, and moreover that that part wasn't a random part but an *extremity*, and it didn't change in an arbitrary way but in a somewhat *natural* way, which is to say, into a closely related object in a canonical and universally memorized sequence (namely, the alphabet).

In short, you would most likely have (unconsciously) encoded this event in your memory in something like the way we stated above: "My rightmost letter got replaced by its successor" (note that this ignores the identity of the letters in the string; that level of concreteness is seen as irrelevant). Later, when you heard *abc* telling its exciting story of being changed into *abd*, you would encode *abc*'s story in the same way, extracting from it the same conceptual skeleton; this explains why your own story from the past (*pqrs* ⇒ *pqrt*) would come leaping out at you when you heard your friend's more recent tale. For you, hearing *abc*'s story of turning into *abd* would nearly be a case of *déjà vu.*

Of course this little episode is just the tip of the iceberg of the me-too phenomenon in the Copycat world. In order to show how the Copycat domain explores and sheds light on the me-too phenomenon (and thus on the mystery of encoding) in deeper ways, we will now examine a number of more complex examples.

How Humans Do Not Perceive Situations

What event would instantly spring to the mind of *tky* when it hears *abc* tell of being changed into *abd*? Surely the time it was changed into *tkz*, no? That seems like a trivial Copycat challenge, almost the same as the question "What should *pqrs* do, in playing the copycat game, if *abc* turns into *abd*?" And yet *tky* is not a nice tight segment of the alphabet like *abc* or *pqrs*, and therefore if *being an alphabetical segment* is part of the way we see and remember *abc*'s story, then *tky* doesn't align at all with the stored memory, and so we would have to punt on the Copycat challenge, answering, "The conceptual skeleton that covered *abc*'s change to *abd* is unable to stretch to cover *tky*, because *tky* is not a string of alphabetic successors." People don't react that way, and that's a good thing: we would be very uninsightful beings if our brains came up with conceptual skeletons that were so narrow and so rigid.

Let's suppose, now, that *iijjkk* was also listening when *abc* told its story of getting changed to *abd*, and that *iijjkk*, too, responded by saying, "Gee, the same thing happened to me one time!" What do you think happened to *iijjkk*? It's most unlikely that you think that it got turned into *abd.* But why is it so improbable that such a thought would ever have crossed your mind? Because we humans do not like looking at events in the world in such a literal-minded fashion; we prefer *fluid* analogies by far.

In our many years of asking people such questions, no one has ever replied, "Oh, that's easy — just change *iijjkk* to *abd*!" People spontaneously move away from the literal level; they systematically seek a higher abstraction. Thus in this case, having seen that *abc* changed only at its right end, virtually all people instantly look just at the right end of *iijjkk*, hoping to "do the same thing" in that spot. It doesn't occur to people to replace the *entire* string *iijjkk* by something else; instead, they seek to replace just a small part of it, since just a small part of *abc* was affected by the original change. In short, people don't see *abc* as having been replaced lock, stock, and barrel by *abd*; rather, they feel that just the *c* was affected. And they don't think of that letter as "the *c*", either, but as "the rightmost letter", even though they may well *refer* to it as "the *c*". What they really mean is not the letter itself, but *the role that it plays.*

So this gives us some clues as to how to do "the same thing" to *iijjkk.* We look at its rightmost letter (that is to say, its *c*), which is *k*, and we replace the *k* by its alphabetical successor, namely *l.* This will give us *iijjkl.* Exactly the same thing!

Do you disagree? We certainly hope so. Or to go back to an earlier way of talking about this, we certainly hope that when *iijjkk* is given the chance to spell out how "exactly the same thing happened to me one time", it doesn't say, "I got changed into *iijjkl.*" To be sure, that would in *some* sense be "exactly the same thing", but it would be an impoverished sense of "the same". We would feel much happier if *iijjkk* said, "The exact same thing happened to me the other day — I got changed into *iijjll.*" Now why does this version seem so much more satisfying to us?

The Inescapable Role of Esthetics

In a sense, the event *iijjkk* ⇒ *iijjkl* is "exactly the same" as the event *abc* ⇒ *abd*, yet in a deeper sense, the event *iijjkk* ⇒ *iijjll* is even *more* exactly the same as *abc* ⇒ *abd.* This is an esthetic judgment, and when simple, basic perceptions are concerned, there is a widely shared sense of esthetics that depends on many factors, including abstraction and frame-blending (discussed below), and that sense is crucial to how we humans perceive the world around us. There are virtually universal mechanisms that guide us in our perceptions of the world, and that's lucky, because it means that in many ways, two people (or even a whole crowd of people) will be likely to see what happens in a given situation in the same way, even though *in theory* they all could have focused on entirely different aspects of the situation, and could therefore have encoded the situation in wildly different ways. In many simple situations, there is a strong natural tendency to see things in just one way and one of the beauties of the Copycat domain is that it helps us put our finger on some of what these nearly universal tendencies are.

What lies behind the answer *iijjll* is an almost irresistible tendency for people to *make perceptual chunks.* It is thanks to gestalt psychology, developed in the early part of the twentieth century and particularly influential in the 1930s, that today everyone takes for granted the importance of such perceptual chunking. But it was not always so obvious. Even if gestalt psychology has often been criticized for merely describing rather than explaining mental phenomena, the importance of its findings has never been questioned. Certain perceptual principles, including that of "continuity", of "good form", and the motto that "the whole is greater than the sum of its parts" (which has to do with our ability to perceive large-scale patterns), come directly from gestalt psychology. For our purposes, what matters most is the idea that certain ways of making perceptual chunks strike most people as natural — indeed, as well-nigh intrinsic — ways of apprehending the world.

Let's return to the case of *iijjkk* ⇒ *iijjll.* Instead of seeing the original string *iijjkk* as a string of six letters, most people tend to see it as a string of *three pairs*: *ii–jj–kk.* This interpretation invites us to map these three pairs onto the three letters of *abc.* Here something subtle happens. Because of the perceptual chunking that took place, the rightmost "letter" of *ii–jj–kk* is no longer the single letter *k* but the *group* of two *k*'s. In other words, the meaning of the word "letter" is extending naturally outwards under contextual pressure. In this case, we recoil at the idea of merely changing the single rightmost *k* into an *l*; instead we change *both* of the *k*'s simultaneously into *l*'s, yielding the string *iijjll.* In sum, we changed the rightmost "letter" to its successor, but in so doing, we instinctively took the notion of *letter* fluidly and abstractly.

The story behind the answer *iijjll* will not be a surprise to you, but the real point is how intimately and inextricably wound up in all of this our sense of esthetics is, and how often people see eye-to-eye on the esthetic qualities of different answers.

Let us be even more explicit here. In a strict sense, there is nothing *wrong* with the answer *iijjkl* (it clearly has a certain kind of logic to it), nor for that matter is there anything wrong with *iijjkd* ("replace the rightmost letter by *d*"), nor is there even anything wrong with the answer *abd* ("replace the entire string by *abd*"). We are not concerned with *rightness* or *wrongness*, *validity* or *truth* here. This is not a black-and-white domain in which there is just one right answer and all other possibilities are dead wrong; rather, there is a spectrum of possible answers, and each different answer has its own logic, and answers vary in their degree of appeal. *Appeal*, not truth, is the name of the game. Rather than being *wrong*, the various answers discussed above might be said to have differing "scores" of subtlety or finesse, depending on how much or how little abstraction was involved, as well as on an elusive "sense of essence". Especially the final pair (*iijjkd* and *abd*) are out of tune with how humans tend to see their world.

The Left Hand Doesn't Know What the Right Hand is Doing

To shed some perspective on all this, we propose a simple analogy challenge. First please hold up your two hands, with their palms facing you, and wiggle your right thumb. Now we'll simply ask your left hand to "Do the same thing."

So you wiggled your left thumb? Good move! But you certainly could have wiggled your left hand's little finger instead, since it is that hand's *rightmost* finger, and since your right thumb is your right hand's rightmost finger. However, what you *could have done* doesn't necessarily have much to do with what people in fact *do*, because we humans are ceaselessly categorizing the world, and we try to do so efficiently and even *elegantly.* We don't usually see a little finger as being analogous to a thumb, unless there are strong pressures to do so — but when they are strong enough, we gracefully yield and make the conceptual slippage, often without the slightest thought. Thus, for instance, if you happened to have rings on your right thumb and on your left little finger, and no other rings anywhere on either hand, it's very possible that you wouldn't have wiggled your left hand's thumb but the finger with the ring on it. And the more similar and salient the two rings were, the more likely you would be to move your little finger. Under sufficient pressure, concepts slip into other, related concepts.

More Me-too Stories

Let's return to the Copycat domain and its miniature me-too stories. It happens that *mrrjjj* was among the group of friends, and it, too, exclaimed, "Exactly the same thing happened to me!" What me-too story do you suppose *mrrjjj* then told?

Most people guess that *mrrjjj*'s story was that one fine day, it was changed into *mrrkkk*. What leads people to make this guess? Everyone agrees that *mrrkkk* beats its "rivals" *mrrjjd* and *mrrjjk* and *mrrddd* hands down (and we won't even mention *mrdjjj*, let alone some others). None of these "rival" answers is *wrong*, no more than it would have been *wrong* for Leonardo to have put a mustache on the Mona Lisa; these are esthetic decisions. Each answer is justifiable in its own way, and each will satisfy some people while leaving others dissatisfied. Their justifications reside at different levels of literality. For example, *mrdjjj* is hands down the most literal of the just-cited answers, as it replaces the *third* letter by a *d* — a level of literality so extreme that we have never seen anyone come up with this answer. Answer *mrrjjd* is slightly less extreme, as it involves replacing not the *third* but the *rightmost* letter by a *d*. Answers *mrrddd* and *mrrjjk* are yet less literal, since for the former, the final *group*, instead of the final *letter*, is replaced by a set of *d*'s, while for the latter, the rightmost letter is replaced by its *successor* rather than by a *d*. Each time one moves to a higher level of abstraction, one finds hidden structures and patterns that tend to be more esthetically pleasing. As the old dictum says, there's no arguing over tastes — but luckily, at least in the tiny Copycat microworld, people tend to agree on what is in good taste and what is not. And most people suggest that the best event for *mrrjjj* to recall would be when it got changed to *mrrkkk*.

And yet what's curious is that *mrrjjj* did not, in fact, recall that event. It did *not* say, "It's just like when I got changed into *mrrkkk*!" It certainly *could* have told this story, but that's not the memory that came to its mind. The story *mrrjjj* told was this: "It's just like when I got changed to *mrrjjjj*" (with four copies of *j*). Now why would this be "exactly the same thing" as *abc*'s story of having gotten changed to *abd*?

Clearly *mrrjjj*, just like *iijjkk*, can be broken up into three natural parts: *m–rr–jjj*. Now, just as with *ii–jj–kk*, a mapping leaps out at the eye between this new tripartite structure and *abc*. And there are some interesting things to observe in *m–rr–jjj*, such as the different sizes of the three pieces. Indeed, there's a group of length 3 at the right end, a group of length 2 in the middle, and… Could that be a group of *length 1* at the left end? Isn't that *m* sitting all by itself a perfectly valid group of length 1? No one could disagree. And so we have a situation whose essence resides no longer at the letter level — the *literal* level, quite literally — but at the *numerical* level. That is, we are coming to see the essence of string *mrrjjj* as the hidden pattern "1–2–3", and this essence has little to do with the string's quite arbitrary component letters *m*, *r*, and *j*, which simply acted as a *medium* carrying us the *message* "1–2–3". In this context, we ignore the letters — the surface is irrelevant here — and we focus on something deeper.

In mapping the tripartite structures *abc* and "1–2–3" onto each other, we will of course see the "3" as the *c* of "1–2–3", and we'll change that "letter" to its successor. Several slippages take place at the same time here, yet all happen effortlessly. We are not looking at the rightmost *letter* of *mrrjjj* (let alone at an instance of the letter *c*) but at the rightmost *number* of "1–2–3", and we are not changing that "letter" to its *alphabetical* successor but to its *numerical* successor, which is of course "4". That is the reason that *mrrjjj* said to the assembled crowd, "And so I got changed to *mrrjjjj*, just like *abc* got changed to *abd*." This esthetics-drenched answer exemplifies conceptual fluidity.

Gilding the Lily

Some people, when asked what might have happened to *mrrjjj*, suggest the answer *mrrkkkk*, based on changing not just the final group's length but also all of its letters, in one fell swoop. What about this double-barreled, perhaps "superfluid", answer?

Well, what would you think if someone, in answering the thumb-wiggling puzzle, simultaneously wiggled both their left thumb and their left hand's little finger? Would that constitute a sensible me-too? Would that amount to "exactly the same thing", in the world of the left hand, as wiggling the thumb alone, in the world of the right hand? We would say no. Taking two rival answers to a single question (in one case, wiggling the left hand's thumb *and* little finger, or in the other case, changing both *j and* 3 to their successors) and blurring them together is, in a word, confused. The answer *mrrkkkk* may have an initial razzle-dazzle, but on further thought it is an incoherent trap. There is no more reason to combine these two answers into one answer than there is to combine French fries and orange sherbet into one dish, simply because one is fond of each of them. To put it another way, there is no good reason to gild the petals of a lily. It's often said that it's pointless to argue over taste, since taste is so personal, but it is also a fact that in some contexts, there is a strong consensus that certain combinations of tastes or of ideas make no sense and are displeasing. Sometimes taste is nearly universal.

There is no such thing as a *proof* that an analogy is good or bad, whether in the Copycat microdomain or in the far wider real world. And so, in our attempt to explain why we see the answer *mrrkkkk* as confused and unsatisfying, we didn't seek an ironclad *logical argument* but instead resorted to a trio of *caricature analogies* (the first involving finger-wiggling, the second involving mixing gustatory delights, and the final one involving a famously fatuous flower-furbishing act), hoping readers would agree with our subjective sense of their relevance to this situation. We sought convincing *esthetic* reasons for rejecting *mrrkkkk* by thinking carefully about the idea of combining two actions, each of which, taken on its own, does a fine job, but which, if fused together, yield something silly. Does our "superfluid" answer not seem superfluous, now?

Seeing a More Abstract Gist than the Gist that was Encoded

Something noteworthy took place when *mrrjjj* recalled being changed into *mrrjjjj*. When hearing *abc* tell its *abd* story, it unconsciously encoded that story as "the rightmost letter got changed to its successor". On the other hand, much earlier, *mrrjjj* had its own experience of being changed into *mrrjjjj*, and at that time it encoded this experience in an abstract fashion as "the rightmost group got extended by one unit". As is obvious, *mrrjjj*'s encoding of its own experience is similar to, but far from identical with, its encoding of *abc*'s story. Life would be very simple if every me-too retrieval episode were due to the two events having been encoded *identically*, but that is a naïve hope.

Let's look at another example of the subtleties of encoding in this domain. It's very unlikely that on first hearing the story of *abc* being changed into *abd*, you would think of the *c* as "a group of *c*'s of length 1", but if you were to do so, then you would probably encode the story *abc* ⇒ *abd* as follows:

> The rightmost group got changed into another group of the same length, with all the letters inside it being replaced by their alphabetic successors.

Although this encoding is undeniably *correct* (provided one is willing to bend over backwards to see each of the three letters inside *abc* as constituting a *group* on its own), it is an unnatural, bizarre, and topheavy description of what happened to *abc*, and no one would ever come up with such a strange description.

But now recall *iijjkk* being changed to *iijjll*. The unnatural encoding just displayed for *abc* ⇒ *abd* now seems perfectly natural for *iijjkk* ⇒ *iijjll*. But this doesn't make it a good encoding for the event *abc* ⇒ *abd*, because it contains extraneous ideas that are irrelevant to that event. A single letter is not perceived as a "group of length 1" unless the perceiver is under intense pressure! Therefore, the strong, natural analogy between the events *abc* ⇒ *abd* and *iijjkk* ⇒ *iijjll* cannot have been mediated by their sharing the exact same encoding. No; different encodings were created at different times (one involving the rightmost *letter*, the other involving the rightmost *group*), but even so, the latter story activated the former, because on some abstract level their gists were sufficiently similar (in each case, one changes the "rightmost thing" into an abstract kind of "successor", whether that "thing" is a letter or a group).

Now let's turn to string *ace*. What came to its mind when it heard *abc*'s story? Hint: it wasn't the time when *ace* became *acf*, nor the time when it became *ade*. Of course either of those events *could* have come to *ace*'s mind, but in fact *ace* recalled the time when it turned into *acg* — when its rightmost letter was replaced by its *double* successor. *This* memory, for *ace*, was "exactly the same" as what happened to *abc*. Now where did the curious concept *double successor* come from? From, of course, the internal texture of the string *ace*, which is analogous to the internal texture of *abc*. Namely, where *abc* is a short chain made out of successor bonds (*a–b* and *b–c*), *ace* is a short chain made out of double-successor bonds (*a–c* and *c–e*) — and so this prompts a natural slippage from the concept of *successor* to the closely related concept of *double successor*. But surely no one could have anticipated this esoteric slippage when looking at the event *abc* ⇒ *abd* in isolation — and yet in this special *ace* context, the pressures to make that slippage are very strong. *Not* to make the slippage (and thus to insist on the greater appropriateness of *ace* ⇒ *acf*) would seem like an unreasonably rigid stance. Once again this is a case where the natural encodings of the two analogous stories are similar but not identical.

The Copycat domain is filled with cases where the encodings of the two stories involved in a me-too reminding episode are not identical but are *analogous* to each other. In other words, even when one jumps up to a fairly high level of abstraction, ridding oneself of nearly all of an event's details and thereby arriving at a tiny, compact summary of its essence — nothing but a gist — the two gists linked by a me-too reminding episode are usually not *identical*, but only *analogous* to each other. And if one insisted on having *identical* conceptual skeletons for two analogous events, one would often be forced to leap to such an artificial level of abstraction that the conceptual skeleton applying to *both* stories would be an absurdly strange-sounding legalistic formula that no one would ever concoct in real life. Furthermore, such a skeleton could never arise as the spontaneous encoding of the first event alone, before the second event had been encountered. That would require clairvoyance.

The profound mystery of how human remindings take place is not solved by positing that we always encode events with marvelously clairvoyant conceptual skeletons that anticipate all possible other events that might ever be analogous to them, even many years down the pike. Something much subtler is involved in the act of encoding memories than just manufacturing "clairvoyant" conceptual skeletons on the fly, since that notion is a chimera.

A key ingredient so far missing from our account is the ability of a crux to evoke *analogous* cruxes in memory. This ability is what allows us to see connections between events that are too dissimilar on their surfaces to have been encoded identically, and yet that still are deeply similar. In short, the secret of making good analogies involves making good but *more abstract* analogies — analogies between encodings, or conceptual skeletons. This may sound like an infinite regress, and thus a hopeless conclusion, but since analogies between cruxes are *more abstract* than analogies between original stories as wholes, "kicking the problem upstairs" to the level of finding analogies between cruxes is in fact a genuine simplification.

The Zaniness of the Letter "Z"

Next we will explore one favorite example of a me-too event in the Copycat domain. It turns out that the string *xyz* was among those listening to *abc*'s tale. What experience was it reminded of? It all comes down to figuring out how to answer this question: "What is 'the *c*' of the string *xyz*, and how should it change?" Well, quite obviously the *c* of *xyz* is the *z* — what else could it be? — and so one's first natural impulse is to change the *z* into its alphabetical successor. Here, though, we run into a snag, since *z* has no successor; it's the last letter of the alphabet. End of the line; everybody off!

However, some people — many, in fact — are not in the least stymied by this snag; undaunted, they immediately propose *a* as *z*'s alphabetical successor, giving the answer *xya*. Now where does this idea come from? Are we explicitly taught in school that the alphabet is a circular, wraparound structure? Of course not. However, as we grow up we all learn about "circular sequences", such as the days of the week, the months of the year, and the hours of the clock. There are also decks of playing cards, where the ace not only is the lowest card but is also higher than the king. Structures very much *like* a circular alphabet are "in the air" all around us, and thanks to their unconscious influence, the answer *xyz* ⇒ *xya* is easily found. In short, *xya* results from importing the concept of *circularity* from various familiar external sources into the microdomain whose alphabet is *not* circular. In so doing, therefore, one tampers with the nature of the microdomain. We might even say that when people carry out such a conceptual importation, they "corrupt" or "contaminate" the pure and pristine Copycat domain by throwing in alien ideas that are extraneous to it.

Despite this caveat, we'll accept *xya* as a legitimate reminding for *xyz* to have had — but what if *xyz* had never had that experience? In that case, what other event(s) in its memory might the *abc* ⇒ *abd* story have triggered? Stated otherwise, if *z* has no alphabetical successor, then what event(s) in the life of *xyz* might be analogous to the event *abc* ⇒ *abd*? And in fact there is an answer that strikes most people, once they have seen it, as being far more elegant than *xya*.

Everybody wants to change the *z* into something else; the question is, into *what*? Well, since taking *z*'s successor seems to be at the heart of what is giving us trouble, we might try to go back and explore alternate interpretations of what happened to the *c*, interpretations not involving the concept of *successorship*. For instance, instead of saying that the *c* changed into its successor, we could say that it changed into a *d*. In that case, *xyz* could perfectly reasonably be reminded of the time when it got changed into *xyd*. That's one possible answer; however, because of the intrusion of the literal *d* into the *xyz* world, it's not very appealing. Are there other more appealing alternatives?

Well, under this situation's unique combination of pressures, we might try to reperceive what happened to *abc* and say, "The letter *c* got replaced by a *d*", where by "the letter *c*", we now literally mean "the instance of the letter *c*", rather than "the string's rightmost letter". In that case, we certainly don't have to worry about the pesky *z* any more. Instead, we want to scour *xyz* for one or more instances of the letter *c*, and then, if and when we find one, we will change it to a *d*. A moment's scouring of *xyz* reveals, however, that there is no *c* in it, and hence no letter to change to *d*. And so, one possible reaction on *xyz*'s part would be to say, "Ah, yes — *abc*'s story reminds me of a memorable occasion one time long ago, when nothing at all happened to me…"

We're far from having exhausted the possibilites. Another thought might be to recall how *mrrjjj* was perceived on an abstract level as 1–2–3, which then, in analogy to *abc*'s becoming *abd*, became 1–2–4. Well, then, why couldn't *xyz* be seen on an abstract level as 1–1–1, meaning three groups of length 1? In that case, it could turn into 1–1–2 on that abstract level, yielding the answer *xyzz* back on the literal level. This doubling of a letter is reminiscent of how stadiums and large theaters often extend the alphabet, going from single occurrences ("A, B, C, D, ... , W, X, Y, Z") to double occurrences ("AA, BB, ..."). But even so, it feels like a bizarre evasive maneuver. If *abc*'s story had been *abc* ⇒ *abcc*, then of course for the story *xyz* ⇒ *xyzz* to bubble up out of dormancy would seem like a perfect me-too. But as we know, that wasn't the story that *abc* told.

In short, no solution given so far seems pleasing, let alone elegant.

The Snag Triggers a More Satisfying Reperception

In trying to take the successor of *z*, we repeatedly stubbed our toe, and this repeated annoyance focused our attention more and more on the fact that *z* has no successor — otherwise put, that *z* is the last letter of the alphabet. Now focusing on the alphabet's *last* letter is but a stone's throw away from focusing on its *first* letter. (As we pointed out in Chapter 5, slippages between opposite concepts are both natural and frequent, and can sometimes give rise to fascinating errors, such as a confusion between *reading* and *writing*, or between *being born* and *dying*, or between *grandparents* and *grandchildren*, and so forth.) Now inside *abc* there is an *a* staring us in the face. What more natural act, then, than to link the first letter of the alphabet, on *abc*'s left side, with the last letter of the alphabet, on *xyz*'s right side?

Thanks to this fresh new analogy between the concepts *first* and *last*, we have uncovered a new perspective on the connection between the two strings — a charming symmetry that is not so easy to spot but that, after the fact, seems as plain as day. This symmetrical mapping of the *a* in *abc* onto the *z* in *xyz* is reminiscent of the finger-twiddling challenge, where, in order to imitate on your left hand the twiddling of your right hand's rightmost finger (its thumb), you decided not to twiddle the *rightmost* finger (the pinky), but the *leftmost* one (also a thumb).

If the *left* end of *abc* maps onto the *right* end of *xyz*, that makes it natural — indeed, compelling — for us to use the left–right reversal consistently, by mapping the strings' *other* ends onto each other as well — so that the *c* of *xyz*, rather than being the obvious, once-irresistible choice of *z*, now becomes the *x*. Notice that this happy choice means that we won't stub our toe. After all, *x does* have a successor — namely, *y*. Lucky us! And indeed, replacing the *x* by a *y* will give us *yyz*. Now there's a sweet new answer!

And yet... Would *xyz* call this event "exactly the same thing" as what happened to *abc*? We just saw how the two changes might be called "exactly the same", but still, something smells fishy. After all, *yyz* very saliently has two identical letters right next to each other, whereas *abd* contains no such pair. In that sense, the two changes seem glaringly unlike each other. It's almost as if the *yy* pair is serving as a warning signal, a red flag, hinting that something crucial was overlooked. And indeed, the insightful idea of left–right reversal, though it was pushed somewhat, was not carried far enough.

Once we've chosen to map *abc* onto *xyz* with their physical directions reversed, then *forwards* motion in the alphabet is implicitly being mapped onto *backwards* motion in the alphabet, and thus implicitly, as part and parcel of this mapping, the concept of *successor* is being mapped onto the concept of *predecessor*. Another way to put this is that when *abc* is read *c*-wards, starting at the *a*, then its fabric is one of *successorship*, whereas when *xyz* is read *x*-wards, starting at the *z*, then its fabric is one of *predecessorship*. And so, from the simple act of seeing the *a* and the *z* as each other's counterparts, a tight little cascade of conceptual slippages has flowed in a natural and (almost) irresistible manner. At the outset we saw *first* slipping to *last*, and then we saw *left* slipping to *right*. And now we've hit the final slippage in the cascade — the slippage from *successor* (the fabric of the *abc* world) to *predecessor* (that of the *xyz* world).

This final conceptual slippage means that instead of wanting to take the *successor* of the *x* in *xyz* (an overly rigid thing to do), we'll want to take its *predecessor*. All of this will give us, in the end, the string *wyz*. This is a surprisingly pretty answer to the question "Suppose *abc* changed to *abd*; can you make *xyz* change in the same way?" Several coordinated slippages have led us to a *fluid* analogy. And all this insight came from "stubbing our toe" on the letter *z* — that is, from the snag due to *z*'s lack of successor.

This symmetry-based answer is, in a certain sense, an even stronger and more convincing analogy than when *pqrs* recalled turning into *pqrt*. That is, *xyz* ⇒ *wyz* is arguably more like *abc* ⇒ *abd* than *pqrs* ⇒ *pqrt* is, strong though that resemblance is. Why is this? Because both *abc* and *xyz* are strings wedged at their respective ends of the alphabet. There's a perfect symmetry here, and the two changes are like perfect mirror images of each other. No flaw is perceivable anywhere in this analogy, and this is what makes it so appealing. Although *pqrs* has quite a bit in common with *abc*, it's not as similar to *abc* as *xyz* is (what can rival a mirror for making symmetry?), even if *pqrs* is clearly much more similar to *abc* than *tky* is.

In Chapter 5, we cited the striking conceptual-proximity slippage made by a grandfather who, as he and his son drove by a cemetery, observed, "This is where all four of your grandkids were born." There we pointed out the very human drive towards internal consistency, pushing the grandfather towards making *two* slippages between opposite concepts, rather than just one. We even suggested that either of the slippages could have brought the other one along "on its coattails". Such a "coattails" characterization would also describe the way in which slippages coordinatedly cascade in this Copycat analogy between *abc* ⇒ *abd* and *xyz* ⇒ *wyz*, although we are obviously talking about *esthetic* coattails rather than electoral ones.

On Dizziness

We wish now to venture momentarily into yet murkier waters, offering the event *xyz* ⇒ *dyz* as a potential analogue to our standard old tale *abc* ⇒ *abd*. What's going on here? To come up with this analogy, a person would have to exhibit a highly implausible kind of confusion, which we will call "dizziness". First they would discover the subtle idea of mapping of the *a* onto the *z*, and then they would notice the spatial reversal that this implies, involving the conceptual slippage *right* ⇒ *left*, which leads to the subtle idea of modifying the *x* instead of the *z* — but then, for some reason, they would drop the ball. Totally forgetting all the subtleties they had just noticed, they would blithely interpret the event *abc* ⇒ *abd* as nothing more than a replacement of a certain letter by the random letter *d*, without any deeper rhyme or reason, and would mindlessly apply this recipe to the *x*. This is a bizarre twist, because genuine initial insights are followed up by an act that totally betrays the spirit of finesse of what was done only moments before. If one can use the phrase "in bad taste" in such an austere domain, this would certainly seem an appropriate time to do so.

In a word, this analogy is *dizzy*, in the sense that the hypothetical person who made it seems truly confused, exhibiting an incoherent mixture of smartness and stupidity, of insight and myopia, of brilliance and dullness. First there is a moment of genuine insight (mapping *a* onto *z*) and then, further into the process, the same high level of insight continues (mapping *c* onto *x*), and yet, at the very last moment ("What kind of insight is called for now, to modify the *x*?"), comes an act of myopic, local literality — seeing simply the raw letters *c*, *d*, and *x*, instead of the roles they play inside their worlds.

Such a dizzy hodgepodge, where high insightfulness is found happily coexisting with mind-boggling literal-mindedness, is implausible, to put it mildly. We have never run into anyone who in all seriousness proposed the answer *dyz*. And yet anyone who carefully explores the subtle angles of this particular analogy problem in the Copycat domain will sooner or later bump into this answer, whatever they may think of it in the end; and if this answer is explained to people, they invariably understand its "logic", even if they find it strange or uncomfortable. We may not *like* this answer, but we can still understand its strange logic, and when we do so, we can even find it humorous.

In this sense, the *dyz* answer to the *xyz* problem is analogous to a joke that circulated in the 1980's about Nancy Reagan, then the First Lady. It was stated that, upon hearing that there was a glut of butter, she perkily declared, "Oh, before it goes bad, we should distribute all the extra butter to the poor people of our country, so that they have something to dip their lobster tails in." The First Lady in the world of the joke understood poverty in a ridiculously implausible manner: while realizing perfectly clearly that poor people have no butter, she totally failed to realize that they *also* don't have lobster tails a-plenty. This joke is reminiscent of the infamous remark by Marie Antoinette, also about poor people, but in the streets of Paris: "Oh, have they no bread? Poor dears! Then let them eat cake!" This is what we mean by "dizziness".

During the heyday of the First Lady joke, a similar joke about the United States and the Soviet Union was making the rounds. An American, bragging to a Russian about our land's great freedom of speech, proudly proclaimed, "We in America are so free that we can march up and down in front of the White House and shout, 'Down with Reagan!'" Unimpressed, the Russian replied, "That's no big deal. We in Russia can do exactly the same — we too can march up and down in front of the Kremlin and shout, 'Down with Reagan!'" The analogy is almost complete, but the ball is fumbled on the one-yard line (much as in *xyz* ⇒ *dyz*).

Frame Blends

We are approaching the rich and important topic of *frame blends*. Rather than defining that term in an abstract fashion, we will do so through a series of very concrete examples in the Copycat microworld, since the phenomenon emerges with great clarity there. Once again we trot out our same old story of *abc* changing into *abd*, and this time (exactly as before) we ask, "Of what event in its past was *iijjkk* reminded?" Here is a set of possible answers:

iijjkk ⇒ *abd*	("Replace the whole string by *abd*")
iijjkk ⇒ *iijjkk*	("Replace the letter *c* by a *d*")
iijjkk ⇒ *iidjkk*	("Replace the third letter by a *d*")
iijjkk ⇒ *iikjkk*	("Replace the third letter by its successor")
iijjkk ⇒ *iijjkd*	("Replace the rightmost letter by a *d*")
iijjkk ⇒ *iijjd*	("Replace the rightmost group by a *d*")
iijjkk ⇒ *iijjdd*	("Replace the rightmost group by an equal-sized group of *d*'s")
iijjkk ⇒ *iijjkl*	("Replace the rightmost letter by its successor")
iijjkk ⇒ *iijjll*	("Replace the rightmost group by its successor")

We've listed these answers in an order that echoes the amount of frame-blending involved, starting with a lot and finishing with none. Take the third answer, for instance, where the third letter — *j* — was replaced by a *d*. This answer seems very maladroit, for the concept *d* feels like an intruder in the *iijjkk* world; that is, it belongs to the *abc* world, near the beginning of the alphabet. The presence of a *d* in the midst of the letters *i*, *j*, and *k* (which constitute their own little world, far away in the alphabet from the *abc* world) seems muddle-headed. But that's not all; the concept *third letter* also seems like an intruder. Borrowed literally from the *abc* world, it has been bluntly thrust into the *iijjkk* world with no regard at all for the fact that *iijjkk* is twice as long as *abc* is, so that the richly pregnant meaning of *third letter* in the *abc* world is utterly lost when the concept is exported in such a literal fashion to the other world. Thus the third answer is the result of a great deal of blending, and esthetically speaking, it suffers for it.

Each of the other answers shown, except for the last one, involves some type of frame-blending, in the sense that some concept belonging to the *abc* world has been borrowed and thrust into the *iijjkk* world, contaminating it. Even the second answer, *iijjkk*, which on its surface exhibits no contamination from the *abc* world, is in fact infected, because the *logic* behind it ("replace the letter *c* by a *d*") is based on the concepts *c* and *d*, which are both alien to the *iijjkk* world.

On the other hand, the last answer, *iijjll*, doesn't borrow any concept from the *abc* world that doesn't belong just as much to the *iijjkk* world. In that sense, the final line constitutes a "purer" form of analogy than all the lines above it do. All the other lines feature some degree of frame-blending, sometimes very blatant and sometimes less so. And there are plenty of other possible frame-blended answers, such as *iijjkk* ⇒ *aabbdd* or even *iijjkk* ⇒ *aabbll*.

Finally, it's amusing to observe that all the answers in our list, except for the first one, are covered by the single frame-blending recipe "Replace the *c* of *iijjkk* by its successor". Of course this formula is highly ambiguous and subjective, since inside *iijjkk* there is no more an instance of *c* than there is a Golden Gate Bridge in Indiana, and yet that wouldn't in the least keep one from speaking of "the Golden Gate of Indiana", which is actually a much less wacky idea than our old friends from Chapter 4, "the Meryl Streep of spitting" and "the Mussolini of mulligatawny".

Strengths and Weaknesses of Frame-blending

We don't wish to portray frame-blending as being intrinsically feeble-minded or confused. Some frame blends, in fact, are carefully designed to exhibit creativity and humor. The *xyz* ⇒ *dyz* story is a fine example of this. In it, the letter *d* was quite intentionally stuffed into a world where it has no business at all, like a bull that has been led into a china shop. This move, deliberately gauche, induces a feeling of complete nonsensicality, and thus of humorousness. One needs quite a sense of fantasy to come up with as off-the-wall an answer as this.

But a frame blend can also be made inadvertently, simply by overlooking something, in which case it is more like a makeshift solution to a real-world dilemma: acceptable but not optimal. Indeed, many people come up with the fairly flat answer *xyd* rather than *wyz* for exactly this reason — they don't examine the strings deeply enough to let the subtle *a–z* connection, followed by its rapid chain reaction of conceptual reversals, jump out at them.

A frame blend can also be made when one simply doesn't possess the knowledge required to see certain aspects of a situation, and so one necessarily leaves them out. If, for example, someone knew the alphabet but didn't know how to count, then the answer *mrrjjj* ⇒ *mrrjjjj* would simply be unavailable to them. On the other hand, with the set of concepts they possess, they could easily come up with *mrrjjj* ⇒ *mrrkkk*. To some observers, this answer may seem filled with insight and elegance, while to others it may look like the most glaring of frame blends (what is the concept of *alphabetic successor* doing in a world whose essence is not alphabetic but numerical?).

Or suppose that someone with a strangely defective kind of vision were looking at the string *iijjkk*, and could just barely make out that it contained six letters but had no idea that any of them were repeated. They might say, "Change the rightmost letter to its successor!", thus defining the event *iijjkk* ⇒ *iijjkl*. To us this looks awkward, but if we take into account the lack of information that gave rise to it, we can understand the oversight. They performed as well as possible, given that their information was partial and incomplete. This is an echo of the earlier idea that people make the deepest analogies that they can, as constrained by their conceptual repertoires.

Although these examples may seem artificial, scientific analogies are often made in just this way — as brave leaps way out into the thickest of fogs. Decades or centuries later, some of those leaps may look silly, because, having been based on precedents well known at their time, they incorporated various naïve assumptions that in the end turned out to be utterly inappropriate to the new situation — but such *ex post facto* judgments of naïveté are a luxury. They are the wisdom that one acquires when one is blessed with, as the phrase goes, "20–20 hindsight".

To be concrete, let's recall the way that, in the seventeenth century, light waves were hypothesized to be very similar to sound waves: they were presumed to be longitudinal compression waves carried by some all-pervasive elastic substance very much like air but even more pervasive and even more tenuous than air. (A sufficiently bold spirit might even dare to describe that elusive substance as "ethereal"!) This conceptual leap was an audacious one and contained far more than a grain of truth, as was revealed centuries later, but in those early days, people's perception of the nature of light was contaminated by incorrect analogical importations from the world of sound — in other words, by a frame blend in which concepts that were to some degree inappropriate were imported from the *sound* side of the bridge.

To be specific, both the idea that light had to be a longitudinal compression wave and the idea that it would need an elastic medium to propagate it were practically perfect carbon copies of what was known to be the case for sound — but in the case of light, both presumptions were eventually discovered to be wrong. Nonetheless, these guesses about light, even if they were later realized to have been slightly off, constituted a great and bold leap in the dark, affording humanity its first inklings of light's true nature. It would be unjustified to criticize such boldness simply because *some* of it eventually turned out to be wrong. Why would anyone have postulated a *transverse* wave propagating through *vacuum* when such a thing was unheard of, and indeed, virtually inconceivable to people at that time? To criticize the natural even if somewhat cautious leap (as seen from today's viewpoint) that *was* made would be nonsensical. After all, even the greatest thinkers of the 1600s couldn't possibly have leapt so far ahead as to anticipate Maxwell's equations and Einstein's theory of special relativity, which are indispensable ingredients in today's understanding of light as a wave.

The greatest scientists of any era venture into the wild unknown by making bold but often naïve analogies about phenomena that they do not fully understand, and almost inevitably, in making these frame-blending analogies, they inadvertently bring along irrelevant baggage that comes from their own limited knowledge. They borrow a neat little "package" of various facts about phenomena that they *do* know, taking (or mistaking) those facts for universal truths, and indeed, some of the facts in the packet will turn out later to be inapplicable to the new phenomena. And yet the best of such analogies contain so many grains of truth that they open up whole new perspectives, despite various incorrect assumptions that have been unwittingly imported from other domains. This kind of partly-insightful, partly-defective analogy-making is a hallmark of all humans, and will be discussed extensively in the next two chapters.

We might add that frame-blending is at times deliberately exploited in literature and other artistic domains for its lively and stimulating qualities. And the activity of reading a book or watching a film, play, or opera takes for granted the act of projection by each reader or viewer into the scenes, identifying with one and then another of the characters. Such mental blends are done rapidly and often without any realization on the part of the viewer, but they are what give to any dramatic work its emotional meaning, since without them the work would merely be a cold, third-person recitation of events.

Fauconnier and Turner's Conceptual Blends

We have brought up the concept of frame-blending because it is an enormously rich source of insight into many phenomena in human cognition, and under the name "conceptual integration" it has been beautifully and richly explored and described by cognitive scientists Gilles Fauconnier and Mark Turner and their colleagues and students. They have shown time and again that frame-blending is found throughout human thought, sometimes using marvelous examples that seem exotic, just as often using examples that are as down-to-earth as can be, but in any case demonstrating the fundamental importance of the phenomenon.

One of the examples analyzed by Fauconnier and Turner in their book *The Way We Think* is a billboard that was put out by the state of California in a campaign to combat smoking. As they describe it, the billboard featured a large photo of a macho cowboy (much like a classic Marlboro man) riding a horse and conspicuously smoking a cigarette, and at the bottom it stated, in big bold letters, "WARNING: SMOKING CAUSES IMPOTENCE." The crucial frame-blending touch was that the cigarette was clearly bent and drooping downwards. A complex mapping was of course intended, in which the drooping cigarette would be seen as analogous to a non-erect penis, and in which the act of smoking would be seen as somehow *causing* both droops at once. It's obvious that the drooping cigarette is a mental contamination of the scene, and that the drooping has simply been borrowed from an analogous situation (a cigarette being in some ways clearly analogous to a penis, Sigmund Freud's famous disclaimer "Sometimes a cigar is just a cigar" notwithstanding). Whether you consider it cheap or clever, there's no doubt that this advertising maneuver constituted a frame blend *par excellence*, and that its clever ploy induced some long-time male smokers to reconsider their habit.

A careful series of examples in *The Way We Think* demonstrates how the understanding of tiny linguistic units that are as innocuous-seeming as adjective–noun combinations — for instance, "a safe child", "a safe beach", and "a safe knife" — are fraught with intricate blendings of one frame into other frames. Fauconnier and Turner show that the analogies needed to produce or understand such phrases are in some ways extremely simple and in other ways very subtle. All these analogies involve counterfactual situations. For instance, the phrase "a safe beach" makes one conjure up (most likely unconsciously) a scenario of a beach where bathers are in some fashion threatened, perhaps by sharks, perhaps by a powerful undertow, and then the actual beach and the hypothetical beach are mapped onto each other in order to emphasize their difference. What could be simpler than a mapping between a beach with an undertow, and the same beach without an undertow? Such analogies seem trivial, and yet modifying the true beach into a momentarily false version of itself requires nontrivial mental operations.

Not only the adjective "safe" depends upon an unconscious blending of frames on the part of both speakers and listeners, but even the most garden-variety of color adjectives — "red", "green", and so on — are shown by Fauconnier and Turner to involve blends, as in such phrases as "red wine", "red pencil", "red fox", "red hair", "red light", and so forth. More surprising yet is the fact that even an isolated noun may require an involuntary and subtle frame blend in order to make sense. For example, understanding the word "dent" necessarily involves a comparison between a norm (a pristine automobile fresh from the factory) and what we might call an "abnorm" (a car that has been deformed by an accident). Without a notion of how the car *should* be, and without the construction of a mapping between the norm and the "abnorm", the concept *dent* would be incomprehensible.

Are Analogies Different from Blends?

In their book, through many dozens of fascinating case studies such as the ones just mentioned, Fauconnier and Turner convincingly document how rife our everyday understanding of our world is with frame blends, or, to put it in our terms, with unconsciously produced analogies between situations in which elements belonging to situation A (or rather, to A's mental representation) may wind up being carried over into (the mental representation of) situation B, and vice versa.

An example described by Fauconnier and Turner as a blend, and by us as an analogy, is the importation of the idea of *desk* to computer screens. We describe this as a use of analogy, while they describe it as a conceptual blend or "blended scenario", and they argue that, although it is *based* on an analogy, it is *not* an analogy. Their point is that there is a *hybrid* structure in people's minds, some of it coming from old-fashioned physical desks, and some of it coming from people's perception of images on screens, with the two being blended in people's minds. We agree with this analysis. But what else is such a hybrid structure, other than an analogy? As we have shown in these past few sections, analogies are often blurry blends resulting from conflating two separate situations in one's mind, mixing their ingredients, and in fact this is what happens in many of the analogies that we make on a daily basis, as Fauconnier and Turner show.

It's not as if the word "analogy" were reserved for "uncontaminated" mappings. Indeed, as we said above, deciding which mappings are "contaminated" and which are "pure" becomes a subjective question, and it depends on the perspective taken on a situation. In order to bring out how and why the existence of blending is not an objective fact but a subjective opinion, we propose the following Copycat analogy challenge: "If *aabc* ⇒ *aabd*, then how should *pqrr* change 'in exactly the same way'?"

One very appealing answer would be *pqrr* ⇒ *pqss*, based on the recipe "Change the rightmost 'letter' to its 'successor' ", where the terms "letter" and "successor" are both allowed to flex naturally in meaning, as a function of the new context. This may well seem like a perfectly "pure" answer, uncontaminated by any blend at all, and indeed gracefully fluid in its double slippage — and yet there is a rival answer that could be argued to be even "purer" — namely, *pqrr* ⇒ *oqrr*.

At first, the answer *oqrr* may seem completely off the wall, but if one pays attention to the curious feature of having a double letter at one end — a feature of both *aabc* and *pqrr* — and *if* one feels that this feature is so salient (this is where subjective judgments come in) that it cries out to be taken into account in establishing a mapping between the two worlds, then one will wind up making a mapping in which *aa* maps to *rr*, which induces a mapping of the *c* onto the *p*. In this new mapping, the concepts *left* and *right* are reversed — and then, exactly as in the *xyz* problem, which we so carefully analyzed above, the concepts *successor* and *predecessor* will also reverse roles (and once again, they'll do so "on the coattails" of *left* and *right*). This very different double conceptual slippage explains the answer *oqrr*, which certain people will find very pure and pleasing.

This new answer casts the answer *pqss* in a completely different light. To some people, *pqss*, which at first seemed elegant and "pure and uncontaminated", will now seem like a frame blend in which the concept of *rightmost letter* was rigidly imported into the *pqrr* world, where it is in fact an unwelcome intruder, and hence the analogy giving rise to answer *pqss* will seem inelegant and "impure and contaminated". And yet there will be other people to whom the idea of paying any attention at all to the double-letter feature of both *aabc* and *pqrr* will seem like an unimportant, irrelevant, and pointless luxury, and to them the answer *oqrr* will merely seem like the precious self-indulgence of an over-intellectual mind. To such people, the answer *pqss* does not involve the rigid importation of a concept from one world to another; it simply "does exactly the same thing" in the second world, and it does so without the least trace of contamination.

We thus see that frame-blending is not an objective quality of an analogy. To the contrary, the choice between slapping the label "blend" and "non-blend" on a given analogy depends on one's esthetic preferences, which are often unconscious, and in any case are incapable of being logically or objectively proven correct or wrong. Esthetic preferences are prejudices that lie deep in the makeup of one's way of looking at the world; they are not facts about the world itself. And so the notion that there is a sharp and clean distinction between frame-blending ("impure") and analogy-making ("pure") is an illusion.

In short, to us, frame blends (or "conceptual blends", in the Fauconnier–Turner terminology) are not *exceptional* analogies but *typical* ones; indeed, they are analogies that have the interesting feature that one can point out one or more aspects of the mental bridge built between the two situations that involve a blur between entities located on both sides of the bridge. As a result, one isn't always sure which side of the bridge one is standing on. Our next example, much in the spirit of Fauconnier and Turner's book, shows this clearly.

A Childish Frame Blend

Scott and his three-year-old daughter Ellie, on a visit to a natural history museum, are looking at a display of a family of antelope-like mammals called "bongos", all of which have been taxidermically stuffed and arranged in a setting that resembles an African savanna. The bongos are standing near what looks like a shallow pond, although on closer inspection it is seen to be just a sheet of transparent plastic. A large male bongo is leaning his head down over the pond and sticking his tongue out, as if he's just about to take a sip of water. Here we tune in on the conversation between human father and daughter:

Ellie: Oh, look at the poor daddy bongo... He's so sad!
Scott: Why is he sad, Ellie?
Ellie: Because he's very thirsty, but he can't drink anything!
Scott: Why can't he drink, Ellie?
Ellie: Because he's *dead*!

Ellie's unexpected reaction is most amusing. Let's try to spell out the unconscious analogy present in her mind (and hopefully in the mind of any observer of the scene). It involves two different scenarios or, as Fauconnier and Turner would put it, "mental spaces". One scenario is the intended *effect* of the display: despite the total immobility of everything in the scene, visitors are supposed to see animals on the savanna and to imagine them as living out their lives in a natural fashion, halfway around the globe. The other scenario is that of the museum display itself: visitors know perfectly well that they are not on a savanna in Africa but in a room in a building in an American city, and that the background scenes, showing other animals, some trees, and a mountain range in the distance, have merely been painted on the wall, and if they look carefully, they can also see that the pond is really just a sheet of plastic. But the paintings on the wall are *analogous* to a savanna scene, and the plastic sheet is *analogous* to a pond.

The bongo family, however, is at a different level. All the bongos are three-dimensional and life-sized, and thus, even when the plastic pond and the painted savanna are seen as fake, the bongos continue to be seen as real. And indeed they *are* real, or at least they are a significant step closer to reality than the other elements of the display, because these bongos were once alive. And so, even when the rest of the exhibit shatters into the falsities that it consists of, the bongos remain as somehow genuine and authentic.

Viewers will effortlessly map the artificial scene (with painted mountains, a plastic pond, stuffed bongos, and no motion whatsoever) onto a scene in Africa, and hopefully they will not allow any of the fake aspects of the artificial scene to travel across the analogical bridge and contaminate the far end of the analogy; that is to say, hopefully nothing from the museum-display side of the analogy will leak over and invade the African-savanna side. This desired way of looking at the display is often called "suspension of disbelief". But one can imagine that despite the best efforts of the museum staff, such contamination might occur in the minds of certain visitors, especially the youngest ones.

And so when we hear little Ellie voice pity for the daddy bongo, we anticipate how she'll reply when her father asks what's keeping him from drinking. Of course she'll say that it's because the pond isn't made of water — it's a *fake* pond, a plastic *non*-pond. The poor daddy bongo! We can see it coming a mile away — Ellie will let the museum-display side of the analogy contaminate the African-savanna side.

And indeed she does, but in a way that catches us off guard, saying, "He can't drink because he's *dead*!" Now where did *that* idea come from? Only a moment ago she said that the daddy bongo was longing to drink. But how can he long to drink — how can he have any feelings at all — if he's dead? If the daddy bongo is indeed dead, then Ellie shouldn't be feeling sorry for him at all, because the whole scene is lifeless, desireless. In short, Ellie is having her cake and eating it too — for her, the daddy bongo is both alive (full of longing, and part of the African-savanna side of the analogy) and dead (insentient, and part of the museum-display side of the analogy) at the same time.

This is reminiscent of the dizzy Copycat answer *xyz* ⇒ *dyz*, where a keen insight is instantly followed by a thought so shallow that it feels like we've just suffered a whiplash. Such modes of thought exude such dizziness that they often will provoke laughter, and when this anecdote is told, it unfailingly does so.

Perhaps this strange, confused-seeming blending of ideas should not surprise us all that much, since children are constantly playing with all sorts of objects, pretending that they are something very different, yet knowing full well that they are not *really* that thing. Children live in this kind of superposition of spaces much more than adults do, and so they are used to the constant back-and-forth between two levels of interpretation of what is in front of their eyes. For a child at play, a wooden block can easily be a knight riding a horse and a wastepaper basket can easily be a castle that the knight is charging. Of course the child knows that they are *really* a wooden block and a wastepaper basket, and can shift into that mental space on a dime (for instance, if the wastepaper basket tips over and a few papers fall out of it and have to be put back in), but while in the pretend-and-play mode, the child has no trouble keeping that knowledge somehow compartmentalized.

But Ellie's remark goes way beyond the usual shifting-on-a-dime that children standardly do; somehow it violates all expectations. It is as if the child playing with the wooden block and wastepaper basket were to say, "The knight is sad, because he wants to jump up on top of the castle, but he can't!" If we were to ask the child why not, we would expect an explanation of this sort: "He can't jump up there because he knows the castle has no roof and he'll fall inside it and be trapped forever in it!" That reply would gracefully blend the wastepaper basket's actual shape and intended function with the play world, where it is a castle. But the child throws us by saying, "He's sad because his horse can't jump at all — it's just a stupid piece of *wood*, and anyway, it's only the size of my *hand*! It could never jump to the roof of a castle!"

Frame Blends Are Analogies; Analogies Are Often Frame Blends

In this book, aside from the examples given above in the Copycat microworld, we have discussed many everyday analogies that, if one goes back and examines them, are easily seen to be frame blends. For example, recall the story in which a person pointed to the seat across the aisle in a train and remarked, "The chatterbox on the trip down was sitting right *there*!" Two frames were being conflated and blended, thanks to an analogy between two trains traveling on different days. More precisely, the directly visible seat in the current train was being mentally inserted into a different train that had been traveling in the opposite direction on a different day. In the interest of efficient communication, truth was being efficiently conveyed through a strategy mixing falsity, analogy, and blending.

Another frame blend occurred in Chapter 4, where we compared an eclipse of the moon, which involves sun, earth, and moon, to an analogous situation with a flashlight, an orange, and a ping-pong ball. What made this a frame blend was when we added the idea of ants on the orange watching the whole event in the "sky". It became even more of a blend when we suggested blowing the scene up by a factor of a hundred million. The resulting image of a gigantic flashlight in outer space, which is pointing at a colossal orange (the earth's size) floating in the void, on whose rind are standing stupendously large ants (a hundred times taller than Everest) staring out into space and watching with awe a huge, darkened ping-pong ball, is a quintessential frame blend.

Another frame blend we discussed was how one decides whether one wants to take a job that has been offered, by mapping oneself onto a friend who has a similar job or who works in a similar place. In the analogy one makes, one inevitably blends some aspects of oneself into one's model of the friend, or some aspects of the friend into one's model of oneself, thus producing a hybrid imaginary individual, neither fish nor fowl.

As we pointed out earlier, frame-blending is not by any means always a "contaminated" or "defective" way of thinking. Hypothetical individuals — mental blends between real people — are in fact common in everyday analogies. For example, we described Mark, who was reading a newspaper article about the swimming competitions at the Beijing Olympics, comparing himself to Michael Phelps, and wondering what he would have done in Phelps's shoes (or lack thereof) — mixing *himself* into the 2008 Olympic games, feeling the water of the pool and the excitement of the competition, adjusting his age and athletic abilities in order to make the imaginary insertion work. In coming up with this analogy, Mark thus created an intricate mental blend of himself and Phelps.

For a more complex example of an analogy that blends frames, consider the case of a woman who, flaring up at her husband for an insensitive remark, picks up a plate, then puts it down, and says, "If I'd been my mother, I *would* have thrown it at you!" The implicit analogy links the current fight to numerous fights that the wife witnessed, as a child, between her parents. She is envisioning her mother in this room (but much younger than she actually is), married to this man (or to a blend of him and her father) and who actually throws the plate. On the other hand, all the pent-up fury stemmed from the fight that just took place, so the hypothetical plate-thrower is as much the woman herself as it is her mother. The natural and inviting analogy between the two marriages, including their contrasts, is being exploited by the wife in order to shed new light on their current fight. By momentarily blending herself with her mother and envisioning this version of herself *actually* throwing the plate, and then by telling her husband about this scenario, the wife has at least managed to let off a bit of steam.

The Dream of Mechanical Translation

The act of carrying a book from the culture in which it was conceived to another culture cannot help but involve a large number of frame blends. Suppose one is translating a novel written in America into Chinese. Not only will all the people in it wind up fluently speaking a language that they don't know, but all the concepts denoted by the words in the story have grown in different "gardens" in the minds of Chinese readers. We need only to think of what happens to such concepts as *city*, *bicycle*, *house*, *store*, *rice*, *river*, *mountain*, *poem*, *writing*, *word*, *eyes*, *hair*, and so forth, when they are transported (through literal translation) from the American culture to the Chinese culture. The central members of each of these categories, making up its "old town", are clearly very different in China and in America. The upshot is that Chinese readers of the novel will automatically and unconsciously bring in certain Chinese preconceptions when they read the novel in Chinese. The places and the events they imagine will subtly blend America (where the events take place) and China (where the concepts grew). And the same thing happens, of course, in any act of translation between any two languages, because some parts of the original work remain constant while other parts are necessarily subject to distortion.

Despite the complexity that we've just described, translation seems like a relatively mechanical activity to some people. And indeed, the idea of mechanical translation was first proposed three score and several years ago, at which time it seemed reasonable and relatively straightforward. For example, one of the founders of the field, the noted mathematician Warren Weaver, wrote the following: "When I look at an article in Russian, I say, 'This is really written in English, but it has been coded in some strange symbols. I will now proceed to decode.'" Weaver's humorous statement expresses the credo underlying machine translation, which is that translation is an act of "decoding" essentially analogous to using a substitution cipher, in which, in order to encode or decode a message, all one needs to do is to replace the message's symbols, one by one, by other symbols, according to a fixed table of correspondences. To be concrete, one might encode a message by replacing every letter in it by its alphabetic successor, which will in general yield an illegible piece of text, such as "Gpvs tdpsf boe tfwfo zfbst bhp"; then, in order to decode such an encrypted message, one would do the reverse transformation, replacing every letter by its alphabetic predecessor, in this case yielding Abraham Lincoln's famous opening gambit, "Four score and seven years ago".

In the early days of machine translation, translation between languages was seen as this same process but just on a larger scale, in the sense that it operated not on letters but on words, and the correspondence table wasn't just 26 items long but was a huge bilingual dictionary, giving, for each word, *the* matching word in the other language. And lastly, it was presumed that, in order to fix up awkwardnesses due to discrepancies between the grammars of the two languages involved, the scrambled word order that would almost surely result could be cleaned up by a complex but straightforward mechanical rearrangement process that took both grammars into account.

This substitution-and-rearrangement process remains the most common philosophy underpinning mechanical-translation efforts even today, except that the units in the semantic correspondence table are often taken at a somewhat higher level than that of individual words — they can include idiomatic phrases and other large chunks (for example, "to be under the weather" might be rendered in French, as an indivisible chunk, by "ne pas être dans son assiette", literally meaning "not to be in one's dish"). Indeed, bilingual data bases containing many millions of corresponding phrases are thoroughly scoured in ultra-rapid fashion in order to allow the computer to find "the best" (in some sense) alignments that will yield the "decoding" into Language B of a phrase, sentence, or any passage originally written in Language A.

We might test the efficacy of this strategy by seeing how the world's most readily available (and perhaps also its most sophisticated) machine-translation "engine" performs on our little Lincolnian phrase "Four score and seven years ago". (The term "engine" is a friendly tip of the hat to Charles Babbage, the great nineteenth-century British computing pioneer, who invented an important predecessor of today's computers, which he called the "Analytical Engine".) We asked Google's engine to translate this phrase into French, and in but milliseconds it shot back at us the words "Quatre points et il ya sept ans". (Strangely enough, "ya" is not a French word, but that's nonetheless what the engine came up with.) To test the program's understanding of its own output, we fed this phrase (with "ya" unchanged) into the engine and threw it into reverse gear; instantly, out popped "Four points and seven years ago" (the clever engine didn't stumble in the least over the non-word "ya"). Anyone can imagine how the translation engine might have gotten "four points" out of the words "four score", but, despite the plausibility of that guess, the engine's output, whether in French or in English, doesn't make any sense.

We carried this experiment out not just once but many times, and discovered that the French output was not stable at all. Sometimes it was better, sometimes not; in any case, the first output we quoted reappeared on more than one occasion. When we asked the same translation engine to render Lincoln's phrase in German, it returned to us "Vier der Gäste und vor sieben Jahren" — "Four of the guests and seven years ago". We have no clue as to how, beginning with "score", it came up with "guests". In Spanish, it gave us "La puntuación de cuatro y siete años atrás" — "The grade of four and seven years back" ("grade" meaning "school grade").

The translation engine has no notion of meaning. It is not trying to *understand* its input, but simply to operate on the marks making it up. In that regard, the operations carried out by the Google translation engine are, indeed, much like the operations of encoding or decoding using a substitution cipher, which don't involve meaning in any way at all. This is exactly the view that Warren Weaver proposed sixty years earlier, his vision having been of an unthinking, meaning-ignoring, mechanical substitution process that would allow one to toggle back and forth between any two languages.

It is difficult to fathom how such a simplistic view of translation could possibly have been put forth by the same person who wrote *Alice in Many Tongues*, a delightful short study devoted, with clear love, to a discussion of how some of the trickier passages in Lewis Carroll's wordplay-filled *Alice in Wonderland* had been rendered by highly creative translators into French, Italian, German, Danish, Swedish, Spanish, Russian, Polish, Hungarian, Hebrew, Swahili, Pidgin English, Japanese, and Chinese. In this book, Warren Weaver pays careful attention to how superb translators handled such complex challenges as parodied verse, wordplay, nonsense passages, and other ways in which form and content are deeply entangled with each other, and with reverence he describes the artistry with which some of the translators found brilliant solutions to these challenges. Over and over again he points out that high-quality translation is anything but mechanical, and in so doing, he implicitly demonstrates as many times that translation in no way resembles a "mechanical decoding" process, but that it requires continual discovery of ingenious new analogies. Indeed, his book's message is that translation depends crucially on drawing deeply from the well of one's life experiences and mental resources in order to come up with appropriate analogues in Language B for strings of characters written in Language A.

Only a few years after the dream of machine translation was hatched, it was already starting to run into profound problems. These problems were articulated by a number of skeptics, of whom perhaps the most vociferous was the logician Yehoshua Bar-Hillel, who had once been one of the field's earliest and most enthusiastic researchers. In the mid-1950's, then, there was already deep skepticism over the idea of translation as a "mechanical" or "algorithmic" process, and such skepticism is still warranted today, as we have just seen.

Good Analogies Make Good Translations

How, then, should one translate Lincoln's famous speech-opener? Suppose we wished to translate it into French, for instance. It helps to know what it means. To begin with, then, since a score consists of twenty items, it would follow that "four score" means "eighty", and that "four score and seven" means "eighty-seven". Can we say this in French? Yes, fortunately, the French, who have a strong mathematical tradition, do have a way of saying "eighty-seven" — namely, "quatre-vingt-sept". To say "years", say "ans", and to say "ago", just put "il y a" at the very front. So putting the little pieces together, we get "Il y a quatre-vingt-sept ans". All done!

There is a problem, though, which is that Lincoln's phrase "four score and seven" is not the usual way of saying "eighty-seven" in English, but a special way of saying it, based on a somewhat odd way of looking at the number and which, more importantly, has a clear poetic resonance. These are both key qualities of Lincoln's turn of phrase, so it would be crucial to preserve both of them in our translation.

Now, by a remarkable coincidence, the French word for "eighty" happens not to be based on "huit" (the French word for "eight"), but on the words "quatre" and "vingt" ("four", "twenty") — so that "quatre-vingts", the standard French word for "eighty", actually means "four twenties"! It would seem that the ideal solution — "Il y a quatre-vingt-sept ans" ("Four-twenty-seven years ago") — fell right out of the sky. What luck! It certainly would not have worked if the Gettysburg Address had been delivered in 1843 (an event that would have been confusing to more than one historian), since "three score and seven" is expressed in French not as "trois-vingt-sept" (that is, not as "three-twenty-seven") but as "soixante-sept" (that is, as "sixty-seven", very literally).

It might at first blush seem that thanks to this stroke of luck, we are in fact done, but this is a hasty conclusion. French speakers almost never notice the "quatre" and the "vingt" inside "quatre-vingts" — no more than English speakers hear the words "for" and "tea" inside "forty", or, for that matter, an allusion to the terror of Mongol invaders inside "hundred". Even if for English speakers, "quatre-vingts" seems to brim with the proper arithmetical meaning and also to exude an archaic, poetic flavor, it does nothing of the sort for French speakers; to them, it is just an ordinary, pedestrian word for the number halfway between 70 and 90. By contrast, for English speakers, Lincoln's phrase requires a tad of conscious calculation and resounds with poetry and nobility. Therefore, if we hope to capture its essence in another language, then we have to avoid, at all costs, the mundane — and this means that what we at first took for a great stroke of luck turns out to have been a cursing in disguise! And so, it's back to the drawing board.

Our goal is to be high-quality copycats, which is to say, to "do what Lincoln did" — and what Lincoln did is to replace the standard English *eight-tens* view of eighty by an exotic, attention-grabbing *four-twenties* view. At this point, it may seem obvious to some that the solution is simply to use a reversal analogy — namely, to replace French's standard *four-twenties* view of eighty by an exotic, attention-grabbing *eight-tens* view. Indeed, this is even reminiscent of that elegant Copycat flip of perspective whereby *xyz* is seen as the mirror image of *abc*. And as a matter of fact, the word "huitante" (meaning essentially "eight-ty"), though exotic in France, is commonly used in Switzerland in lieu of "quatre-vingts", and in some French-speaking locales the even more exotic word "octante" is also a dialectal way of saying "eighty", though today it has almost fallen out of currency. So what about using one or the other of these unusual French words for "eighty"? Well, unfortunately, the phrases "Il y a huitante-sept ans" and "Il y a octante-sept ans", rather than sounding poetic and uplifting, come across to most French speakers as simply quaint. As would-be translations of Lincoln's noble phrase, they are both extremely wanting. In this case, the often clever idea of a reversal analogy doesn't pay off.

Clearly, then, we will need to search around for a deeper French analogue to Lincoln's phrase. What might be the French counterpart of the poetic phrase "four score and seven years ago"? Without recounting all our failed forays, we can simply say that we began by exploring the idea of "seven dozen and three years ago", which uses the common French word "douzaine". This, however, did not seem analogous enough to the original because, among other things, it involved too much conscious calculation, although it did at least activate the closely related thought of using the French word "dizaine" (which is analogous to "douzaine", but features only ten items). And once again, although "il y a huit dizaines et sept ans" sounded silly to our ears, it reminded us of the word "trentaine" (which also is like a dozen, but involves *thirty* items), so that "eighty-seven" could be expressed as "trois trentaines moins trois" — "three thirties minus three". At this point, however, our translation was becoming too much like an arithmetic problem, or even a mild tongue-twister. Hardly our goal!

Eventually, it dawned on us that French has the word "vingtaine" as well — based on "vingt" and meaning a collection of twenty items (in other words, a score). Thus we finally hit upon the translation "Il y a quatre vingtaines d'années, plus sept ans". This was far more promising, but still, its flavor was not sufficiently analogous to that of the original, since the explicit mention of the arithmetical operation "plus" was too heavy-handed, as if Lincoln had said "four score *plus* seven years ago". In the end, though, we were able to raise the loftiness of the tone by making some minor adjustments, as follows: "Voici quatre vingtaines d'années, et encore sept ans..." ("Four score years ago, and yet seven more..."). We were quite proud of our collaborative find.

Another idea was suggested to us by a translator friend who took advantage of the poetic word "lustre". Although most native speakers of French think of it as simply meaning "a long time", it can also mean a five-year chunk of time. This fact allowed our friend to render the president's lustrous words by "Voici seize lustres et encore sept ans..." ("Sixteen lustres and seven years ago..."), which exudes a rather grand and lofty flavor. (Of course he could have said "Seventeen lusters and two years ago...", but it's not clear that that's an improvement.)

Altogether, then, through a slow process of carefully honed analogy-making and analogy-judging, we eventually managed to recreate some of the high-sounding flavor of Abraham Lincoln's immortal phrase, while sidestepping various superficially enticing traps along the way.

Potential Progress in Machine Translation

The preceding anecdote confirms the pervasive thesis of Warren Weaver's book *Alice in Many Tongues*, which is that to translate well, the use of analogies is crucial. In order to come up with possible analogies and then to judge their appropriateness, one must carefully exploit one's full inventory of mental resources, including one's storehouse of life experiences.

Could machine translation possibly do anything of the sort? Is it conceivable that one day, computer programs will be able to carry out translation at a high level of artistry? A couple of decades ago, some machine-translation researchers, spurred by the low quality of what had then been achieved in their field, began to question the methods on which the field had been built (mostly word-matching and grammatical rules), and started exploring other avenues. What emerged with considerable vigor was the idea of *statistical* translation, which today has become a very important strategy used in tackling the challenge of machine translation.

This approach is based on the use of statistically-based educated guesswork, where the data base in which all guesses are rooted consists of an enormous storehouse of bilingual texts, all of which have been carefully translated by human experts. A typical example of such a data base is the proceedings, over several decades, of the Canadian Parliament, which are legally required to be made available in both English and French. Such a data base is a marvelous treasurehouse of linguistic information, if only one can figure out how to exploit it.

The basic idea of statistical machine translation is to choose among the many possible meanings of a "chunk" (that is, a word or several-word segment) in a piece of input text (*i.e.*, a text to be translated) by exploiting the *context* in which the chunk appears in the given passage. Suppose, for instance, that the engine is translating from

English to French. The English chunk to translate may appear in many thousands of diverse contexts in the English side of the bilingual data base, but only a small number of those thousands of contexts (say, two dozen) are likely to be found sufficiently "similar" to the original context (where "similarity" is judged by a complex statistical calculation). This phase of narrowing-down on the basis of statistical similarity is the crux of the matter. In the human-translated bilingual data base, each of these relatively few English-language contexts comes aligned with a corresponding French-language context. The translation problem would thus seem to have been reduced to simply zeroing in on the corresponding chunk in these few French contexts. Unfortunately, though, this vision is too optimistic. In general there won't be just one precisely corresponding chunk in the French contexts; there may be quite a few rival candidate French chunks, and so, to get a good candidate, educated guesswork (*i.e.,* further statistical calculations, the details of which we will skip) is called for. The long and the short of it is that in this computationally intensive fashion, which takes advantage of vast amounts of human-translated text, the French chunk that is "most probably equivalent" to the English chunk is pinpointed and is inserted into the outgoing stream of French words.

One way of describing the translation algorithm that we've just sketched is that, through a sophisticated and highly efficient set of computations, it repeatedly makes *analogies between pieces of text* in the two languages involved. This sounds nothing if not promising, but the proof of the pudding is in the eating, and so we will now proceed to sample the pudding. In order to do so, we'll take a careful look at a short piece of French text in order to see how two extremely different machine-translation programs dealt with it — one using the old strategy, and one using the new strategy. The passage we will examine is taken from an obituary of the novelist Françoise Sagan, written by the literary critic Bertrand Poirot-Delpech, and which appeared in the highly respected national newspaper *Le Monde* in September of 2004. The paragraph we selected is written in elegant and evocative but standard French, readily understood by any literate native speaker. We did not choose it for its difficulty; indeed, its density of "traps" for a translator is no higher than that of any typical article in *Le Monde.*

Below we give the original French, followed by the translation furnished by Google's translation engine shortly after the obituary appeared. At that time, the Google engine was based on the original "Weaverian" machine-translation philosophy — namely, first via lookup in a very big on-line dictionary, followed by enhancement using grammatical "patching".

Original paragraph from Le Monde, September 2004:

Parfois, le succès ne fut pas au rendez-vous. On a beau y penser très fort, le bon numéro ne sort pas forcément. Sagan prenait les échecs d'auteur dramatique comme les revers de casino, avec respect pour les caprices de la banque et du ciel. Il faut bien perdre un peu, pour mieux savourer la gagne du lendemain. Qui ne l'a pas vue « récupérer » en quelques quarts d'heure les pertes de toute une nuit ne peut comprendre comme c'est joyeux de narguer le sort.

Google's translation engine, September 2004:

Sometimes, success was not with go. One thinks of it in vain very extremely, the good number does not leave inevitably. Sagan took the failures of dramatic author like the reverses of casino, with respect for the whims of the bank and the sky. It is necessary well to lose a little, for better enjoying gains it following day. Who did not see it "recovering" in a few fifteen minutes the losses of a whole night cannot include/understand as they is merry of narguer the fate.

It is obvious that the "decoding" technique — the technique that lay behind the original optimistic vision of machine translation — was hopelessly inadequate to the task, since the output that the translation engine yielded is pretty much nonsensical to an English speaker.

It is ironic that the only French word that the Google translation engine considered ambiguous in this passage was the word "comprendre", for which it gave two possible interpretations separated by a slash, as if to suggest that were the only spot in the whole paragraph where a translator might have some doubts as to how to word things properly in English. (The French word "narguer", found near the end and roughly meaning "to flout", was apparently not in the engine's on-line dictionary, so it was simply left in French.) This example gives a sense for the quality of machine translation in the fall of 2004.

But now let us fast-forward to the spring of the year 2009. At that point, Google's translation-engine developers had radically switched strategies in favor of the new idea of statistical machine translation, so their new engine had little in common, other than its name, with its former incarnation. Given the inadequacy of the old method, which we have just witnessed, this would seem like a wise decision. To cut to the chase, here is the output text that the totally revamped translation engine yielded:

Google's translation engine, April 2009:

Sometimes, success was not there. It was nice to think very hard, the proper number does not necessarily spell. Sagan took the failures as a dramatist such as backhand casino, with respect to the whims of the Bank and the sky. It must be losing a little, better enjoy the gains overnight. Who did not see "recover" in a few minutes lost a whole night can not understand how happy it is the sort of taunt.

This is English, in some sense, but it is still English that makes no sense. (What on earth is *backhand casino*, for instance? And why does "the proper number" not necessarily *spell*? And why is, or was, it "nice" to think very hard? And how happy can *the sort of taunt* get, anyway?) In sum, there is little visible improvement here over what was produced in 2004.

To round out this section, we give one last translation of the paragraph in the Sagan obituary in *Le Monde*. This human-produced translation comes from our joint pen, and in fact it is what we would have offered had we been requested to anglicize Poirot-Delpech's obituary of Sagan for publication in an American newspaper.

Careful human translation:

Sometimes things just didn't work out right; no matter how hard she wished for it, the dice simply wouldn't come up her way. But Sagan always took her failures as a playwright much as she took her gambling losses, acknowledging the arbitrary whims of the house and of divine fate. After all, everyone has to lose now and then, so that the next day's victory will taste all the sweeter. And if you never saw her win back a whole night's losses, often in well under an hour, you just can't have any idea of the glee she took in laughing in the face of destiny.

On reading this hand-done translation along with the two machine translations given earlier, one sees that humans and translation engines are not playing on the same turf — in fact, they are not even playing at the same game. Any decent human translator has a rich storehouse of first-person and vicarious experiences in all areas of life. Human translators can imagine in detail a scene that was sketched with just a few brushstrokes and can pick up on subtle allusions, they have an excellent mastery of the grammar of both languages, and they are past masters at expressing themselves fluently and idiomatically. Just like the original passage, this translation paints a vivid picture of the complicated strategems a certain human being had for dealing with life's setbacks.

How could one possibly translate without making use of one's knowledge? For example, in the very limited context of weather reports, it will in general be a safe strategy to translate "ciel" as "sky", but in the wide world of human affairs, the different possible meanings of "ciel" include (but certainly are not limited to) the sky, heaven, the heavens, the air, mid-air, climes, blueness, a vault, a canopy, a ceiling, the atmosphere, clouds, the firmament, the stars, space, the cosmos, the universe, Providence, God, divine will, destiny, predestination, and on and on. If we know for sure that we are dealing only with meteorological phenomena in the most mundane of contexts, then blindly replacing "ciel" by "sky" will probably do (although even in a weather forecast, someone could always throw a curve ball), but in a completely open-textured domain where any possible idea might come up or be evoked in some standard or brand-new metaphorical manner, then all bets are off. In our translation of the paragraph taken from the Sagan obituary, we happened to opt for "divine fate", but in some other context we might have chosen "heavens to Betsy", and in yet other contexts perhaps "the orange glow" or "the stars above" or "the celestial vault" or "light blue" or just plain old "azure". And the word "ciel" is by no means exceptional; it is quite typical.

We've argued, in this book, that choices of words and longer lexical items are carried out by an unconscious process of analogy, and this holds in translation between two different languages no less than in monolingual speech production. But the analogies involved in carrying out human translation are not of the sort exploited by current machine-translation techniques, which make surface-level analogies between linguistic "chunks" based on statistical information that can be extracted, using intense calculation, from human-translated bilingual data bases, as we described above, rather than on understanding the *ideas* that are being talked about, which of course are the entire reason that pieces of text are created.

In a human translator's mind, the evocation of words and phrases in the target language will take place much as when they summon up words and phrases of their native language in a conversation without any kind of input from a foreign language. This evocation process consists of analogy-making once again — the type of analogy-making that gives rise to the most mundane acts of labeling, the type of analogy-making that zeroes in on the *mot juste* or the *locution juste* or the *proverbe juste*, the type of analogy-making that makes someone say "probab-*lee*!" as opposed to just "*prob*-ably", the type of analogy-making that makes someone call a certain room a "study" or an "office", based on their prior life experiences, the type of analogy-making that makes Cheryl repeatedly call her husband "Chuck", which is her brother's name, the type of analogy-making that makes phrases like "sour grapes" and "the left hand doesn't know what the right hand is doing" pop instantly to mind in certain types of circumstances, and so forth and so on.

In the specific case of the obituary paragraph, when we humans read "les revers de casino", we do not think of *backhand strokes* in games with rackets, but of *setbacks in gambling*, because we are familiar with the kinds of things that go on in casinos. When we humans read "les caprices de la banque", knowing that the context is one of casinos, we don't think about *financial institutions* where we save our money but about the casino's own *storehouse of money*, because we understand what gambling is about. And when we humans read about "les caprices du ciel", we don't think of the *whims of the atmosphere*, nor about whims of the *starry sky*, nor of the *color blue*; we think of the *mysterious heavenly forces* that someone might imagine as governing the rolls of dice and roulette wheels.

And this is because all the foregoing French phrases trigger *ideas* in our human minds, rather than merely triggering counterpart English phrases, and the evoked ideas fit together into large patterns that trigger large-scale *memories*, which, only at that stage of the game, evoke English *words*. The process of translation depends crucially on the intermediate phase in which memories and concepts are triggered — an unavoidable phase usually called "understanding". And this process involves putting together all the pieces of a sentence in a carefully coordinated manner, which means exploiting all the indications that grammar gives us about how the ideas fit together in a sensible pattern. No translation worthy of the term can afford to ignore the meaning of the text to be translated, and meaning can be grasped only if complex grammatical constructions are taken into account, which means making a precise linguistic analysis of the text, which today's translation engines are unable to do.

Consider just one example from our obituary paragraph. The last sentence opens as follows: "Qui ne l'a pas vue « récupérer » en quelques quarts d'heure les pertes de toute une nuit...". There arises the question of why an "e" has been tacked onto the past participle "vu" of the verb "voir" ("to see, to watch"). This is a question of French grammar, and its answer is that whenever a past participle is preceded by its direct object, it must agree in gender and in number with that object. That little "e" is therefore telling us that the direct object of the verb "voir" is feminine and singular. It carries a crucial meaning! It tells us that a thing or a person of feminine gender is being watched (or rather, since the verb is negated, that this thing or person is *not* being watched). And indeed, the text is talking about failing to see an "entity" (possibly an animate one) that was quickly recovering from the losses of an all-night gambling session. We instantly realize that this "entity" is in fact none other than Françoise Sagan. The past participle's feminine singular ending has clued *us* in, but both the 2004 and 2009 engines were clueless. To ignore grammar is to invite disaster.

Could a good mastery of French grammar conceivably have helped the translation engines? That is a rhetorical question, whose answer is "yes", of course. Translation depends on understanding, and understanding depends on grammar, because grammar tells us how the smaller pieces fit together to make a large, coherent structure.

Various and Sundry Challenges in Translating this Book

The preceding section's purpose was not to deride efforts at machine translation. Indeed, the blinding speed with which virtually anyone today can, entirely for free, get a glimpse of what is going on in a piece of text written in a language of which they don't know a single word, or perhaps even a single symbol, is very impressive. Thanks to the efforts of many researchers over many decades, anyone today can, in but seconds, get a feel for what a Website written in, say, Arabic, Estonian, Hindi, Icelandic, Korean, Malay, Maltese, Romanian, Swahili, Thai, Turkish, or Vietnamese is talking about, and sometimes the one-second miracle is coherent for fairly long stretches. This accomplishment is both astonishing and humbling. But for all its impressiveness, it is not to be confused with what human translation is about.

Genuine translation — translation that merits the label "translation" — is indeed about analogy-making on all levels imaginable, from the most minuscule grammatical ending of a word to the entire overarching cultural context in which the text and the events and notions of which it speaks are embedded. In further support of this thesis, we will now briefly touch on some of the issues in analogy-making that have arisen in the translation of this book, since it is appearing simultaneously in both French and English, and in order to produce these two versions, we, its authors, have had to grapple with innumerable nontrivial translation challenges.

Take the second paragraph of Chapter 1, for instance. In the French version (which was the source for that particular paragraph), a picture is painted of what a rider of the Paris métro is taking in, and this picture is familiar to every French reader. We could have converted that passage into English in a non-transculturating manner, in which case the corresponding two sentences of Chapter 1 would have run like this:

> We see ads everywhere, we think vaguely about the names of the stations as they go by, and at the same time we are absorbed in our own thoughts. We wonder when we'll find a free moment to go to the bank, we think about the health of an old friend, we are upset by a headline in the newspaper that some man sitting near us is reading, which speaks of a terrorist attack in the Middle East, we inwardly smile at the jokes in the advertisements on the walls, we try to make out the words of the song that the accordionist who just stepped into our car is playing…

But we chose not to translate it in that fashion, not wanting to disorient our American readers from the first moment. Of course our American readers are likely to know, either personally or indirectly, the Paris métro. Nonetheless, it struck us that it would seem distinctly *foreign* if the book started out in a foreign country, and we didn't want to give our book that tone. Indeed, that is why we started out the *French* version of our book in the Paris métro — we wanted it to feel familiar to our French-speaking readers. Naturally, we were shooting for "exactly the same" effect in English; and especially since our book is about analogies, it was self-evident that we had to find an *analogous* way to start our book in English.

The most obvious analogy would have been to transplant the scene to the New York subway system, in which case most of what we wrote above would have worked just fine, although we might have changed the accordion to another instrument. The problem we felt with that idea was that riding the New York subway is not a universally shared experience of all Americans. It feels "New-Yorkocentric" in a way that the French version did not feel Paris-centric. Paris plays such a central role in France that scenes taking place in Paris feel generic to all French readers, whereas to many American readers, scenes in New York City exude an unfamiliar feel. Moreover, riding the subway is a rather unusual, perhaps even exotic, experience for many Americans. In that case, what in the life of a typical American corresponds to taking the Paris métro? It could of course be driving a car somewhere, but we preferred to keep the image a little closer to the original, thus in the domain of public transport, and so we chose the idea of sitting in an airport somewhere — not a specific airport, but just a random, generic one. Accordingly, here is what we came up with:

> We see ads everywhere. We think vaguely about the cities whose names come blaring out through loudspeakers, yet at the same time we are absorbed in our private thoughts. We wonder if there's time enough to go get a frozen yogurt, we worry about the health problems of an old friend, we are troubled by the headline we read in someone's newspaper about a terrorist attack in the Middle East, we sniff the enticing odors emanating from the nearby food court, we are puzzled as to how the little birds flying around and scavenging food survive in such a weird environment...

Now who would ever have suggested, *a priori*, that a handful of hungry birds fluttering about, hither and thither, in the concourse of some random airport is "the American translation" of an impoverished accordionist who has just boarded a random car on a random line of the Paris métro? Is this really a case of translation? To be sure, our answer is "yes". To us it was crucial that our book have "the same feel" to native speakers of both languages. To attain this effect demanded, on occasion, not just bland vanilla translation, but rich chocolate mint-chip transculturation. Had we merely converted the opening of Chapter 1 directly into English, leaving it in the Paris métro, as in the text displayed at the bottom of the previous page, or even had we transposed it to the New York subway, it would have been a wooden, almost mechanical kind of translation (although light-years more sophisticated than today's machine translation).

Could we not have elected, after having chosen an airport scene over a subway scene in the American version, to go back to the original passage in French and to rewrite it so that it takes place in an anonymous French airport? Indeed, *shouldn't* we have done that? Well, had we done so, we would have gained some uniformity, but we would have sacrificed elegance. Transculturation struck us as the best choice here. But transculturation is not the choice we would have made if we had been translating a *novel* whose opening scene was set in the Paris métro. To transculturate a novel is to move it completely from one cultural setting to another — it is to uproot it, to create a counterpart story in a different land. That is a much more radical shift than merely transculturating an illustrative paragraph or two every so often in a book of nonfiction. Our book in particular has nothing inherently to do with France or America; it has to do with human cognition, which transcends cultures, and in it we illustrate cognition's mechanisms in contexts of many sorts, feeling free to dream up scenes that fit the culture of our readers.

For instance, in Chapter 1 we used the concept *mother* in both languages, but we replaced Zinédine Zidane and his sport, soccer, by Tiger Woods and *his* sport, golf. In Chapter 2 we had to replace nearly every example involving a compound word or an idiom, since they were totally local, and generally we wound up using examples that were completely unrelated on the superficial level, but nonetheless they illustrated the same cognitive phenomena.

In Chapter 5, speech errors are given a lengthy treatment, and any speech error is a unique event intimately linked not only with a particular language but also with a particular individual's brain. Rather than transcribing such an irreproducible event into a different language, we relied principally on examples that were directly observed in the language we were writing in. But we were fortunate in that, over the years, we had collected sizable corpora of errors in both French and English, and so the translation strategy there was more akin to transculturation. On the surface, totally different speech errors were used as illustrations, but at a deeper level, the mental mechanisms that we described were all the same. For that reason, this important part of our book reads very differently in French and English, and yet in their essence, the two discussions really do say "exactly the same thing".

Another striking example of transculturation comes from Chapter 5, and involves the way we found of rendering in French a telephone conversation between two Americans. One of them, a jazz musician, had placed a classified ad in a newspaper in order to sell his old cornet, and the other one had called him to inquire about it. However, the potential buyer was under the impression that the entity in question was a used Dodge Coronet. Despite this considerable disparity in topic, the two had a conversation that lasted a minute or more that seemed to each of them to make perfect sense. Since this misunderstanding came about as the result of a phonetic resemblance between words that are not familiar to most French speakers, a straightforward translation into French of the dialogue would have been heavy-handed, and would in fact have ruined the example for French readers. The analogous conversation that we chose instead for the French version of the chapter was taken from a television advertisement in which two people are having a conversation over lunch about a documentary film concerning "emperors". The one who has seen the movie knows it was about emperor penguins, and that's what he has in mind (the ad explicitly jumps back and forth between the private mental images of the two people), while the one hearing about it for the first time is imagining a crowd consisting of clones of the French emperor Napoleon in various scenarios, starting in the steppes of Russia and winding up in Antarctica. Neither one, of course, has any idea for at least a minute of the fact that they are talking past each other. The extended ambiguity in the French dialogue has very much the same flavor as that of the English dialogue, which makes the translation work very nicely. And yet it might seem rather odd that in passing from American soil to French soil, a musical instrument turned into a penguin and a used car turned into a monarch.

In the original version of our lengthy list of caricature analogies in the current chapter, one featured a French speaker mocking another French speaker for a nonstandard pronunciation of the number word "cinq", meaning "five". The mocker, to cast doubt on the acceptability of this usage, spontaneously came up with an analogous but clearly sillier usage involving the number word "six". This quip was too language-specific to be translated, but by chance, we happened to have observed a strikingly similar case in English, this one involving two bisyllabic verbs, "invite" and "accept", of which the former, with shifted stress, can be used as a noun but somewhat questionably, whereas the latter, similarly stress-shifted, makes an unarguably silly-sounding noun. The two anecdotes were parallel at the level of their essences, but clearly "invite" and "accept" are not the standard translations of "cinq" and "six" — no more than "birds flying around in an airport concourse" is the standard translation of "un accordéoniste dans le métro".

Chapter 3 included perhaps the most complex of all the translation challenges in the book — two poems from the pen of our friend Kellie Gutman, each of which describes one side of the analogy linking Dick and his bottlecaps to Danny and his ants. Originally, Kellie was inspired to write the poem about Dick at Karnak simply because it captured an amusing episode involving her husband during a cruise they made up the Nile. However, since witnessing that episode had triggered in Doug's mind the far-off memory of his infant son Danny at the edge of the Grand Canyon, and since that reminding incident was featured in our book, we asked Kellie if she could write a parallel poem about the earlier episode. She took up our challenge with verve, using exactly the same poetic form, perfectly matching the syllable counts on corresponding lines and also preserving the precise pattern made by feminine (bisyllabic) and masculine (monosyllabic) rhymes. When we were writing Chapter 3 and were starting to frame the passage about the Dick/Danny analogy, we knew that Kellie's two poems would make excellent accompanying pieces and thus decided to include them. This fact created a major obstacle — not only were both poems composed in playful English, but they obeyed several detailed formal constraints. Both despite and because of its linguistic virtuosity, Kellie's performance called for faithful translation on all levels in the French language.

We will not describe the details of the two pairs of poems, since that could occupy several pages, but below we simply exhibit the final stanza of all four poems.

At last, the North Rim: strange striations
with shades evoking exclamations —
unless you're Danny... Then you treasure
the leaves and bugs! While grownups measure
the grandeur of vast rock formations,
you play with ants — a simpler pleasure.

In Karnak's heat, our guide expounded
on gods and temples, while surrounded
by columns far too grand to measure.
We contemplated them with pleasure,
but as we gazed on high, dumbfounded,
Dick stooped to pluck a humbler treasure.

Le Grand Canyon enfin s'révèle !
Falaises rocheuses, couleurs trop belles —
à moins d'avoir taille trop modeste...
or là, tu scrutes, à l'aide de gestes,
fourmis et feuilles à p'tite échelle,
et ça t'procure une joie céleste...

Au Temple de Karnak, le guide
louait les hauts piliers splendides
(et nous brûlions), quand, d'un beau geste,
s'agenouillant de façon leste,
Richard saisit, d'une main humide,
un p'tit trésor bien plus modeste.

Among the constraints governing the two final English-language stanzas is the fact that all twelve of their lines end in *feminine* rhymes (meaning two-syllable rhymes in which the penultimate syllable is stressed and carries the rhyming action, and in which the final syllable is unstressed and is identical in both words — *e.g.*, "pleasure" and "measure"). In French the concept of *feminine rhyme* exists but is slightly different: it requires that the final word of each of the two lines involved should end in a so-called "mute 'e' ", which, despite its name, is not in fact totally silent but is pronounced, albeit just slightly (for instance, "modeste" and "céleste" form a feminine rhyme in French). Kellie's conscious decision to use exclusively feminine rhymes as she crafted the twelve lines in these two stanzas in English wound up, a couple of years later, forcing all twelve lines of the corresponding French stanzas to end in mute "e" 's.

As if this wasn't enough, Kellie playfully chose to use exactly the same set of rhyming words in lines 3, 4, and 6 in both of her stanzas — but in one poem their order was "treasure", "measure", "pleasure", while in its analogical mate their order was cyclically permuted: "measure", "pleasure", "treasure". This subtle cross-poem pattern posed yet another thorny translational hoop through which to jump.

In the end, however, the challenge was met, and all the various hoops were jumped through simultaneously. The meter and the rhyme pattern were preserved throughout, syllable counts were preserved, the feminine-rhyme/masculine-rhyme distinction was preserved, the cross-poem sameness of the rhymes on lines 3, 4, and 6 was preserved, the cyclic permutation of those three rhymes was preserved, and last but not least, linguistic playfulness was preserved. A careful look at the three feminine rhymes on lines 3, 4, and 6 of the French version shows that in the lefthand stanza the sounds "deste", "gestes", and "leste" are used (the "s" at the end of "gestes" is just as silent as the comma that follows it), while in the righthand stanza these same sounds show up cyclically permuted — namely, as "geste", "leste", and "deste".

Poetry translation necessarily involves a large number of searches for complex analogies because one is laboring under the combination of many simultaneous pressures, some of which are explicit and quite sharp (such as a precise syllable count or the constraint of rhyming), others of which are anything but explicit and are subject to endless interpretation (such as the meaning of a word or the tone of a passage). The number of pathways that are tentatively explored is enormous, and the number of compromises made is also large.

But the product that emerges from all this searching, adjusting, and compromising does not have to be inferior to the original poem. The writer of the original was no less subject to multiple pressures, and hence necessarily made a vast number of choices that were also compromises. Compromise is the name of the game whenever constraints are involved, as they are in poetry, and a certain sort of creativity is also a frequent outcome of the combination of pressures under which one finds oneself working. Thus compromise can even at times be a source of quality. And in this context, creative compromise, arrived at through elegant analogies, is what we are talking about.

Who is Manipulating Whom?

In this chapter, we've shown how analogies are constantly manipulated by people, using such examples as caricature analogies, explanatory analogies, analogies that help people make personal or larger-scale decisions, analogies situated in the microworld of Copycat, analogies featuring various amounts and sorts of frame-blending, and analogies allowing translation and even transculturation to take place.

The preceding chapter showed how analogies sometimes trick us, manipulating us without our being even in the least aware of it; this chapter, by contrast, has shown how analogies, not content merely to lurk behind the scenes and pull strings surreptitiously, often emerge into broad daylight, where they are then at our mercy. We humans are thus not always mere puppets; indeed, sometimes we are puppeteers who deliberately build or choose one analogy or another, most frequently in order to communicate with others, but sometimes simply to make a situation clearer to ourselves. So in the end, who is manipulated, and who is the manipulator?

To choose one analogy over another is to favor one viewpoint over another. It amounts to looking at things from a particular angle, to taking a specific perspective on a situation. An insightful analogical take on a situation gives you confidence in your beliefs about the situation while also revealing new facts about it. A teacher, a lecturer, a lawyer, a politician, a writer, a poet, a translator, or a lover may pass hours or days in search of the most convincing analogy, like a goldsmith crafting a beautiful chalice for maximal effect. Such individuals work very hard and very consciously to induce in their listeners or readers the same point of view, or the same emotion or feeling or judgment, as they have.

On the other hand, sometimes everything happens in just a fraction of a second, as is the case for caricature analogies that people just blurt out. Although these, too, are conscious analogies dreamt up for a specific purpose, they are often less finely crafted than the ones just described, because the person who comes out with a caricature analogy often has no idea of how they came up with that particular one, as opposed to many others that might have worked just as well. And then there is the swirling sea of unconscious analogies constantly churning below the surface of our minds, forming and unforming without cease, pushing our thoughts "hither and skither" at all moments — and yet we are blithely unaware of them. These are the manipulating analogies that we discussed in Chapter 5. Together, then, these two chapters have painted a picture of a continuum stretching between thoughts that we push around and thoughts that push us around, but no matter where one looks along this continuum, one finds the process of analogy-making as the operative principle.

Although we have called some analogies "manipulated" or "carefully crafted", we could still try to look beneath their surface to find their hidden sources. And as we've seen throughout this chapter, consciously crafted analogies owe their existence to spontaneous unconscious analogical links. This means that even when you think it's you who are pulling the strings, the fact is that you are merely a puppet of whose strings you are unaware. You feel that you are deliberately creating an analogy to advance a certain point of view, but actually it's the other way around: your point of view comes from a myriad of hidden analogies that have given you a certain perspective on things.

Thus, the baseball announcer who spontaneously said, "Trying to get a hit off of Sandy Koufax is like trying to drink coffee with a fork" came up with this colorful image because unconsciously he was seeing one thing (a bat) coming very close to but slipping right by another thing (a ball), and this hidden abstract conceptual skeleton then gave rise to an analogical bridge linking the actual situation (swinging at a fastball and missing it) to an imaginary humorous situation (coffee slipping off the tines of a fork). In short, though we may tell ourselves that we are royally pushing analogies around from the heights of our conscious thrones, the truth is otherwise: we are really at the mercy of our own seething myriads of unconscious analogies, much as a powerful ruler is really responding to the collective will of their people, because if they were regularly going against their people, they would soon be dethroned. And thus the powerful "leader" is unmasked and revealed to be merely a very perceptive follower.

CHAPTER 7

Naïve Analogies

❧ ❧ ❧

Three Anecdotes

Timothy is watching his father shave one morning. He observes his father moistening his face, spreading some shaving cream across it, using the blade, and rinsing. From his four-year-old vantage point, Timothy categorizes the scene as best he can. The idea that a razor, so different in appearance from a pair of scissors or a knife, might be able to cut something does not cross his mind. On the other hand, Timothy knows very well that certain substances dissolve in other substances, such as sugar in hot water. He is therefore absolutely convinced that the shaving cream dissolves his father's stubble, and that the razor's sole purpose is to wipe away the shaving cream once it has done its job.

Janet is on a local mailing list, and one day she received a message from a chatty neighbor sharing this news: "This morning I enjoyed watching a bunch of titmice feasting on the insects on the branchlets of our tulip tree. There's such an abundant crop of insects this year that I think it will attract a large population of insect-eaters like titmice and chickadees." Janet was puzzled by the image of teeny mice scurrying about on the branches of a tree, since she had never witnessed any such thing. When she came to the phrase "titmice and chickadees" she was puzzled yet more, since it seemed odd that mice and birds would happily coexist on the branches of a tree. All at once it hit her that titmice are not in fact teeny mice, but are birds, just like chickadees.

Professor Alexander is bidding good-bye to a younger colleague who is leaving for Germany for a month. He says, "When you arrive, please send me your email address, won't you?" Seeing the look of perplexity on his colleague's face, Professor Alexander bangs his hand against his forehead. "What am I saying? Obviously, your email address isn't going to change at all!"

The age difference between Timothy and Janet is about the same as between Janet and Professor Alexander, but despite these large gaps, the same cognitive phenomenon is at work in the young child, the young adult, and the older adult. All three were taken in by tempting analogies, which, just like the categorical blinders discussed in Chapter 5, led them into error. In the cognitive-psychology literature, one finds all sorts of terms for this phenomenon, including "preconceived notion", "spontaneous reasoning", "naïve reasoning", "naïve theory", "naïve conception", "tacit model", "conceptual metaphor", "misconception", and "alternative conception". Although these terms are not all interchangeable, they do have in common one key thing, and we will call that core notion "naïve analogy".

The idea is that an unfamiliar concept (such as *shaving cream*, *titmouse*, or *email address*, in the three anecdotes) is apprehended plausibly, although inaccurately, through a natural-seeming analogy with a prior piece of knowledge (here, knowledge about hot liquids, mice, and postal addresses). Such analogies allow a person to make at least some sense of the new situation by likening it to something familiar, and yet it is all done in a spontaneous, unconscious, automatic way, without the person's least awareness of making an analogy.

This stands in stark contrast to the standard image of analogy-making as a process of deliberate construction of mappings between situations. Naïve analogies lead directly to conclusions without there being any consideration of other options, and without any uncertainties or doubts arising. Thus the shaving cream is taken for granted as a dissolving substance by Timothy, the hungry titmice as a type of tree-borne rodent by Janet, and the email address as a place-specific address by Professor Alexander. The presence of these analogies is never felt explicitly, however.

Just like other acts of categorization, naïve analogies lead one to a perfectly reasonable (and thus self-consistent) interpretation of a situation, but they unconsciously assume that one is dealing with a typical member of the selected category. However, the situation may well involve an atypical member or even a non-member of the chosen category, in which case the conclusions reached will be irrelevant and useless. Thus if an email address were a postal address (the most familiar type of address to Professor Alexander), then the question he asked would have been totally reasonable, because the colleague was indeed going far away. Similarly, if a titmouse were indeed a very small rodent, then it would have been reasonable to be surprised by an image of such animals scampering about on tree branches, and Janet's confusion at this image would have been perfectly comprehensible.

As for young Timothy's categorization of his father's shaving scenario, that too, was very reasonable, given his prior knowledge. For an adult, the category *shaving* presumes that there is a blade that will cut some hair and that there is a lotion whose purpose is to make the cutting easier and to reduce the blade's chances of nicking the skin. For an adult, it's obvious and unquestioned that what's doing the cutting is the razor blade. But Timothy saw it quite differently. He was quite right in thinking that if the shaving cream dissolved the small hairs, then some sort of spatula might be useful in getting the shaving cream off his father's face. Even if this interpretation might make adults smile, there was nothing particularly childish about the thought process. No matter how wrong it might have been, it is not an iota less self-consistent than the adult's vision of the process. One can even think of it as a rather ingenious invention, for after all, if such a marvelous hair-dissolving cream did exist, then all our razors would soon be museum pieces.

Generally speaking, naïve analogies have a certain limited domain in which they are correct, and which justifies their existence and their likelihood of survival over years or possibly even decades. This domain of validity can be narrow or broad. This is the case, for example, for children who personify animals. A grasshopper has much in common with a person: it is alive, breathes, moves about, reproduces itself, is mortal, and can be wounded; up to that point, the naïve analogy is perfectly useful. However, unlike what six-year-old children generally think, a grasshopper will not be sad if the person who is taking care of it disappears, and this illustrates one limitation (among many) of the analogy.

As for the naïve analogies at the chapter's start, their strengths and limitations are easy to see. For example, the analogy between a postal address and an email address is valid in a number of ways: both are pieces of data that are structured hierarchically, moving from local to global information (an email address starts with something like a personal name; then comes an at-sign; then something that corresponds roughly to a street address; then a dot and finally something vaguely akin to a state name or country name), and which can be given to certain people and kept secret from others; both are associated with "boxes" where mail accumulates and can be accessed; both are subject to occasional changes; and so forth. For this reason, the analogy is shared by nearly everyone, and of course it is implicit in the shared term "address".

But the analogy has its limitations, too. To send something electronically, the sender needs to have an email address, whereas no such thing is needed in order to send something by post. When one sends an email, usually a copy is automatically kept by the sender, in contrast to postal mailings. An electronic message arrives almost instantly, while a postal shipment may take a few days. If one moves, one gets a new postal address but one can keep one's email address. And so forth. It is therefore understandable that it might fleetingly occur to a person to ask about the new email address of a friend who is moving to a new place (though this is far more likely to happen to a novice email user than to a seasoned one). This is where the limit of validity of the analogy becomes clear.

The naïve analogies made by Timothy and Janet are far more idiosyncratic than the one made by Professor Alexander. To figure out just how common Timothy's naïve analogy is, one would have to make a careful study of how children understand the process of shaving. As for Janet's confusion, it probably strikes you as rather quaint that an adult might envision a titmouse as being a very small mouse that scampers about on tree limbs, but if such naïveté makes you smile, keep in mind that we all live in glass houses, for we have all fallen into traps of the same sort from time to time, by making overly rapid and inappropriate categorizations. Let's take a look at an example from classical popular literature.

Many readers will be familiar with Æsop's fable "The Ant and the Grasshopper", and some will know the seventeenth-century French poet Jean de La Fontaine's rhyming version thereof, called "La cigale et la fourmi" (literally, "The Cicada and the Ant"), of which the opening lines run as follows (in our own translation):

All summer long, without a care,
Cicada sang a merry air,
But when harsh winter winds arrived,
Of food it found itself deprived:
It had no wherewithal for stew:
No worm or fly on which to chew.

The last two lines are unlikely to give most readers pause, but in them, in fact, there lurks a mistaken assumption. To bring this out into the open, let's explore a small variant of them. Suppose La Fontaine had instead written, "It had no wherewithal for stew: / No horse or cow on which to chew". In that case, readers would almost certainly be thrown by the incongruous image of a mere insect having failed to build up a stock of barnyard animals on which to feed. And readers would have been even more disoriented had La Fontaine written, "It had no wherewithal for stew: / No shark or whale on which to chew" or else "It had no wherewithal for stew: / No stick or stone on which to chew." Such lines would have instantly aroused suspicion and bafflement.

Although the closing lines of La Fontaine's actual poem appear to lack any such incongruity, that impression is wrong, since it turns out that cicadas are not carnivorous, and so they have no use for flies or worms. The innocuous-seeming assumption that cicadas feed on small creatures of about their own size is simply erroneous. The famous French biologist and science writer Jean-Henri Fabre, in his autobiographical memoirs entitled "Entomological Souvenirs", observed that cicadas have only a sucking tube with which to nourish themselves. In their larval stage, they get their nourishment from the sap of roots, and when they reach maturity, they suck sap from the branches of various trees and bushes. The plausible-seeming image of what cicadas eat is simply based on a naïve analogy with people or farm animals or other types of insects. La Fontaine's poem would have been far more faithful to the true nature of cicadas if it had run this way:

All summer long, without a care,
Cicada sang a merry air,
But when harsh winter winds arrived,
Of food it found itself deprived:
It hadn't stocked one sip to lap
Of what cicadas crave: thick sap.

If the great La Fontaine was so naïve, perhaps we should not judge Timothy, Janet, or Professor Alexander too harshly.

Naïve Analogies, Formal Structures, and Education

As people move into higher realms of abstraction, naïve analogies, with all their strengths and weaknesses, inevitably become trusted guides. The strengths of such analogies derive from their easy availability in long-term memory, in the form of efficient and ready-to-use mental structures. And their weaknesses stem from the fact that in certain contexts they are misleading. Naïve analogies are like skiers who sail with grace down well-groomed slopes but who are utterly lost in powder. In sum, naïve analogies work well in many situations, but in other situations they can lead to absurd conclusions or complete dead ends.

What the study of naïve analogies tells us about the human mind is of paramount importance for education, and this chapter is therefore oriented to some extent towards the educational payoffs connected with our ideas. A certain number of entrenched ideas about what it means to *learn* and to *know* will be called into question, and some new directions for education will be suggested. Below we list three key ideas that we will discuss in this chapter and the following one.

First of all, ideas that are presented in school classrooms are understood via naïve analogies; that is, children unconsciously make analogies to simple and familiar events and ideas, and these unconscious analogies will control how they will incorporate new concepts.

Secondly, naïve analogies are in general not eliminated by schooling. When teaching has an effect on a student, it usually just fine-tunes the set of contexts in which the student is inclined to apply a naïve analogy. The naïve notion does not displace the new concept being taught, but coexists with it. Both types of knowledge can then be exploited by a learner, but they will be useful in different contexts. And this is fortunate, since banishing naïve analogies from people's minds would be extremely harmful. For example, looking at the world from the point of view of a professional physicist in everyday situations would often be hopelessly shackling. A physicist who sees a glass start to fall floorwards doesn't need to wheel out Newton's laws of motion and his universal law of gravitation in order to figure out what's about to happen. Reaching out to catch the glass that's about to be shattered is a straightforward consequence of pre-Newtonian, non-technical world knowledge. And if two astronomy students are walking hand in hand on the beach admiring the pink-and-orange sunset, they most likely are not in the least thinking about the fact that it's the earth that's turning rather than the sun that is descending, and it's most probable that they are enjoying the beautiful colors and the romantic feelings in much the same way as any other couple would.

Finally, a formal description of a given subject matter does not reflect the type of knowledge that allows one to feel comfortable in thinking about the domain. Humans do not generally feel comfortable manipulating formal structures; when faced with a new situation, they favor non-formal approaches. Learning is thus the building-up not of logical structures but of well-organized repertoires of categories that themselves are under continual refinement.

Familiarity and Entrenchment

Familiarity is crucial in analogy-making for the simple reason that, in order to deal with an unknown situation, one intuitively feels more secure about extrapolating what one knows well than what one barely knows. We don't mean that such choices are made consciously; they take place below one's level of awareness. Thus unconscious analogical processes dominate the way we interact with our environment, forming the very basis of our understanding of the world and the situations we find ourselves in.

Quite obviously, we are not equally familiar with all the things that surround us. Certain notions seem totally natural to us, and others very little so. The notions of *addition*, *equality*, *adjective*, *verb*, *continent*, and *planet* strike us as familiar, while the notions of *partial differential equation*, *Fourier series*, *topological space*, *spinor*, *lepton*, *electrophoresis*, and *nucleotide synthesis* exude considerable strangeness for most of us. This is the case not only for abstract notions of that sort, but also for concrete objects. For most of us, rockets are less familiar than cars, and household robots are less familiar than computers. A category's familiarity has to do with how much one has been exposed to it, with the amount of knowledge one has of it, and with the degree of confidence one has in one's knowledge about it. One feels more comfortable with cars than rockets because one has seen many more of them, because one knows much more about them, and because one has greater clarity about how to get into and out of them, about what their control devices (steering wheel, brake, etc.) will do, and so forth. One is also, for similar reasons, far more familiar with gravity than with electromagnetism.

Familiarity has been studied in certain psychological experiments that explore the effect of the degree of entrenchment of a category. For instance, cognitive psychologist Lance Rips showed that a new piece of knowledge is more readily transferred from a *typical* member of a category to an *atypical* member of the same category than the reverse. For example, if participants in an experiment are told that robins are susceptible to a certain disease, they are likely to conclude that hawks, too, might catch this disease, whereas in the reverse situation, in which they are told that hawks are often afflicted with a certain disease, the chance is much smaller that they will infer that robins will suffer from it as well. The greater familiarity of the category *robin* than that of *hawk* is the source of this asymmetry. This means that the more entrenched a concept is (here, *robin* being more entrenched than *hawk*), the more likely it is to act as the source of an analogy.

A set of experiments conducted by Susan Carey, a developmental psychologist, on children of various ages and on adults as well, led to similar conclusions. Participants were told that all the members of a certain category — *dogs*, for instance — have a certain property, such as possessing an internal organ called an "omentum". They were then asked if members of other categories, such as *humans* or *bees*, were also likely to have this internal organ. For very young children, but not for adults, the existence of such an organ was more easily transferred from *person* to *dog* than from *dog* to *person.* The explanation for this is that in the minds of young children, the concept of *person* is more entrenched than the concept of *dog.* Adults were perfectly happy extending possession of such an organ to any species of animal whatsoever, as long as they had been told that two very different kinds of animals (such as dogs and bees) both possessed it. Young children's thought processes were different; they would decide whether or not to extend a property to a new kind of animal depending on that animal's perceived proximity to either of the two species (*i.e.*, depending on the strength of the analogy in question), and they did not use the superordinate category *animal* to make the analogy, because that abstract category was not sufficiently familiar to them.

Everyday Concepts Versus Scientific Concepts

A narrow vision of learning that we would be happy to see fully eliminated from educational programs presumes that knowledge acquired in school is independent of everyday knowledge. This philosophy would encourage the teaching of new ideas without any reference to everyday concepts, except for the most rudimentary notions, which are impossible to separate from everyday experience. Such a limited view of learning is based on a conception of the mind according to which we would keep facts and ideas that we pick up in school in a separate mental compartment from facts and ideas that we pick up in daily life, and year after year we would increase our school-based knowledge much as we would build a brick house: each brick added would be supported by previously installed bricks, and would in turn support new bricks.

Such a vision is appealing in some ways, since it would simplify the design of school curricula. It would make teaching easier, treating each discipline as an island disconnected from other disciplines, and trying, within any given discipline, to decompose its notions in a logical fashion and to arrange them in a strict, natural order. Teaching any complex idea would thus involve teaching a set of simpler ideas in the same domain, each of which would in turn be taught through yet simpler ideas in the same domain, and so forth.

It is a long-lived myth in the world of education that there is a watertight boundary between two types of knowledge: everyday knowledge, presumed to grow on its own with no need for formal teaching, and formal knowledge, conveyed in schools and presumed to be communicable independently of everyday knowledge. In this naïve view, school is seen as a magical shortcut that allows ideas arduously developed by humanity over thousands of years to be transmitted in just a few years to a random human being.

Another belief that exerts considerable influence on school curricula, particularly in science, is that scientific knowledge is best conveyed in precise, formal terms — especially through mathematical formulas — and that precise formalisms correspond to the way in which knowledge ought to be absorbed by beginners and in which it is manipulated by experts. It is presumed that the initially large gap between a printed formalism and one's internal mental representation gradually approaches zero as one comes closer and closer to expertise.

While we're on the subject of distances approaching zero, the following rather prickly formula expresses the idea that the function $f(x)$ is continuous at the point x_0:

$$\forall\varepsilon\ \exists\delta\ \forall x\ |x - x_0| < \delta \Rightarrow |f(x) - f(x_0)| < \varepsilon$$

(The upside-down "A" and the backwards "E" are shorthand notations for the words "for all" and "there exists", and the arrow "⇒" can be read as "if... then...".) Should one presume that all people who are comfortable with mathematics think of continuity in exactly these terms? Do people fluent in calculus really always imagine and mentally manipulate Greek letters, and is this what continuity means for them? Are all their prior intuitive notions of continuity totally dispensed with after they have absorbed this formula? If so, then in order to educate experts, or even to convey technical notions of this sort to ordinary students, the right way would be to transmit just such formulas, since they presumably embody the most distilled and precise essence of the notions. In particular, one would want the notion of *continuity* to be seen by students as synonymous with the above-displayed formula.

However, this view of education unfortunately conflates the actual way that experts think with the use of dense formalisms designed to capture subtle notions as rigorously and unambiguously as possible. It is the result of a long-standing philosophical assumption, reinforced by the broader culture, which is that logical thinking is superior to analogical thinking. More specifically, this view comes from misguided stereotypes of analogical thinking, which maintain that dependence on analogy, although possibly useful when one is just starting out in a field, is basically childish, and that analogies should rapidly be shed, like crutches or training wheels, when one gets down to brass tacks and starts seriously thinking in the domain. A related half-baked stereotype of analogical thinking is that it is like an untamable wild horse, so unpredictable and unreliable that it must be shunned, even if it might once in a while provide a spark of true insight; thus analogies belong not to the realm of reason but to that of "intuitions", which, being irrational, cannot and should not be taught.

The fluid way that a scientific notion is realized and "lives" in the minds of people who deeply understand it is a very different thing from a formal and rigid symbolic notation, which has been carefully devised to be as concise and rigorous as possible, and the two should not be conflated. As we have just seen, the notion of *continuous function* has a precise formal definition, and yet continuous functions form a mental category having blurry boundaries and members with different degrees of typicality, and such judgments will vary from one mathematician to another. A famous historical case illustrates this fact. Toward the end of the nineteenth century, functions continuous at every irrational point and discontinuous at every rational point were discovered, and this behavior was so unexpected that some renowned mathematicians labeled them "pathological" and strove to banish this "scourge" from mathematics, whereas other mathematicians were extremely excited by the revelation of such a rich new area for research. This episode hardly jibes with the image of mathematics as a frozen body of precise and absolutely objective knowledge. In the following chapter we'll see that even a category as familiar as *number* has blurry boundaries and members with varying degrees of typicality, and seasoned mathematicians can hold different opinions and can argue vehemently about what is and what is not a member of the category.

The unfortunate but widespread conflation of austere formal definitions with the psychological reality of concepts in human minds has had worrisome repercussions in education. One consequence is that it tends to make the educational establishment lose sight of what ought to be its primary goal — the construction of useful, reliable categories. Another consequence is that it tends to favor teaching methods based on prickly formalisms and rigorous deductions rather than methods that use analogies to build up suitable families of categories in intuitive ways.

Our view is very different from one in which logic is seen as central. Indeed, as we stressed in Chapter 4, expertise builds up as categories are acquired and organized. Rather than depending on formal perceptions of situations, people have the ability to treat novel situations as if they were familiar, thanks to categorization. To acquire knowledge in a domain is to build up relevant categories. Analogical thinking is the key to understanding new situations and to building up new concepts, and this holds at all levels, ranging from the shakiest beginner to the most fluent expert. The difference between those two is not their style of thinking — logical for the expert and analogical for the beginner — but the repertoire of categories that they have at their disposition, and the way those categories are organized.

For example, in the case of continuity of a function at a given point x_0, do experts really think in terms of Greek letters and flipped and rotated roman letters? Hardly. Instead, they have vivid visual imagery about the concepts concerned. The epsilons and deltas in the formula are merely helpers in translating that imagery. Thus if one wants the value of $f(x)$ to be very close to that of $f(x_0)$ — in particular, if one wants their discrepancy not to exceed the tiny number ε (this desideratum is expressed by the notation "$|f(x) - f(x_0)| < \varepsilon$") — then continuity requires that there should exist some little zone centered on the point x_0 and whose two edges are at a distance of δ from the center (this zone is expressed by the notation "$|x - x_0| < \delta$"), and throughout this zone the desideratum should be the case.

Thus if one has a mathematically trained mind and is thinking about what continuity means, one envisions a small rectangular box centered on the point $(x_0, f(x_0))$ in the plane. Continuity then will mean that no matter how close one wishes $f(x)$ to be to $f(x_0)$ (this desire translates into the image of a box whose *height* is very tiny, so that all y-values in it are *vertically* close), one can always make the box so *narrow* that the desideratum will hold everywhere inside it. The phrase "no matter how close one wishes" is the visual translation of "$\forall\varepsilon$" (pronounced "for all epsilon"), while the phrase "one can always make the box so narrow" is the visual translation of "$\exists\delta$"(pronounced "there exists some delta"). In the end, it all comes down to the idea of zooming in on the graph of the function at arbitrarily small scales, and thus it has to do with such familiar things as magnifying glasses and microscopes and walking towards objects so that what is blurry at a distance comes into focus. These are the kinds of everyday experiences in which a mathematical sophisticate's understanding of continuity is rooted. Thus we see that to understand the rich concept of *continuity* is to have built up this kind of imagery, and the epsilons and deltas are merely tools, used briefly and discarded quickly, which help one to realize that goal.

In sum, notions taught in schools and colleges are internalized by students not formally, but by means of naïve analogies, which is to say, by means of analogies with familiar categories. These categories can come from any domain, and they tend to come from domains that are not covered in school. Thus the presumption that a given area of knowledge is self-contained is erroneous, because students will inevitably connect every new notion in that area with experiences they've had in other areas of life. Moreover, a naïve analogy, once it has taken root in a beginner's mind, will not go away as time passes; it will stick in the mind as school goes on, year after year, because everyone has a deep need for simple, basic intuitions. Indeed, some naïve analogies are so persistent that one might begin to doubt whether schooling can have any effect at all in certain areas. The goal of transmitting to average pupils subtle ideas that humanity as a whole took thousands of years to discover and absorb is admirable, but it is not something that just happens by itself.

We will see that certain notions that are usually thought of as extremely simple and that are taught in elementary school are, unfortunately, understood through naïve analogies having quite limited realms of validity, and these often-misleading analogies remain entrenched through middle school, high school, and college. Their subliminal effect lives on, long after education should theoretically have thoroughly drummed them out. Thus naïve analogies, thanks to their great robustness, lie behind the thought processes of well-educated adults, even experts; this casts doubt on the idea that education gets rid of them. The stereotypical vision, by contrast, is that people who have deeply absorbed scientific ideas swim in a soup of formally defined notions that obey logical laws; analogies play no part in this vision. That, however, is just wishful thinking. To be sure, scientific education does convey ideas that go beyond everyday experience, and in particular they tend to be more abstract, but scientific ideas are not learned any differently from other ideas — they, too, are rooted in naïve analogies, and this is a universal fact, whether one is talking of beginners, intermediates, or experts.

Where Novelty and Familiarity Walk Hand in Hand

There is no question that the development of computers and related technologies has given rise to a major revolution on our planet. Those who remember the day when there were no computers in homes or businesses, and when there was no World-Wide Web to which to connect such devices, have children who are filled with wonderment at the idea that people could get along at all in such a primitive world. It thus makes sense to suggest that computers were *the* great innovation of the twentieth century. It also might seem that when something so new and so completely unprecedented takes over so widely, it would involve such a radical break with prior concepts that completely new categories would have to be introduced, categories not grounded in any kind of analogy at all.

If computer technologies are so hugely innovative, don't they have to do much more than merely use various analogies to clothe familiar old notions in fancy new wrappings? Actually, the key question is the exact opposite — namely: how could anyone imagine a great burst of creativity taking place without its pioneers building left and right on familiar notions? And how could the lay public possibly absorb the tsunami of radically new ideas without basing their new skills on analogous skills involving notions that were already deeply rooted? The simple truth is that only through homey analogies based on familiar, everyday categories can an average person relate to all the revolutionary new technologies; such analogies are anchors keeping people from getting totally lost in a huge sea of technical terms. The following memo is a good case in point.

> Please find my "Recent stuff" folder and make a copy of the document in it, then put it in my "Paid bills" folder, and also send it to my personal address. Then could you clean up my messy desktop? Would you also look in my "Miscellaneous" folder and toss out anything in it that's outdated or irrelevant? When you're done, please empty the trash, close all the windows, and turn everything off. Thanks a lot!

This note, had it been written in the 1970s, would have referred to a wooden desk, some cardboard folders, documents and sheets of paper, a metal wastebasket, glass windows, and a postal address, and there couldn't possibly have been any confusion as to what was being talked about. Today, however, the paragraph is highly ambiguous, because the supplies and furniture to which it refers have both real and immaterial embodiments, which are so closely analogous to each other that it's impossible to tell which type of stuff is being referred to.

We thus see that computer technology — the domain that represents the greatest break between this century and the previous one — relies on scads of analogies with old-fashioned things, and these analogies are so tight that a perfectly realistic little scenario can be concocted in which there is no way to tell which century's technology is being referred to. Actions that can be performed on physical objects have their counterparts in the virtual world. The desk, the files, the folders, the trash all exist in both worlds and just about anything that can be done electronically can also be done in the old-fashioned way. Even if people today read *documents* that cannot be crumpled, placing them in *folders* that cannot be torn, stacking those on *desktops* that are not made of wood, sending them to *addresses* that have no particular geographical identity, opening and closing *windows* that have no handles, and filling and emptying the *trash* with just a few flicks of their fingers, they are always doing so by analogy with the most familiar situations they know.

The reason that computers have revolutionized our society but not our vocabulary is that these very powerful devices have all been grafted onto familiar categories and have borrowed their verbal labels wholesale. These days, everyone speaks about virtual technologies using terms that wouldn't in the least have disoriented our grandparents. Just as in the good old days, we open our mailbox to get our mail, we send messages to addresses, we send and receive documents, we visit sites, we chat with friends, we do searches for things we want to find, we consult pages, we make new links, we browse, we toss useless files into the trash, we open and close folders and windows, and so forth.

To be sure, there are certain new expressions that would have completely thrown anyone who heard them thirty years ago, such as "hosting a site on a server", "surfing the Web with one's browser", and "installing a firewall against hackers", but even so, these droll expressions, too, bear witness, in their own way, to the fact that all of our electronic activities are still cast in terms of the physical world around us. The humorous juxtapositions audible in these phrases reveal that as a species, we humans are totally dependent on familiar categories in order to adjust to brand-new realities. If *hackers* are using the *Web* to transmit *viruses* hidden inside *spam*, then who's to say we shouldn't use *firewalls* to protect ourselves from such shenanigans?

A more systematic exploration of the lexicon that has grown up around the Web and electronic technologies only serves to confirm our thesis that extremely familiar, everyday physical categories are overwhelmingly the most standard and relied-upon source of analogies for brand-new phenomena. Here, for instance, are roughly one hundred terms all of which could be found in dictionaries published long before computers played any role in society, but which, if read today, give the sense of being technological:

> account, address, address book, animation, application, archive, attach, bit, bookmark, bootstrap, browser, bug, burn, bus, button, capture, card, chat, chip, clean up, click, close, compress, connection, copy, crash, cut, delete, desktop, disk, dock, document, drag, dump, entry, erase, figure, file, find, firewall, folder, font, forum, gateway, hacker, highlight, history, home, host, icon, image, input, install, junk, key, keyboard, library, like, link, mail, mailbox, map, match, memory, menu, mouse, move, navigate, network, open, output, pad, page, palette, paste, peripheral, point, pop up, port, preference, preview, print, process, program, quit, read, reader, record, save, screen, scroll, search, select, send, server, sheet, shopping cart, shortcut, site, sleep, style, surf, tab, thumbnail, tool, trash, turn on/off, virus, wall, web, window, workplace, worm, write, zoom...

One might at first be inclined to suppose that many of these terms are deceptive — that is, that they are not really based on helpful analogies to older ideas, and that their new and old meanings are no more intimately related than are the two different meanings of "race" ("a running competition" and "a subspecies") or the two different meanings of "number" ("a quantity" and "more anesthetized"). But this idea is far from the case: the semantic connections linking the new technological meanings to the meanings that originated in far older domains are in every case quite straightforward. Aside from a couple of terms such as "mouse" and "chip", where the analogy between the everyday notion and the technical one is little more than a visual resemblance, what indisputably lies behind each of these familiar terms is an abstraction linking the new and the old uses.

Sometimes one finds much charm in definitions that were drawn up in a long-gone era when today's technologies could not have been imagined by anyone. For example, the 1932 Funk & Wagnalls New Standard Dictionary, previously quoted in Chapter 4, defines the words "browse", "folder", and "hacker" as follows:

browse. *trans. verb* To feed upon, as twigs, grass, etc.; nibble off; also, sometimes, to graze; as, the goat *browsed* the hedge.

> The fields between
> Are dewy-fresh, *browsed* by deep-uddered kine.
>
> TENNYSON *The Gardener's Daughter*, stanza 3.

intrans. verb To eat the twigs, etc. of growing vegetation; graze.

> Wild beasts there *browse*, and make their food
> Her grapes and tender shoots.
>
> MILTON *Psalm XXX*, stanza 13.

folder. 1. One who or that which folds. Specif.: (1) A flat knife-shaped instrument for folding paper. (2) A map, time-table, or other printed paper so made that it may be readily spread out. (3) A leaf, as one containing a map, larger than the other leaves of a book into which it is folded and secured. (4) A folding-machine. (5) A folding sight on a firearm. (6) An addition to a sewing-machine which folds the material before it is sewn.

hacker. A tool for making an incision in a tree to permit the flow of sap.

It's rather astonishing to think that less than a decade before the invention of the digital computer, in an era when radios and movies, cars and airplanes, telephones and record players already existed and were even commonplace, the three concepts above were conceived of in ways that today strike us as incredibly concrete, limited, and quaint; indeed, if someone from back then could visit us today, it would take us a little while to explain to them the modern technological senses of these words. There is no doubt that the old and new meanings are cousins, and yet it would take a nontrivial crash course and some major leaps of the imagination to spell out some of the analogies that bridge the gap, but once one sees the common abstract core, it is very clear.

Consider, for instance, this possible contemporary definition of "hacker": "A person who invades a data base to permit the illicit flow of information". The analogy with the 1932 definition has deliberately been made salient, and yet the conceptual gap is still a huge mental stretch, somewhat reminiscent of the vast conceptual gap between California's rural and picturesque Santa Clara Valley in the post-Depression years, dotted with scenic orchards and small farms, and what it became just a few decades later: the ultra-modern bustling metropolitan area known as Silicon Valley, jam-packed with high-tech firms, criss-crossing freeways, upscale housing developments, and more Thai, Indian, and Chinese restaurants than you can shake a stick at.

Your Trip Has Been Placed in Your Shopping Cart

The emergence of modern computer-oriented meanings for many words not only has given rise to new categories, but has helped to make the abstract essences behind the terms in the above list become clearer (as was discussed in Chapter 4). As we already saw in the case of the word "desk", the original meaning of a term gives rise to an analogical extension, allowing the old concept to apply to virtual (or "software") entities; this extension in turn simplifies and refines the old category, in the sense that some of its once-central qualities are now seen as superficial and thus merely optional. For instance, the virtual concept of *address* has made us understand that *physical location*, standardly tied tightly to a postal address, is not crucial to this concept, but that what matters, in fact, is *accessibility through a symbolic label.* When a message is sent to an electronic address, what matters is simply that it should reach the person concerned, and the geographical whereabouts of said person are of as little import as whether a desktop is made of wood or pixels or whether it has drawers or not. In short, the recent appearance of a digital version of the category *address* has made us see that geography is not relevant to the term, even though up till then, geography had been absolutely indispensable to the notion.

Certain curious phrases that crop up now and then on Web sites, such as "Your trip has been placed in your shopping cart", reflect this tendency of categories' essences to be revealed ever more clearly as categories get analogically extended. Thus an on-line shopping cart shares with a physical shopping cart the property of accumulating potential purchases, about which one can change one's mind at any moment until one comes to the "checkout stand". In real life (*i.e.*, in "the good old days" before computers became a central reality in our lives), it would seem absurd, to say the least, to speak of "placing a trip in one's shopping cart", but what would be absurd for the old and narrow categories of *trip* and *cart* need not be so for the new and broader categories, because once a *trip* has become a *potential purchase*, and once a *shopping cart* has become a *repository for potential purchases*, then the phrase "to place a trip in a shopping cart" seems perfectly innocuous — indeed, quite sensible.

Finding a new category that gracefully combines the pre-technological and the technological versions of a concept often requires one to jump to a relatively high level of abstraction, as is the case, for example, for *firewall*, *hacker*, *peripheral*, and *port.* Thus in the on-line world, a *firewall* is a protection against *hackers*, whereas in a more traditional context, the phrase "a firewall against hackers" has little or no meaning. Similarly, in a computer context, it makes perfect sense to speak of "plugging a peripheral into a port", whereas in the pre-computer world, such a phrase would merely have sounded like meaningless gibberish.

And yet there are other categories, such as *address book*, *keyboard*, *move*, *screen*, *delete*, and *send*, which scarcely seem to have been extended at all to incorporate their new aspects. One might think that these categories have emerged unscathed from all the technological upheavals, that they have withstood the earthquake without being touched in the least. However, the truth of the matter is a bit subtler than that. Thus, for instance, an electronic *address book*, unlike its paper forebear, is not a concrete object. One "writes" entries in it not with a pen but by typing (or possibly by speaking aloud!), and it allows one to do certain things in a flash that would be very laborious with an old-fashioned address book, such as finding a person given their telephone number.

Asking whether a virtual address book *belongs to the category* called "address book" or simply *is analogous to* an old-style address book is a question that needn't be answered, because the dichotomy it presumes is a false one. Belonging to a familiar category and being analogous to a familiar thing are not black-and-white matters, and should not be thought of as opposites, or as excluding each other; both have sliding scales (or shades of gray) that depend on both perceiver and context; indeed, *strength of category membership* comes down to nothing but *strength of analogousness.* The gradual and natural extension of technological terms provides an excellent illustration of the fact that analogy and categorization are just two sides of the same coin.

What about the concepts expressed by verbs such as "move", "erase", and "send"? It might seem that their meanings haven't budged an inch as a result of the computer revolution. And yet, would you say that "moving" a file from one folder (on a hard disk) to another is *exactly the same thing* as moving a paper file from one cardboard folder (in a wooden drawer) to another? Or would you say that highlighting (on a screen) the set of pixels forming an alphabetic character and then hitting the "delete" key is *exactly the same thing* as quickly and forcibly rubbing the pink end of a pencil back and forth across some marks (on a piece of paper) until they are barely visible any longer? And would you say that sending an electronic message is *exactly the same thing* as sending a letter via "ordinary" mail? It's easy to forget that when one sends a material letter, one has to relinquish it physically, whereas when one sends a message electronically, the original remains on one's hard disk. And so we are reminded once again that even when it comes to terms that, on first blush, seem completely unextended by their computer versions, the truth is that the categories in question have indeed been broadened, and it is only thanks to analogy-making with previously familiar everyday categories that these extensions could take place.

The Best Interface is No Interface at all

Do the hundred-odd terms given a few pages back, all resulting from analogical extensions, coexist with all sorts of other new terms that did *not* come from analogical extensions? The fact is that whenever a new technology comes along, the standard way of devising a new set of terms that work naturally with it is to borrow pre-technological terms and to rely on the predictable naïve analogies that most people tend to make. Anyone who doubts this should just listen to computer people talking and take note of terms that didn't exist fifty years ago. They will discover that everyday down-to-earth words are ubiquitous, while terms that are unique to the new technologies and that are nowhere to be found in old dictionaries, such as "motherboard" or "pixel", are not all that frequent. To be sure, if one were to transcribe a technical conversation between two computer specialists, one would find a rich harvest of acronyms and other narrow technological terms, just as one would for any specialized discipline, but there can be no doubt that ordinary people, in speaking about their computers, are constantly exploiting terms that hark back to a day long before anyone could have dreamed of the abstract uses to which those terms would be put, decades or centuries hence.

Thus concepts from the world of computers now permeate our daily lives because our down-to-earth concepts, through hundreds of naïve analogies, have permeated the technology itself. These naïve analogies, building as they do on extremely familiar categories rooted in mundane daily activities, allow us to endow complex and mysterious technological entities with all sorts of simple and unmysterious properties, and they do so at minimal cognitive cost. Terms based on naïve analogies catch on easily because they naturally bring out qualities that otherwise would be highly elusive.

Thus thanks to our naïve analogies, we all speak with ease and accuracy of "placing objects on the desktop", of "inserting documents into a folder", of "opening or closing a window", of "moving, copying, or throwing away a file", and so forth — and we didn't need a course or an instruction manual to gain this skill. Moreover, this down-to-earth lexicon for describing the behavior of abstract technological entities (such as *desks*, *files*, *windows*, and *documents*) is not just a rough-and-ready set of linguistic tools, serving solely to help novices get their feet wet but then to be summarily dropped; to the contrary, even the most technically savvy people speak in this concrete manner. They, too, open and close windows, files, and folders — and when using such phrases, they feel they are expressing themselves perfectly straightforwardly and non-metaphorically.

Experts in the field of human–machine interface design have stated that the best possible interface should be invisible — indeed, that "the best interface is no interface at all". Such assertions, stressing the importance of natural and intuitive interfaces, mean essentially that designers should always try to exploit analogies to familiar things. Only if this is done will the interface become "transparent", which means that users will feel as if they are manipulating everyday objects, a feeling that frees them up to concentrate on their main goals. Interfaces designed in this felicitous manner convert the computer into an easy-to-use tool for accomplishing a particular type of task. Instead of working up a sweat figuring out how to use the tool, one simply concentrates on the task itself.

Interfaces carefully based on analogies to familiar activities do not suffer from the great awkwardness of poorly-designed interfaces. Donald Norman, who is not only a distinguished cognitive psychologist and error-collector, -classifier, and -modeler, but also a pioneer in human–machine interface design, has stated this idea succinctly: "The real problem with the interface is that it is an interface. Interfaces get in the way. I don't want to focus my energies on an interface. I want to focus on the job... I don't want to think of myself as using a computer, I want to think of myself as doing my job."

The Naïve Side of Naïve Analogies

Although indispensable, naïve analogies that help us to relate to new technologies have their limitations, because members of the new category will sometimes behave differently from those of the older, more familiar category. In such cases, the naïve analogy is likely to lead one down a garden path. After all, when one depends on a naïve analogy, one does so, by definition, naïvely — that is, lock, stock, and barrel. For better or for worse, the naïve analogy is one's only guide — and on occasion it will mislead. In a word, we are back again in the land of categorical blinders.

When the differences between a virtual category and its old-fashioned analogue do not involve the categories' most central aspects, there generally is no problem. For example, it's obvious that virtual desktops, files, and folders have no volume or mass, cannot get dirty, and aren't subject to wear and tear. This is because categories such as *virtual file* and *virtual folder* are immaterial, just as is the category *virtual desk* (or *desk*$_2$).

It's another matter, though, when the discrepancies involve central aspects of the familiar category; the naïve analogy can then give rise to serious confusions. For instance, at one time Apple systems required users who wished to eject a disk to drag its icon into the trash. Many users balked at doing so: it struck them that there was a fair chance that taking such an action would delete all the data on the disk. The analogy behind this reaction was so natural and irrepressible that even experienced users couldn't help but feel a twinge of uncertainty when dragging their disk into the trash, as if this operation, no matter how many times they'd done it before, was still just a tiny bit risky: "Uh-oh… Could it be that this time all my files will get erased and will be lost forever?" It's as if they were imagining that the computer itself was thinking by analogy, and that it might get confused like a human and, by error, throw all the disk's contents away. ("Oops! Sorry about that! I got a little distracted and when I saw you'd dragged the disk's icon into the trash, I just tossed everything on it out. Silly me! I'd forgotten that for disks I'm not supposed to do that, but should just eject them.")

In light of this common fear, should one infer that Apple's user-interface designers suffered from a fleeting "senior moment" when they decided that dragging the disk icon into the trash was the natural way to say, "Please eject the disk"? Not really. They were simply presuming that users would easily jump to a higher level of abstraction than *obliteration* when they dragged the disk icon into the trash — say, to a concept such as *getting rid of something no longer relevant*. However, as it turned out, this presumption overestimated the typical user's mental fluidity. When the designers finally realized this, they removed the ambiguity in subsequent versions. Nowadays, whenever the icon for a disk is brought near the wastebasket icon, the latter magically mutates into a different icon denoting ejection rather than destruction.

Another striking example of the failure of a naïve analogy involves the virtual desktop. Usually, the hard disk is represented as sitting *on* the desktop (or possibly in the workspace, which is itself on the desktop). But the fact of the matter is that all the data in memory are stored on the hard disk, and this includes the entire desktop. There is thus an apparent paradox, with the disk being *on* the desktop and the desktop being *inside* the disk. What sense does it make for A to be on top of B, while B is at the same time inside A? This shows that at times naïve analogies cannot fully do the job they were intended to do. To be sure, we humans can live with small inconsistencies on our computers just as we do in life in general, but sometimes this little paradox does cause genuine confusion, as when a user wants to locate the desktop in the computer's memory. Interface designers, after recognizing the possible confusion due to this naïve analogy, eventually made a patch. Today the hard disk is no longer shown as being located *on* the desktop; instead, it is simply accessible through it. Still, it took designers some twenty years to take care of this small problem.

The World of Computers Yields Analogy Sources for Itself

Not all naïve analogies designed to help people use computers more easily are rooted in pre-computer experiences, because today computers are familiar enough that some of their best-known properties can themselves be exploited as sources for naïve analogies. Indeed, something that was once understood only by analogy can eventually become familiar enough that it can act in turn as the source for new analogies. This happens not only with technological devices, but in all aspects of life. For example, sound waves, which were first understood by analogy to water waves, became in turn, many centuries later, the analogical basis for understanding light waves.

The world of computers is thus starting to yield sources for its own analogies. For example, the notion of a *floppy disk*, which for many years was the standard device on which one saved all one's files, was originally understood by analogy to a vinyl record. But once floppies had become familiar to all computer users, thanks to their widespread use in the 1980s and 1990s, they became the source for new analogies. And thus even today, the icon that stands for saving a file is frequently a stylized picture of a floppy. This is ironic, since writing a file onto a floppy disk is almost unheard of today; floppies have long since been supplanted by internal and external hard disks, CDs and DVDs, flash drives, and so forth (each of these, after a brief day in the sun, gave way to newer technologies). Although the floppy-disk icon is still found in some software, it is a remnant of a bygone era — a bit like the stylized pictures of ancient bicycles with huge front wheels that were sometimes used, not too long ago, to indicate bike lanes in the U.S. — and it's a safe bet that this icon will soon go the way of floppies themselves. As the objects themselves are no longer around, the concept of *floppy disk* is approaching extinction. Children who see the square icon don't know where it comes from, and the recent tendency is to make the icon for file-saving look like a hard disk instead.

When the Virtual World Helps Us to Understand the Real One

Because of their constant and increasing presence in our lives, computers and related technologies have recently turned into a rich source of categories that, through their great familiarity to us, can serve as rich new sources for analogies. This is a curious twist, since computers, for most of their brief existence, have standardly been explained through analogies to phenomena in the physical world, but today, the reverse is happening: that is, physical things are coming to be described through analogies to phenomena in the world of digital technology. For example, a recent television ad for an SUV crowed, "Think of it as a search engine helping you to browse the real world!" Who would have predicted such a reversal of roles? This tendency will surely increase, ushering in unpredictable changes in society.

Take the concept of *multitasking*. This was a clever invention of the 1960s allowing a computer to execute several distinct processes concurrently by breaking each process into tiny steps and doing a single step of process #1, then a step of process #2, and so on, thus interleaving the various processes so finely that, to all appearances, they are all being carried out simultaneously. But in our lives today, the concept of *multitasking* is routinely applied to human beings and their activities. Thus, sentences like "As a single mom, believe me, I'm constantly multitasking!" and "Every morning on my way to work, I sip my mocha, yak on the phone, savor the scenery, and listen to music, all while driving my car — I'm such a multitasker!" are standard parlance. Indeed, the computer origin of the term has begun to fade out of view. Here are definitions of it taken from two dictionaries a couple of decades apart:

Webster's New World Dictionary (1988):

> *Computer science*: the execution by a single central processing unit of two or more programs at once, either by simultaneous operation or by rapid alternation between the programs.

The American Heritage Dictionary (2011):

> **1.** The concurrent operation by one central processing unit of two or more processes.
> **2.** The engaging in more than one activity at the same time or serially, switching one's attention back and forth from one activity to another.

As this comparison shows, the term was purely technical in the 1980s, whereas today many people use it fluently for everyday activities, having no idea that it came from computer science or even has *any* technical application! There was a transition period where the analogical extension was explicitly felt by people who knew they were stretching the concept, but after a while, the stretching had been accomplished, and the term, stripped of its original technical connotations, entered the public lexicon.

Another computer term that has been imported into daily life is "to interface", which originally meant adapting two pieces of hardware or software so that they would work together seamlessly, but which is now used in such nontechnical phrases as "The gay community needs to interface much more with the black community."

The term "core dump" was used for decades to mean a printout of the entire contents of a computer's main memory (once called "core"); this was a somewhat desperate measure that could help in pinpointing very recalcitrant bugs. But it was analogically extended to daily life, with the result that today nontechnical people say things like, "Sorry for going on and on so long — I didn't mean to give you a brain dump!" What is preserved is the abstract idea of visibly or audibly outputting a huge amount of information that normally is invisibly stored in some kind of memory device.

Another computer concept that has recently enjoyed considerable popularity as the source of casual conversational analogies is *cut-and-paste*. Thus, a television newscaster describes a political candidate's speeches as being "cut-and-pasted from her previous speeches", a newspaper describes attempts to cut-and-paste Silicon Valley into various European countries, and a book reviewer criticizes a new book by saying, "This book is just a cut-and-paste of other books on the same subject; I learned nothing new from it."

The notion of *debugging a computer program* is yet another fertile source of imagery for everyday life. Thus a salsa dancer says, "I'm working on debugging my Latin hip motion — my hips always move in the wrong direction", while a Chinese teacher says, "You really have got to debug your tones — they're all mixed up!"

And finally, a few miscellaneous examples of the insidious manner whereby computer terminology worms its way into everyday discourse. One business executive confides to another: "I'd really like to have your input on this matter." A medical-school student sighs: "Every so often, I just need to disconnect from this crazy routine." An advertisement exhorts: "Reboot your brain with a caffeine nap!"

Taken together, these examples provide another pillar of support for the thesis of the first two chapters that word choices are made via analogies, and that word-choice analogies are usually (though not always) made unconsciously. Take the computer-science term "multitasking", for instance. In the first few years of its existence, the idea of applying it to some kind of *human* behavior was a choice available only to computer-savvy people, and for them the computer-to-human analogical bridge was built with deliberateness and delight. Pushing a word's range outwards is fun (because making inventive analogies is fun). Eventually, however, this extended sense of the term leaked, by osmosis, into the much broader community of non-computer-savvy people, and at that point, using the term didn't require building an analogical bridge linking a human behavior to an arcane computational trick (after all, these speakers knew nothing of that trick); the analogical bridge was simply to *human* behaviors that had been labeled by the catchy new term. But no matter which set of speakers we focus on — technically savvy or technically non-savvy — we see that it's always analogy in the driver's seat, always analogy that is handing words to speakers, and often handing them words on such convenient silver platters that, to them, their word choices seem to have taken place instantly, naturally, effortlessly, and without any help from analogy-making. That silver platter was just magically *there*. Of course, that's an illusion — just another case where the human mind, not surprisingly, fails to be aware of the seething activity constantly going on below the surface of its familiar linguistic behavior.

Technomorphism — an Analogue to Anthropomorphism

Sometimes it's not the choice of a computer-oriented *word* that betrays the tendency of technical ideas to slip into everyday thinking, but simply a computer-oriented *habit* that unconsciously pops up and tries to insinuate itself into an unfamiliar new situation. After all, when one is constantly dealing with technology and using the Web on a day-in day-out basis, one can't help starting to see this familiar old world with fresh new eyes. This can happen, for instance, when one is trying to find a favorite passage in a book. While randomly flipping through its pages, seeking the desired passage much like someone seeking a needle in a haystack, one can easily grow very frustrated, since one is painfully aware that if one only had an on-line version of the book, it would be a piece of cake to find the desired passage.

The following anecdotes illustrate the very human and very natural — indeed, irresistible — tendency for computer-based concepts to pop to mind in everyday situations as if they applied, when in fact they don't apply at all (or at least not yet!).

A woman was driving her ten-year-old son, a video-game addict, to another city. After several hours, the boy started complaining of being bored to death. His mother replied, "Would you please stop your whining? I've got to concentrate on the road." Her son shot back, "But it's not fair — you get to *play* the whole time!" For the boy, who had *virtually* driven *virtual* cars for several years, the act of *real* driving was exactly like playing a video game, and thus to him, it seemed that his mother had the good fortune to be having lots of fun while he was unable to do so.

A little girl walked into a room where two digital photo frames were sitting on a buffet, each of them periodically flashing one image, then another, then another, and so on. Between them was an old-style frame containing a standard still snapshot. She commented, "Look, Daddy, the frame in the middle is broken!" For her, the old-style frame was automatically seen as a member of the category *digital photo frame*, and as such, it was clearly a defective member of the category.

An eight-year-old girl was in her family's car when a heavy rain started to come down as they were driving through the countryside. She remarked, "Watch out, Dad, you can't see anything on the screen any more!"

A computer addict confessed that when he was sitting at his computer and a fly landed on the screen, his knee-jerk reflex was to try to get rid of it by clicking on it and then dragging it off the screen.

When you're at home and can't find an object, you often wish you could simply search for it just as one can search for things on one's hard disk or on the Web, or for that matter on one's cell phone. If only you could type in a couple of key words, you feel, you could instantly retrieve any lost article! Take Alice, for example. She was very groggy when she woke up, and badly needed her morning coffee. While drinking it, she got a phone call from her mother, so she set the cup down. When she hung up, she'd forgotten where she'd placed it, so she blithely picked up her cell phone and started typing the word "coffee" in the "Search" slot of her list of contacts. Then there's Bob (another example we found on the Web), who for five straight days had been working all day long at his computer, trying in desperation to finish a grant proposal before the deadline. At some point, he misplaced his glasses and realized that he couldn't squelch the desire to type a couple of words into a search engine to find them. Examples of this sort are a dime a dozen on the Web.

Unfortunately, no tool at that level of sophistication exists to help us locate lost items in the physical world, and for that reason the on-line world's concept of *search*, which by now is second nature to nearly everyone, has become the source of analogies in the physical world, rather than the other way around. It is very tempting to think that objects that we've misplaced over and over again, such as our keys, our wallet, our checkbook, or our glasses (or their case), should be just as easily coaxed to chime out their presence to us as is our misplaced cell phone. In short, we'd like to be able to summon anything that we can't find and have it instantly chirp back at us, telling us just where to find it.

By now, the Web and cell phones have given us the sense that pretty much anything should be within easy reach at any time. Wherever we are located, just about anything that matters to us is within clicking range, emailing range, cell-phoning range, or text-messaging range. This gives us the illusory feeling that anything of any sort should be available to us instantaneously.

The great ease of obtaining things from afar these days has had the effect of reinforcing the sense of presence of people who are in fact absent. Some people, for example, have started to feel ill at ease about making the slightest critical comment about anyone, any time their cell phone is nearby — as if the person being criticized might somehow overhear the comments even without any call having been initiated. This scenario is a bit reminiscent of living in an apartment with very thin walls, where one never knows for sure whether one might be overheard by others in neighboring apartments. Along much the same lines, certain people, when in a chat room on the Web, often start whispering to physical people who are physically near them, as if the virtual occupants of the virtual room could overhear every word they say aloud. In short, some computer addicts develop the jittery feeling that the physical world all around them is populated by virtual people who are able to overhear conversations about themselves.

One very useful feature of computers is that they offer us all sorts of chances to undo mistakes that we have made. Thus when we are using a word processor or a photo editor, it seems only reasonable that we should instantly be able to reverse any action that we've done — including massive deletions carried out by accident — and we get very used to such luxuries, taking them for granted. The habit then spills over into other domains of life and becomes an expectation, and when we realize there is no "undo" button in nearly any of them, we become frustrated. Here are two anecdotes illustrating this tendency:

> A student who was cooking a cake had put too much flour in her dough, and she would have liked to go back a few moments in time and undo her mistake. All of a sudden, she caught herself wishing that she could just push the "undo" key that she was so accustomed to.

> A teen-ager who was plucking hairs from her eyebrows suddenly realized that she'd done more than she intended, and she thought to herself, "Oh, no problem, I'll just revert to the older version."

Not only the act of "undoing" but many other types of frequent computer actions can become such strongly ingrained habits that they wind up shaping the way people see and behave in the material world, as the following set of anecdotes demonstrates:

> A teen-ager was looking at a photo in a magazine in which she recognized some people but not others, and she said, "My first reflex was to wonder, 'How come they're not tagged?' Then I remembered I was looking a *magazine*, not a Facebook page."

A man reported that one time when driving he found himself using two fingers on his rear-view mirror in an attempt to blow up the image. A woman replied that "the same thing" had happened to her but on her flat-screen television, and yet another said "exactly the same thing" had happened to her when she was looking at her reflection in her bathroom mirror.

An avid moviegoer said that she wanted to go to the restroom during a film and for a split second she found herself trying to grab her mouse to stop the film momentarily.

Another film buff confessed, "When I was in the movie theater, I tried to jiggle my mouse a little bit to see how many minutes were left in the movie."

A schoolboy said that during a quiz, he wasn't sure whether he had spelled a certain word right, so he waited for a few seconds to see whether what he'd just written in pencil on paper was going to get underlined in red.

A college student described how she was cramming for an exam and had finally, with great effort, figured out how a certain complex biochemical reaction worked. At that point, she was sitting in front of her computer, and she mindlessly hit the pair of keys that she always would hit when saving a file.

These days, amusing mental contaminations due to these types of crossed wires involving computer concepts and pre-technological concepts are a dime a dozen, and there are Web sites aplenty "where" (if we may analogically borrow this concept from the physical world) people gleefully report and laugh at their own gaffes of this sort.

Some Equations Are More Equal than Others

Now that we have taken a careful look at the ways that naïve analogies originating in recent technology have insidiously invaded our lives, we can turn back to the field of education, focusing specifically on the role that naïve analogies play in how children pick up basic mathematical concepts in school.

Is the equation "3 + 2 = 5" completely clear? Is there just one way to understand it? Do all educated adults understand the equals sign in the same way? Theoretically, an equation symbolizes a perfect equivalence or interchangeability; that is, an equals sign tells us that the two expressions flanking it stand for one and the same thing. The notion of equality, when described this way, seems so simple and straightforward that it would seem hard to imagine any other way of interpreting it. And yet there is another side to the notion of equality, and it comes out of a naïve analogy that we will call "the operation–result analogy".

In this alternate interpretation, the left side of an equation represents an *operation*, while the right side is the *result* of the operation. This is a naïve analogy in which equations are tacitly likened to processes that take time, and it crops up in situations that have nothing to do with school or mathematics, and which influence everyone, including young children. For instance:

point at + cry = obtain a desired object
vase + knock over = shards of glass on the floor
mud + hands = mess
DVD + DVD player + remote control = watch a movie
chocolate + flour + eggs + mix + bake = cake
cheese + lettuce + tomato + bread = sandwich
$3 + 2 = 5$

Here, the equals sign is a symbol that links some sort of action in the world to its outcome, and it can be read as "gives" or "yields" or "results in". When seen that way, "$3 + 2 = 5$" is not the statement of an equivalence at all; rather, it expresses the idea that the process of adding 3 and 2 results in 5.

The ideas of *interchangeability* and *operation–result* are different. The second point of view clearly embodies an asymmetrical conception of equations, in which the two sides play different roles, one side always standing for a process and the other always representing its outcome. To write "$5 = 5$" would be incompatible with this viewpoint, since no process is indicated. Likewise, writing "$7 - 2 = 8 - 3$" is also troublesome, since now there is no result. And lastly, writing "$5 = 3 + 2$" would be disorienting, because the operation and its result occupy the wrong sites. Indeed, many first- and second-graders understand equality in just this fashion, insisting that "$5 = 3 + 2$" is "backwards", and that "$7 - 2 = 8 - 3$" makes no sense because "after a *problem* there has to be an *answer*, not just another problem". Some even balk at "$5 = 5$", replacing it with something such as "$7 - 2 = 5$".

The operation–result naïve analogy guides children before they encounter the concept of *equivalence*, because the notions of a *process* and its *result* are familiar even to toddlers. These notions are close cousins to the notions of *cause* and *effect*, as well as to the idea that certain means have to be used to reach certain ends.

Although today's children may acquire a fairly decent understanding of equality in elementary school, coming up with the symbol "=" took a long time for humanity as a whole. A symbol for equality in mathematics first appeared only in the year 1557, in a book by the Welsh mathematician Robert Recorde. He wrote:

> I will sette as I doe often in woorke use, a pair of paralleles, or Gemowe lines of one lengthe, thus : ═══, bicause noe 2. thynges, can be moare equalle.

The word "gemowe" means "twins", and the "twinnedness" of the upper and lower horizontal lines was intended to symbolize the general idea of equality. The fact that a symbol for equality took such a long time to occur to anyone, even though mathematics had existed for at least two millennia, reveals that it is far from a self-evident notion.

Although for many adults today the idea that "equality equals equivalence" may seem obvious in a mathematical context, it doesn't follow that the operation–result view of *equality* has disappeared from their minds. In fact, people often write down, and read aloud, equations in a way that reflects their unconscious understanding. For instance, in reading "4 + 3 = 7", many people will say "four plus three makes seven", whereas for "7 = 4 + 3" they might say "seven is the sum of four and three". If education always resulted in equations being seen as statements of interchangeability, then by the end of high school, the operation–result view of the equals sign would surely have disappeared for once and for all. The order of the two sides in an equation would be completely irrelevant, and both ways of writing an equation down would elicit exactly the same commentary. However, this turns out not to be the case. Let's take a look at some specific cases, starting out with some that are very remote from mathematics.

Naïve Equations in Advertisement

It's standard practice for advertisements to appeal to the child inside each of us rather than to the budding mathematician. Here are a few examples of "equations" culled from real ads:

buy two items = 50% off on the second one
buy a pair of glasses = get a pair of sunglasses for free
buy a loyalty card = free home delivery for a year
buy a TV set = a DVD player for just $1
buy any pizza = get another one free

Just to convince ourselves that interchangeability is not the idea behind these equations, let's flip them around. As you will see, the resulting "equations" sound utterly silly, even nonsensical.

50% off on the second one = buy two items
get a pair of sunglasses for free = buy a pair of glasses
free home delivery for a year = buy a loyalty card
a DVD player for just $1 = buy a TV set
get another one free = buy any pizza

As these examples reveal, people's first glimmerings of understanding of the equals sign come from the naïve analogy of an operation followed by a result, and even if the concept of interchangeability gains some ground in the course of twelve years of education, the *operation–result* viewpoint is never fully eradicated. It can always be coaxed out of dormancy when the right cues are presented. We thus see that education does not eliminate the first naïve ideas about a mathematical notion — even one that we tend to think of as completely trivial because it is taught in elementary school, when children are only six or seven years old. In childhood and even when one is fully grown, the naïve view coexists with a different view, which is instilled at school but which is also dependent on a pre-existent and familiar notion: that of *same thing* (that is, *identity*). The transition from the earliest viewpoint (operation–result) to the more sophisticated one (identity) does not obliterate the earlier viewpoint, which remains triggerable, and on which one frequently relies on a day-by-day basis, sometimes even in scientific contexts, such as mathematics or physics, as we shall now see.

Of Equations and Physicists

For physicists, the most fundamental formula of classical mechanics is doubtless Newton's second law, which describes how a force affects the motion of an object. The basic idea of this celebrated law is compatible with the naïve analogy that says that one side of an equation should represent a *process*, with its other side representing the *result* of that process. In this case, the process (ideally occupying the left side of the equation) would be the action of a force of size F on a mass of size m, and the result (ideally on the right) would be an acceleration of size a imparted to the mass. Rendered symbolically, this yields the equation "$F/m = a$". Unfortunately, though, Newton's law is virtually never written this way. Instead, it is almost always cast as follows: "$F = ma$". This famous formula is quite confusing to many students, since neither side of it cleanly symbolizes either the process or the result. The alternative notation "$F/m = a$" encodes the naïve analogy much more clearly, and would therefore be easier for students to relate to, but it is seldom if ever found in textbooks. From a purely *logical* standpoint, these two versions of Newton's law are completely equivalent and interchangeable, but from a *psychological* and *pedagogical* standpoint, they certainly are not.

Luckily, physicists are often sensitive to such psychological pressures, and most of the time they try hard to cast their equations in the form of clean and clear cause-and-effect relationships, with one side giving rise to the other side. Take, for instance, the first of Maxwell's four equations for electromagnetism:

$$\text{div}\, \boldsymbol{E} = 4\pi\rho$$

where $\boldsymbol{E}$ represents an electric field, ρ represents electric charge density (basically, a description of how much electric charge there is in each point of space), "div" stands for a certain operation in differential calculus called the "divergence", and π is the familiar circular ratio 3.14159...

This formula is universally seen by physicists as saying, "A certain distribution of electric charges in space (the *cause*) always gives rise to a certain pattern of electric fields in space (the *effect*)." For historical reasons, however, the cause (the charge distribution) is conventionally placed on the *right* side of this equation and the effect (the electric field) on its *left* side, thus reversing the usual operation–result order. Why do physicists always write it in this flipped fashion? That's hard to say, but basically it's just a harmless "professional deformation". In any case, Maxwell's first equation intuitively embodies a physical cause-and-effect relationship, with the cause on the right side and its effect on the left side. (In fact, all four of Maxwell's equations embody similar cause-and-effect relationships, and they all have this same kind of right-to-left causal flow.)

There is, however, another way of looking at Maxwell's equations. For concreteness' sake, let's once again consider the first one, as shown above. It says that if you calculate the divergence of the electric field, you will obtain the charge density. Now such a calculation can also be seen as a kind of cause-and-effect or process-and-result relationship, wherein certain quantities are fed into a calculating machine that churns for a while and eventually outputs new quantities. Seen this way, the "cause", or initial event (namely, the feeding of input values into the computing device), is always on the left side, while the "effect", or subsequent event (namely, the number that the device spits out), is always on the right. So now we have a left-to-right causal flow!

But one must keep in mind that this is only a *mathematical* kind of causality, meaning you can *calculate* the charge density if you're given the electric field everywhere in space. However, as we pointed out above, the equation can also be read as a *physical* kind of causality, asserting that if you arrange electrical charges in a certain way in space, you will always find that a specific pattern of electric fields surrounds them: in short, the charges *produce* the fields. When the equation is read in this latter way, the causality flows from right to left (*i.e.*, from charges to fields). And that's how physicists view this equation, whether they do so consciously or unconsciously. Indeed, it would strike a physicist as absurd if someone were to say that Maxwell's first equation means that an electric field spread out all over space gives rise to a tiny electric charge sitting somewhere. That would sound as backwards as saying that a strong stench wafting all through a neighborhood will give rise to a frightened skunk crouching under a bush (note the use of a caricature analogy here).

In summary, the equals signs in Maxwell's equations can be understood either as expressing *physical* causality (a physical cause giving rise to an effect), when they are read from right to left, or as expressing *mathematical* causality (a calculation giving rise to a result), when they are read from left to right. And Maxwell's equations are in no way exceptional. Physicists always try to manipulate their equations so that they will have this quality — namely, with causes on one side and effects on the other. Doing so is certainly not logically necessary, but it contributes greatly to clarity. For example, here are two alternative ways of writing Maxwell's first equation that are both perfectly correct yet would make physicists scratch their heads in puzzlement and ask, "What's the point of writing it *that* way?"

$$\text{div}\,\boldsymbol{E}\,/\,2 - 2\pi\rho = 0 \qquad\qquad \text{div}\,\boldsymbol{E}\,/\,4\rho = \pi$$

Indeed, these equations both cloud up the crux of the law, which is the fact that one phenomenon gives rise to another.

In short, physicists, no less than other people, have a weakness for, and also derive benefit from, the naïve analogy likening equations to cause-and-effect relations.

Does Multiplication Always Imply Getting Bigger?

For some concepts that one learns in school, there is an early naïve analogy that is very helpful, but there is no other familiar category that helps one develop the concept more deeply. In such cases, the naïve analogy will very likely be one's only means for grasping the concept, and it will retain this primary role even after many years of education. In such cases, refining the concept so that it becomes more general and flexible will be far harder. It so happens that multiplication and division, two of the most basic notions in mathematics, are cases of this sort.

Addition, subtraction, multiplication, and division are taught in elementary school and are presumed to have been fully understood by middle school. Since so many other mathematical notions are built on them, they are often called the four basic arithmetical operations. These classic notions feel as if they are part of the cultural heritage of every member of our society, and any adult claiming to have no idea what multiplication or division is would be looked at askance. Taught all the way through childhood, these notions should be clear as a bell to high-school and college students. And yet the belief that these operations have been mastered by most adults is an illusion. The next few sections illustrate how this can be so.

Let's consider multiplication. We have found, in surveys of many quite advanced university students (we don't mean advanced math majors), that if they are asked, "What is the most precise possible definition of multiplication?", they are generally very satisfied with either of the following two definitions that we suggest:

> Multiplying is repeatedly adding a value a certain number of times.
>
> Multiplication is taking *a* times *b*, which means adding *b* up *a* times.

It's hard to find anyone who disagrees in any way with these definitions, and virtually no one sees any way to improve upon them. We have also asked groups of advanced university students to supply definitions themselves, and exactly the same themes reappear. It always comes down to the idea that multiplication means, by definition, adding a given number over and over to itself, counting how many times it is done, even if the formulation is not always as concise or clear as the two definitions offered above. For example:

> A multiplication is the iterated addition of a given number a specified number of times.
>
> Multiplying means adding a given figure to itself as many times as one is told to do so.
>
> To multiply is to add a particular number to itself as many times as the other number tells you to do so.
>
> Multiplication is a calculation in which one is told how many times one should add a given quantity to itself.

On the Web, definitions of this sort abound. One site proposes: "Multiplication is thus nothing but an addition in which the numbers being added up are all equal to each other. This is why we say that it amounts to repeating the multiplicand as many times as there are units in the multiplier."

For a bit of historical perspective on the question, one can take a look at definitions along these lines proposed by professional experts. Thus in 1821, the renowned French mathematician Étienne Bezout wrote, in his *Treatise on Arithmetic Intended for Sailors and Footsoldiers*: "Multiplying one number by another is summing up the first of these as many times as there are units in the other."

Well, now... is this view of multiplication as repeated addition really as indisputable as it would seem from all the above? Hardly! As a matter of fact, this view is a naïve analogy that falls far short of the target, and in the long run, it is almost sure to lead anyone who relies on it into confusion.

First of all, this view of multiplication requires that one of the two values be a positive whole number, since otherwise "as many times" has no meaning. What would it mean to speak of adding a number to itself 2½ times or 1/3 of a time, let alone $\sqrt{2}$ times or π times? And yet, requiring one of the factors in a multiplication to be an integer should raise suspicions, since everyone knows that multiplying two non-integers is not forbidden; indeed, in school we all learn how to do it, and pocket calculators don't balk at all at multiplying any two numbers they are given. What on earth would the expression "$\pi \times \pi$" mean if at least one of the two factors had to be an integer?

The next stumbling block lurking in this definition is the common belief that when one adds b to itself over and over again, the result will always be greater than b. We would not merely expect the result to be *somewhat* greater than b, but, by definition, a times greater than b. Dictionaries confirm this naïve idea, as does everyday speech. Indeed, the words "multiply" and "multiplication" suggest a clear image of growth, never an image of shrinking (even though, as we pointed out in Chapter 4, things that shrink can be said to be "growing smaller"). Thus rabbits are said to multiply, quickly resulting in overcrowding; in good times, one's assets multiply, making one wealthier; in bad times, risks multiply, making one less secure; and so on. The prefix "multi" also tends to make one think of growth, as in words like "multinational", "multicolored", "multilingual", "multimillionaire", and so forth. However, this preconception runs into a brick wall when one is instructed to multiply by a value less than 1, as doing so yields a result *smaller* than the multiplicand. This is incompatible with repeated addition.

There is still more trouble. The best-known property of multiplication is that it is commutative — that is, for any numbers a and b, it is always the case that $a \times b = b \times a$. Why this should hold for every pair of numbers a and b is not at all obvious if multiplication is conceived of as repeated addition. In fact, the naïve analogy would suggest that multiplication is intrinsically *asymmetric*, since it treats the multiplicand and the multiplier differently: the former is repeatedly added to itself, while the latter counts how many times the operation is carried out. This certainly does not fit the image of commutativity, in which the two numbers play totally interchangeable roles. Since the naïve analogy gives no insight into this key property, a child (or an adult!) may be baffled by the fact that a added to itself b times always gives the same result as b added to itself a times. To be sure, one can enrich one's notion of multiplication by arbitrarily tossing in the fact of commutativity like icing on the cake, but the naïve analogy of multiplication as repeated addition makes this fact seem mysterious rather than natural.

These minor stumbling blocks turn into serious obstacles when pupils who depend on the naïve analogy are given word problems to solve. For instance, when middle-school students in England were given the problem "If one gallon of petrol costs 2.47 pounds, what is the price of 0.26 of a gallon?", only 44% of them recognized that this is in fact a multiplication problem. The remaining 56% took it to be a *division* problem (namely, 2.47 divided by 0.26)! And thus a multiplication problem that should be very easy even for elementary-schoolers stumped roughly half of the middle-schoolers.

What happens if one changes the numbers in this problem? If one merely replaces "0.26" by "5" and asks the question again ("If one gallon of petrol costs 2.47 pounds, what is the price of 5 gallons?"), then 100% of the middle-schoolers solve it correctly. This discrepancy is due to the fact that the first problem doesn't meet the naïve analogy's image of *adding a number repeatedly*, since the idea of adding a number to itself 0.26 times makes no sense. On the other hand, using the naïve analogy of repeated addition works just fine in the modified problem (2.47 + 2.47 + 2.47 + 2.47 + 2.47). Discrepancies between participants' performances on the two problems reflect the fact that the naïve analogy is of no help in the first one, yet is appropriate in the second one.

Adding Thrice and Fifty Times are Different Kettles of Fish

It's enlightening to compare the preceding findings with some experiments carried out in Brazil. The participants were teen-aged boys who had dropped out of school and were making a living as street vendors. The following simple problem was given to a group of them:

> A boy wants to buy some chocolates. Each chocolate costs 50 cruzeiros. He decides to buy 3 of them. How much money will he need?

The same problem was also given to a different group, except that the two numbers were interchanged, as follows:

> A boy wants to buy some chocolates. Each chocolate costs 3 cruzeiros. He decides to buy 50 of them. How much money will he need?

To all readers of this book it will surely be trivially obvious that each of the two boys will have to fork over 150 cruzeiros, even if one of them winds up with far fewer chocolates (at least in number) than the other one. When we read the two problems, they appear equally easy; the scenario is the same, they involve the same numbers (50 and 3), and they both involve the same arithmetical operation: multiplication. But were they equally easy for the two groups of street vendors? Not in the least.

The first problem was handled pretty well by most: 75% got it right. The second problem, by contrast, was not solved by any of the street vendors. The reason behind this discrepancy is relatively simple; it comes down to reliance on the naïve analogy of repeated addition. To solve the first problem, all one needs to do is add 50 + 50 + 50, to get 150 cruzeiros. This is just two additions, and it involves only very simple facts: first, that 50 + 50 = 100, and second, that 100 + 50 = 150. The variant problem, however, is another ball of wax entirely. To compute the answer, one has to carry out a very long process of iterated addition — namely, $3 + 3 + 3 + ... + 3 + 3 + 3$, involving fifty 3's. As one can easily imagine, this is not a challenge that an elementary-school dropout would be very likely to be able to handle.

This might seem the moment to shower praise on our educational system, thanks to which we educated adults are all instantly able to solve a problem that to school dropouts seems impossibly hard. We all know in a flash that 50 x 3 equals 3 x 50, end of story. Given this contrast with the 0% success rate of the school dropouts, one might be tempted to conclude that schooling very effectively gets across the true nature of multiplication. However, things are not that simple.

There is no disputing the fact that schooling teaches us that the two numbers in a multiplication can be interchanged — we all know that multiplication is commutative, that $a \times b = b \times a$ — and we all carry out such switches without a moment's thought. However, carrying out multiplications using one's knowledge of commutativity doesn't mean that one's understanding of multiplication, as an adult, goes far beyond that of elementary-school children. Indeed, a quick informal survey reveals that almost no one, aside from serious math enthusiasts, knows *why it is the case* that, for instance, 5 x 3 equals 3 x 5. Middle-school students, high-school students, even university students are generally unable to say *why* the two numbers in a product are switchable. How, then, do they justify to themselves the idea that five threes equals three fives, or symbolically, the fact that $3 + 3 + 3 + 3 + 3 = 5 + 5 + 5$?

Most people, if asked this question, will answer readily that one can check this out in any specific case ("Just go get a calculator and try it out for whatever pair of numbers that you wish!"). Some people will state it more as an axiom: "In multiplication, you have the right to switch the two factors"; others will baldly assert, almost as if some kind of magic were involved, "That's just how it is" or "Well, it's known to be a fact." In short, for most well-educated adults, since multiplication is conceived of as repeated addition, its commutativity appears simply as a kind of miraculous coincidence, lacking any clear explanation or reason.

The above-cited treatise on arithmetic by Étienne Bezout provides a somewhat wordy and obscure justication for the commutativity of multiplication. If one's vision of multiplication is rooted in the naïve analogy of repeated addition, then asking why the order of the factors makes no difference in multiplication amounts to asking why two different repeated additions give the same answer, and there is no obvious symmetry between the two operations involved. Bezout tries to resolve this dilemma, but his words are not terribly clear:

> As long as one considers numbers as abstract entities — that is, as long as one ignores the units attached to them — it makes little difference which of two numbers to be multiplied is taken as the multiplier and which as the multiplicand. For example, 3 times 4 is nothing but the triple of 1 taken four times, while 4 times 3 is the triple of 4 taken one time. Now it's self-evident that 1 times 4 is the same thing as 4 times 1; and one can apply the same reasoning to any other number.

Most adults consider the fact that $a \times b$ always equals $b \times a$ to be a very useful but unexplained coincidence, which simply is empirically true. They "understand" the commutativity of multiplication in much the same way as they "understand" why bicycles don't topple over and why airplanes can stay airborne: simply because they've seen such things for most of their lives and have long since forgotten that these phenomena are mysteries that crave explanation. And so, although education certainly drills into students the rote fact that multiplication is commutative, it fails to instill a deep *understanding* of multiplication's nature; instead, it leaves them dependent on their initial naïve analogy.

What about Division?

For division, too, there is a widespread naïve analogy that dominates people's thought, profoundly affecting how people conceive of the operation. Although one would tend to think that division is perfectly understood by anyone who has gone through school, this is an illusion.

Here is a very simple "home experiment" that reveals the hidden presence of this naïve analogy, demonstrating how concrete and standard it is. There is no trap here, just a couple of straightforward challenges. The first challenge is a warm-up exercise: Invent a word problem involving division — that is, a problem whose solution requires just one operation of division. Hopefully, this will pose no problem for any reader. Here, for example, are a few division word-problems that were invented by university students:

- 4 friends agree to share 12 candies. How many candies will each one get?
- 90 acres of land is going to be divided into 6 equal parcels. What will the area of each parcel be?
- A mother buys 20 apples for her 5 children. How many apples will each child get?
- A theater has 120 seats arranged in 10 rows. How many seats are in each row?
- A teacher buying food for a class picnic filled 4 grocery carts with a total of 20 watermelons. How many watermelons were in each cart?
- It takes 12 yards of cloth to make 4 dresses. How many yards does it take to make one dress?

Each of these problems involves a starting figure to be divided by something, and the result of the division is always *smaller* than that starting figure. Thus the first problem reduces 12 candies down to 3, the second problem reduces 90 acres down to 15, the third problem reduces 20 apples down to 4, and so forth.

This observation is already significant, since it shows that when people are asked to invent division problems, they come up with situations where the key idea is "making smaller". Just as the standard image of multiplication as repeated addition locks in the belief that multiplication necessarily involves *growth*, here it seems that there is a naïve analogy at work that locks in the image of division as involving *shrinking*. And this supposed fact about division jibes perfectly with the way the word "division" is used in everyday speech. When one speaks of dividing X up, one sees X as being broken into pieces, with each piece obviously being smaller than X itself. The 1988 edition of *Webster's New World Dictionary* confirms this vision, defining division as follows: "a sharing or apportioning; distribution". Moreover, division is often associated with the notion of weakening. For instance, the slogans "united we stand; divided we fall" and "divide and conquer" imply that if an entity is divided into pieces, it will be weaker than the original entity.

Well, that was the first of our two small challenges. The second one is simply to invent one more division problem, subject to one extra constraint: the answer must be *larger* than the starting figure. Readers, to your marks!

As you probably have noticed, this slight modification of the assignment changes everything. Observe, for instance, that none of the problems in the preceding list meets this constraint. We assume that most of our readers experienced a jump in difficulty between the first and second challenges. Whereas inventing a word problem involving division is a piece of cake for nearly everyone, inventing a division problem where the answer is *bigger* than the starting number is generally not easy at all. It requires a bit of mind-stretching, and for many people it simply is beyond their reach. After all, how can *dividing* something possibly result in something that's *larger*?

The reactions of typical university students to the second challenge are quite diverse. Some students are categorically negative: the challenge is simply impossible. For them, *division* is by definition incompatible with the idea of *making larger*. Therefore, instead of inventing a word problem meeting the requirement, they explain why the task makes no sense:

- "Can't be done. Division always makes things get smaller."
- "When you have a certain value at the outset and you divide it up, you necessarily have less at the end, so it's not possible."
- "Division means sharing, and with equal-sized shares. So each person gets less than what there was at the start. Therefore, it's impossible to invent a division problem where someone winds up with more than there was at the beginning."
- "Impossible, because dividing means cutting something up into pieces. To get more, you have to *multiply*, not divide!"
- "No way, because whenever you divide something, you always reduce it!"

Some other students acknowledge that division problems can indeed have the requested property, because from school they recall the fact that dividing by a number between 0 and 1 has this effect. However, they are convinced that this kind of formal mathematical operation doesn't correspond to any situation in the real world, and so they assert that there can be no word problem that meets the requirement. At least they can't think of any. Here are some comments along these lines:

- "I could say '10/0.5', which gives 20, but that's just a *calculation.* You can't make up a corresponding *word* problem, because in the real world you always divide by 2, 3, 4, and so on. That is, you always divide by numbers bigger than 1."
- "Yes, it's possible — for instance, '5/0.2' — but I can't think of any actual situation that this formula would describe."
- "Any time you divide by a quantity less than 1 you get a larger answer, but I can't think of any real situation where it works like that."
- "When you divide something by one-half, you get more, sure — but the thing is, it's not *possible* to divide anything by one-half!"

Then there are some students who invent various problems that seem to them to work, but they cheat in one way or another, because the problems they give don't match the assignment. For example:

- "Rachel has 20 bottles of wine. She sells half of them at 8 dollars apiece. How much money does she get?"
- "Eric had 8 marbles. In a game, he won half again as many. How many marbles did he wind up with?"

Despite all these protests, it is perfectly possible to devise a division word-problem whose answer is larger than the starting number. Some people find good examples:

- "How many half-pound hamburgers can I make with 4 pounds of meat?"
- "If I have 3 days to prepare for an exam, and it takes me 1/5 a day to read a book, how many books can I read before my exam?"
- "I have 10 dollars, and a chocolate mint costs a quarter (of a dollar). How many chocolate mints can I buy?"
- "How many scarves can I make out of a 3-yard roll of cloth if each scarf requires 3/8 of a yard?"

It turns out, however, that to come up with a problem such as these last four is quite hard. Among 100 undergraduate students, roughly 25 came up with a problem of this type, while the other 75 couldn't do so, and were split into roughly equal-sized groups associated with the three types of failures quoted above. And so we see that an arithmetical operation that in theory should have been completely mastered in elementary school still gives a great deal of trouble to adults, even university students. Could it be that division problems lie so far back in their past that they've forgotten what they once knew about division? Well, no, because the same challenge was set to 250 seventh-graders, all of whom had been studying division for the previous three years, and so for them this kind of challenge was very fresh in their minds (indeed, they had studied problems involving divisors smaller than 1 for at least one full year), and yet it turned out that over three-fourths of them said that it's impossible to invent a situation where division gives a larger answer, and of the 250, only one single student invented a word problem that correctly met the challenge.

Why is it So Hard to Dream up Such Problems?

It's a common belief that when situations are concrete, people think more clearly, but this challenge shows that concreteness is no guarantee of clear thinking. The kinds of problems invented by university students in both parts of our little test featured essentially the same kinds of everyday items (cakes, candies, glasses of water, books, scarves, and so forth), and they were set in the same kinds of environments (kitchens, schools, trips, shopping, and so forth). What, then, is the nature of the conceptual gulf between solving the first challenge, which virtually everyone was able to do, and solving the second challenge, which so few people could do?

The explanation is that the two challenges belong to two quite different categories of problems. They do not rest on the same naïve analogy. To be specific, the problems dreamt up in response to the first challenge, which didn't ask for a larger answer than the initial value, were all problems involving *sharing.* The examples we quoted above were selected in order to give readers some variety, but in truth, two-thirds of the problems invented were extremely routine, always involving sharing the same kinds of things — relatively uniform everyday objects — among the same kinds of recipients — children, siblings, or friends. From the examples cited, it's obvious that the division word-problems that people spontaneously come up with nearly always involve the concept of sharing, and more specifically, the splitting-up of a certain quantity into a number of precisely equal shares. The most typical case involves countable items (candies, apples, marbles) shared among people, and the word "sharing" often shows up explicitly in the problem's statement. Nonetheless, there are more abstract kinds of sharing that show up in a few of the word problems suggested.

In such cases, one has to imagine a more abstract manner of sharing than merely distributing a given set of objects to a given set of people. It might still involve the distribution of entities, but not to human recipients — say, the sorting of cookies into bags, or the arrangement of chairs into rows. It can also involve non-countable substances, such as flour, water, sugar, or land, which get split up into several equal-sized portions. Here there is no sharing in the marked or narrow sense of the term — that is, a counting-out of items, similar to dealing cards out to players in a card game — but there is still sharing in a more general or unmarked sense of the term, in which a whole is divided, through some process of measurement, into smaller chunks. But in any case, none of the responses given by students to the first challenge, whether they involved the marked or the unmarked sense of the concept of sharing, was a division whose result was *larger* than the initial quantity. And this is no surprise, because the nature of sharing is that it makes something *smaller.* Sharing involves breaking an entity into smaller parts, with each recipient necessarily receiving less than the whole that was there to start with. A part cannot be larger than the whole from which it came.

By contrast, in word problems that successfully meet the second challenge, a different naïve analogy operates behind the scenes — that of *measuring* something. (In mathematics education, such problems are said to involve "quotative division".) Division problems of this type can always be cast in the form, "How many times does *b* fit into *a*?" This is a measuring situation, in the sense that *b* is being treated as a measuring-rod with which *a*'s size is being measured. If the size of *b* is between 0 and 1, then there will be more *b*'s in *a* than the size of *a*, which means that the result is bigger than the initial size. For example, the calculation 5/0.25 can be phrased: "How many times does 1/4 go into 5?" The answer, 20, is of course larger than 5. What all this shows is that if a division problem is of the *sharing* sort, then its answer can't be larger than the starting value, but if it is of the *measuring* sort, then its answer can be larger.

It turns out that from a historical and scholarly point of view, measuring is a more fundamental way of looking at division than sharing. The definition of division given by Bezout in his 1821 treatise is quite explicit: "To divide one number by another means, in general, to find out how many times the first number contains the second." Indeed, the etymology of the terms involved in division reflects the view of division as a measuring process. As readers will recall from elementary school, the result of a division is called its *quotient.* (As Bezout explained it: "The number to be divided is the *dividend*; the number by which one is dividing is the *divisor*; and the number that tells how many times the dividend contains the divisor is the *quotient.*") The English word "quotient" stems from the Latin word "quotiens", which is a variant of "quoties", meaning "how many", and which derives from "quot", a word that refers to the counting of objects. In sum, today's terminology echoes the conception of division as measurement, since "quotient" means "how many times".

Bezout is aware that seeing division as measurement is not the only possible point of view, but he wants his readers to act as if it were: "One's goal in doing a division is not always to find out how many times one number is contained in another number; however, one should always carry out the operation as if this were indeed one's goal." This shows that the view of division as being primarily a kind of *sharing* did not come from mathematicians, for they tend to favor the view of division as measurement or counting. To the contrary, the origins of the naïve analogy of division as *sharing* lie outside of mathematics. As we mentioned earlier, dictionaries tend to define "to divide" in its everyday sense along the following lines: "to separate into parts; split up; sever; to separate into groups; classify; (*Math*) to separate into equal parts by a divisor" (this taken again from the 1988 edition of *Webster's New World Dictionary*).

Is Division Mentally Inseparable from Sharing?

The experimental results we've just described show that for most people, division is understood through the naïve analogy of sharing; after all, most people find the first challenge very simple and invent word problems that involve sharing, while the second challenge, which is easily handled if one simply uses the analogy of measuring, is much harder for most people. Although children spend years learning about division in school and are thus presumed to have mastered this basic operation by the end of middle school (and adults are assumed to know division yet better), it turns out that people of all ages have trouble thinking of division other than through the naïve analogy that equates it with sharing.

Most people use the term "division" not to describe a concept that they learned in school, but to describe a category of situations that was part of their lives before they started school — *sharing*. When sharing comes up in a mathematical context, they have learned from school to use the term "division" instead. In other words, most people think that "division" is just a technical term to denote the concept of sharing, especially when a calculation is called for, and that's all there is to it. When one is in math class, sharing has a fancier name, just as in certain arenas of life people use various special terms to designate familiar concepts, even though such terms don't lend any particular insight. Thus one learns that when one is at the opera, it's better to say "aria" than "song", and likewise, when one has truck with wine connoisseurs, one soon gets used to hearing about the "bouquet" rather than the "smell" of the wine; one also gets used to the fact that one's doctor will tend to speak of "apnea" rather than of "having trouble breathing", or of "hypertension" instead of "high blood pressure".

To summarize, although it is tempting to think that schooling teaches people the full-blown concept of division, thus allowing them to throw away the naïve analogy of *sharing* like a no-longer-needed crutch, the truth is that the crutch remains the central way of understanding division — it merely disguises itself by donning the more impressive-sounding mathematical label of "division".

Mental Simulation in the Driver's Seat

To solve either of the following two word problems is a challenge as easy as they come:

> Paul had 27 marbles. Then during recess, he won some, and now he has 31. How many marbles did he win?
>
> Paul lost 27 of his 31 marbles during recess. How many does he have left now?

Both of these problems are solved by carrying out exactly the same operation — namely, subtracting 27 from 31. At first glance, they thus seem identical in terms of what is going on mentally when we solve them, but let's set aside the formal operation by which we solved them; instead, let's try to *visualize* these situations in our mind's eye — that is, we'll try to mentally simulate each of them. What happens?

In the first case, it's easy to imitate what happened by counting on one's fingers or in one's head. Paul's marble count moved up from 27 to 28 ("1"), then to 29 ("2"), then to 30 ("3"), and finally it reached 31 ("4"). The solution takes four simple steps.

The second case, however, is very different. This time, starting from 31, one has to move downwards 27 steps: first to 30 ("1"), then to 29 ("2"), then to 28 ("3"), then to 27 ("4"), then to 26 ("5"), … , and after a long time one will finally hit 4 ("27").

We thus see that these two word problems, although they're both solved by the same *formal* operation (31 – 27), are not imagined or mentally simulated in the same fashion at all. One process involves just four easy counting steps, while the other takes 27 steps, which, to make matters worse, involve counting backwards.

This contrast should recall a similar one from earlier in the chapter — namely, that of the teen-aged street vendors in Brazil. As we saw then, the product of 50 and 3 can be mentally simulated either as 50 + 50 + 50 or as 3 + 3 + 3 + ... + 3 + 3 + 3, depending on how the problem was stated; here, likewise, the subtraction "31 – 27" can correspond to two very different mental simulations, one very short and one very long.

Let's now compare the following four word problems:

1. If we break a stack of 200 photos into piles of height 50, how many piles do we get?
2. If we break a stack of 200 photos into 50 piles, how many photos are in each pile?
3. If we break a stack of 200 photos into 4 piles, how many photos are in each pile?
4. If we break a stack of 200 photos into piles of height 4, how many piles do we get?

The first two are solved by the division "200/50", and the last two by "200/4". That is quite obvious. But are some of these problems easier or harder than others? Perhaps it seems as if we are asking if the division "200/50" is easier (or harder) than the division "200/4". If so, the first two problems would be easier (or harder) than the last two. But things are trickier than that, as a bit of mental simulation will show.

Let's try to envision the first situation: *If we break a stack of 200 photos into piles of height 50, how many piles do we get?* We first imagine a tall stack of 200 photos; now we want to break it into smaller stacks of height 50. In order to find the answer by simulation (as opposed to doing it by formal division), we take the number 50 and add it to itself until we get 200. 50 plus 50 makes 100, and then another 50 makes 150, one last 50 to make 200. *Four* 50's altogether — that's our answer.

Now let's try to imagine the second situation: *If we break a stack of 200 photos into 50 piles, how many photos are in each pile?* Formally speaking, this problem also involves the division "200/50". We had a stack of 200 photos and we divided it up into 50 smaller ones. We need to add up *some* number 50 times, but we don't know *which* number. In fact, not only do we have to do 50 additions to find the right answer, but we may have to do it a bunch of different times, guessing about what to add to itself! 2 + 2+ 2...? After much toil, we get 100, and see that "2" was wrong. 3 + 3 + 3...? Again a lot of toil to wind up with the wrong answer. 4 + 4 + 4...? Well, this time, if we add right, we'll get 200. But solving this problem is not a piece of cake.

So although the first two problems are both solvable by the same division (200/50), from the point of view of seeing what's going on in one's mind's eye, the first is much easier than the second.

Now let's look at the third one: *If we break a stack of 200 photos into 4 piles, how many photos are in each pile?* Here we break our tall stack into four shorter stacks. How many photos in each short stack? This is very similar to the second situation, where we want to do a repeated addition in order to reach 200, but this time we only need to add the mystery number to itself *four* times, instead of 50 times. On the other hand, we have to guess at the mystery number's identity. But if we're clever, we may hit on "50" without too much trouble. For instance, if we recall from everyday life that 50 + 50 = 100, then we can quickly figure out that 50 + 50 + 50 +50 = 200, and we're done.

Finally, the fourth problem: *If we break a stack of 200 photos into piles of height 4, how many piles do we get?* We know we're dealing with repeated addition of the number 4, but the question is: *how many additions?* We know that 4 + 4 + 4 + ... + 4 + 4 + 4 = 200, but the mystery is how many copies of "4" there are in this sum. This is tough, because we'll need to keep track of two things in our head at once — firstly, how many copies of "4" we have added up so far, and secondly, what the running tab is. We thus see that mentally simulating the third situation is far easier than mentally simulating this one.

To recapitulate, it turns out problems #1 and #3 are fairly easy to simulate in one's mind's eye, while problems #2 and #4 are challenging. If we go back to our earlier distinction between division as *sharing* and division as *measuring*, we see that two of these problems are easy to simulate mentally: there is an *easy sharing problem* (#3, where 200 photos are shared among 4 piles) and there is an *easy measuring problem* (#1, in which a 200-photo stack is measured using big stacks of height 50). There are also two problems that are very hard to simulate mentally: there is a *difficult sharing problem* (#2, where 200 photos are shared among 50 piles), and there is a *difficult measuring problem* (#4, where a 200-photo stack is measured using small stacks of height 4).

Let's rephrase this in another way. One problem involving the division "200/4" is *easy* to simulate mentally (#3: 200 photos shared among 4 piles), while another problem that involves exactly the same division is *hard* to simulate mentally (#4: "Measure a 200-photo stack using piles of size 4"). Likewise, one problem involving the division "200/50" is *easy* to simulate mentally (#1: "Measure a 200-photo stack using piles of size 50"), while another problem involving exactly the same division is *hard* to simulate mentally (#2: 200 photos shared among 50 piles).

Here is a short summary of what can be said about these four word problems, all of which are very similar in form (each involves a division whereby a pile of photos is broken into smaller piles), yet differ greatly in how one conceives of them (some involve sharing, some involve measuring) and also in terms of their difficulty for a solver who is mentally simulating them (some are easy, some are hard):

1. If we break a stack of 200 photos into piles of height 50, how many piles do we get?
   [200/50; easy measuring problem]
2. If we break a stack of 200 photos into 50 piles, how many photos are in each pile?
   [200/50; difficult sharing problem]
3. If we break a stack of 200 photos into 4 piles, how many photos are in each pile?
   [200/4; easy sharing problem]
4. If we break a stack of 200 photos into piles of height 4, how many piles do we get?
   [200/4; difficult measuring problem]

Let's presume that when subjects solve such a problem, they do so by making a mental simulation rather than by instantly carrying out an arithmetical calculation. If that's the case, then the first and third problems should both be *easy*, while the second and fourth should be *difficult*; moreover, knowing which arithmetical calculation is involved (200/4 or 200/50) does not tell us how hard the problem is.

This way of looking at word problems stands in marked contrast to the traditional view, which takes the formal arithmetical operation needed to solve a problem as a gauge of the problem's difficulty. In place of this, the new perspective highlights the spontaneous way in which the situation is framed — that is, the analogy that allows one to solve the problem in a very direct way, namely by counting in one's head. This point of view shows why different word problems, even if they are equally down-to-earth, and even if they all involve exactly the same *formal* arithmetical operation, can nonetheless have extremely different levels of difficulty. The surprising findings that we described earlier concerning Brazilian teen-aged street vendors tackling multiplication problems now seem to apply more generally, both to other kinds of word problems and to other groups of people. The key variable is seen to be the simplicity of the mental simulation that will yield the correct solution.

Rémi Brissiaud, a developmental psychologist, has done pioneering research into these ideas in the context of learning to do arithmetic, and he has written very innovative and efficient new mathematics textbooks inspired by his discoveries. In collaboration with him, we have studied how seven-year-olds who are just beginning to learn the basic arithmetical operations tackle different kinds of word problems. Our results show a clear distinction between problems that tend to be solved by mental simulation and ones that tend to make children resort to formal arithmetical operations. In cases where the mental simulation is not unwieldy, it is always preferred. This principle is illustrated by the following examples:

Paul has 10 boxes containing 4 cookies apiece. How many cookies does he have in all?

Paul has 4 boxes containing 10 cookies apiece. How many cookies does he have in all?

The first problem, if solved by simulation and not multiplication, requires adding 4 to itself 10 times. In such a simulation, children might imagine Paul at the grocery store, taking boxes one by one off the shelf and placing them in his shopping cart. Thus it would go as follows: "Box #1 (4 cookies); #2 (8); #3 (12), … , #9 (36), #10 (40)". This process requires one to count and to keep a running tab at the same time, and for children just learning how to add, repeatedly adding up all these 4's is far from easy. Mental simulation is hard here. By contrast, the second problem is solved by mental simulation quite handily. It takes just four additions, and what's more, they're all easy; in fact, each sum along the way echoes the counting number just preceding it, as follows: "Box #1 (10 cookies); #2 (20); #3 (30); #4 (40)".

Our experiments have shown that children are much better at solving word problems in which mental simulation comes easily than they are at solving problems in which it does not. Not only does this hold *before* the relevant formal arithmetical operation has been taught to them (not too big a surprise!), but it also holds *after* it has been taught (this, by contrast, is quite surprising). We observed that even two years after the relevant arithmetical operation has been taught, if mental simulation provides a short solution path, the problem is solved much more easily than via formal calculation.

Some of our findings fly in the face of received ideas about the relative difficulty of arithmetical operations. For instance, subtraction is usually considered to be an easier operation than division, and is thus taught in schools already roughly at age 6, whereas all mention of division is put off for another two years or so. And yet our experiments have shown that children who have supposedly mastered subtraction but have heard nary a word about division manage to solve certain division problems (those that can be done via mental simulation) better than they can solve subtraction problems in which mental simulation is inefficient. Here is an example of what we are talking about:

> Jill passes out 40 cookies to her 4 children. How many cookies does each child get?

This division problem is much more easily solved by children at the above-described stage than the following subtraction problem:

> Paul has 31 marbles. He gives 27 to his friend Peter. How many does he have left?

This is not simply due to the fact that 40 breaks easily into 10 + 10 + 10 + 10, because the problem "Jill has 40 cookies and wants to make little packets of 4 each. How many packets will she make?" turns out to be far harder than the one given above, involving 40 cookies given out to 4 children, although both have the same answer (namely, 10).

Our experimental findings show that whenever it's possible, children opt for using analogies to real-world situations rather than making formal arithmetical calculations. If a word problem can be conceived of in such a way that formal calculations can be bypassed, then simulation is the pathway that children tend to follow. The formal technique will be wheeled out only when there is no alternative — that is, in situations where mental simulation would be inefficient, either because it would require too many steps (adding up ten 4's) or because it would require the use of arithmetical facts about which the child is still a bit shaky (*e.g.*, "4 + 16 = 20").

The Influence of Language on Naïve Analogies

Does *sharing*'s dominance over *measurement* as a naïve analogy for division mean that the former is a simpler concept than the latter? Is *sharing* such a natural and familiar idea that it automatically and irrepressibly jumps to mind as a ready-made analogue for division? And is *measurement* such a rare and unfamiliar idea that it is unlikely to be used as an analogue for division? Is this why *sharing* enjoys the lion's share of mental imagery for division?

The answer is no. The predominance of sharing as the naïve analogy for division is not due to intrinsic simplicity, but merely to an accident of language. Division is unconsciously associated with sharing in the minds of most speakers of English because the English word "division" has both a mathematical meaning and an everyday meaning, and connotations of the everyday meaning inevitably spill over into the technical meaning; as a consequence, the naïve analogy of *division as sharing* overwhelms that of *division as measuring*.

Suppose there were an arcane mathematical notion called "surgery" (indeed, it exists). If you were told that surgery sometimes involves smoothly tying things together and other times involves tearing things asunder, it seems likely that, thanks to your prior familiarity with medical surgery, the naïve analogy that you would unconsciously exploit in trying to make sense of the notion would tilt more towards *tying together* than *tearing apart*. A similar story can be told about division and sharing. Suppose that hundreds of years ago, the English word assigned to the mathematical concept of *division* had not been "division" but "measurement". Had that been the case, then children today, on first hearing about the arithmetical notion called "measurement", would tend to create a very different primary naïve analogy for it. And the idea of sometimes getting a larger answer than what they started with (that is, a/b being sometimes greater than a) wouldn't strike them as strange or confusing in the least.

In sum, the predominance of the naïve analogy "division is sharing" doesn't imply that envisioning a measurement ("How many B's will fit inside A?") is cognitively more demanding than envisioning an act of sharing ("If I cut A into B parts, how big will each part be?"). Indeed, quite to the contrary, we've seen that a measurement problem such as "How many 50's are there in 200?" is much easier than a sharing problem that involves exactly the same numerical values: "Dole out 200 candies to 50 kids!"

What we learn from this example and similar ones is that the prevalence of *sharing* as the naïve analogy for division is not because sharing is easier to imagine than measuring; it is because there is an unconscious bleed-through of the everyday meaning of "to divide" into the technical term "to divide", and this semantic contamination gives a big head start to *sharing* as the source of the naïve analogy, even if there are many cases where *measurement* would be a more apt analogy. In other words, it's easy to solve "How many 50's are there in 150?" in one's head ("50 and 50 and then 50 again — that makes 150, so the answer is 3"), but to realize that one has just solved a *division* problem is not easy at all, because the usual feeling of carrying out a division is pervaded by the everyday sense of that word, which has no connection with the idea of measuring anything.

What Schooling Leaves Untouched in Our Minds

Does what we learn in school profoundly affect how we see situations? Does school teach us to think "formally" about situations? By this, we mean acquiring the ability to zoom straight to the abstract core of a situation, not deflected by its concrete details. We all do this in some aspects of daily life — that is, we routinely ignore many aspects of certain situations, though we are fully aware of them intellectually. Thus, we know but we forget that our closet door was once part of a tree, that Adolf Hitler was once a baby, that the meat on our table was not long ago inside an animal grazing in a field, and so forth. Even if we accept the *truth* of these facts, we systematically ignore them, so that it's fair to say that we simply don't see an ex-animal in the steak, nor an ex-tree in the door, nor an ex-baby in photos of the Führer. This is not stupidity but intelligence.

In the same way, we pay no attention to all sorts of properties of the objects that surround us. Who would think of using a painting hanging on the wall as a tray on which to carry the dirty dishes into the kitchen, or as a bulletin board onto which we could post a bunch of family photos, or as a throw rug that might decorate our floor? And yet in theory, and in case of extreme need, any of these uses might come to mind, perhaps even seeming eminently reasonable. Only when one is extremely angry or frightened does it ever occur to one that a candle, a plate, a vase, a glass, a statuette, a chair, and a mirror are all potential weapons, and this "forgetfulness" is as it should be.

Categorization involves taking a certain point of view, and once one has chosen a category for something in one's environment, that act tends to suppress the perception of all sorts of properties that are irrelevant to the chosen category. Who ever wonders if the hamburger in their bun came from a male or a female cow? And who would ever care if it's a left or a right shoe when one is so starved that one is desperate to eat it? In short, we are constantly abstracting and thus constantly ignoring thousands of potentially observable facets of things and situations. Once again, this is not stupidity but intelligence. Does school teach us to use this kind of "intelligent forgetting" or "intelligent ignoring" more systematically, especially in mathematics?

When most people are given a mathematical word problem — even a very simple one — they have great trouble ignoring some of its irrelevant aspects. Instead of treating such a problem in a *formal* manner, they tend to be influenced by some of its salient concrete features. Even if someone eventually discovers the abstract mathematical structure in a word problem, that recognition never fully overrides the person's more spontaneous initial view of the situation; various concrete aspects get blurred in with more abstract ones. Our ability to perceive mathematical situations formally — that is, in such a way that our thinking is not contaminated by some of their irrelevant, surface-level aspects — is very limited.

We will illustrate this using one of the easiest possible cases for an adult — multiplication, which, as we stated earlier, is one of the first operations taught in school, and which is a concept that few people feel they have anything more to learn about. To make it even simpler, we'll look only at cases of multiplication involving positive integers. We compare three situations:

> I go into a store where every item costs 4 dollars, and I buy 3 pens. How much money do I spend?
>
> For each of my 4 children, I buy 3 pens. How many pens do I walk out with?
>
> For each of my 4 children, I buy one blue pen, one green pen, and one red pen. How many pens do I walk out with?

Seeing things purely formally would involve instantaneously filtering out the store, the narrator, the pens, the colors, the children, and the money, and jumping directly to the bare-bones idea that "These problems all involve multiplication of 4 and 3", and then simply carrying out that operation. Casting the problems this way should involve no asymmetry between the two factors, because if these problems are seen on a purely abstract level, they are all just multiplication problems. And indeed, given the above problems, few us would wonder, "Is this a '4 x 3' situation or is it a '3 x 4' situation?" We would instead simply tend to think, "Multiply 3 and 4 together." Or at least this is one's first introspective impression when, as an adult, one tackles these problems. But do we really perceive, in our mind's eye, nothing but a "Multiply 3 and 4" situation in all three cases? Do we really see and do exactly the same thing when we solve these three problems? Let's take a closer look at them.

In the first problem, where every item costs 4 dollars, I buy one pen for $4, then another, and then another. Thus in the end, I spend 4 + 4 + 4 = 3 x 4 = 12 dollars.

In the second problem, where I'm buying pens for my children, I buy 3 pens for each child. Thus I buy 3 pens for the first child, 3 for the second, 3 for the third, and 3 for the fourth child. Altogether, then, I buy 3 + 3 + 3 + 3 = 4 x 3 = 12 pens.

In the third situation, there are two possible perspectives. From the first perspective, we focus on what each *child* will get — that is, we see things much as in the second problem. Each child gets 3 pens, and that makes 4 cases of "getting 3", which means 3 + 3 + 3 + 3 = 4 x 3 = 12 pens.

From the second perspective, I focus on *colors* rather than on children. I'm buying 4 blue pens (one for each child), 4 green pens, and 4 red pens. Therefore, since there are 3 colors concerned, I am buying 4 + 4 + 4 = 3 x 4 = 12 pens.

If a person solving one of these word problems immediately perceived the abstract idea that it involved *multiplication*, then the way in which the problem happened to be concretely embedded in the world should have no effect on the order of the factors. By contrast, if these word problems are subliminally perceived through the filter of the naïve analogy *repeated addition*, as opposed to being perceived formally as *multiplications*, then the naïve analogy should guide the way the problems are solved, and the solutions that people find should be influenced by the preceding considerations.

To test this hypothesis, we asked older elementary-school students and also university students to solve these problems without using multiplication. It turned out that nearly everyone, children and adults alike, used repeated addition, which is no surprise, and moreover that the additions that were chosen depended crucially on how the problems were stated, which may be more of a surprise.

Thus roughly 90% of the subjects used the addition "4 + 4 + 4" for the first problem, and roughly the same percentage solved the second problem using the addition "3 + 3 + 3 + 3". For the third problem, involving colored pens, roughly 50% of the subjects went for "3 + 3 + 3 + 3" (thus seeing it in terms of children), roughly 40% of them went for "4 + 4 + 4" (thus seeing it in terms of colors), and the remaining 10% didn't group anything together explicitly; that is, the repeated addition they saw was "1 + 1 + 1 + 1 + 1 + 1 + 1 + 1 + 1 + 1 + 1 + 1". Of course those subjects still made unconscious groupings — either "(1+1+1) + (1+1+1) + (1+1+1) + (1+1+1)" (each group representing a *child*) or else "(1+1+1+1) + (1+1+1+1) + (1+1+1+1)" (each group representing a *color*).

This finding clearly supports our prediction, which is that given a word problem, people will try to solve it not just by perceiving its formal structure, but also by doing their best to find, based on the way the problem is worded, an analogy to repeated addition. In summary, to solve these word problems, people tend to mentally simulate the situations described in them.

If, on the other hand, the situation had been spontaneously perceived "formally", meaning that the abstract concept *multiplication* had been instantly evoked in subjects' minds, then (since using multiplication had been explicitly banned) the repeated addition "4 + 4 + 4" would have been suggested by all the subjects for all the problems, because that sum involves the least computation, and there would have been no good reason to opt for the repeated addition "3 + 3 + 3 + 3", since it uses more summands.

To summarize, we have shown that even in a bare-bones mathematical situation, people are very seldom able to ignore all of its superficial, concrete aspects and to home in on just its abstract formal structure. For better or for worse, people are influenced by how situations are concretely described, by their familiarity with similar situations, and by the naïve analogies that these situations evoke naturally.

Sometimes Situations Do Our Thinking for Us

Here is a problem to solve, which contains no hidden trick:

> Lawrence buys an art kit for \$7, and also a binder. He pays \$15 altogether. John buys a binder and also a T-square. He pays \$3 less than Lawrence did. How much does the T-square cost?

Of course you've already gotten the correct answer, but that's not the main point here. We would instead ask you to think carefully about this problem and try to find the most streamlined, efficient way to solve it, showing exactly how the shortest, simplest solution would work, step by step.

Usually, people suggest a solution that involves three calculations, as follows:

Price of the binder: 15 − 7 = 8 dollars.
Price paid by John: 15 − 3 = 12 dollars.
Price of the T-square: 12 − 8 = 4 dollars.

This a perfectly correct way to solve it. But now, how about tackling the following problem, again looking for the shortest route to the solution:

> Laurel took ballet lessons for 7 years and stopped at age 15. Joan started at the same age as Laurel but stopped 3 years earlier. How long did Joan take ballet?

Once again we are interested in the most economical way of solving this problem, and in seeing what exactly are the steps that must be taken. We thus ask you to indulge us once again in trying to find and spell out the minimal pathway to the solution.

One idea is to carry out essentially the same steps as in the preceding problem:

> Age at which Laurel (and hence also Joan) started ballet: 15 − 7 = 8 years.
> Age at which Joan quit: 15 − 3 = 12 years.
> Total time that Joan took ballet: 12 − 8 = 4 years.

Although this pathway to the solution is totally correct, another pathway might have come to your mind. If Laurel and Joan started taking lessons at the same age and Joan stopped 3 years earlier than Laurel did, then Joan took ballet 3 years less than Laurel did, so she took lessons for 7 − 3 = 4 years (we were explicitly told that Laurel took lessons for 7 years). This pathway involves just *one* arithmetical operation!

The existence of this alternate route to the solution of the second problem suggests to most people that the two word problems differ fundamentally from each other, since the ballet-lesson one can be solved in just one step, whereas no similar shortcut exists for the school-supply problem. Is this really true, though?

What would it mean to use the subtraction "7 − 3" in the context of the *first* problem? To be sure, it gives the right answer — 4 dollars — but does it *mean* anything? Most people who are given this problem tend to think it doesn't, or that if it is a meaningful thing to do, then it would take a long time to figure out why, and it's not worth it. And yet, the situation can be described as follows: "Both Lawrence and John bought a binder plus some other item"; this view leads one to a solution using just one single operation. One of them paid 3 dollars less than the other one paid. Therefore, the difference between what the two boys shelled out is due totally to the *other* item. Hence the price of *John's* other item (the T-square) must be 3 dollars less than the price of *Lawrence's* other item (the art kit): 7 − 3 = 4.

What's remarkable, when one compares the solutions of these two problems, is that fewer than 5% of elementary-school children and fewer than 5% of adults (in fact, of highly educated adults — namely, university students and schoolteachers) find the direct one-subtraction solution to the school-supply problem, whereas roughly 50% of the children and also 50% of the adults spontaneously find the one-subtraction solution to the ballet-lesson problem. And so, although all that's required here is to carry out very trivial subtraction operations in extremely concrete situations, nonetheless the angle of attack that yields a one-step solution is almost never found in the first context while it is very often found in the second context.

What Does It All Mean?

It's no accident that the very same one-operation method can be used to solve both problems, as both involve situations that could be handled by using the following formal rule, which has the feel of a theorem one might find in a set-theory textbook:

> If two sets overlap, then the difference between their sizes equals the difference in the sizes of their non-overlapping parts.

If we apply this rule to the school-supplies problem, it tells us that the difference between what Lawrence paid and what John paid must be equal to the difference between the non-overlapping parts of their purchases — that is, the difference between Lawrence's art kit and John's T-square. If we apply the rule to the ballet lessons, it tells us that (since Laurel and Joan had equally long periods with *no* lessons), the difference between their ballet-quitting ages is equal to the difference between their ballet-lesson periods. If this formal rule were learned and fully absorbed by everyone, then we would expect that both problems would be always solved by the single-subtraction method. As we have seen, though, nothing of the sort happens. How come?

Simply because the formal rule is not part of most people's mental repertoire. Even people who discover the one-operation method for the second problem are unlikely to be aware of any such rule. Rather, they just allow the problem itself to direct their thoughts. If they come up with a one-step solution, it's because that is what they are naturally led to. Each situation is defined, in a person's mind, by the categories it effortlessly evokes, and that perception, rather than the application of any formal rule, is what guides the person's thinking.

The ballet-lesson problem is thus perceived as follows. If two events start at the same moment, and one of them lasts N time-units less than the other, then it will end N time-units before the other one does. This is so patently obvious to us all that the sentence tends to sound like a mere tautology, a vacuous triviality. Let us nonetheless restate it slightly differently: if two events start simultaneously, then the difference between their *durations* equals the interval between their *cutoff moments.* People's perception of time is so deeply anchored and they so intuitively understand this basic principle — *taking less time means ending earlier* — that they often recall the statement of the ballet-lesson problem in a distorted fashion.

To be specific, if students who have read the statement "Joan started ballet lessons at the same time as Laurel but took them for three years less" are asked to write it down by memory, they often do not reproduce it correctly, writing instead: "Joan started ballet lessons at the same time as Laurel but quit three years earlier." Their deduction is so deeply fused with their perception of the situation that they don't see it as such; they are unaware of having transformed the sentence in committing it to memory.

Indeed, the transformation of the problem from the initial phrasing ("Joan took three years less") to the final phrasing ("Joan finished three years earlier") converts a difference between two *lengths of time* into a difference between two *temporal stopping points.* In the isomorphic school-supply problem, this would be comparable to someone converting the three-dollar price difference between Lawrence's total outlay and John's total outlay into the difference between just the art kit and the T-square. However, in our experiments we have never run across any subject who read the phrase "John's total outlay was three dollars less than Lawrence's" and subsequently wrote it down by memory as "the T-square costs three dollars less than the art kit".

What we see here is that prior knowledge about relations involving time ("three years less" can be converted into "three years earlier" or *vice versa*) allows people to solve the temporal word problem in one single step. In this sense, one could quite reasonably claim that it's the *situation* that's "doing the thinking" for the subjects. Finding the one-step solution is not a consequence of having mastered some general set-theoretic rule such as we quoted at this section's outset.

As is probably quite clear, students who solve the ballet-lessons problem do not rely on its abstract, formal structure (as expressed in the above-stated "theorem"); they have no need at all to evoke the abstractions of set theory in order to solve it. When they carry out their reasoning, they don't perceive the time intervals involved as *sets* (if they did, each set would contain infinitely many infinitesimal moments!), nor do they perceive the common age at which Laurel and Joan started their lessons as the *overlap* of two sets (their "intersection", in set-theoretic language), nor do they see the lengths of time that the two girls took lessons as the *non-overlapping* parts of sets, nor do they see the girls' ages when they stopped their lessons as the *sizes* of two sets. In fact, it would take careful intellectual work to recast this problem in set-theoretical language, because set theory is not the framework in which humans naturally perceive it, and that's why efficient solving of the problem by a person should not be taken as showing that the formal rule (the "theorem") was correctly applied. Rather, the act of perceiving the ballet-lesson situation in terms of familiar time-categories does the bulk of the work for the student, and there is no need whatsoever to code the situation into an arcane, abstract technical formalism such as set theory.

The diagrams on the facing page show three ways of conceiving these isomorphic word problems. The diagram at the top shows how the school-supplies problem is imagined by nearly everyone to whom it is given. The typical assumption is that in order to figure out the price of the T-square, you have to subtract the price of the binder from John's total outlay (which itself is found through a subtraction), and the binder's price is found by subtracting the price of the art kit from Lawrence's total outlay — thus three subtractions *in toto* seem necessary. In this diagram, one doesn't see that the difference between John's and Lawrence's total outlays is identical to the difference between the prices of the T-square and the art kit. That key idea, a prerequisite to solving the problem in a single step, is missing, and so three arithmetical operations seem to be needed.

The middle diagram shows the way that many people very naturally envision the problem of the ballet lessons. They begin with the idea that Joan and Laurel took up ballet at the same age, and so the difference between the *lengths of time* they took ballet has to equal the difference between their *ages* when they quit. This idea is screamingly obvious in the diagram, and that explains why the one-step solution is often found for the ballet-lessons problem. (Let's not fail to note the frame blend on which this diagram is tacitly based, and which Gilles Fauconnier and Mark Turner would delightedly point out — namely, the fact that we are imagining aligning the two girls' lives on a single horizontal time axis. Aligning their lives means placing their births at the same spot on the time axis, and as a result of this maneuver, their first lessons will also coincide.)

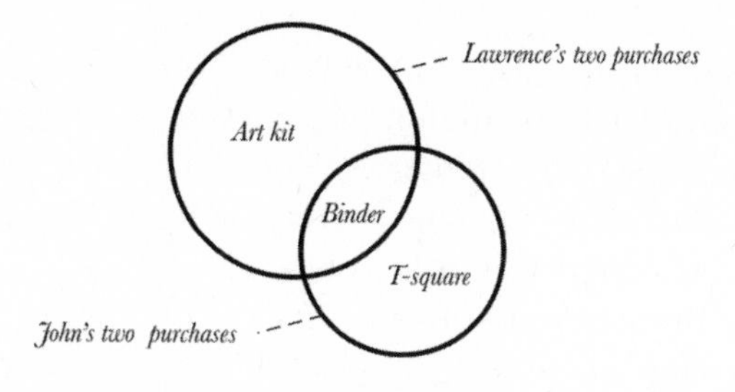

Diagram for the school-supplies problem, which usually takes 3 operations

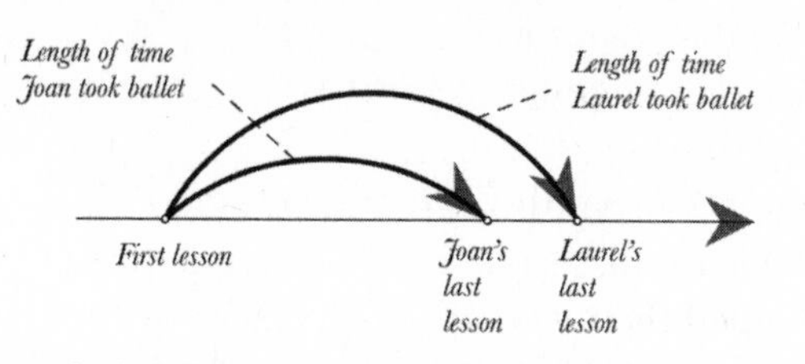

Diagram for the ballet-lessons problem, which often takes just 1 operation

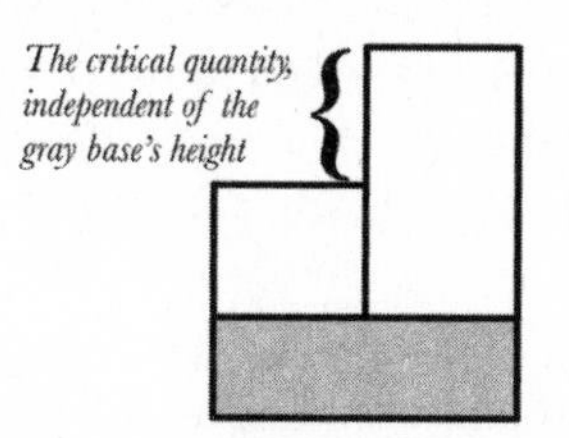

Diagram for both problems, revealing their deep structure, and taking just 1 operation

Each of the two upper diagrams (the top one, involving three operations, and the middle one, involving just one) was tailored to fit one specific problem, and it is not apparent that they have much in common with each other. Looking at the upper two diagrams alone, one might think that the problems they represent are of very different sorts; in fact, this is what most people claim who try to solve them both. But the bottom diagram not only shows how the two problems can be seen in a single unified manner; it also reveals how a one-step method works just as naturally for the school-supply problem as for the ballet-lesson problem. While the "theorem" enunciated above was abstruse and difficult to grasp, this visual encoding of the two problems in one single picture reveals their isomorphism (*i.e.*, the fact of having the same underlying structure), as well as how they both can be solved with just one arithmetical operation.

These three diagrams show how the way a person visualizes a word problem can either bring out or hide a pathway to the solution. The bottom diagram could be said to be more abstract than the upper two in that it unifies them in a single diagram, but on the other hand, the image of two boxes standing on the same pedestal is very concrete, and as soon as these problems are cast in a form that involves a shared "pedestal" (the binder, in the first problem, and the starting age, in the second one), the abstruse idea expressed in the "theorem" suddenly becomes crystal-clear, as it has been fleshed out in a concrete manner, using simple, everyday images, such as boxes resting on a shared pedestal. Teaching a young student to see the "pedestal" in these two word problems imparts an elegant insight that is unavailable to most untrained adults.

A key challenge for educators is thus to take into account the way people manage to adroitly sidestep the formal encoding of situations by exploiting the way their familiar categories, built up over years of interactions with the world, work. Although most teachers are quite aware that the way in which a problem is "dressed" can profoundly affect its difficulty, educators have not yet figured out how to make the art of dressing mathematics problems into a powerful teaching tool. To achieve this would be a great advance, but of course doing so constitutes a great challenge as well.

A Naïve Analogy that has Ill Served Psychology

Do you find it hard to see naïve categorizations such as *multiplication is repeated addition* and *division is sharing* as constituting *analogies*? Despite the multiple arguments we've mustered and the many situations we've dissected with a fine-tooth comb, perhaps a little voice inside you keeps insisting, "I'm sorry, but categorization and analogy-making are just not the same thing. Taking two notions that initially seem very far apart and then building a mental bridge between them because one sees that they have certain abstract qualities in common is a profoundly distinct type of act from seeing something and merely recognizing that it belongs to a familiar category, such as *table*."

Why do so many people, perhaps even you, have an inner voice that so strongly resists the thesis that the building of analogies between things is just the same activity as the assigning of things to categories? How come your inner voice hasn't gradually calmed down and grown silent over the course of reading this book? How come it hasn't listened to all the reasons that we had hoped would convince you of our thesis?

The answer — a rather ironic one — is that our thesis itself explains why so many people have so much trouble accepting it: namely, the belief that analogy-making and categorization are separate processes springs from none other than a certain naïve analogy about the nature of categories. This naïve analogy, which has the dubious honor of having seriously held back progress in the field of analogy and categorization for a long time, has already been cited in this book. Here it is, once again:

Categories are boxes, and to categorize is to put items into boxes.

This is the everyday, down-home view of categorization. Let's think about it a little bit.

If categories really *were* boxes and if there really *were* a reliable, precise mechanism for assigning things to their boxes, then it would make eminent sense to distinguish between two types of mental process. First would be *categorization* — a rigorous, exact algorithm reliably placing mental items in their proper boxes; second would be *analogy-making* — a subjective and fallible technique for dreaming up fanciful, unreliable bridges between mental items that do not enjoy a contents-to-container relationship.

However, as research in psychology has shown, and as we have stated throughout this book, the vision *categorization = placing things in their natural boxes* is highly misleading, for categorization is every bit as subjective, blurry, and uncertain as is analogy-making. A categorization can be outright wrong, can be partially correct, can be profoundly influenced by the knowledge, prior experiences, prejudices, or goals, conscious or unconscious, of the person who makes it, and can depend on the local context or the global culture in which it is made. In addition, categorizations can be just as abstract as analogies, can be nonverbal as well as explicitly verbal, can be shaded, and so forth.

Recent work, in fact, makes this point so obvious that the old view of categories as boxes is now usually called "the classical approach to categories", because in cognitive science there is scarcely anyone around any longer who still puts stock in it; today the classical approach tends to be looked upon as simply a quaint historical stage in the development of a far more sophisticated theory of categories and categorization.

And yet, the "categories are boxes" naïve analogy is still seductive, and leads us all to fall victim to the nearly irrepressible belief that all objects and situations we encounter have a privileged category to which they belong, and which constitutes their intrinsic identity. Let's recall the object Mr. Martin purchased; though it subsequently bounced merrily back and forth among such motley categories as *fragile object*, *dust-gatherer*, *spider carrier*, and *home for tadpoles*, it always seemed to remain *in truth* just one thing — namely, a *glass* — and this label constituted its genuine, intrinsic identity. The naïve unspoken analogy "categories are boxes" implies that each item in the world, just like Mr. Martin's glass, has a proper box to which it belongs, and that this connection between a thing and its natural box is universally shared in all people's heads, and finally that this is simply the nature of the world, having nothing to do with thinking or psychology. In this view, not only would category identity exist, but it would be precise and objective.

Over the years, the insidious motto "categories are boxes" has underwritten much scientific research reinforcing the belief that there is a clear-cut distinction between analogy-making and categorization. Indeed, the view of categorization that reigned in cognitive psychology for decades, though expressed in far more sophisticated terms, was essentially indistinguishable from this motto. That view was a restatement of the motto in technical terms borrowed from mathematical logic. It portrayed each mental category as possessing a set of "necessary and sufficient conditions" for membership. An entity belonged to the category *if and only if* it had all those properties. Thus each category-box was thought of as being precisely defined and having rigid, impermeable walls. In this theory of categories, there was no room for degrees of membership, nor for contextual effects on membership in categories. Only in the mid-1970s, when psychologist Eleanor Rosch published her seminal series of articles on categorization, was this erroneous but nearly universally accepted theory at last discredited.

And so, after finally being liberated from the motto, how do today's researchers view analogy-making and categorization? Has a clear consensus emerged on what, if anything, makes the two different? Well, let's listen to some of the top authorities. On the one hand, Thomas Spalding and Gregory Murphy write, "Categories let people treat new things as if they were familiar"; on the other, Mary Gick and Keith Holyoak state, "Analogy is what allows us to see the novel as familiar". If there is a distinction between these two characterizations, it eludes us! Or, to cite Catherine Clement and Dedre Gentner, "In an analogy, a familiar domain is used in order to understand a new domain, especially to predict new aspects of this new domain", whereas for John Anderson, "If one establishes that a given object belongs to a certain category, then one can predict a great deal about the object." Once more, these descriptions of supposedly different processes make them sound as alike as Tweedledum and Tweedledee.

The great overlap of these experts' definitions confirms (if confirmation was needed) the tight relationship of analogy-making and categorization, and provides grist for the mill we are defending — namely, that the idea that analogy-making and categorization are separate processes is illusory. Still, someone who wished to play devil's advocate might argue that by overhauling the definitions of analogy-making and categorization, one might be able to show that they are indeed *two* processes and not just *one*. In fact, after Chapter 8 there is a dialogue that carefully explores this possibility, and we hope that that dialogue, in addition to closing our book, will also close the book on this issue.

However, if you, even after having read this far in our book, still feel reluctant to accept our thesis, rest assured that you are in excellent company, for even experts in the fields concerned fall victim all too often to this same naïve vision. Indeed, we — the authors of this book, the most fervent proponents of our thesis — find ourselves from time to time falling into the very trap we've tried so hard to warn our readers about! Yes, we too fall occasionally for the tempting illusion that categories are boxes. The following way of putting it would probably have warmed the cockles of Aristotle's heart:

> People make naïve analogies.
> Analogy experts are people.
> ______________________________
> Therefore, analogy experts make naïve analogies.

Specialists of any ilk, be they experts in psychology, mathematics, physics, or any other field, do not belong to a different species from ordinary people; they make naïve analogies not only in their daily lives, but also in their professional lives, even ones that involve the concepts with which they are the most proficient.

In short, we sympathize with those readers who still have some doubts about the identity of categorization and analogy-making. It is, after all, a counterintuitive view, and a little voice inside, prompted by a beguiling naïve analogy, continually whispers, "It's wrong! It's wrong!" Nonetheless, we harbor fond hopes that in the remaining pages of our book, we might still get some doubters to swing around to our view.

In any case, speaking of the remaining pages, it is high time that we moved from the often troubling world of naïve analogies to the ever-admirable role of analogies in scientific discovery. And yet in so doing, will we really leave naïve analogies far behind?

CHAPTER 8

Analogies that Shook the World

❧ ❧ ❧

The Royal Role of Analogies in the Realm of Rigor

In Chapter 7, we saw that students who encounter new mathematical ideas for the first time lean heavily on analogies at every step, doing their best to keep from taking tumbles in the abstract world into which they are tremulously treading. We also saw that naïve analogies have a way of sticking in people's minds for a long time. Indeed, naïve mathematical analogies often last a lifetime in the minds of non-mathematicians, tending to lead their makers into dead ends, confusions, and errors. To avoid this fate, one has to gradually refine one's category system as one is exposed to mathematical notions having ever higher levels of sophistication and abstraction. But what about professional mathematicians? Do they, too, rely on naïve analogies in order to keep from stumbling left and right, or would such a vision of their professional lives itself be a result of making too naïve an analogy between beginners and experts?

To some people, it might seem far-fetched to imagine that any role at all might be played by analogical thinking in the professional activity of a mathematician. After all, of all intellectual domains, mathematics is generally thought of as the one where rigor and logic reach their apogee. A mathematical paper can seem like an invincible fortress with ramparts built from sheer logic, and if it gives that impression it is no accident, because that is how most mathematicians wish to present themselves. The standard idea is that in mathematics, there is less place for intuitions, presentiments, vague resemblances, and imprecise instincts than in any other discipline. And yet this is just a prejudice, no more valid for mathematics than for any other human activity.

Anecdotes as Antidotes

The great French mathematician Henri Poincaré devoted much thought to the nature of scientific creativity. In a commentary on mathematicians, he wrote:

> Could anyone think… that they have always marched forward, one step after another, without having any clear idea of the goal they were trying to reach? It was necessary for them to guess at the proper route to get them there, and to do so they needed a guide. This guide is primarily analogy.

This observation about mathematical thinking strikes your authors as being spot-on — but we don't expect readers to take Poincaré's word for it. In support of his thesis, and as antidotes to any possible skepticism about the role of analogies in mathematics, we now offer a bouquet of anecdotes.

For starters, let's go back to the early part of the sixteenth century, to a time when European mathematicians were struggling with the challenge of solving third-degree polynomial equations (of the form "$ax^3 + bx^2 + cx + d = 0$"), otherwise known as "cubics". For more than fifteen centuries, the formula that gave the general solution of *second*-degree polynomial equations (that is, quadratic equations, which have the form "$ax^2 + bx + c = 0$") had been known, and at the heart of this now world-famous formula there was a square root — that of a certain quantity (specifically, $b^2 - 4ac$) calculable from the three coefficients (a, b, and c) of the specific quadratic in question. For the few people who cared about such arcane matters, it was therefore natural to wonder if *something of the same general sort* might not also be the case for cubic equations.

But why would a mathematician entertain such a blurry, imprecise thought? The answer is simple: any serious mathematician would suspect that these two equations, so similar in form, must be linked by a hidden connection — an *analogy*. More specifically, one would expect that there would be a general formula for the cubic and that it would contain, at its core, the *cube root* of some quantity analogous to $b^2 - 4ac$, and that this quantity would involve the *four* coefficients (a, b, c, and d) of the given cubic equation.

Such an analogy is irresistible here, just as were the *me-too* analogies described in Chapter 3. Mathematicians, even the most exceptional ones, are also ordinary human beings, and without any conscious thought, they automatically anticipate that there will be an analogy between, on the one hand, the *two* solutions of any *quadratic* equation, which one can calculate by extracting the *square root* of a special number determined by the equation's *three* coefficients, and, on the other hand, the *three* solutions of any *cubic* equation, which one can calculate by extracting the *cube root* of some special number determined by the equation's *four* coefficients. This little guess, sliding a couple of times from *two*-ness to *three*-ness, and also once from *three*-ness to *four*-ness (which in itself comes from a mini-analogy: "4 is to 3 as 3 is to 2") seems like an utter triviality, but without very simple-seeming conceptual slippages of this sort, which crop up absolutely everywhere in mathematics, it would be impossible to make any kind of progress at all.

Let's return to the story of the solution of "the" cubic equation (the reason for the quote marks will emerge shortly). It all took place in Italy — first in Bologna (Scipione del Ferro), and a bit later in Brescia (Niccolò Tartaglia) and Milan (Gerolamo Cardano). Del Ferro found a partial solution first but didn't publish it; some twenty years later, Tartaglia found essentially the same partial solution; finally, Cardano generalized their findings and published them in a famous book called *Ars Magna* ("The

Great Art"). The odd thing is that, as things were coming into focus, in order to list all the "different" solutions of the cubic equation, Cardano had to use thirteen chapters! Nowadays, by contrast, the whole solution is covered by just one formula that can be written out in a single line, and which could easily be taught in high schools. What lay behind such diversity, of which we no longer see any trace today?

The problem was that no one in those days accepted the existence of negative numbers. For us today, it's self-evident that the coefficient in the third term of the equation $x^3 + 3x^2 - 7x = 6$ is the *negative number* -7. It jumps right out at us, since we are completely used to the idea that a subtraction is equivalent to the addition of a negative quantity. We could rewrite the equation as follows: $x^3 + 3x^2 + (-7)x = 6$. For us who live five centuries after Cardano, these two equations are trivially interchangeable. The conceptual slippage on which their equivalence is based is so minute that we don't even perceive it at all. But for the author of the vast tome on the third-degree equation, the concept of *negative seven* simply didn't exist. For him, the only legitimate way to get rid of a subtraction in a polynomial (that is, a term with a negative coefficient) was to move the misfit term to the other side of the equation, thus yielding a different but related equation — namely, $x^3 + 3x^2 = 7x + 6$ — all of whose coefficients are *positive.*

The upshot of all this is that before he could handle all the different cases of the general problem of cubic equations, Cardano had to move terms around so that there were no more subtractions anywhere, thus obtaining new equations that had only plus signs (and therefore positive coefficients). As it turns out, this procedure gave rise to thirteen different types of cubic equation, each one being — in the eye of specialists of the time — *essentially different* from the twelve others. And thus, in order to publish the general solution to "the" cubic equation (now the *raison d'être* for the quote marks should be apparent!), Cardano had to write thirteen chapters, each of which contained a complicated recipe covering one of the thirteen types of cubic. All in all, Cardano's book on cubic equations, *Ars Magna,* was a long, heavy, and formidable tome, but its reception was, nonetheless, positive, shall we say.

From a contemporary viewpoint, what Cardano did is comparable to someone who invents thirteen kinds of can-openers, each one working for just one type of can. It was a great feat, but what was lacking was an umbrella formulation, laying bare the hidden unity lying behind all this apparent diversity. That is, what was missing for "the" cubic equation was its universal can-opener. But this goal was unthinkable until someone recognized that all these different equations *were,* in fact, just one equation.

Indeed, although all thirteen of Cardano's recipes were somewhat different, there were nonetheless striking similarities between them — analogies, that is, that inspired his successors to try to combine them all into just one formula. However, in order for such a unification to come about, some concept was going to have to stretch, expand, or bend. In this case, the concept in question was the most basic of mathematical concepts — that of *number.* The unification of Cardano's recipes for solving cubic equations depended on a conceptual extension, and a quite significant one, which would allow thirteen different types of algebraic recipes, as seen by as highly skilled a mathematician as Cardano, to melt down into just a single one.

Obviously, what was needed was a new conceptual leap, this time extending the category *number* to include negative numbers. This was by no means an easy step to take. Ever since the ancient Greeks, it had been known that there were very simple equations that lacked solutions, such as $2x + 6 = 0$. The idea of giving such equations solutions had been considered but was always rejected (at least in Europe). Cardano himself understood that "fictitious" numbers (as he referred to them) could satisfy such an equation, but he rejected the idea with disdain. To him, the concept of *negative three*, being in no way visualizable, was an absurdity, somewhat like the concept of an object that violated the laws of physics. Such an idea might be stimulating to the mind, but it had to be recognized as absurd, because there was no way of actually realizing it in the world. Since Cardano was unable to associate negative numbers with any kind of entity in the real world, he labeled them "fictitious" and discarded them.

Nonetheless, his successors — most especially Raffaello Bombelli in Bologna — were powerfully driven to find the elusive unity in Cardano's troublingly diverse set of thirteen chapters, and in the end they wound up accepting the notion of *negative numbers* on the same level of reality (or nearly so) as positive numbers. This move yielded an enormous simplification in the solution for the cubic equation, as the thirteen families were gracefully fused into a single family, and the thirteen recipes associated with them were also fused into a single, compact recipe.

This is an excellent illustration of the recurrent theme in earlier chapters that the human mind is forever driven to *transform* its categories, not just to use them as givens, and that intellectual advances are dependent on conceptual extensions. In this particular case, welcoming negative numbers into the fold was a decision that grew out of *a desire for unification*, and it led to such a gratifying simplification that no one would have wished to go back to the previous stage, with its long list of special cases.

Nonetheless, the welcoming of negative numbers into the category *number* was not immediate or universal. Even 250 years later, the English mathematician Augustus De Morgan, a central figure in the development of symbolic logic, was still resisting, as this passage from his 1831 book *On the Study and Difficulties of Mathematics* shows:

> "8 – 3" is easily understood; 3 can be taken from 8 and the remainder is 5; but "3 – 8" is an impossibility; it requires you to take from 3 more than there is in 3, which is absurd. If such an expression as "3 – 8" should be the answer to a problem, it would denote either that there was some absurdity inherent in the problem itself, or in the manner of putting it into an equation…

DeMorgan's comment is reminiscent of what a seven-year-old girl once said to one of us when she was a participant in experiments on subtraction errors. To explain why she'd written "0" at the bottom of a column containing the numerals "3" and "8", she said, "If I had three pieces of candy in my hand and I wanted to eat eight, I'd eat the three I had and there wouldn't be any more left." Despite the passage of several centuries, the sizable age gap, and the immense amount of mathematical sophistication separating our two commentators, their reactions still share a common essence.

Further on in the same chapter of his book, De Morgan gives a slightly more concrete example, and comments as follows on it:

> A father is 56 and his son 29 years old. When will the father be twice his son's age?

De Morgan translates this word problem into the equation $2(29 + x) = 56 + x$, where x is the number of years that will have to transpire before the described moment arrives. Then he solves the equation, easily obtaining the value of -2 for x. But for him, this result is absurd. What could it mean to say "−2 years will pass"? What could "−2 years from now" mean? Absolutely nothing — or perhaps even less! He then explains that we shouldn't have fallen into the trap lurking in the words "the father *will be*". Instead, we should have thought of x as the number of years that *have passed* since the doubling, which would have led us to write down the equation $2(29 - x) = 56 - x$, whose solution is $x = 2$, which corresponds to the fact that two years ago, the father's age was twice his son's age.

Only at this point is De Morgan happy, admitting that the idea "−2 years will pass" is equivalent to the idea "2 years have passed". He thus does accept the idea that a *numerical value* can be negative, but not that a *length of time* can be negative. All this would lead one to think that De Morgan had no qualms about negative numbers within pure mathematics, even if he didn't think they applied to the real world — and yet, a little later in his book, when he deals with the quadratic equation, instead of considering it as one single, unified problem, he breaks it up into six different families of equations, insisting (in perfect Cardano style) that all three of its coefficients must be positive! De Morgan thus finds that there are *six different* quadratic formulas, rather than one universal one. And all of this nearly 300 years after Cardano!

De Morgan's qualms reveal that the extension of any concept, driven by the mental forces of analogy and by a quest for esthetic harmony based on unification, is a gradual and subjective process, and that even the most insightful of minds within a domain can balk at certain extensions that, some time later, may strike other minds as being as innocent as baby lambs.

In the preceding chapter, we showed that naïve analogies made in early years of school, such as *multiplication is repeated addition* or *division is sharing*, continue to influence the reasoning of adults, including students in college. In this chapter, we have seen that some highly regarded mathematicians, such as De Morgan, could vacillate on the nature of the concept *number*, swinging back and forth between welcoming and barring negative numbers. For him, the legitimacy of the entities being manipulated depended on what they represented in the situation described in words.

Today, the notion of negative numbers seems commonsensical, even bland, which shows how tightly linked the concrete everyday world is to the abstract mathematical world. We are all familiar and comfortable with negative temperatures in the winter, with basements and subbasements whose floor numbers are negative, and with credits on bills, which are indicated with minus signs, meaning that you owe the company a negative amount of money, which is to say, the company owes you a positive amount of money. Children who encounter negative numbers in natural contexts of this sort have no trouble absorbing the general idea. And thus, over time, what was once a most daring intellectual insight turns into a commonplace, unreflective habit.

Enter Complexities

Raffaello Bombelli, after he had fully accepted the existence and reality of negative numbers (around 1570), found himself forced to confront an even greater mystery, which flowed directly out of his acceptance of such strange numbers. The source of the problem was that the formula for the solution of the cubic equation sometimes required taking *square roots of negative numbers.* Bombelli understood the multiplication of signed numbers better than anyone else, having been the first person ever to state the rules for it. He knew that multiplying two negative numbers always gives a positive result, as does multiplying two positive numbers, and thus that there is no number (today we would say "there is no *real* number") whose square is a negative number. In short, all squares are positive (or zero), and so negative numbers do not have square roots.

All this would be fine, but the problem was that square roots of negative numbers cropped up in the formula for the solution of the most innocent cubic equations. For example, the formula giving the solutions to the equation "$x^3 - 15x = 4$", one of which is $x = 4$ (as can easily be checked), was a long algebraic expression inside of which was found (not just once but twice) the square root of -121, which is to say, $\sqrt{-121}$. What to do, when faced with such seeming nonsense? And yet Bombelli knew beyond a shadow of a doubt that this long, curious, and troubling algebraic expression somehow had to stand for the extremely familiar and very real number 4.

Seeing this paradox as a subtle hint, he took the courageous step of accepting the mysterious square root point blank, manipulating it just as he would any other number, using the standard rules of arithmetic (such as the commutativity of addition and multiplication, and so forth). Having adopted this new attitude, he discovered that the solution formula, using the mysterious quantity $\sqrt{-121}$ twice and thus seeming to embody a piece of mathematical nonsense, did indeed act formally like the real number 4. To be more concrete, when he plugged the strange expression into the polynomial $x^3 - 15x$ — that is, when he formally cubed it and then subtracted 15 times it from the result — he discovered, to his amazement, that all the frightening square roots of -121 either were squared, simply giving -121 and thus losing their fearful fangs, or else canceled each other out in pairs, leaving him in the end with just the number 4, exactly as the cubic equation's right-hand side said it should. Bombelli thus realized that he could manipulate $\sqrt{-121}$ just as he would manipulate any "genuine" number, and this analogy with more familiar numbers made the new quasi-number quasi-acceptable for him. From that moment on, Bombelli began to accept the square roots of negative numbers, although he didn't have the slightest idea what they actually *were.*

As mathematicians gradually got used to the fact that these mysterious expressions acted in many ways just as ordinary numbers do, that they were not going to lead people into paradoxical waters, and moreover that they enriched our collective understanding of the world of mathematics, the objecting voices slowly faded away and the mathematical community opened up to them, although not unanimously. Here, for example, is how Gottfried Wilhelm von Leibniz, co-inventor with Isaac Newton of the infinitesimal calculus, described, in 1702, the numbers that René Descartes had dubbed "imaginary": "an elegant and marvelous trick found in the miracle of Analysis: a monster of the ideal world, almost an amphibian located somewhere between Being and non-Being." And even Leonhard Euler, the Swiss genius who deserves enormous credit for putting the theory of complex numbers on a solid footing, declared, on the subject of the square roots of negative numbers, that they are "not nothing, nor less than nothing, which makes them imaginary, indeed impossible."

In any case, despite the protestations of lesser and greater minds, imaginary numbers slowly took hold over the course of the next couple of centuries after Bombelli's early explorations, thanks largely to the discovery of a way of visualizing them as points on a plane, which led to an elegant geometrical interpretation of their addition and multiplication. (This crucial development is vividly described in Fauconnier and Turner's *The Way We Think.*) It would be hard to overstate the importance of geometrical visualization in mathematics in general, which is to say of attaching geometrical interpretations to entities whose existence would otherwise seem counterintuitive, if not self-contradictory. The acceptance of abstract mathematical entities is always facilitated if a geometrical way of envisioning them is discovered; any such mapping confers on these entities a concreteness that makes them seem much more plausible.

N-dimensional Spaces

For example, the notion of *square number* owes its name to the geometrical image of a square on a plane, all of whose sides have the same length, and whose area is therefore the product of that length by itself. In like manner, a *cubic number* was originally understood as the volume of a physical cube: the product of three equal lengths. But no one dared to go beyond the case of the cube — at least not if such a quantity was going to take its name from a visualizable object. Thus the arithmetical operation written as "5 x 5 x 5 x 5" was perfectly fine, so long as one didn't try to attach any *geometrical* interpretation to it. (This is reminiscent of division by a number between 0 and 1, discussed in the previous chapter; most people can carry out the operation *formally*, but only a small minority of adults, even university students, ever come to understand division clearly enough that they are able to dream up a real-world situation that such an operation could represent.) When the audacious analogical suggestion was made that the fourfold product 5 x 5 x 5 x 5 might stand for something that was somewhat like an area or a volume, but associated with a *four*-dimensional space, people objected strenuously, feeling that this violated the essence of what space was. Even at the start of the nineteenth century, that is how many mathematicians protested. As this shows, a reliance on concrete analogies to make sense of abstract mathematical operations is not limited to children or non-mathematicians. The lack of spatial analogues to such arithmetical expressions thus kept the concept of *dimension* from being extended beyond three, even for mathematicians, until well into the nineteenth century.

Once the ice had been broken, however, it didn't take long for the general idea of *N*-dimensional spaces — 4, 5, 6, 7,... — to be accepted, thanks to tight analogies between the theorems that hold in higher-dimensional spaces and those that hold in familiar low-dimensional spaces. Indeed, one of the best ways to imagine four-dimensional space is to try to put oneself in the place of a poor *two*-dimensional being striving mightily to visualize a *three*-dimensional space. One easily recognizes oneself in this creature valiantly struggling against its own limitations; and just like that more limited being, one uses one's analogy-making skills to extend one's mental world. One thinks, "My limitations are just like its limitations, only slightly more elaborate!", and one attempts to transcend one's *own* limitations by imagining how that two-dimensional being would transcend *its* limitations. Once this barrier has been broken, the analogy is so strong that there is no going back. Pandora's box is open and one jumps readily from four to five dimensions, then six, and so on, all the way to infinity.

"What?! A space with an *infinite* number of dimensions? Balderdash!" Thus reacted many mathematicians at the end of the nineteenth century, yet such objections would just bring smiles to the lips of their counterparts today, for whom the idea seems self-evident. In fact, this is just the tip of the iceberg, for after the work of the German mathematician Georg Cantor, it became a commonplace that there is not just *one* infinity, but *many* infinities (of course, there are an infinite number of different infinities).

Spaces with a *countably* infinite number of dimensions (this is the smallest version of infinity, and mathematicians would say that it is the cardinality of the set of natural numbers — that is, of the set of all whole numbers) are called "Hilbert spaces", and for theoretical physicists, quantum mechanics "lives" in such a space; that is to say, according to modern physics, our universe is based on the mathematics of Hilbert spaces. This connection to the physical world lends plausibility to the notion of infinite-dimensional spaces.

Let's not forget that *between* the integers there are plenty of other numbers (for example, 1/2 and 5/17 and 3.14159265358979..., etc.), and mathematicians in the early twentieth century who were interested in abstract spaces — especially the German mathematician Felix Haussdorff — came up with ways to generalize the concept of dimensionality, thus leading to the idea of spaces having, say, 0.73 dimensions or even π dimensions. These discoveries later turned out to be ideally suited for characterizing the dimensionality of "fractal objects", as they were dubbed by the Franco–Polish mathematician Benoît Mandelbrot.

After such richness, one might easily presume that there must be spaces having a negative or imaginary number of dimensions — but oddly enough, despite the appeal of the idea, this notion has not yet been explored, or at any rate, if it has, we are ignorant of the fact. But the mindset of today's mathematicians is so generalization-prone that even the hint of such an idea might just launch an eager quest for all the beautiful new abstract worlds that are implicit in the terms.

How Analogies Gave Rise to Groups

The enrichment of the real numbers by the act of incorporating i, the square root of -1, was a great step forward, because the two-dimensional world that was thus engendered — the *complex plane* — turned out to be "complete" in the sense that any polynomial whose coefficients are numbers lying in the complex plane always has a complete set of solutions within the plane itself. To state it more precisely, any complex polynomial of degree N has exactly N solutions in the complex plane; one never needs to reach beyond the plane to find missing solutions. One can imagine that at this stage mathematicians might well have joyfully concluded that they had at last arrived at the end of the number trail: that there were no more numbers left to be discovered. But the predilection of the human mind to make analogies left and right was far too strong for that to be the case.

The discovery of the solution of the cubic by the Italians in the sixteenth century inspired European mathematicians to seek analogous solutions to equations having higher degrees than 3. In fact, Gerolamo Cardano himself, aided by Lodovico Ferrari, solved the quartic — the fourth-degree equation. Even though there was no *geometric* interpretation for an expression like "x^4", the purely formal analogy between the equation $ax^3 + bx^2 + cx + d = 0$ and its longer cousin $ax^4 + bx^3 + cx^2 + dx + e = 0$ was so alluring to Cardano that he could not resist tackling the challenge. And in short order Cardano and Ferrari, using methods analogous to those that had turned the trick for the cubic, came up with the solution. However, there were some curious surprises lurking in the formula they discovered.

Most strikingly, there was no fourth root anywhere to be seen, although the natural (but naïve) analogy would lead anyone to expect one. Instead, there was a *square root*, and underneath the square-root sign there was a complicated expression in which was found *another square root*, and then underneath the inner square-root sign there was another complicated expression containing a *cube root*, and finally, underneath the cube-root sign there was another long expression that contained *another square root* — all in all, a fourfold nesting of radicals! Who could have predicted such a curious, complicated nesting pattern? Why did only square roots and cube roots appear? Why no fourth root? Why were *four* radicals involved, and not two or three or five — or twenty-six? And why in the order "2–2–3–2" rather than, say, "2–2–2–3"? For that matter, why not "3–3–3–2"? The unexpected and unexplained pattern of these nested radicals in the solution formula for the quartic equation suggested that the general solution for the Nth-degree polynomial must contain some deep secret. And thus was launched the quest for the general solution to polynomial equations of any degree.

Despite the intense efforts of many mathematicians for more than 200 years, nothing worked even for the fifth-degree equation, let alone for its higher-degree cousins. But finally, toward the end of the eighteenth century, the Franco–Italian mathematician Joseph Louis de Lagrange began to intuit the nature of the subtle reason behind this striking lack of success, although he was unable to pin it down precisely. Lagrange saw that there were tight relationships between the N different solutions of the

*N*th-degree polynomial — so much so that these numbers, though not identical, acted indistinguishably in certain respects. Like Tweedledee and Tweedledum, if they were interchanged, there was a sense in which no effect was observable. This meant that there was a new kind of formal symmetry involving such numbers. Lagrange looked into what happens if one permutation of an equation's *N* solutions was followed by another, then yet another, and so forth. He thereby opened up the theory of "substitutions", planting the seed of what turned out later to be the theory of groups, one of the linchpins of modern mathematics.

Lagrange's first ideas about substitutions were extended by Paolo Ruffini in Italy and Niels Henrik Abel in Norway, among others. In 1799, Ruffini published a proof that the fifth-degree equation was in fact *unsolvable* in terms of radicals — that is to say, using square roots, cube roots, fourth roots, and so on. To prove that a goal in mathematics was *unattainable* was almost unprecedented at that time. Unfortunately, in a couple of passages in Ruffini's proof there were some gaps, and for that reason, few of his colleagues took his result seriously. It took another thirty years until Abel published a more complete (though still not flawless) proof of the same result. Finally, in 1830, the very young French mathematician Évariste Galois put the crowning touch on all this work by writing an article that spelled out the precise conditions under which a polynomial equation would or would not be solvable using radicals.

At the heart of Galois' work was the notion of a *symmetric group*, which is to say, the set of all permutations of a finite set of entities (in this case, the *N* solutions of a given polynomial equation). Among these permutations are *swaps*, for instance, in which just two entities are interchanged ($A \leftrightarrow B$), and which are thus analogous to reflections of a human face in a mirror. Then there are *cycles*, such as $A \rightarrow B \rightarrow C \rightarrow A$. This cycle, which carries us back to our starting point after three steps, is analogous to a rotation of an equilateral triangle through 120 degrees. Any physical object whose appearance doesn't change under a reflection or a rotation has some kind of symmetry, and Galois understood the crucial importance of the symmetry that was possessed by the abstract "object" constituted by the set of all *N* solutions of an *N*th-degree polynomial. In making this connection, Galois was consciously generalizing the idea of the symmetries of a physical object, extending it to collections of abstract algebraic entities. The "reflections" and "rotations" of the set of all *N* solutions to a given polynomial were transformations that could be applied one after another, and they reminded Galois of a series of arithmetical calculations carried out one after another.

This reminding soon gave rise to the central analogy that guided Galois in his quest. Having written down a "multiplication table" of all the different permutations of the solutions of a particular polynomial, he proceeded to study the patterns in his table. For example, in the case of the most general quartic equation, Galois knew that the group of symmetries of its four solutions consisted of all the different ways of permuting four indistinguishable objects (*ABCD, ABDC, ACBD,* ...... *DBCA, DCAB, DCBA*). There are 24 such permutations, so the multiplication table of this group has 24 rows and 24 columns. Galois suspected and then showed that the secrets of polynomial equations resided in hidden patterns in the multiplication tables of such groups of permutations.

The history of groups is far too long to tell here, but for us what is important is the idea that the structure of groups themselves became a new domain for research in mathematics. Specifically, Galois discovered that inside a group there are often smaller groups — "subgroups" — and inside subgroups there can be subsubgroups, and so on. He saw that there could be jewels nested inside jewels going many levels down: an incredible feast for his inquisitive young mind!

Once mathematicians had understood and absorbed Galois's highly novel ideas, they collectively passed a point of no return. They made the leap of moving away from the study of concrete, visualizable objects, like the regular polyhedra, and towards the study of more abstract entities such as rotation groups (which reflected the hidden symmetries of concrete objects) and then substitution groups (which reflected the hidden symmetries of abstract entities). And it was the teen-aged Galois who, around 1827, first understood the tight connection between the nesting pattern of subgroups of the group of permutations of the solutions of a polynomial equation and the possibility of solving the given equation with radicals. As the twentieth century unfolded, groups started to pop up in every field of physics, like wildflowers in spring meadows, thus showing the profound prescience of the intuitions of this genius who died at just twenty.

Fields, Rings, *N*-dimensional Knots...

The very same teen-aged Galois also came up with the theory of *fields*, which are groups that have two different operations, one being analogous to addition in its familiar sense and the other to multiplication. The key connection linking these two operations is the distributive law: $a \times (b + c) = a \times b + a \times c$. (Note: The symbols "x" and "+" do not stand for the usual operations of multiplication and addition, but for analogous operations that act on the entities belonging to the field.) Once again we see that the jump to a new kind of entity (sets of things having *two* operations linked by a distributive law) came thanks to an analogy with an already-familiar entity (ordinary numbers). The most familiar of all fields is the infinite set of real numbers, and another example of a field — smaller though still infinite — is given by the rational numbers (a subset of the reals). Certain other familiar sets of numbers, however, such as the whole numbers, are not fields, because if one tries to carry out division, one is led outside of the set (for instance, 2/3 is not a whole number).

Out of the blue, drawn forward solely by his sense of beauty, Galois made a new analogical leap. He observed that fields, like the groups of symmetries that he had just invented, did not have to have an infinite number of members. He saw that there could be finite fields as well, such as the set of integers from 0 through 6, which is a seven-element field, provided that the "sum" of two elements m and n is defined to be the remainder after the ordinary sum $m + n$ is divided by 7. Thus 4 + 5 would not be 9 but just 2 — and similarly, 4 x 5 would not be 20 but just 6 (the standard term for these operations is "addition and multiplication modulo 7").

In any field, whether infinite or finite, one can write down polynomials and ask about the solutions of these polynomials. All the natural questions discussed above in connection with "real" polynomials arise once again in these new worlds. Just as in the field of real numbers, there can exist polynomials that have no solution inside the field in question. Évariste Galois, at age 17, took a marvelous flight of fancy, deciding that a field lacking such a solution could be enriched by adding to it one or more "imaginary" elements modeled on the classic "imaginary" number i, which solved the quadratic equation $x^2 + 1 = 0$, which had no solution in the classic field of real numbers.

What the young Galois dreamt up opened the door to a new style of mathematics characterized by a constant drive towards greater abstraction. One takes a familiar phenomenon — say, the set of prime numbers 2, 3, 5, 7, 11, 13, 17, 19, ... — and one tries to pinpoint its most essential essence, its deepest crux. Then one tries to locate this exact essence in other structures that are less familiar. A good example is given by the "Gaussian primes", which are complex numbers of the form $a + bi$, with a and b being ordinary integers, and which share with the classical prime numbers the quintessential property of having no divisors except themselves and unity. The study of such generalized prime numbers is a rich vein of research in contemporary mathematics.

Not long after the tragic death of Galois, mathematicians came up with a new kind of algebraic structure that was similar to a field but in which division was not always possible. The idea lying behind these new structures, called "rings", came from the attractive — indeed, well-nigh irresistible — analogy that said that every field had to contain a special subset that played the role played by the integers in the classic field of the rational numbers. Whereas you can always divide any rational number by any other to get a third one, this does not hold for integers — dividing 5 by 3 does not yield another integer. Similarly, in a ring, division of two arbitrary elements is not always possible. Thus, given any specific ring, one could imagine its set of "prime numbers" — those entities that are not the product of any elements of the ring except themselves and unity. The study of the prime elements of rings attracted many researchers in the second half of the nineteenth century, inspiring them to define structures and phenomena that were ever more abstract and that seemed ever more distant from "real life". And yet many of these ultra-abstract phenomena turned out, decades or centuries later, to be profoundly implicated in the laws governing the physical universe, and so in the end they turned out to be every bit as real as the more familiar mathematical notions that everyone learns in school.

The study of groups furnishes a good example. Galois noticed that under certain special conditions, a group can be "divided" by one of its subgroups (of course we are not speaking of the familiar arithmetical notion of division, but of a far more abstract analogue — an *unmarked* sense of division, to use the term we used in Chapter 4). This process gives rise to what is called a "quotient group".

In order to get some sense of what Galois's new kind of abstract division is like, imagine a 100-floor skyscraper all of whose floors are identical. We now wish to "divide" the building by one of its floors. What we will get is a skeletal skyscraper, a tall vertical structure no longer having 100 floors but simply 100 dots. *Et voilà !* This abstract structure is the "quotient skyscraper", the result of "dividing" the original skyscraper by its generic floor. If you are able to carry this rarefied notion of division from the world of skyscrapers back to the much more abstract world of groups, then you will have some feel for what "dividing" a group by one of its subgroups is all about. This process of division by a subgroup can be repeated until one finally obtains a chain of "prime factors" of the original group. Thus one is led back to the familiar notions of *factorization* and *prime number*, or at least their very distilled essences, now instantiated in this abstract and austere context.

The theory of knots, launched by Leonhard Euler in 1736 and ever since then profoundly enriched by thousands of researchers, furnishes another good example of this kind of push towards abstraction via analogy. At a certain point, the concept of *factorization* was extended into the world of knots (and here, we really do mean knots on a string — a very concrete idea, for once!), which meant that a complicated knot could be seen as the "product" of two simpler knots, and so forth. This process of reducing complex knots to simpler and simpler knots then led, naturally, to the idea of *prime* knots. However, no sooner had this idea been suggested than it was inevitably generalized to intangible knots inhabiting N-dimensional spaces, and all at once things became completely unimaginable to anyone except to those few hardy souls who are able to survive in the forbidding world of such high abstractions.

As we've tried to show, modern mathematics is pervaded by an unrelenting pressure to move towards ever more abstract ideas, but unfortunately the ubiquity and intensity of this pressure are almost impossible to convey to non-mathematicians. The *modus operandi* of mathematical abstraction is pretty much as we have described it above: you begin with a "familiar" idea (that is, familiar to a sophisticated mathematician but most likely totally alien to an outsider), you try to distill its essence, and then you try to find, in some other area of mathematics, something that shares this same distilled essence. An alternative pathway towards abstraction involves recognizing an analogy between two structures in different domains, which then focuses one's attention on the abstract structure that they share. This new abstraction then becomes a "concrete" concept that one can study, and this goes on until someone realizes that this is far from the end of the line, and that one can further generalize the new concept in one of the two ways just described. And thus it goes...

Mechanical Mathematical Manipulations: Also the Fruit of Analogy

Before we take our leave of mathematics, we would like to comment on the use of standard techniques for manipulating mathematical expressions. Take, for example, Cardano's shifting of a term from one side of an equation to the other, by which he got the term to change sign. This is a general strategy or technique that children today are taught in beginning algebra classes. For instance, to solve the equation $3x - 7 = x + 3$, one carries the variable "x" across from the equation's right side to its left side while changing its sign, and (analogously) one carries the number -7 across from the left side to the right side while changing its sign. The upshot is the equation $2x = 10$, and here one divides both sides by 2, thus winding up with the solution $x = 5$. But what, if anything, do such routine algebraic operations have to do with analogy-making?

A great deal! Such operations come from analogies that reside at the lower end of the creativity spectrum, very much like perceptual acts such as looking at an object sitting on a table and mentally calling it a "a paperweight". For someone to be able to recognize objects and to assign them labels, their memory has to be efficiently organized, allowing new experiences to evoke, by analogy, linguistic labels that long ago were attached to old experiences. Such labels seem so self-evident that one can easily fall prey to the seductive illusion that one is doing nothing other than mechanically allowing the *intrinsic* labels of the various items in the situation to emerge explicitly.

The very same illusion arises in routine mathematical situations as well. The recognition, when one is in a brand-new mathematical situation, that it is crying out for a certain standard memorized technique is much like the everyday recognition that one is in a new situation that is crying out for a standard linguistic label, or for a standard canned kind of behavior. Recognizing that the equation $3x - 7 = x + 3$ is crying out for two of its terms to be transposed to the other side (and for their signs to be changed), and then recognizing that the resulting equation, $2x = 10$, is in turn crying out for division of both its sides by 2, is no different, in principle, from recognizing that a suitcase "cries out" to be lifted (an "affordance", as defined in Chapter 6), or that a certain situation calls for hitting a nail into a wall with a hammer.

There are all sorts of standardized techniques for manipulating algebraic equations, such as the ones just mentioned; indeed, mathematics abounds with routine techniques at all levels of abstraction. Among these are various ready-made, off-the-shelf types of logical argument, such as the famous schema called *reductio ad absurdum*, which basically says that if you wish to prove that an idea X is true, see what happens if you presume that the opposite of X is the case, and if this assumption leads to an absurdity (more specifically, a self-contradiction), then X must be correct. This mode of reasoning is a standard, often-used ingredient in every mathematician's toolkit. Likewise, one of many standard techniques for doing integrals in calculus is that of making "trigonometric substitutions". One recognizes that a certain situation "smells" as if it needs such a substitution, and so one goes for it. The fact that it eventually takes on a recipe-like quality, however, does not disqualify it from being a case of analogy-making, since if one has a good deal of mathematical experience, the formulaic manipulations that one uses all the time were committed to memory a long time earlier. Now, thanks to one's ability to categorize situations efficiently, these mathematical techniques are evoked "automatically" — that is to say, so effortlessly that it feels as if analogy plays no role. But this is simply the same illusion that says that the paperweight on the table evokes the label "paperweight" independently of one's analogy-making capacity.

How many times have you played Monopoly or a similar game? Probably very often. And of course your style of play today is consistent with, and indeed comes from, all those times when you played it before. You know when you feel comfortable taking risks buying pieces of property or railroads or houses or hotels, and you know when it feels too dangerous. Although such decisions occur in a much more concrete world than that of a mathematician, they have much in common with a mathematician's decisions about whether to try using a certain technique or not in a given situation.

Exploring the technique might waste a lot of valuable time and effort, but it might pay off in the end. Is it worth it? It all depends on how the unfamiliar and nebulous new situation "smells" to the seasoned mathematician. These are subtle qualities that come only from long experience and from having slowly built up a refined repertoire of categories. And as in Monopoly, so in mathematics.

One day, two math graduate students were arguing about a famous unsolved problem in number theory. One of them was insisting that the highly elusive and mysterious distribution of prime numbers must play a crucial role in the matter; the other argued that *that* couldn't possibly have any relevance at all. As it happens, neither student had any intention to devote years of their life to this thorny problem, and so their clashing opinions on what was of the essence versus what was irrelevant were merely idle words without import; but if someday they both decided to tackle this challenge independently, then their diametrically opposite intuitions about its nature would have profound impacts on the directions that their research would follow. Thus whichever of them had "sniffed" the true crux of the problem would wind up following far more promising pathways than the other one would. And what would have given them these powerful, make-or-break intuitions (whether they were correct or incorrect)?

Analogies in mathematics range widely in sophistication. At the lower end of the spectrum is the evocation of standard canned recipes such as we have been discussing. At the upper end lie the strokes of genius by great mathematicians, such as the importation of the idea of *imaginary number* into the realm of finite fields, or the search for prime numbers (or prime knots!) in highly abstract structures. Perhaps mathematics seems at times to be merely a symbol-manipulating game, but it is important to realize that even the most rote-seeming manipulation of symbols relies on analogy.

In summary, mathematics involves the making of analogies at all points along the spectrum, ranging from the blandest of memorized symbol-shunting recipes all the way up to the most vertiginous generalizations of ideas that are already highly abstract. In the sections preceding this one, we concentrated on quite sophisticated analogies in mathematics because this chapter is principally about what lies behind scientific creativity and "discoverativity". It's understandable that people may not at first see lowly symbol-manipulation recipes as requiring analogical thought, but this is only because they are letting themselves be subliminally guided by a naïve stereotype of analogy-making itself. This ironic fact is par for the course, though; after all, people seldom recognize at first how much analogy-making is involved in the mundane-seeming acts that result in the evocation of words and phrases. Once they get used to the idea, however, it becomes self-evident. And the same idea applies to routine techniques in mathematics.

Physics and Logical Thinking

We now shift our attention from mathematics to physics. In cartoons one often sees a wild-haired physicist standing before a blackboard overflowing with equations and esoteric symbols. Such images suggest that the prototypical thought processes of any physicist consist in the manipulation of symbols according to the laws of mathematics. High-school and college physics courses encourage this vision, as they abound in precise formulas for all sorts of sophisticated and elusive notions such as *magnetic field*, *kinetic energy*, and *angular momentum*. To be sure, all these formulas are perfectly correct, but they are nonetheless misleading, because they give the impression that physics is an axiomatic discipline in which, starting with a few basic principles, one can use the most rigorous logic and from them deduce all the key results of mechanics, thermodynamics, electromagnetism, the theory of waves, and so forth. This is the standard image of research in physics for the public at large, for intellectuals, and even for a good number of physicists — but it is, in the end, a myth.

Although many textbooks written by physicists paint physics as a purely deductive science, a good number of others effectively counter that image. And some of the greatest physicists have even tried to elucidate the pathways of thought that once led them to their most important discoveries. They invariably tell stories of analogies that came to them — analogies establishing links between just-discovered, fresh, unexplored physical phenomena and older, well-explored, and clearly understood physical phenomena.

And so, would the difference between a great and a mediocre physicist have to do with the degree of wildness and craziness of the analogies that each is willing to entertain? Would a genius in physics be someone who deliberately sets out to explore bizarre and far-fetched analogies between extremely distant concepts, someone who purposefully tries to make implausibly wide leaps of imagination connecting concepts that nobody would have dreamt had the slightest relation, all based on small pieces of evidence that other physicists would simply laugh at? And contrariwise, would the mediocre physicist be a timid individual who explores just a small neighborhood of ideas right under their nose, and who is never tempted to make bold mental leaps?

Such an image of genius versus mediocrity, though in some ways appealing, is not our view. Geniuses do not deliberately set off with the goal of concocting a wild-sounding analogy between some brand-new phenomenon, shimmering and mysterious, and some older phenomenon, conceptually distant and seemingly unrelated; rather, they concentrate intensely on some puzzling situation that they think merits deep attention, carefully circling around it, looking at it from all sorts of angles, and finally, if they are lucky, finding a viewpoint that reminds them of some previously known phenomenon that the mysterious new one resembles in a subtle but suggestive manner. Through such a process of convergence, a genius comes to see a surprising new essence of the phenomenon. This is high-level perception; this is discovery by analogy.

Albert Einstein, Analogizer Extraordinaire

Some years ago, this book's senior author, who many years earlier had gotten his doctorate in physics and who had always been fascinated by the pathways of discovery in that discipline, decided to develop a lecture on the role of analogies in physics. Long before starting to prepare the talk, he impulsively chose as his title "The Ubiquity of

Analogies in Physics", and so, when he actually tackled the topic in earnest, it was with a sense of great relief that he discovered that each branch of physics is indeed as densely riddled with analogies as he could ever have hoped. Better yet, at the heart of nearly all the great discoveries of the last three centuries, he found that there lay a crucial, decisive analogy. Here is a small sampling from the list:

- gravitational potential: analogous to a hill — Lagrange and Laplace, 1770
- electric potential: analogous to gravitational potential — Poisson, 1811
- magnetic potential: analogous to electric potential — Maxwell, 1855
- four-dimensional space–time: analogous to three-dimensional space — Minkowski, 1907
- electrons as waves: analogous to photons as particles — de Broglie, 1924
- quantum-mechanical matrices: analogous to Fourier series — Heisenberg, 1925
- quantum-mechanical waves: analogous to classical waves — Schrödinger, 1926
- quantum commutators: analogous to classical Poisson brackets — Dirac, 1927
- "isospin" of related particles: analogous to quantum spin states — Heisenberg, 1936
- weak nuclear force: analogous to electromagnetic force — Fermi, 1931
- strong nuclear force: analogous to electromagnetic force — Yukawa, 1934
- vector bosons: analogous to photons — Yang and Mills, 1954

It thus appeared that just about every breakthrough by the greatest physicists, such as Newton, Maxwell, Dirac, Heisenberg, Fermi, and many others, had been the fruit of one or a number of analogies intuitively "sniffed" by its discoverer. Indeed, it seemed as if the lecturer's early-chosen title had been confirmed far more powerfully than he had ever thought was possible. Curiously, however, he had deliberately shied away from looking at the work of Albert Einstein, as he was nagged by a fear that the genius of Einstein might have been so profound that he was able to do completely without those irrational leaps of the sort made by "ordinary geniuses", and that Einstein might instead have relied exclusively on the purest of logical reasoning to arrive at his revolutionary ideas. Was it possible that Einstein's exceptionally deep thinking style might undermine the thesis of the talk that he was going to give?

It happened that a few years later, the same speaker was invited to take part in a conference celebrating the centennial of the year 1905, which is known as Einstein's *annus mirabilis* — his "miraculous year" — since that was the year that he revolutionized physics with a series of five unparalleled articles. For this special occasion, the speaker decided to take the bull by the horns and to look carefully into the pathways of Einstein's thought in the hopes of finding at least a handful of analogies in them. And *mirabile dictu*, to his great relief, he found that analogies abounded in the thought processes of Einstein, every bit as much as in those of any other physicist.

There was just one difference — namely, that often what at first seemed to be a shaky analogical leap between two phenomena would, years later, turn out to be the unification of two phenomena that up till then had been thought of as totally separate. Over and over again, two things were revealed to be really just one thing. Could it be that Einstein always knew from the very start that he had discovered an *identity* as opposed to a mere *analogy*? In fact, no. Even the genius of Einstein could not foresee that his newborn intuitive hunch that two physical phenomena were *related* — his new analogy — would someday wind up totally *unifying* them. From his perspective, he was merely offering a promising analogy whose consequences would unfold with time. But in Einstein's case, unification happened so frequently that one is forced to say that this is one of the most salient traits of Einstein's thinking style. By contrast, most other physicists find analogies that, though revealing a fruitful network of similarities between phenomena, do not lead to a deep unification.

To make this idea a bit more concrete (via a caricature analogy), suppose that a chemist of a few centuries ago, after observing a number of similar characteristics of coal and diamonds, had made the wild suggestion that these two totally different-seeming substances were linked by some kind of hidden analogy. What a thrill it would have been for this chemist to learn one day that diamonds and coal are just two superficially different forms of one single chemical element — namely, carbon!

Low-level and High-level Einsteinian Analogies

Before we step aboard the train of grand Einsteinian analogies, we wish to make it clear that it was not solely in the highest and most exalted flights of his scientific imagination that Albert Einstein resorted to making analogies. Just like every human being, Einstein perceived the world around him by making analogies on many levels, all the way down to the tiniest mental connections.

A rather obvious example of an Einsteinian analogy comes from a reminiscence he wrote, late in life, about why he choose, as a teen-ager, not to become a mathematician: "I saw that mathematics was split up into numerous specialties, each of which could absorb the short lifetime granted to us. Consequently I saw myself in the position of Buridan's ass, which was unable to decide upon any specific bundle of hay." The image is amusing, especially when one notes that Albert Einstein is comparing himself to a ravenous but confused donkey that is utterly immobilized because it is surrounded by a number of tempting things to eat, and which, in the end, perishes because it can't bring itself to make a choice. The analogy isn't bad — indeed, it has charm — but it certainly didn't take an Einstein to make it!

And there are smaller and humbler analogies aplenty lurking unnoticed in this same pair of sentences. To take just one, Einstein describes each specialized area of mathematics as potentially "absorbing" one's entire short lifetime. Any reader will effortlessly conjure up an image of a professional life utterly "sucked up", "eaten up", "swallowed up", or "devoured" by the need to master a vast field of knowledge. Almost surely this choice of a verb by Einstein was not the result of a prolonged linguistic rumination but was made casually and instinctively, as are most word choices when one writes or says anything. One can also point to many other word choices in Einstein's writings that have a similar flavor — for instance, his choice of the verb "to fall in love" in the sentence "I really began to learn [to play the violin] only when I was about 13 years old, mainly after I had fallen in love with Mozart's sonatas."

The point is that *all* of Einstein's word choices, whether he was writing, speaking, or just thinking, were likewise the fruit born from myriads of tiny, evanescent analogies, as is the case for all humans. Whenever Albert Einstein saw a shoe and the word "Schuh" came to his mind (or the English word "shoe", once he had moved to America), or whenever he saw someone ambling down the street and the word "walk" came to his mind — in any such case, he was building an analogical bridge between a current percept and a stored structure that had been built up in his memory, no less than in yours or ours, by untold numbers of previous analogies made all through his lifetime.

To be sure, such tiny, mundane Einsteinian analogies are far from being "analogies that shook the world", and if a chapter advertised as being largely about analogies made by the great Albert Einstein did nothing but discuss occasions on which he made such forgettable pronouncements as, "Where are my old shoes?" or "Oh, look, there's my old friend Gödel walking home", readers would be forgiven for feeling that they had been sold a bill of goods, so to speak. But we felt it important to bring out the fact that even the great Einstein's thought was utterly pervaded by unspectacular, throwaway, workaday analogies that allowed him to live. But from here on out, when we speak of "Einstein's analogies", we will mean his few dozen *great* analogies, not his millions of mundane ones. With this out of the way, we now head straight for the big game.

A Crazy "Swimming Pool Table" Analogy Emits Quanta of Light

Toward the end of the nineteenth century, one of the great mysteries in physics was that of the famous "blackbody spectrum". A *spectrum* is a graph that shows what proportion of the energy carried by a bunch of waves is associated with each different wavelength, while a *black body* can be thought of as a hollow cavity whose walls are held at a specified temperature, and which is filled with electromagnetic waves that are bouncing against the walls and crisscrossing the empty space — the vacuum — in the cavity, in the manner of ripples crisscrossing on the surface of a swimming pool. The most familiar type of electromagnetic wave is visible light, which includes all the colors of the rainbow, but the notion also includes ultraviolet, infrared, X-rays, gamma rays, microwaves, radio waves, and so forth. The essence of a *black body* thus resides in the co-presence of electromagnetic waves of many wavelengths, continually bouncing off the walls of the cavity just as ripples reflect off the walls of a swimming pool.

This analogy between a black body and a swimming pool will in fact be very useful to us. If you toss an object — a wedding ring, a bowling ball, a grand piano — into a swimming pool, the splash will create circular waves having various wavelengths, which will then be reflected off the pool's walls. Roughly speaking, the lighter the object, the daintier will be the waves (both in their amplitude, or height, and in their wavelength). A physicist might well ask, "Given the mass of the object thrown into the water, what will be the dominant wavelength of the waves it causes?" Along the same lines but more sophisticated would be the question, "Given the mass of the thrown object and also some particular wavelength, how much of the total energy of the agitated water do the waves at that wavelength have?" One easily imagines that for each different object tossed in, there will be a specific wavelength that will be dominant in the pool, and that waves having much longer or much shorter wavelengths will hardly be triggered at all by the splash. A graph showing the distribution of energy among different wavelengths would constitute the spectrum of waves on the pool.

Much the same holds for a black body, except that there is no palpable medium that wiggles up and down or back and forth, like the water in the pool; the vacuum through which the electromagnetic waves course, crisscrossing each other all the time, contains no matter (whether solid, liquid, or gas), by the very definition of the word "vacuum". The waves themselves are ghostlike entities, consisting of electric and magnetic fields oscillating in space in time, which have the ability to make charged particles move. An electromagnetic wave is thus much like a fluctuating gravitational field; it is something that pervades empty space and whose value at any point in space changes according to a "wave equation", or possibly a set of intertwined equations — in this case, Maxwell's equations, discovered in the early 1860's.

The black body's temperature corresponds, in our analogy, to the size of the object tossed into the pool, which gives rise to the ripples on the surface. The hotter the black body's walls are kept, the more electromagnetic energy there will be in the cavity, and, perhaps surprisingly, the shorter will be the dominant wavelength of the waves. This may seem counterintuitive, since it contradicts what one would guess on the basis of the swimming-pool analogy. After all, a sofa heaved into a pool will make very long-wavelength ripples, while a pebble will make short-wavelength waves. But we won't worry about the conflict with the water-wave situation; physicists at the time were well aware of these phenomena in blackbody radiation. They were very familiar with the bell-curve shape of the graph of the blackbody spectrum, as determined through many careful experiments. They simply had no coherent explanation for that shape.

Admittedly, in 1900, the German physicist Max Planck had discovered an elegant mathematical formula that, for any given temperature, exactly reproduced the experimentally found graph of the blackbody spectrum, but at the outset he had no justification for his formula. It was as if he had pulled it out of his hat, like a rabbit. Then, after several months of obsessive toil, he showed how one could deduce his formula from a strange and arbitrary-seeming assumption about the atoms in the walls of the cavity — namely, that their vibrational energies were limited to taking on just certain special values, determined by a new constant of nature, to which Planck gave the name "h", which later was dubbed "Planck's constant". An atom, instead of being able to vibrate with any arbitrary amount of energy (as all physicists would have expected), could only assume energies that were exact multiples of a certain very small chunk or "quantum" of energy, as Planck called it. The idea of energy coming in such chunks struck physicists, even Planck himself, as very weird; however, everyone realized that Planck's formula, based on this idea, matched the experimental results perfectly.

For macroscopic creatures such as humans, the fact that the atoms in the walls couldn't take on arbitrary amounts of vibrational energy would be unnoticeable, as the magnitude of h was so tiny. Indeed, if h were equal to zero, then all possible energies would be allowed and the quantum restriction would not hold. However, the graph that could be calculated from the assumption that *h* equaled 0 didn't look at all like the true blackbody spectrum, as found in experiments; contrariwise, Planck's theoretical graph, which he had calculated using his tiny but non-zero constant, matched the mysterious blackbody spectrum with total accuracy. These developments certainly had a promising feel to them, but for most physicists of the time, Planck's theory of the "quanta of excitation" of the atoms in the cavity's walls seemed too arbitrary, and for that reason almost no one took it seriously. Even Planck felt very uncomfortable. Although he had led himself to the quantum pond, he did not want to sip from it.

This lull provided a most opportune moment for another genius to step into the scene — a young clerk in the Swiss patent office in Bern. The great masterstroke of Albert Einstein, for indeed he is the person we mean, was to spot a key parallel between a black body and a different system that also has a spectrum determined by its overall temperature. In particular, we are alluding to an *ideal gas* trapped inside a container. Why did Einstein make this analogy, which no one else — or almost no one else — had imagined? And why did he trust it so deeply? We will shortly give some speculations.

Curiously enough, until the *annus mirabilis*, no totally convincing proof had been found of the existence of atoms (whether in solids, liquids, or gases). To be sure, certain audacious scientific spirits, such as Ludwig Boltzmann, from Austria, and James Clerk Maxwell, from Scotland, had conjectured that all gases consisted of myriads of small particles constantly bashing into each other, as well as bouncing off the walls that contained them, and from this "ideal gas" assumption they had been able to derive certain formulas that matched the empirical observations of *real* gases to an incredible degree of precision. Although this was a strong piece of evidence in favor of the atomic hypothesis, a good number of physicists, chemists, and philosophers remained skeptical.

It will help us here to draw another explanatory analogy — this time between an ideal gas and a frictionless pool table. On the pool table we imagine hundreds of tiny balls that have been set in motion by a violent explosion (such as the "break" at the beginning of a pool game), and that now are all madly bouncing off the table's walls and off of one another. If someone were to tell us the amount of energy of the initial explosion, then we might wonder what the dominant speed (*i.e.*, the most common speed) of the balls on the table would be, once things had settled down into a more or less stable state. More ambitiously, we might ask what the *distribution* of speeds of the balls will be. It might seem surprising, *a priori*, that there is a precise answer to such questions, but in fact there is. And in an analogous fashion, there is a precise formula — the so-called "Maxwell–Boltzmann distribution" — that gives, for any particular value of kinetic energy you wish, the percentage of molecules of gas that will have that energy (given the temperature of the gas). The location of the peak of this graph reveals the dominant kinetic energy of the gas molecules.

Albert Einstein had a hunch that these two types of system — the black body and the ideal gas — were deeply related despite their surface-level dissimilarities. In both cases, there is a container filled with energy, but beyond that, what would make anyone suspect that these two systems were deeply linked? Let's shift the question back to the more familiar territory of our two analogues — the swimming pool and the pool table.

It then becomes a question about the likelihood of there being a profound connection between the rippling surface of a swimming pool agitated by a splash, and hundreds of tiny balls bouncing about in a frenzied manner on a pool table, all set in motion by a sudden explosion. Both situations are filled with random motion and take place on horizontal, flat surfaces, but those very superficial facts would hardly seem to add up to a strong reason to make anyone suspect that a deep relationship links them.

Thus, for the vast majority of physicists at that time, Einstein's analogy between the ideal gas (here likened to a seething billiard table) and the black body (here likened to a natatorium's undulating surface) seemed utterly implausible. So why did Einstein see things differently? First of all, as he stated in the first of his articles of 1905, he had noticed a curious *mathematical* similarity linking the two formulas giving the energy distributions (for the blackbody spectrum, Einstein used a formula discovered by the German physicist Wilhelm Wien before Planck found his more precise one, and for the ideal-gas spectrum, he used the formula of Maxwell and Boltzmann), and this suggested to him that the *physical* similarity of the two systems might easily go well beyond the surface. All one can say here is that Einstein had an eagle eye; he almost always knew how to put his finger on just what mattered in a situation in physics.

It is fascinating to note that Wilhelm Wien, in his search for a formula for the blackbody spectrum in the mid-1890s, had had the excellent intuition — closely related to Einstein's intuition some ten years later — to try using an analogy he had "sniffed", linking the blackbody spectrum to Maxwell and Boltzmann's ideal-gas spectrum. It is thus no coincidence that Einstein refound Wien's analogy when he looked at the two formulas at the same time, for Wien's formula was rooted in the Maxwell–Boltzmann formula. For Wien, however, the analogy between the two systems was solely *formal*; it did not suggest to him that the two systems had a deep *physical* link, and so in his mind he did not pursue it nearly as doggedly or as profoundly as did Einstein.

One other factor that might have contributed to Einstein's faith in his analogy between the physics of the ideal gas and that of the black body (not just between the mathematical formulas for their spectra) was the fact that only a few months earlier, he had found and deeply exploited an analogy between an ideal gas and another physical system — namely, a liquid containing colloidal particles whose nonstop, apparently random hopping-about could be observed through a microscope. This analogy had allowed him to argue persuasively for the existence of extremely tiny invisible molecules that were incessantly pelting the far larger colloidal particles (like thousands of gnats bashing randomly into hanging lamps) and giving them their mysterious hops, known as "Brownian motion". It is thus probable that two distinct forces in Einstein's mind — the mathematical similarity of the formulas and also his recent Brownian-motion analogy — gave him great trust in his analogy between a black body and an ideal gas.

In any case, building on the bedrock of his latest analogy, Einstein undertook a series of computations, all based on thermodynamics, the branch of physics that he thought of as the deepest and most reliable of all. First he calculated the entropy of each of the systems and then he transformed the two entropy formulas so that they would look as similar as possible to each other; in fact, at the end of his ingenious manipulations, they wound up exactly identical except for the algebraic form of one simple exponent. This provocative maneuver made it clear that the two systems were far more intimately related than Wilhelm Wien had ever suspected.

In the key spot in the formula for the ideal gas's entropy, the letter "N" appeared, standing for the number of molecules in the gas; in "the same" spot in the formula for the black body's entropy, the expression "$E/h\nu$" appeared. (The letter "h" stands for Planck's constant, and the Greek letter "ν" — "nu" — for the frequency of the electromagnetic waves, always inversely proportional to their wavelength.) Einstein had thus compressed the entire distinction between these two vastly different physical systems down into one tiny but telling contrast: an integer N in one case, and the simple expression "$E/h\nu$" in the other.

But what did this precision pinpointing of their difference mean? Well, $E/h\nu$ represents the act of dividing up the total energy E (a large number of ergs, an erg being a standard energy unit) into many minuscule chunks all having energy $h\nu$ (a tiny fraction of one erg). This ratio tells how many small chunks make up the larger chunk; and thanks to the cancellation of the units (ergs in both numerator and denominator), it is a "pure" number: its value is independent of the system of units used. Einstein's analogy now plays a key role, telling us that this number in the blackbody system maps onto the number of molecules N in the ideal gas. The dividing-up of E into identical pieces all having size $h\nu$ (a "measurement" of E, to echo the term used in Chapter 7 for one of the naïve analogies to division), was an unmistakable clue, for Einstein, that the radiation in the cavity was composed, as is a gas, of discrete particles. For any given wavelength, all "light chunks" carried the same tiny load of energy.

Even for its finder, this was a monumentally shocking idea, because to him, just as to all physicists of his day, *electromagnetic radiation* was synonymous with *light* (along with light's cousins having longer and shorter wavelengths), and Einstein was very aware, as were all his colleagues, that the ferocious battle between advocates of *light as corpuscles* and advocates of *light as undulations* had finally been conclusively won, a century earlier, by the undulatory side. Furthermore, ever since then, thanks especially to Maxwell's fundamental equations, discoveries in physics had reinforced over and over the view of light as continuous waves and not as discrete particles. How, then, could a corpuscular view of light possibly stage a comeback a hundred years after its demise? And yet, this is exactly what seemed to be happening, thanks to a very simple analogy.

Einstein, recognizing that there was nothing to do but accept the image so clearly suggested by his analogy, came to the staggering hypothesis, flying in the face of the most solidly established facts, that the electromagnetic radiation in a blackbody cavity consisted of small corpuscles — small packets of energy, analogous to the N molecules in an ideal gas (and, to show how far ahead of his times Einstein was, we point out that at that moment in physics, even the existence of atoms and molecules was still considered suspect by some skeptical holdouts!). Each of these mysterious "lumps of radiation" would necessarily possess exactly the energy $h\nu$, which thus had to be the *minimal* amount of energy associated with the frequency ν. Called "light quanta" by Einstein, such particles are known today as "photons".

Light Quanta Are Scorned While Sound Quanta Are Welcomed

Although not in the least controversial today, Einstein's bold suggestion in 1905 that light must consist of particles was harshly and unanimously dismissed by his colleagues. Later in life, he declared this hypothesis, based on but the shakiest of analogies, to be the most daring idea of his entire career; indeed, it was so daring that it unleashed, among his colleagues, a barrage of scorn and hostility whose magnitude, duration, and ferocity he surely could not have anticipated.

In the conclusion of his light-quantum article, the young "Technical Expert, Third Class" (the lowest rank at the Swiss patent office) had both the cleverness and the courage to suggest three possible experimental ways to confirm or refute his theory, thus taking the risk of handing weapons to his enemies, with which they could potentially shoot him down! In particular, the second of his suggestions involved looking at the photoelectric effect, in which, when electromagnetic radiation (such as light) falls on a piece of metal, some electrons come flying out of the metal. It was an odd little effect but was considered of no great moment for physics, and had been observed for the first time, but only very crudely, in 1887 by the German physicist Heinrich Hertz, in a series of experiments in which he conclusively demonstrated the existence of electromagnetic waves, thus brilliantly confirming Maxwell's equations.

Einstein realized that his theory of light quanta yielded precise predictions for the photoelectric effect. In particular, in a very simple equation, it predicted the rate of ejection of electrons as a function of the wavelength of the incident light, and this prediction was in stark contradiction with predictions based on Maxwell's universally accepted equations. Einstein could not know, nor could any other physicist of the time, what would be revealed by precise measurements of the photoelectric effect, but it was clear to him that such experiments would be decisive and might lead to a great battle, because if his prediction turned out to be correct, the world of physics would be forced to reject Maxwell's equations as the basis of electromagnetism. This was among the most paradoxical moments in the entire history of physics, for Hertz's experiments, which had so triumphantly *confirmed* Maxwell's equations, were also the source of the tiny anomaly that now threatened to *undermine* those very equations. However, the investigation that Einstein suggested in his conclusion was very difficult to carry out, and it took quite a number of years before the experiments yielded clear results.

In 1905, though, no one paid the least attention to the light-quantum hypothesis, as everyone but Einstein was completely convinced of the validity of Maxwell's equations. Light was made of waves; that was that. To doubt it was simply insanity. Even Max Planck, who had dreamt up the idea of *quanta of energy of vibrating atoms*, proclaimed that the new hypothesis of *quanta of light* was senseless. (It is of note that Planck, some years earlier, had also declared that the hypothesis of atoms was senseless, but by 1905 he greatly regretted having done so.) Despite his colleagues' unanimous scorn, the young Einstein had an unshakable faith in his own ideas, and was not discouraged. (Actually, calling them "his colleagues" is a bit of a stretch, since until 1908, Einstein was merely an amateur physicist, his official job being that of Technical Expert in the patent office.)

In 1907, Einstein pushed his quantum ideas yet further. He proposed a new analogy that built both on Max Planck's idea of *energy* quanta in vibrating atoms and on his own idea of *light* quanta. This analogy had to do with sound waves inside solids. Essentially, Einstein came up with the idea of *sound quanta*, although he never used this terminology. (Today, the quanta making up sound waves are called "phonons", echoing "photon"; they play a key role in the physics of matter.) With his new way of conceiving of vibrations inside solids, Einstein was able to resolve a major mystery concerning the heat capacity of solids. This time, most curiously, the world of physics, even as it disdainfully continued to reject *light* quanta, unanimously accepted the validity of Einstein's explanation of the heat capacity of solids, based on *sound* quanta, and in 1909 the Dutch physicist Peter Debye deepened Einstein's theory and created a very powerful theory of heat capacities, which physicists quickly and warmly welcomed, all while still giving the cold shoulder to Einstein's light-quantum hypothesis.

A strong friendship and great mutual respect developed between Albert Einstein and Max Planck, and in 1913, the latter nominated Einstein for membership in the Prussian Academy of Sciences, which was one of the most distinguished scientific societies in the world. In his nomination letter, Planck sang Einstein's praises, but when it came to the subject of light quanta, which Einstein had continued to champion, Planck commented, "That he may sometimes have missed the target in his speculations, as, for example, in his hypothesis of light quanta, cannot be held too much against him, for it is not possible to introduce really new ideas even in the most exact sciences without sometimes taking a risk."

In the decade from 1906 to 1915, the distinguished American physicist Robert Millikan carried out a long and very careful series of experiments on the photoelectric effect. From the start, he was convinced that Einstein's ideas on the subject were worthless, since they directly contradicted the century-old finding, due to Thomas Young in England and Augustin Fresnel in France, and spectacularly confirmed in 1887 by Heinrich Hertz in Germany, that light consists of waves, and this fact precluded particles of light. For Millikan as for nearly everyone else, the idea of light being *both* particulate and wavelike was inconceivable. Nonetheless, his experiments wound up confirming Einstein's predictions perfectly, which plunged Millikan into deep cognitive dissonance. In a major book summarizing his work, published in 1917, Millikan admitted that his results supported Einstein's revolutionary predictions to the hilt, but he insisted that one should beware of Einstein's "reckless" ideas about light because they had no theoretical underpinning. Put otherwise, although Einstein's conjectural explanation of the photoelectric effect had furnished impeccable predictions, one should give it no credence because it had not been rigorously derived from previously known physical laws. Millikan even had the temerity to declare in his article that Einstein himself no longer believed in his own "erroneous theory" about light (a pure speculation on Millikan's part, without the slightest basis in fact).

To add insult to injury, although the 1921 Nobel Prize in Physics was awarded to Albert Einstein, it was not for his theory of light quanta but "for his discovery of the law of the photoelectric effect". Weirdly, in the citation there was no mention of the ideas *behind* that law, since no one on the Nobel Committee (or in all of physics) believed in them! Light quanta had been unanimously rejected by the members of the community of physicists, even the most adventurous among them. For example, the following year, Niels Bohr, the great Danish physicist and admirer of Einstein, in his acceptance speech for his own Nobel Prize, which had just been awarded to him for his contributions to quantum theory, brusquely dismissed Einstein's ideas about the corpuscularity of light as "not able to throw light on the nature of radiation".

And thus Albert Einstein's revolutionary ideas on the nature of light, that most fundamental and all-pervading of natural phenomena, were not what won him the only Nobel Prize that he would ever receive; instead, it was just his little equation concerning the infinitely less significant photoelectric effect. It's as if the highly discriminating Guide Michelin, in awarding its tiptop rank of three stars to Albert's Auberge, had systematically ignored its chef's consistently marvelous five-course meals and had cited merely the fact that the Auberge serves very fine coffee afterwards.

Vindication of Einstein's Boldest Analogy

The turning point when light quanta at last emerged from the shadows came only in 1923, when the American physicist Arthur Holley Compton astonished the world of physics with his experimental discovery that when an electromagnetic wave approaches an electrically charged particle (an electron in an atom, for instance), it transfers to the particle some of its kinetic energy and momentum, but does not do so as Maxwell's equations predicted. In fact, Compton found that the wave–particle "collision" that takes place in such a situation obeyed the long-known mathematical rules of collisions between *two particles*, with the energies of the incoming and outgoing waves matching exactly what Einstein had predicted in his 1905 paper about light quanta. And thus, at long last, light became particulate!

It still took three more years for the catchier word "photon" to be coined by the American chemist Gilbert Lewis, but in any case, today the notion of a photon — that is, a "wave packet" of light — is a completely familiar denizen of the physics world, and no physicist would dream of denying its reality.

It thus took almost twenty years before the idea of light quanta, the fruit of an analogy conceived in 1905, was taken seriously by physicists — and even after the Compton effect, it still took a bitter battle before the idea was universally accepted. Today, oddly enough, this story is hardly remembered; indeed, most contemporary physicists have the erroneous impression that this first of Einstein's five great articles in 1905 was written solely in order to explain the "famous" photoelectric effect, basing it all on Max Planck's idea that the atoms in the walls of a black body can only take on quantized amounts of vibrational energy. But that is *not* what Einstein's article was written for. Indeed, in 1905 the photoelectric effect was so new and so unexplored that there were not enough data to call for a precise explanation. And thus, in his article, Einstein didn't propose an *explanation* of a famous, well-charted effect; rather, he made a precise *prediction* of the behavior of a barely-known effect, suggesting in a very clear way how his prediction might be tested; however, all of this occupied but two pages near the very end, since his article's *main* topic was the radical idea of light quanta, which had very little in common with what Max Planck had hypothesized in 1900 (as Planck's violent rejection of the idea shows). In sum, Einstein's light-quantum article was nothing but the dogged pursuit of a subtle analogy linking a black body to an ideal gas. Once he had glimpsed this analogy, Einstein went way out on a limb, placing all of his chips on it, in a move that to his colleagues seemed crazy, and then he patiently waited nearly twenty years before being vindicated by Compton's experiments.

This saga, rather troubling but at the same time enlightening, beautifully illustrates Einstein's ability to put his finger on the true essence of a physical situation where his colleagues either saw nothing of special interest or saw only a fog without any recognizable landmarks. For us, the story of this analogy constitutes an example of human intelligence at its very finest.

The Marvelous Conceptual Slippages of Albert Einstein

What is the most famous equation in the world? The most plausible candidate, other than "1 + 1 = 2", would surely be "$E = mc^2$", the celebrated formula by which Albert Einstein revealed a profound but unsuspected relationship between the concepts of *mass* and *energy*. In the next several sections of this chapter, we will concentrate our attention mainly on the process by which the Technical Expert, Third Class gradually deepened his understanding of the meaning of his discovery. It took him two full years — from 1905 till 1907 — to come to see the unsuspected depths hidden in these five little symbols. How this conceptual evolution took place in Einstein's mind is a fascinating but surprisingly little-known story.

But one must begin at the beginning — that is, the origins of $E = mc^2$. To set the stage, we need to describe how the mechanisms of analogical category extension and vertical category leaps can be used in scientific discovery. Both types of process played key roles in the intellectual style of Albert Einstein, and together they carried him to fantastic destinations.

Using Analogy to Extend Concepts in Science

To illustrate the scientific role of analogical category extension, we will consider for a moment the *annus minimus* ("minimal year") of Doctor Ellen Ellenbogen. It was in 1905 that Doctor Ellenbogen, who was not yet employed as a physician but rather as Dishwasher, Third Class in a restaurant in Bellinzona, Switzerland, made not several, alas, but just one medical discovery, and a very modest one, at that. To be specific, shortly after Doctor Ellenbogen had read an article about a marvelous yet very simple treatment that had been recently discovered by Doctor Knut von Knie for an acute *knee* disease, it occurred to her that Doctor von Knie's method might well be also applied to afflicted *elbows*. Here are the words with which, many years later, the Dishwasher Third Class explained her bold mental leap:

> That a treatment of such great simplicity [namely, Dr. von Knie's] should work with such efficacity for *one* part of the human body [the knee], and yet be utterly inefficacious for *another* part of the human body [the elbow], is *a priori* not very probable.

This explanation, although clear, does not allude to the rather salient *resemblance* between elbows and knees, which played a critical role in Doctor Ellenbogen's discovery. This is a regrettable oversight, since one might well wonder if she thought that Doctor von Knie's treatment for knee ailments might not also work equally well in combating diseases affecting the eyes, ears, stomach, kidneys, and so forth. But the truth is that Doctor Ellenbogen never made any such mental leaps, which suggests that she did not see enough of a resemblance between knees and stomachs (for example), or between knees and eyes, to lead her to guess that she might extend the virtues of Doctor von Knie's miraculous cure to those organs.

Once one has seen that the concept of *knee* can be analogically extended outwards, yielding the more general concept of *knee-plus-elbow* (which is the extension found by Doctor Ellenbogen), then this wider category exists in its own right in one's mind. This kind of addition to a conceptual repertoire typifies the process of conceptual broadening by analogy. We will call this kind of broadening *horizontal.*

By contrast, in a *vertical* category leap, one would make an upwards move from the concept of *knee* to the more abstract concept of *joint*, which, in the minds of most adults, would subsume the concepts *knee* and *elbow* (and in addition, the concepts *ankle*, *shoulder*, *knuckle*, and so on). So let us now imagine that the brilliant Doctor Gregorius Gelenk starts with Doctor von Knie's treatment for knee diseases and makes a mental leap allowing him to announce a uniform treatment for a varied group of illnesses that afflict all the different joints, including ankles, knuckles, and so forth. In this case, we would be dealing with a *vertical* jump: the result of following a pre-existent link between *knee* and *joint* in Doctor Gelenk's personal repertoire of concepts.

Such horizontal categorical broadenings and vertical category leaps are natural and unsurprising — the bread and butter of human thinking. But on occasion, such broadenings and leaps can also be insightful and admirable. Suppose Doctor Zygmund Zeigefinger discovers a cure for an ailment of the *index* finger, and his colleague Doctor Renate Ringfinger modifies this cure so that it works also for the *fourth* finger. It would certainly be shocking, would it not, if the Nobel Prize in Medicine were awarded to Dr. Ringfinger for her "remarkable breakthrough"? The analogy on which such a "breakthrough" would be based is far too obvious to deserve such a great honor. People would protest, "Come on! Two fingers are as alike as two peas in a pod!"

On the other hand, suppose Doctor Zora von Zehe adapted Doctor Zeigefinger's cure so that it also worked for *toe* diseases. It goes without saying that we would applaud her contribution more than that of Doctor Ringfinger, but we would still be bewildered were she to receive a Nobel Prize for her work.

Finally, suppose that Doctor Hartmut Herz, in a bold moment of inspiration, had the sudden insight that there might be a connection between (of all things) the index finger and the human heart, and that then, by slightly tweaking certain aspects of

Doctor Zeigefinger's cure, he stumbled upon a miraculous cure for certain *coronary* diseases. In such a case, we would have no problem understanding why the Nobel Committee had seen fit to award a Nobel Prize in Medicine to Doctor Herz.

Language can play a catalytic role in such situations; for example, the French terms "doigt" ("finger") and "doigt de pied" ("toe", but literally "foot finger") make it clear *a priori* that we are dealing with essentially the same thing in both cases. The finger–toe analogy is thus extremely simple and natural for French speakers, whereas for English speakers there is no silver platter on which the analogy is delivered to them. However, for easily understood *visual* reasons, it is still quite obvious. In neither language is just one word used to denote both elbows and knees, and since the resemblance of elbows to knees is perhaps slightly subtler than that of fingers to toes, it would be less probable for someone to connect these concepts. On the other hand, if in German the word for "elbow" were "Armknie" (regrettably, it is not), we would expect there to be a somewhat larger amount of unconscious crosstalk between the two concepts in the minds of German speakers than in the minds of English or French speakers.

And lastly, if in German the human heart, by some odd quirk, were called "der Brustzeigefinger" (that is, "the chest-index-finger"), and if Doctor Herz were a native speaker of German, then his mother tongue could have furnished him with an intuitive hunch that a cure for a disease affecting index fingers might be adaptable to the heart (that is, to the chest's index finger — at least from his German-speaking viewpoint). The idea of such a compound word in German is not totally far-fetched, by the way, since, as readers may recall from Chapter 2, the German for "glove" is "Handschuh", or even "Fingerhandschuh", as contrasted with "Fausthandschuh", a compound that, when taken apart, means "fist-hand-shoe" or "fist-glove" — which is to say, "mitten".

When pathways between concepts are handed to one in advance — for example, by extremely salient physical resemblances (as between fingers and toes), or by related linguistic expressions ("hair" is a clear example in English, since the hair on one's head and the hair on one's body are denoted by one and the same word, whereas in French they are denoted by two extremely different words — namely, "cheveux" and "poils" — which break a single anglophone category into a pair of francophone ones) — then the corresponding horizontal category extensions via analogy will be obvious and inviting. Much the same holds for vertical category leaps. If in one's mind there already exists a connection between a specific concept and a more general concept, then the shift in perspective involved in making the leap from one to the other, such as seeing knees or elbows as *joints*, is a simple and natural act.

Category Broadenings as Sources of Special Relativity

We now return to the genesis of $E = mc^2$. Any given entity belongs simultaneously to an unlimited number of categories. Nonetheless, in daily life one often has the illusion of dealing with an entity belonging to just one category. In general, our surroundings often strike us as being clear and unambiguous, as if there were just one correct, objective way to perceive them; indeed, it's this illusion that allows us to live. If we had to take into account, at every moment, the boundless number of categories that the situations we run across might belong to, we would constantly be spinning our wheels in utter mental confusion and we would be incapable of taking any action.

Labels and stereotypes found and stuck on in a flash are indispensable, but the flip side is that they also limit us tremendously. How does one find one's way efficiently in the space of all possible categorizations without taking forever to do so? Where is the happy medium in daily life, or in scientific thinking, between finding a quick-and-dirty categorization and putting one's finger on the perfect one in a given context? In a new situation, how can one know, when one is trying to pinpoint its essence, whether one should settle for the fastest and easiest categorization that comes to mind? How can one recognize situations in which expanding horizontally outwards, thus constructing a *broader category*, or else jumping vertically, thus reaching a *higher level of abstraction*, would be a wise move (or an unwise one)?

When one reads works by Einstein himself, as well as his more scientifically oriented biographies, it is clear that the great physicist frequently ran into just such dilemmas. Indeed, this is one of the most salient traits of his intellectual style. For example, in 1905, in coming up with special relativity, Einstein made a very fecund category extension based on what might seem to be the most innocuous of analogies.

Taking a fresh look at an old and fundamental principle known to us today as "Galilean relativity", he extended it outwards in the most innocent-seeming fashion, suggesting that it held not only within the domain of *mechanics*, but also within the larger domain consisting of *mechanics together with electromagnetism.* (Mechanics is the earliest branch of physics, and it deals solely with the movement of tangible bodies in space — thus with speed, acceleration, rotation, gravity, friction, orbits, collisions, springs, pendulums, vibration, tops, gyroscopes, and so on — but it does not include optics, electricity, or magnetism, let alone nuclear forces; none of these branches of physics were known in Galileo's day.)

More precisely, the principle of Galilean relativity said: "Given two frames of reference moving at a constant relative velocity, there is no mechanical experiment whatsoever that will distinguish one from the other." To make this more concrete, if one is inside an airplane that is flying in a straight line at a fixed altitude at a speed of 500 miles an hour, the principle of Galilean relativity states that no mechanical experiment carried out inside the plane will be able to reveal that it is not standing stock still in a hangar — or conversely, if the plane is sitting on the ground, no mechanical experiment performed inside it will be able to reveal that it isn't streaking along at 500 miles an hour. We all know that when we are flying in the sky at a great speed, as long as the speed is constant, we can pour ourselves a glass of water without taking into account the fact that we are moving. It will feel exactly as if we are perfectly still. And the same holds, of course, when we are in a train moving along a straight stretch of track at a fixed speed, whatever that speed may be.

We can easily imagine experiments of various sorts that one could carry out inside a plane or train to see if their outcomes do or do not depend on the vehicle's state of motion. Mechanical experiments might involve pouring water into a glass, spinning a gyroscope on a table, swinging a pendulum from the train car's ceiling, making weights bob up and down on springs from which they dangle, sliding and colliding hockey pucks on a frictionless surface, rolling a ball down an inclined plane, floating a helium balloon above our heads, and so on. And as it turns out, such phenomena look and feel exactly the same inside a parked plane or train and a smoothly moving one, thus fully confirming Galilean relativity.

But what can be said about optical and electromagnetic phenomena in trains and planes? We have all, while traveling, turned lights on and off, looked at ourselves in mirrors, checked what time it is by consulting our digital watches, used laptops and video games, and so forth. All these devices seem to us to work "just like normal" — every bit as normally as does water poured from a pitcher into a glass. That is, in a plane or train they look just as they look when we are sitting in a chair in our living room. The devices that are involved — glasses, watches, telephones, computers — are combinations of simpler things such as lenses, mirrors, prisms, batteries, bulbs, coils, magnets, currents, and so forth, and the physical laws governing this class of things are those of optics and electromagnetism.

With this prelude, let's return to Albert Einstein in his *annus mirabilis* in Bern. In that year, as he pondered the principle of relativity formulated by Galileo, Einstein asked himself why this profound-seeming principle should be limited to just *mechanical* experiments. If observations of a pendulum, a spring, or a gyroscope did not allow one to figure out whether one was in motion or not, then why should observations of a candle, a magnet, a mirror, or an electrical circuit have a better chance at doing so? Einstein saw no reason that they should. He was pushing Galileo's principle outwards in his mind, generalizing it by analogy, but only in the gentlest of fashions.

It's not clear, incidentally, whether it's better to call this a horizontal category-broadening act or a vertical category leap, because one can see it either way. On the one hand, all that Einstein did was to replace the phrase "any kind of *mechanical* experiment" by the more general phrase "any kind of *mechanical or electromagnetic* experiment"; in this sense it seems like a horizontal extension justified by a simple analogy between one area of physics (mechanics) and another (electromagnetism). This is similar to a student saying, "I'm not going to take the electricity-and-magnetism course next semester, because this semester's mechanics course was so hard for me."

On the other hand, one might also say that Einstein replaced the idea "any kind of *mechanical* experiment" by the more abstract idea "any kind of *physical* experiment" — that is, he made a leap from a narrow concept to a wider, more general one that encompasses it. This would be like a discouraged student saying, "I'll never take another physics course again, because this semester mechanics was such a bear."

And how did Einstein himself see his analogical move? Well, he once described his feelings at the time in the following manner:

> That a principle of such broad generality [namely, Galilean relativity] should hold with such exactness in *one* domain of phenomena [namely, mechanics], and yet should be invalid for *another* [namely, electrodynamics], is *a priori* not very probable.

At first one might be inclined take these words as describing a horizontal broadening — an analogy-based extension from just *mechanics* to the *union* of mechanics with electromagnetism. (This should remind readers of Dr. Ellenbogen's new way of curing an elbow disease, based on her *horizontal* analogical extension of the treatment of a knee disease, moving outwards from just *knees* to the *union* of knees and elbows, which are clearly close cousins.) But couldn't one equally well hear Einstein's sentence as declaring that he had such deep *a priori* faith in the uniformity of physics that he was willing to bet that Galileo's principle holds *for all imaginable areas of physics*, not just for mechanics alone? (This alternative way of seeing Einstein's act reminds us of Dr. Gerhard Gelenk's *vertical* generalization of the treatment of knee diseases, wherein he changed perspective from just *knees* to the more abstract category of *joints.*) In sum, in this case, as in many others, we see that there is no sharp line of demarcation between vertical category leaps and horizontal category extensions.

In any case, whether it was a vertical or a horizontal mental move, Einstein's extension of the Galilean principle of relativity wound up profoundly undermining much of the physics of the preceding three centuries. And yet this revolution emerged from the act of paying attention to the trivial-seeming similarity between experiments in a train (or any similar reference frame) that were limited to mechanics, and experiments that might also involve electromagnetism. Einstein's intuition told him that any such distinction was unnatural, since, in the end, any conceivable experiment in any conceivable branch of physics belongs to the single unified tree of physics. Of course, this kind of sixth sense for how and when a category can be broadened is mysterious, and is one of the deepest of all arts.

A Two-headed Flashlight Loses a Tiny Bit of its Mass

The consequences of category-broadening by analogy, applied to the principle of Galilean relativity, were enormously deep and led Einstein to a rich network of ideas whose names are familiar to anyone interested in science today, such as the relativity of simultaneity, time dilation, the contraction of moving objects, the non-additivity of speeds, the twin paradox, and so on. But those ideas, fascinating though they are, are not our focus. We wish now to come back, as promised, to the equation $E = mc^2$, which, oddly enough, was nowhere to be found in Einstein's first article on relativity. That thirty-page article, published in the summer of 1905, contained plenty of other equations whose consequences were unprecedented and revolutionary, but it lacked the little tiny equation that became the indisputable emblem of relativity.

For most people, experts and non-experts alike, the equation $E = mc^2$ is so tightly linked with the notion of Einstein's theory of relativity that imagining relativity with its signature equation completely absent would seem as strange as imagining the 1927 Yankees without Babe Ruth, or the town of Pisa in 1100 A.D., before its signature tower had ever been dreamt of. And yet the truth of the matter is that Einstein did not discover the now-celebrated equation until some months after his first relativity article appeared. He deemed his new finding interesting enough to warrant another article, which appeared in November of that same year (thus just barely squeezing in under the wire of the *annus mirabilis*), and which was just two pages long.

Strictly speaking, the famous equation didn't appear in that article, either, since the way Einstein saw fit to express his new discovery was through words rather than through an equation; however, those words were tantamount to saying "$E = mc^2$". And then, two years further down the pike, he realized that his second relativity paper concealed some highly important implications that he hadn't at all suspected when it was published. And so in 1907, he published yet a third article, at last spelling out the full meaning of the symbols "$E = mc^2$". It was this article that grabbed the world's imagination, because its conclusions not only were counterintuitive and scientifically far-reaching, but also had profound potential implications for society.

We will consider these developments in chronological order, starting with the very short article of November, 1905. In it, Einstein imagined an object that could simultaneously emit two flashes of light in opposite directions (say east and west). So let us imagine a flashlight with bulbs at both ends. Since a flash of light possesses some energy, and since energy is always perfectly conserved by all physical processes (mechanical, electromagnetic, and so on), our two-headed flashlight will necessarily *lose* some energy — namely, the total energy carried off by the two departing flashes. From the point of view of energy, one has to pay for producing light! All this is quite obvious.

The key step Einstein took here was the (nearly) trivial idea of looking at the two-headed flashlight from another frame of reference — specifically, a moving frame of reference, such as a train moving at 30 miles an hour, let's say westwards. According to special relativity, observers sitting in the train have the right to consider themselves stationary and to claim that *the flashlight is moving eastwards* at 30 miles an hour (and *always* at that speed, since their frame of reference — the train — has a fixed speed). For the train's passengers, the two flashes necessarily undergo the Doppler effect.

For those readers unfamiliar with it, the Doppler effect merits a brief digression. It holds for any kind of wave, including light and sound waves. In the case of sound, it's the shift that one hears each time an ambulance approaches, passes by, and then recedes into the distance: just at the moment it drives by, its siren seems suddenly to sink to a lower pitch. To those inside the ambulance, nothing changes, of course, but for people standing on *terra firma*, it's quite another story. Why does this surprising sonic shift take place?

Imagine a pond into which a stone has just been tossed. From the spot on the surface where the stone plunged into the pond and is now sinking, circular ripples go spreading out. Now toss in a cork floating somewhere on the pond's surface. Soon enough the concentric ripples will reach the cork, one after the other, and they will start making it bob gently up and down at a regular frequency. This bobbing cork is analogous to the vibrating eardrum of a person who hears the siren from *within* the ambulance: the vibration clearly has a fixed frequency.

But now imagine, by contrast, a toy motorboat speeding across these same circular ripples, first heading straight toward the center of the concentric circles (the waves' source), and then continuing onwards towards the far bank. While it is moving toward the center as the ripples expand, the toy boat bobs up and down *more frequently* than the cork does (for the boat, the circles seem to be coming out to meet it), but once it has crossed the circles' center (located just above the sinking stone), the toy boat has to catch up with the ripples that are now fleeing from it, and so it meets them *less frequently* than before, meaning that it bobs up and down less quickly than before. This is an aquatic Doppler effect: the felt frequency of the ripples suddenly falls, just at the moment when the boat passes their source.

Likewise, in the ambulance situation, the perceived frequency (*i.e.*, the pitch) of the siren suddenly falls as the ambulance rushes by the observer on the street. The Doppler effect generally says that, if an observer is moving with respect to the source of some waves (or conversely, if the source is moving with respect to the observer), then the frequency of the waves, as perceived by the observer, will depend on the relative speed of the two reference frames. Of course it was a nontrivial analogical extension to generalize the original effect from sound waves to other types of waves, such as light waves and ripples on a pond — but that's another story. Suffice it to say that the Doppler effect as applied to electromagnetic waves was a fairly new idea at the turn of the twentieth century, and the third-class patent clerk in Bern, though he didn't invent the notion, took great advantage of it.

Indeed, Einstein calculated the Doppler effect for the double flashlight using his own theory of special relativity, freshly minted just a few months earlier. He imagined himself in the frame of reference where the flashlight was moving at a constant speed (in other words, the frame in which the train is stationary), and he carried out relativistic Doppler-effect calculations that gave him the energy of each of the two flashes of light that sped off simultaneously. By adding these energies together, he got the total energy lost by the flashlight. He was able to use this sum to calculate how much *kinetic* energy the moving flashlight had lost at the instant when the rays were emitted, which should have been exactly zero, since the flashlight had just kept on moving at a constant clip. But it wasn't exactly zero — it was just a tiny bit different from zero. Einstein's Doppler-shift calculations revealed to him that the moving flashlight had to have *lost* some kinetic energy by sending off two flashes of light.

This result was extremely peculiar. It was obvious that to produce light, the flashlight had to give up some *electrical* energy (in its battery), but why would it also give up some of its energy of *motion* (which is given by the standard formula "$Mv^2/2$", the capital M of course denoting the flashlight's mass, and v denoting its velocity)? We know that the flashlight doesn't slow up in the least! By fiat, it is moving at a *constant speed.* (Recall that in the first frame of reference it is perfectly stationary; it's only the observers on board the train who see it as moving, because their frame of reference is gliding down the tracks. And given that their frame is gliding at a perfectly *constant* speed, and that the flashlight is stationary with respect to the ground, the "moving" flashlight never loses or gains a speck of speed, as seen by train-bound observers.) How then can the steadily-moving flashlight have lost even the tiniest fraction of its energy *of motion*? Let's devote a moment's thought to this humble riddle, a humble riddle whose solution shook the world.

If, upon releasing the two flashes of light that carry total energy E, the flashlight loses even the tiniest amount of its kinetic energy, then the just-cited formula for kinetic energy "$Mv^2/2$" tells us that either the flashlight's *mass* M or its *velocity* v must have suddenly diminished at the moment of emission. But as we just mentioned, the train has a constant speed, which means that the flashlight, as seen from the train, also has a constant speed. Thus v is unchanged. We therefore have no choice: the only thing that could possibly have become smaller is M, the flashlight's mass, and according to Einstein's Doppler-shift computations (which we will not spell out here), the tiny amount of mass that the flashlight loses, which we'll denote by lowercase m, is equal to E/c^2. (It's crucial not to confuse the flashlight's *total* mass M with the negligible quantity of mass m that it loses when it gives off the two rays of light.)

The Definition of the Concept of Energy

Anyone who follows Einstein's (rather simple) calculations must agree with him that an object that gives off electromagnetic radiation will necessarily lose some mass — namely, an inconceivably tiny quantity of mass that depends on the amount of energy E carried off by the radiation. Why tiny? Because the energy of the light itself (E, which is the fraction's numerator) is negligible, and the fraction's denominator c^2 is incredibly huge — after all, it's the square of the speed of light, which is to say, the square of 299,792 kilometers per second (that is, the square of 1,079,252,849 kilometers per hour). And when one divides an already microscopic energy — that of the two flashes — by this gigantic quantity, the result will necessarily be infinitesimal.

The fact that one is multiplying a mass by a speed squared here (mc^2) might surprise a nonscientist, but it doesn't surprise physicists, for ever since Galileo, Kepler, and Newton, physicists have grown accustomed to the idea that the laws of nature involve algebraic expressions — often powers (most often squares or cubes) of quantities that are directly observed. Indeed, to anyone who has ever taken any physics at all, the formula "K.E. = $Mv^2/2$", giving the kinetic energy of a moving object with mass M and velocity v, is both familiar and unsurprising.

So let's come back to the quantity E/c^2, which Einstein had just identified as being relevant to this situation. What is surprising in this quantity, then, is not the nature of the algebraic expression itself, featuring an energy divided by a velocity squared, the result of which will necessarily have the units of mass — but its *meaning*. The amazing thing is firstly that this m represents the mass lost by our energy-emitting object, no matter what its original mass M was, and secondly that the relationship between the sizes of m and E is mediated by a special and universal constant of nature — namely, the speed of light. *This* is what was truly new and strange, not the algebraic structure of the formula (an energy divided by a velocity squared), which in itself contains no surprises. In summary, it's the *idea* suggested by this formula — the idea of *energy possessing mass* — that should catch people totally off guard, not its mere *algebraic form*, which is rather ho-hum to anyone who realizes, as physicists already had realized for three centuries, that energy always has the units of mass times velocity squared.

What is the formula actually telling us, then? Well, we are now going to try to reconstruct Einstein's own thinking process on this subject, starting with his first article on the famous formula, which came out in the fall of 1905, and in which he described just the tip of the iceberg, and finishing with the publication of his follow-up article in 1907, in which he finally revealed the iceberg's entirety.

Energy and Mass

In his two-page article in the fall of 1905, Einstein showed that any object that emits energy in the form of light loses thereby a small — in fact, unimaginably small — amount of mass. This conclusion caught physicists off guard, but the public at large paid it no attention at all, since infinitesimal changes in an object's mass, whether counterintuitive or not, have no potential use to society. To return momentarily to our caricature analogy involving Pisa and its tower, the appearance of this first article about $E = mc^2$ was like the appearance, in 1173, of an elegant new stone tower in the center of Pisa — a tower that stood straight up, just as towers should. In those days, an Italian town with a tall tower gained a bit of prestige, but not an enormous amount of it. Though towers were impressive structures, they were pretty commonplace. Likewise, the two-page article in the fall of 1905 didn't attract huge amounts of attention.

We shall come back to Pisa and its tower very shortly, but in the meantime, let us consider what happens to the tiny bit of mass that a radiating object loses. Does it just poof out of existence without a trace, or do the departing flashes of light carry it away with them? It is tempting to localize the missing mass in the rays, and thus to conclude that the light in flight *weighs* something. (By this, we mean that if one were to catch the light inside a box with mirrored inner walls between which it will bounce, and then if one were to place the box on a scale, one would obtain a microscopically higher reading than for an identical box with no light in it.) But such a conclusion is based on the idea that if some mass seems to have vanished, then it must have *gone* somewhere. In other words, the conclusion that the fleeing light rays must be carrying off some mass with them follows from the belief that *mass is indestructible*, or, stated another way, that in all physical processes, there is a law of *conservation of mass*, just as there is a law of *conservation of energy*. (Notice the words "just as", which suggest that mass and energy behave in analogous ways. This analogy will become crucial to our discussion.) If there is such a law for mass, then clearly the departing flashes of light would have to be carrying off the mass lost by the object. (Where else could the mass go? Isn't loot likely to be carried off by the thief?) But this is rather puzzling to a human being, because we are all imbued with the image of light as an insubstantial, ghostly entity — in some ways as the diametric opposite of matter. How, then, could light weigh anything?

In any case, *any process of radiation inevitably entails a loss of mass by the radiating object*, the precise amount of which is given by Einstein's famous formula. Once again we stress that the heart of Einstein's first discovery linking energy and mass is not the precise *value* for the loss, which is specified by his mathematical formula, but rather the statement in italics, above. But this was just the first act; it was not this initial finding but other, deeper meanings of the equation, discovered in the ensuing two years, that finally rendered it so enormously famous.

Banesh Hoffmann's Special Way of Looking at Einstein

The physicist and mathematician Banesh Hoffmann was a collaborator of Einstein's during the 1930's, and in 1972 he published an exemplary biography of Einstein. That book, *Albert Einstein: Creator and Rebel,* is remarkable for the limpid fashion with which it conveys the inner workings of the mind of the great thinker. Certain passages in it give a sense for the subtlety of the analogies with which Einstein gradually homed in on the essence of this discovery that is expressed by just five symbols. Rather paradoxically, the essence of the discovery is also *masked* by those five symbols, because an equation in physics is not self-sufficient, in the sense of explaining itself; an equation just sits mutely on a page. It's up to physicists to decipher its meaning, or rather, its various meanings at different levels, because there can be several levels of meaning, even for a very tiny equation.

For example, the equation "$E = mc^2$" is often stated without any clear context. In such a situation, what do the letters "E" and "m" stand for? What energy and what mass are meant? Are they always attached to the same spot and the same moment of time? To be more precise, does the equals sign mean that the mass is *accompanied* by a certain energy, or that it actually *is* an energy, or that it *yields* an energy, or that it *results from* an energy? Does this equation mean that some energy can *transform* into some mass (or vice versa, or both)?

The answers to questions of this sort are by no means self-evident. They do not effortlessly jump off the page, nor is mathematical skill the magic key to their answers. Even today, very few non-scientists know how to interpret these symbols, and there are a good many physicists whose understanding of them is at times a bit shaky as well; the fact is, this simple-seeming equation's meaning is elusive. Even its discoverer had to mull it over for a couple of years in order to fathom its full depth.

In order to try to understand Einstein's intellectual pathway between 1905 and 1907, let us begin with the following passage by Banesh Hoffmann, which describes a key moment in the article that Einstein published in the fall of 1905:

> With his instinctive sense of cosmic unity he now tosses off a penetrating and crucially important remark: that the fact that the energy is in the form of light "evidently makes no difference".

In other words, once Einstein had formally derived the counterintuitive result, he was perfectly happy to ignore his own derivation and to jump to the conclusion that it must also hold in far more general circumstances than those that allowed him to discover it. In particular, Einstein wrote that exactly the same result must hold in any situation in which an object releases energy *in any form at all* — thermal energy, kinetic energy, sound waves, and so forth.

Now this is a classic vertical category leap by Einstein, supposedly justified by the modest word "evidently" (which, incidentally, would have been better translated by Hoffmann as "obviously"). However, Einstein's calling it "evident" or "obvious" does not legitimize the leap, for it is an extremely bold leap of generalization, owing nothing to logical or mathematical reasoning or to algebraic calculations. This leap comes solely from a physical intuition that all processes of energy release have so much in common with each other that if a given result has been rigorously established for *one* type of process, then it must hold for *all* such processes. In other words, it comes from an analogical belief that, in this type of situation, all forms of released energy are equivalent. This first broadening by Einstein of the meaning of his equation was thus the idea that any object, whenever it releases an energy E of any type whatsoever, loses a minute amount of mass equal to E/c^2.

Actually, before this, Einstein made one prior extension of his equation's meaning. It came from a smaller, more modest leap — a leap involving a conceptual reversal. He declared that any object, whenever it *absorbs* an amount E of incoming energy, *gains* an amount of mass equal to E/c^2. This mental turnaround constituted a nontrivial analogy: the object, instead of giving off some energy, absorbs some, and instead of losing some mass, gains some. In other words, Einstein saw that there was not a profound difference between the new-found phenomenon running forwards in time and running backwards in time. Such a conceptual reversal, although it may seem extremely simple, doesn't just step forward all by itself; somebody has to *imagine* it. And even such a simple mental turnaround can on occasion elude deep thinkers, even "Einsteins" (we'll give an example very shortly) — but this particular conceptual reversal did not elude this particular Einstein.

These descriptions of the process of emission or absorption can be seen as a *causal* interpretation of the equation. As we said earlier, in his first article on these ideas, Einstein didn't write out the now-famous equation with algebraic symbols; he expressed his discovery solely in words by saying, "If a body emits energy E in the form of radiation, then its mass diminishes by E/c^2." This sentence describes an event (emission or absorption of some energy) that inevitably gives rise to a consequence (loss or gain of mass). Much as in the case of some equations discussed in the previous chapter, this sentence amounts to an asymmetric reading of the equation, in which one side is seen as the *reason* behind the other side, but where causality running in the reverse direction is not imagined.

This, in broad brushstrokes, is the meaning that Einstein saw in $E = mc^2$ in 1905. That meaning, although already a very surprising and provocative idea, is not nearly as far-reaching as the final understanding that he reached in 1907. In the course of his ponderings between 1905 and 1907, the equation itself didn't change in any way; all that changed was the interpretation that Einstein attached to its five symbols. In principle, any physicist of the time could have read Einstein's 1905 article, could have reflected on it for two years, and could have arrived at all of its consequences — and yet, to no one else did these ideas occur. What went on that was so different and special in Einstein's mind?

A New and Strange Type of Mass

In order to understand the mental obstacles that Einstein had to overcome, one must try to enter into the mindset of the physicists of that period. The existence of atoms was still not certain in 1905, and if they existed at all, their nature was entirely mysterious. Einstein believed in them with near-certainty, just as he believed in the vibrations of the atoms in a solid as the explanation of heat, even if he wasn't able to envision what the atoms themselves were like. But how could Einstein (or any other physicist of his time) imagine a radiating object, such as a flashlight, losing some of its mass? How could such a bizarre event possibly happen?

For example, would it lose some of its constituent atoms? If so, where would they go? Or would they just suddenly cease to exist? Or else, could some (or all) of its constituent atoms become a smidgen less massive while staying the same in number? In that case, by what mechanism could a single atom lose some of its mass? Was it possible that the fundamental particles (like electrons, which had just been discovered in 1897 by the English physicist J. J. Thompson) might have *variable* masses rather than fixed ones? On the other hand, if the object lost none of its atoms, and if each constituent atom retained all of its original mass, then how could the whole object possibly lose any of its mass? This was a genuine enigma.

All such questions hinged, of course, on how physicists in those days imagined mass. And as to that, there isn't any doubt: they saw it just as we all do intuitively, even today, over a hundred years later — namely, as a fixed property of any *material* object, ranging from clocks to clouds to dust motes to atoms, but not applicable to an intangible notion like a jiggle, a ripple, a rumble, or a tumble, because such verb-like phenomena are merely *patterns of motion* of some matter, and have no weight. As mass was considered a *fixed property* of a material object, it certainly couldn't just poof into or out of existence. Indeed, an object's mass couldn't change at all — unless bits of it broke off and sailed away, like a cigarette giving off tiny particles of smoke that invisibly disperse into the surroundings. But even then, the sum of all the little invisible masses would have to equal the starting mass; this seemed (and still seems) self-evident. The total mass couldn't grow or shrink; it was an invariant, and thus conserved, quantity.

Einstein's new equation put him in a sticky wicket, therefore, because everyone grasps the distinction between material objects and immaterial phenomena, and yet the new equation seemed to be saying that a material object could — in fact, *had to* — lose or gain mass as a result of losing or gaining energy, despite the fact that, to all appearances, energy is anything but a material object. An example of what the new equation implied would be a hot object that is cooling off, giving off a bit of its heat to its surroundings. This object must also, according to Einstein, be losing a bit of its mass. This amounts to saying that some mass is associated with heat, but let's recall that for Einstein, as for most physicists of the time, "heat" was synonymous with "vibration of atoms", which meant that he was forced by his own beliefs to the surrealistic idea that *the vibration of atoms inside a solid contributes to the solid's mass,* with the bulk of its mass residing, of course, in the atoms themselves, seen as material objects.

We are thus led to a dichotomy, in the mind of Einstein and anyone who accepts his conclusions, concerning the notion of mass: on the one hand, there is the familiar type of mass, which we will henceforth refer to as *normal* mass, and which corresponds to the standard everyday notion of "mass of a material object", and on the other hand, there is another type that we will call *strange* mass, which corresponds to the counterintuitive new notion revealed by the famous equation. (Einstein himself made the same distinction in his 1907 article, using the terms " 'true' mass" and " 'apparent' mass".) This breakdown of mass into two types, although unexpected, is imposed on us by the equation; it cannot be avoided. A cloud and a clock obviously possess *normal* mass, since they are both made of *stuff* (that is, atoms), while light, sound, and heat have none; the latter three, however, all possess *strange* mass. Of course, the cloud and the clock will also possess some strange mass, because they contain some heat, and as we said above, heat is imbued with strange mass.

Any ordinary material object thus possesses lots of normal mass and a tiny bit of strange mass, and their relative proportions can change with time. For example, a flashlight that gives off light for a long time will gradually lose its strange mass, thus becoming ever so slightly lighter. What then happens when it has fully exhausted its strange mass (or "runs out of juice", in colloquial terms), and all that's left is its *normal* mass? Well, as we all know, at that point the flashlight will no longer be able to emit any electromagnetic radiation, unless a new battery is installed. The fact that it still contains plenty of *normal* mass would seem utterly irrelevant to its flash-producing capability, because there is no interconvertibility between normal and strange mass. That is to say, it would seem that *we can't draw on a flashlight's abundant reserves of normal mass* to get it to give off light. All of one's experience leads one to think that only by drawing on its very small supply of *strange* mass can an object emit energy (such as light, in the case of a flashlight).

The following allegory may help us to convey more clearly the distinction between these two types of mass. Jan has the wherewithal to purchase the most essential items in her life, but her modest bank account does not allow her to indulge in luxuries. Some time ago she inherited a huge mansion with palatial grounds, worth at least several million dollars — but in her mind, this type of possession doesn't belong to the same category as her day-to-day money; no matter what its official value might be in dollars, she doesn't conceive of her residence as a spendable liquid resource — she sees it merely as a frozen, solid one. For Jan, the two types of possession have nothing to do with each other; it's as if there were a rigid mental barrier separating the concepts. It would never occur to her to sell a few acres of her gardens, let alone her mansion, in order to purchase a luxury item or to take a vacation. In Jan's mind, whereas her meager liquid assets flow easily (indeed, that's why they are called "liquid assets"), her real-estate assets are completely frozen and inaccessible; if she needs money, she never thinks of the latter at all. But then one day, several months after her non-payment of a substantial bill, Jan receives a worrisome letter stating that in a few days some of her belongings will be forcibly seized by law enforcement officers. And then, all at once, something clicks in her mind…

One can easily translate our allegory into the language of mass and the conceptual dichotomy Einstein had discovered. The idea is simply that the seemingly uncrossable mental barrier between strange mass (= liquid assets) and normal mass (= frozen assets) is *not* in fact uncrossable after all, but can be crossed provided that there is enough pressure (= the threat of a repossession) to make the idea leap to mind.

However, for Albert Einstein in 1905 and for the readers of his first article on the idea of $E = mc^2$, the conceptual barrier between *normal* mass and *strange* mass was completely impenetrable. How could it not have been so? Consider a boulder, for instance (a quintessential example of *normal* mass). It's one thing to imagine that the boulder's internal stock of heat (a quintessential example of *strange* mass) will gradually diminish as the boulder emits infrared radiation that warms up its environment. But who would ever have suspected that the boulder *itself* could all at once vanish from the universe, resulting in the shooting-off of much more intensive rays of light? If one is not under severe pressure, one does not jump to embrace wild and woolly scenarios such as that; one does not spontaneously offer a warm welcome to notions that violently clash with a lifetime of experience, not to mention with the collective wisdom of humanity.

In the Copycat analogy problem "*abc* ⇒ *abd*; *xyz* ⇒ ???", no one thinks of the elegantly symmetric answer *wyz* without first having been lured down the pathway of taking the successor of *z* and banging up against that barrier. Only after all one's initial simple and intuitive ideas have failed does one start trying out more radical ideas. In short, it takes a great deal of mental pressure in order to trigger a radical conceptual slippage, and not least among the contributing pressures is one's sense of esthetics.

Here is what Banesh Hoffmann wrote about this very subtle transitional stage in Albert Einstein's thought processes:

> In his paper of 1905 Einstein said that all energy of whatever sort has mass. It took even him two years more to come to the stupendous realization that the reverse must also hold: that all mass, of whatever sort, must have energy. He was led to this by æsthetic reasons.
>
> Why should one make a distinction in kind between the mass that an object already has and the mass it loses in giving off energy? To do so would be to imagine two types of mass for no good reason when one would suffice. The distinction would be inartistic and logically indefensible. Therefore all mass must have energy.

This passage is eloquent and insightful, but it's also rather curious, for its two halves almost seem to contradict each other. Whereas the first paragraph states that it would be very hard (even for Einstein) to transcend the intuition that there are two different types of mass, the second paragraph (which is attempting to give us a privileged, first-person view from within Einstein's own mind) claims that there would be *no good reason* for believing in a distinction between these two types of mass. But actually there *was* a very strong reason for such a belief; it was in fact Einstein's own equation that had created a schism within the formerly monolithic concept of *mass.* Viewed in this way, Hoffmann's short passage constitutes a perfect summary of Einstein's inner mental trajectory over two years. Its first paragraph alludes to Einstein's initial glimpse of a new kind of mass in 1905, as well as his lack of full understanding of it; its second paragraph indicates that this imperfect understanding gave rise to such serious tension in Einstein's mind that he was eventually forced to make a daring esthetics-driven extension of that initial notion (that is, of *strange mass*), getting rid of the conceptual schism and thereby re-establishing conceptual unity, thus leading to a harmonious new state of understanding from which all mental tension had been banished.

We shall now try to put this mental trajectory under a magnifying glass. The fact that light, heat, and sound (etc.) all possess mass (even if it's just an extremely slight amount) implies that we are dealing with *a new type of mass* of which no one had ever dreamt before. An object that gives off energy and in so doing loses a tiny amount of its mass does not lose any of its solid or normal constituents; it loses something radically new — it loses a different *type* of mass. In short, for anyone who understands the equation $E = mc^2$ (and this certainly includes its discoverer), there is a very intense pressure to imagine two extremely different types of mass — the familiar type (normal) and the new type (strange), which is associated with outgoing or incoming energy. To be more specific, the tiny mass carried off by the light rays comes from *strange* mass that was lost by the battery. No *particles* in the battery were lost or destroyed, however; every one of its corpuscles (atoms, molecules, whatever) remained intact. Since the *normal* mass of the battery is never affected by any process of emission (or absorption) of light, this gives rise to the image of a rigid barrier between these two types of mass (informally put, "they don't talk to each other").

Even Albert Einstein took two years to arrive at the conclusion that the impermeable conceptual barrier that was suggested — indeed, *forced* — by his equation did not actually exist. His "instinctive sense of cosmic unity", as Hoffmann dubbed it, eventually led him to the radical notion that nature's internal consistency — that is, the uniformity and simplicity of the laws of physics — required that any material object (*i.e.*, any normal mass), whether an electron or a cannonball, should be able to "melt" into strange mass carried off by escaping light rays, much as does the inert energy stored in a battery, or much as the frozen assets latent in an estate might turn into liquid cash.

This was truly a shocking idea, because it meant not only that solid, massive physical objects *could* literally dematerialize and vanish (or, if we run the scenario in reverse, that such objects could materialize out of nowhere), but also that any such metamorphosis would necessarily be accompanied by the sudden, simultaneous appearance (or disappearance) of a phenomenal amount of energy. Indeed, it was the phenomenal amounts of energy involved that made the newly-revealed full meaning of Einstein's equation stunning and even surrealistic.

What Machinations Took Place Behind the Scenes in Einstein's Mind?

What went on in the hidden recesses of Einstein's mind that brought him, after two years of thought, to this most disorienting idea, for which there was, at the time, no experimental evidence at all?

To begin with, there is every reason to believe that Einstein saw the light rays leaving the flashlight as carrying not only energy but also mass (both of which had been "subtracted" from the flashlight). This amounts to the idea that the strange mass, rather than just poofing out of existence when it left the flashlight, *mutated* from one form to another. Before the two flashes were produced, the strange mass resided in the chemical bonds inside the flashlight's battery, whereas after the flashes' release, it resided in the vibrations of the electromagnetic waves making them up (and thus, if one were to weigh a mirror-lined box in which the rays had been captured and were bouncing back and forth, one would find that the box weighed ever so slightly more than it had before the rays were captured in it).

This fluidity of strange mass — the fact that strange mass can glide back and forth between different forms — could not have failed to remind Einstein of the fluidity of energy (for energy, likewise, is constantly gliding from one form to another), and such a connection would have come to his mind all the more easily given that his equation had revealed an unexpected link between mass and energy. But even if a given type of strange mass could easily mutate into other types of strange mass, it still seemed totally self-evident that *normal* mass could *not* mutate. As we stated above, one never sees boulders or other solid objects (or liquids or gases, for that matter) simply vanishing into flashes of light, or springing magically out of them. Material things are made of tangible *stuff*, and as such they seem to belong to a different class of things from intangible energy and its "strange mass". This sharp distinction makes for a rigid, uncrossable barrier inside the concept of mass, as described earlier, dividing it into two subspecies that are not interconvertible.

Like all physicists of the time, Einstein was intimately familiar with the principle of conservation of energy — the solidly confirmed fact that energy can change form but without ever increasing or decreasing. Countless experiments had shown that heat (thermal energy) could be converted into movement (kinetic energy) of macroscopic objects (for example, of a piston in a cylinder) and vice versa (rubbing something warms it up), and that chemical energy in a battery can be converted into electromagnetic energy, and so on. The technology of the day relied on this fundamental principle.

Einstein had an unswayable faith in the law of conservation of energy; now, all of a sudden, he found himself face to face with a similar new conservation law — namely, the conservation of strange mass. That is to say, strange mass, much like energy, could apparently glide from one form to another without increasing or decreasing. For example, if a crystal absorbed some radiation, a bit of *electromagnetic* strange mass (that is, a ray of light) would suddenly go out of existence and at the same moment a bit of *thermal* strange mass would instantly come into existence; likewise, in an act of radiation, the reverse transformation could take place. We can therefore imagine that in Einstein's mind, thanks to the analogy between the laws of conservation of energy and conservation of strange mass, there was starting to exist a tight analogical link between the concepts of *energy* and *strange mass.*

So far, we have completely neglected to mention an extremely important type of energy — namely, *potential energy*, suggested in 1799 by the French physicist Pierre

Simon de Laplace. This is perhaps the most peculiar and unintuitive form of energy, since it depends solely on the positions of objects relative to each other, but peculiar or not, it plays a crucial role in the conservation of energy. For example, a ball rolling down a hill gains kinetic energy while losing potential energy (which is proportional to its altitude), and vice versa: if it rolls uphill, then it loses speed (and therefore kinetic energy) and all the while it gains potential energy. Another example of how potential energy plays a key role in the conservation of energy is furnished by a spring. In its neutral state (neither stretched or compressed), a spring has no potential energy, but the act of compressing or stretching it gives it some. As long as one holds the spring tight, preventing it from snapping back to its neutral state, its energy remains potential, but at the moment of release, this positional energy is converted into kinetic energy, with the *total* energy remaining perfectly constant at every instant of the process.

Thus energy, like mass, seems to come in two very different varieties: on the one hand, there are all the *dynamic* forms of energy that have to do with movement — heat (jiggling of molecules), waves, rotation, movement through space, etc. — and on the other hand, there is *static* or *potential* energy, which seems very different, because it has nothing to do with movement, just with position. Recalling our financial allegory, we might be inclined to dub the first variety *liquid* energy, since it always involves something that flows, whereas the second variety, potential energy, exists in the absence of any kind of motion, which could encourage us to dub it *frozen* energy.

This splitting-up of the concept of energy into two varieties — liquid and frozen — can't help but remind us of our splitting-up of the concept of mass into two varieties — strange and normal (and we mustn't forget that this second dichotomy was imposed on us by Einstein's equation). The analogy is clear; indeed, it cries out to be made. But despite its salience, this analogy leads us to a problem, for already in Einstein's day, people had known for roughly a century that the two varieties of energy (dynamic and static) are fully interconvertible (otherwise conservation of energy would not hold), whereas we have just been insisting that the two varieties of mass (strange and normal) are *not* interconvertible. If they were, then an iron atom or a pearl or a boulder could just poof out of existence, provided that it left in its wake the proper amount of strange mass — that is, some heat, some sound, some light... But that never happens — or at least this is what any sane person would naturally think. In sum, then, as far as *mass* is concerned, it seems that there has to be a watertight partition between the two varieties, while as far as *energy* is concerned, there is no partition at all between the two varieties. And therefore, our budding mass–energy analogy goes up in smoke. What a shame!

But here is where Einstein's "instinct for cosmic unity" comes into play. As Banesh Hoffmann put it, to insist on the existence of a watertight partition between strange and normal mass "would be to imagine two types of mass for no good reason when one would suffice. The distinction would be inartistic and logically indefensible." If we take Hoffmann's word for it, then, Einstein must have said to himself in 1907, in essence, "My unflagging faith in nature's uniformity leads me to conclude that it must be possible for an ordinary lump of matter possessing *normal* mass to be converted into a quantity of *strange* mass or vice versa, even though nothing of the sort has ever been seen anywhere." This moment of deep inspiration for Einstein, triggered by his esthetic of simplicity, would be analogous to Jan's epiphany when, faced with the threat of repossession, she broke the invisible financial barrier and imagined the previously unimaginable idea of converting her frozen assets into liquid assets.

But what could have prodded Einstein to break the analogous barrier in the concept of mass, which had seemed so definite and so firm? What metaphorical "repo person" came knocking one fine day and put sufficiently intense mental pressure on him? An esthetics-based longing for cosmic unity alone couldn't have done it, because as we said above, it was simply *self-evident* that there was no interconvertibility between the two varieties of mass. Clocks, blocks, and rocks *never* evaporate into flashes of light, sound waves, or anything else. They just sit there, inert and immutable. All this was clear as day. What, then, might have led Einstein to see things otherwise?

Recall how Banesh Hoffmann summarized Einstein's state of mind concerning mass and energy in 1907. If we rephrase that quote using the terminology of this chapter, Einstein would be thinking essentially the following: "Normal mass somehow has to possess energy because it is essentially the same thing as strange mass, and the latter, according to the equation I derived two years ago, possesses energy. Analogy thus forces me to generalize, and so I conclude that *all* types of mass possess energy." The analogy clearly resides in the words "it is essentially the same thing as", but once again we have to wonder *why* Einstein would have been confident of such an analogy, given the vast difference between how one conceives of normal mass and strange mass, and given that there was nary a shred of experimental evidence for the idea that locked up inside every single piece of ordinary, innocent-seeming matter were vast hidden reserves — indeed, inconceivably enormous reserves — of energy.

The key hint for Einstein could well have been potential energy, for as we pointed out above, potential energy is reminiscent of normal mass. While other forms of energy involve movement, potential energy is inert. Likewise, while strange mass involves movement, normal mass is inert.

Up to this point, the analogy between potential energy and normal mass is strong, but in Einstein's mind, it would have been weakened by the fact that *all* forms of energy, including potential energy, are interconvertible, whereas for mass, the notion of interconvertibility applies only to one side of the partition, pointedly excluding normal, "frozen" mass. This is a most disturbing asymmetry — but for that very reason, it is most provocative! Why should there be an impermeable membrane separating strange from normal mass, if the analogous membrane within the concept of energy is perfectly permeable? This is the key question leading to the key breakthrough.

It would clearly be a wild leap in the dark to propose that normal, "frozen" mass also can participate in the fluidlike phenomenon of conservation of total mass. It would lack any justification except a deep esthetic desire for unification, reinforced of course by a suggestive analogy — namely, the fact that potential energy participates in the conservation of total energy. But no matter how suggestive the analogy, to make such a leap would be reckless, because it would oblige one to believe in the wild idea of lumps of mass poofing into and out of existence — an unheard-of kind of event at that time.

Furthermore — and this made the notion even more surrealistic — Einstein was most aware that, because of the enormous multiplicative constant c^2 in his equation, the metamorphosis of even the most insignificant quantity of normal mass into strange mass would make an inconceivably huge amount of energy materialize seemingly out of nowhere (although it would actually have always been there, just hidden out of sight in innocent-seeming lumps of matter, thus strongly analogous to chemical potential energy lurking silently and invisibly in chemical bonds). This release of gigantic reserves of hidden energy would allow the development of stupendous sources of energy, not to mention stupendous weapons. If one day this kind of metamorphosis could be carried out, the world would be profoundly changed.

In sum, the new interpretation of $E = mc^2$ amounted to a daring leap into wild science-fiction scenarios. And yet, though it was based on nothing but an intuitive esthetics-based analogy, this is exactly the leap that Einstein made in print in 1907, thereby opening the door to a revolutionary vision according to which a material object having normal mass could be converted into other, intangible forms of mass, thereby freeing up vast amounts of hidden energy that had been locked up inside it as a kind of potential energy. 1907 is thus the year in which the metaphorical new tower of Pisa started to lean, thereby attracting a great deal of attention. From that moment on, the soon-to-be-cliché phrase "Einstein's relativity theory" became inseparable, in the public's imagination, from the equation $E = mc^2$.

In 1907, however, there didn't exist the tiniest shred of experimental evidence for Einstein's extension of the original meaning of his equation. Only many years later — in the year 1928 — thanks to a subtle fusion of relativity and quantum mechanics, was the idea of *antiparticles* (such as the positron, antiparticle of the electron) proposed on theoretical grounds by the English physicist P. A. M. Dirac, and a few years after that, the sudden and total mutual annihilation of two stationary lumps of matter — specifically, an electron and a positron — was experimentally observed in a process that gave rise to just two photons (the elegant and indispensable word "photon" had finally been coined, in 1926, by Gilbert Lewis) zipping away from each other at the speed of light (by definition!), and undulating with exactly the amount of electromagnetic energy that, when divided by c^2, equaled the sum of the two late particles' normal masses. In other words, this experimental discovery showed that ordinary matter having *normal* mass (in this case, the electron and the positron, which could be thought of as mutually annihilating "nano-boulders", so to speak) could indeed suddenly cease to exist, as long as it was simultaneously supplanted by a burst of radiation energy possessing exactly the same amount of *strange* mass. Thus, after twenty-five years had passed, experimental confirmation finally arrived for Einstein's risky leap that had been based solely on an analogy grounded in esthetics.

We find it revelatory, as does Banesh Hoffmann, that Einstein's dramatic conclusion in 1907 (namely, that *mass always contains energy*) was nothing but the flip side of his 1905 discovery (namely, that *energy always possesses mass*). It's as if, after writing down his equation, he at first read it in only one direction ("$m = E/c^2$" — that is, "an inconceivably tiny amount of mass is possessed by any standard-size portion of energy"), and then finally realized, after two years, that it could be read in the other direction as well ("$E = mc^2$" — that is, "an inconceivably huge amount of energy lurks hidden in any standard-size portion of mass"). This shows that even for the most audacious of spirits, it sometimes takes a great deal of time and intense concentration, not to mention analogy-driven cognitive dissonance, to carry out what might seem, after the fact, to be the most elementary of conceptual reversals.

From 1905 to 1907 in a Nutshell

Below we offer a summary of the many-voiced symphony of ideas about energy and mass in Einstein's mind that eventually led to his breakthrough in 1907, resulting in a far deeper understanding of the meaning of the equation that he had first written down in his *annus mirabilis.*

Ideas inherited from previous eras…

- There are two fundamental varieties of energy: *dynamic* energy, due to the movement of objects and to the oscillation of waves, and *static* (or *potential*) energy, due to the relative positions of objects.
- Either variety of energy can be converted into the other.
- All physical processes conserve the total energy in the given system; the same holds for the system's total mass.

Ideas that Einstein came up with in 1905…

- Whenever any object emits a ray of light, it loses not only a quantity of energy E but also a microscopic quantity of mass, which is given by the equation $m = E/c^2$. Analogously, if a ray of light is absorbed by an object, the object acquires not only some energy but also some mass, given by the same equation.
- A ray of light carrying some energy E must also carry some mass m, once again given by the same equation.
- Conjecture by analogy: not only electromagnetic waves but *any* form of dynamic energy possesses mass. Thus, whenever an object acquires (or loses) a quantity of dynamic energy E, it acquires (or loses) an infinitesimal quantity of mass m, once again given by the same equation.
- Conjecture by analogy: this holds not only for dynamic energy but also for static energy.
- The mass of an object consists of two fundamental varieties: its *normal* mass, which is due to the matter the object is made up of, and its *strange* mass, which is due to the energy it contains.
- Since the basic particles composing an object do not mutate during the emission or absorption of energy, the object's normal mass never varies.

- All the energy contained in an object possesses strange mass; conversely, any strange mass contains energy, the exact amount being given by the equation $E = mc^2$. By contrast, the *normal* mass of an object plays no role in the mass–energy relation, and so the equation $E = mc^2$ applies only to *strange* mass.

A mass–energy analogy starts to form…

- Mass and energy are alike in that both of them are conserved by all physical processes; moreover, the equation $E = mc^2$ connects a given quantity of energy to a corresponding quantity of mass in a simple, natural fashion. Mass and energy are thus analogous entities — indeed, they are intimately related.
- There is a very inviting resemblance between *static* energy and *normal* mass (since both are unrelated to movement), and likewise there is an inviting resemblance between *dynamic* energy and *strange* mass (since both are due to movement). These two resemblances constitute the heart of the incipient mass–energy analogy.

At the same time, a lack of symmetry gives rise to cognitive dissonance…

- Energy (since it is not composed of particles) is endowed with *strange* mass, but it has no *normal* mass. Also the reverse holds: any object's *strange* mass is endowed with invisible energy, sitting quietly in reserve until it is released, but this does not hold for the *normal* mass of the same object (that is, normal mass possesses no energy).
- There is thus an "internal partition" in the concept of mass, separating normal mass from strange mass; because of this partition, the two are not interconvertible. However, this internal partition in the concept of *mass*, keeping two varieties forever apart, has no counterpart as far as *energy* is concerned (all forms of energy being interconvertible). This mass–energy mismatch is a serious blight on the incipient analogy linking the two concepts.

Thanks to a hypothesis that restores "cosmic unity", the cognitive dissonance is dissipated…

- Since there is no partition separating different types of energy, and since there is a promising analogy linking energy to mass, then if one truly believes in this analogy, it becomes conceivable that mass, just like energy, might *not* be divided by an internal partition, but that its two varieties (normal and strange) might be interconvertible.
- This idea, if true, would imply that normal mass, no less than strange mass, constitutes a reservoir of energy, and that (under special circumstances of an unclear nature) it can transform into strange mass (or vice versa). This would imply that an object could (under these special circumstances) completely poof into thin air, as long as its normal mass were instantly transformed into an equal quantity of strange mass.
- The amount of energy associated with the "poofing out of existence" of an object having mass m (or more precisely, the conversion of normal mass into strange mass) is given by the equation $E = mc^2$, and would therefore be astonishingly large, even if the object itself were extremely lightweight.

Clearly, this is a very subtle story, and in our attempt to make all of its many stages vivid, we struggled hard. A big part of the challenge was to find the optimal pair of contrasting English adjectives to convey the key dichotomy between the two varieties of mass. We entertained quite a few possibilities, including "corpuscular / vibrational", "permanent / volatile", "unusable / usable", "solid / liquid", "concrete / abstract", "tangible / intangible", "classical / Einsteinian", "hard / soft", "corporeal / ghostly", and even "lumpy / wiggly". We also considered Einstein's own terms ("true mass" and "apparent mass"), but as he used those terms only one single time, they were far from canonical. And so in the end we settled on "normal mass" and "strange mass". This was a difficult decision, because each of the contrasting pairs that we tried on for size had both virtues and defects: that is, each pair suggests (or comes from) a slightly different analogy with familiar situations, and thus it brings out certain subtleties of this mysterious distinction that are not brought out by other pairs.

Intangible flavors of this sort are what guide a physicist instinctively toward new hypotheses. In this particular situation, Albert Einstein was pushed and pulled in various directions by numerous unspoken, probably largely unconscious, analogies, and finally, after two years, he imagined what he had been unable to imagine in 1905. The allegory of Jan, who, under intense pressure, suddenly had the breakthrough realization that her massive, solid mansion was in principle every bit as lithe and liquid as her bank account was, is an explanatory caricature analogy that helped us to convey, in highly concrete terms, the flavor of Einstein's 1905–1907 esthetics-permeated ponderings.

To sum up, there are different and interconvertible types of mass, just as there are different and interconvertible types of energy; we thus learn that mass is every bit as protean and as ever-changing as is energy. This link between mass and energy is the astonishing analogy lying hidden in the five symbols of this most celebrated equation.

The Analogies of Einstein and the Categories of Physics

We earlier alluded to a special quality of Einsteinian analogies, which is that they often turned out not just to take advantage of similarities but to create deep unifications. Consider how Einstein discovered special relativity. The key step was making a trivial-seeming analogy from mechanics to all other branches of physics. Probably most physicists of Einstein's day, had they been handed the principle of Galilean relativity on a silver platter and told, "Generalize this principle!", would have been able to make the same analogy and would have reached Einstein's generalization of it (although whether they would have realized its far-reaching consequences is another question). But the key fact is that physicists back then were *not* mulling over the limits of the principle of Galilean relativity, so no silver platter was proffered to them. The idea of generalizing Galileo's principle had to be coaxed out of the woodwork. Einstein saw that this simple and fundamental centuries-old principle lay at the crossroads of a number of important problems in physics, and that it was crying out for generalization. In contrast, other physicists were focusing on what *distinguished* electromagnetism from mechanics, instead of seeking a shared essence that could *unite* those two branches of physics.

Similar remarks can be made concerning Einstein's 1907 analogy linking two seemingly different types of mass. Some people might even object that this was not an analogy, because (at least according to the common stereotype) analogies are always just partial and approximate truths, whereas in the case of $E = mc^2$, the link Einstein found between what we've called "strange" and "normal" types of mass turned out to be a complete and precise truth, revealing them to be merely two facets of one single phenomenon. Well, such an objection might seem tempting, but it is off base.

The fact is, the idea started out life as a typical analogy, tentative and shaky — the fruit of a long and patient quest by Einstein to unify two concepts that, in the minds of his very few colleagues who took these kinds of questions seriously, were clearly distinct. Two decades later, however, experimental findings showed that this apparent distinction had to be dropped in favor of a single, more extended concept. The reason is that Einstein's irrepressible instinct of cosmic unity had hit the bull's-eye once again, revealing a new, broader concept built deeply into nature.

The Principle of Relativity and Accelerated Frames of Reference

We will now take a look, albeit brief, at general relativity, at whose basis there are, once again, some Einsteinian analogies that were initially perceived as bold leaps of an idiosyncratic intuition, if not as wild speculations, but which later, once they had been repeatedly shown to be correct, were retroactively perceived as *eternal truths of nature* rather than as merely one individual's subjective and uncertain speculations about some kind of similarity.

What greatly troubled Einstein, as he looked back at his theory of special relativity (which at the outset was not called "special", since at that time it was not part of a more general theory), was that it applied only to frames of reference that were moving constantly and smoothly — that is, without any acceleration. His extension of Galileo's principle of relativity, proposed in 1905, and later dubbed by him the Principle of Special Relativity, said that *for certain types of frames of reference*, it is impossible to tell, using internal experiments, whether one is at rest or not. These frames of reference were those moving at a constant velocity. But internal experiments *can* distinguish, without any problem, between *some* kinds of frames of reference — namely, between accelerating and constant-velocity frames. For example, if you are inside a car that is accelerating rapidly, you *cannot* pour a glass of water just as you would do in your kitchen or in an airplane flying at a constant speed, because the water, as seen by people sitting in the car, will not fall vertically downwards, but will follow a curved arc whose shape is determined by the direction of, and the strength of, the car's acceleration. Anyone inside the car can conduct this very simple experiment, which clearly reveals that the car is not moving with a fixed speed but is accelerating.

Most physicists of Einstein's day would have said that all this shows is that the principle of relativity has its limits and cannot aspire to cover *all* frames of reference — just frames moving with constant velocities. But Einstein could not stop scratching his head over this frustrating situation. He felt that his newly extended principle of relativity should somehow be able to be extended yet further, so as to cover more frames of reference than just those that were not accelerating — in fact, he felt that a truly general principle should be able to cover *all* frames of reference. This strange faith of Einstein's, which flew in the face of the most self-evident facts about the way the world works, was rooted in a profound and nearly inexplicable intuition.

In his mind, Einstein imagined an infinite universe that was totally empty but for one sole sentient observer. This perceiving being considers itself to be stationary, and thus has no sensation of dizziness. After all, as it looks around, it sees nothing but empty space. On the other hand, what if the being were *spinning* in this completely empty universe — would it feel dizzy? Well, what does it mean to speak of "something spinning in a totally empty universe"? Or conversely, what if our observing being was *not* spinning, but the rest of the universe was spinning around it, like a merry-go-round spinning around a stationary pillar? Would the observing being then feel dizzy? And finally, is it in theory possible to distinguish between these two scenarios? Are they identical or are they different? Such thought experiments force us to ask whether it is possible to determine which of the two — the observer or the rest of the universe — is spinning. It would seem that for the two scenarios to be distinguishable, there would have to be a *preferred* frame of reference, sometimes called an *absolute* frame of reference, or "God's frame of reference", or the frame of reference of a hypothetical "ether".

In his early years, Einstein was constantly haunted by philosophical questions like these, and in the wake of his discovery of special relativity, he found the idea of an absolute frame of reference so distasteful that he rejected it out of hand. (Einstein was particularly inspired by the writings of the Austrian philosopher and physicist Ernst Mach — ironically, one of the staunchest disbelievers in atoms! — in which Mach dreamt up hypothetical universes of this sort and carefully studied their consequences, which had led Mach to the conviction that the idea of absolute motion — a notion due to Newton — makes no sense.) Einstein nourished the hope of discovering some way to incorporate even accelerating frames of reference into his principle — in other words, he wanted to show that acceleration, much like speed, is not absolute but depends on the frame of reference that one chooses.

Special relativity implied that what was perceived as motion at a constant speed by the observers in one frame of reference could validly be perceived as perfect immobility by the observers in some other properly chosen frame. Einstein wanted to generalize this idea; he wanted an analogous principle to hold for all types of motion, including accelerated motion. His hope was that if, from the viewpoint of one set of observers in one particular frame of reference, some object was accelerating, it would be possible to find *another* frame of reference in which all observers would say the object was perfectly still. To put it in another way, he hoped that the laws of nature as perceived by an accelerating observer would be identical to the laws of nature as perceived by an observer who was at rest.

Despite this idealistic hope, Einstein was keenly aware of all sorts of phenomena, such as the pouring of a glass of water inside a speeding-up or slowing-down car, that *do* allow one to distinguish accelerating frames of reference from non-accelerating ones.

The undeniable conflict between the unbending reality of nature and his strong intuitions pushed Einstein to focus intently on the nature of acceleration. Very few physicists are driven to seek answers to such pithy and essential questions as "What is acceleration?" or, more specifically, "Must it be the case that something accelerating for *one* observer will be accelerating for *all* observers?" But it was typical of Einstein to do just that — to tackle with unbounded stubbornness questions that concern matters that seem so primordial and so pervasive that nearly anyone else would have wondered what use it could possibly be to worry about them.

Whenever one studies classical mechanics (as Einstein did at ETH, the Swiss Federal Polytechnical School in Zürich), the mathematical form of the laws of physics from the vantage point of an accelerating reference frame is always covered at least briefly. The most salient fact one learns in such an overview is that any observer in an accelerated frame who insisted that the frame was *not* accelerating would have to posit a mysterious "extra force" acting on all objects. Without such an extra force, there would be no way to account for the anomalous movements of the objects in the frame.

Imagine, for instance, that the passengers in a city bus collectively decided, on some odd whim, to declare that their often-accelerating, often-decelerating bus never moved but was always perfectly still. In order to account for the strange phenomena that they witness (*e.g.*, water following a curve rather than falling straight down when poured from a pitcher, not to mention their own frequent sensation of being jerked forwards or backwards), they would have to posit a mysterious extra force sometimes pulling things towards the front of the bus, other times towards the rear. Of course, they wouldn't need to refer to any such force if they acknowledged that their bus is sometimes speeding up and sometimes slowing down, but if they refused to adopt that point of view, then this extra force would have to be included in the laws of physics that they formulate to explain the phenomena that they observe.

Forces like this, which show up only in accelerating frames of reference, are called *fictitious forces*, and it happens that all fictitious forces have a special mathematical property — namely, if such a force acts on an object that has mass m, then it will necessarily be proportional to m, no matter how the frame is accelerating and no matter how the object is moving (think of a tennis ball being dropped inside a car just as the driver slams very hard on the brakes). This proportionality to mass gives rise to an interesting consequence, which is that if one releases several objects at the same instant in an accelerating frame of reference, they will all follow perfectly parallel trajectories. For instance, if a ball, a bell, and a bowl are all released into the air at exactly the same moment inside a car that is accelerating, then all three of them will describe identical-looking curves as they fall. This means that if they start out in a tight cluster, then they will remain in a tight cluster; at the end of their fall, their cluster will be just as compact as it was at its start.

Einstein had understood all this at ETH, just as did all his classmates. But the moment that their doctoral exams were over, all the young physicists were happy to let all of this book-learning sink into oblivion, since the formulas involving fictitious forces in accelerating frames are, as a rule, quite convoluted; indeed, there is no reason to carry out calculations in such a frame, since one can always first describe the situation from the point of view of another observer moving at a constant speed and then do the calculations in that frame of reference. (In the case of the accelerating bus or car, this observer could be a smoothly-riding bicyclist, or a pedestrian standing at a corner.) It's easy to see why young physicists gladly left all these complexities behind. And yet, whenever someone is obsessed by some idea, even the most deeply-buried of memories can suddenly bubble up, almost magically. In order to tell the story of one such sudden retrieval in Einstein's mind, we first need to say a few words about gravitation.

Applying Relativity to Gravity by Analogy

One of Einstein's obsessions, once he had discovered special relativity, was to integrate it somehow with gravity, which, for him, as for any physicist of the period, was a force that bore a tight analogy to the electrical force between two charged particles. One key difference was that for gravity, the force is always attractive, while for two electrically charged objects, it can be either attractive or repulsive, depending on whether the electrical charges of the objects are both positive, both negative, or of opposite signs. Like charges repel, and unlike charges attract. Despite this difference, the analogy between the two forces was clear and compelling. In the case of gravitation, the force between two unmoving objects having masses m and M and separated by distance d is given by the famous Newtonian formula "$m \times M / d^2$", whereas for electricity, the strength of the force between two unmoving objects having charges q and Q and separated by distance d is given by the formula "$q \times Q / d^2$", which was discovered exactly a century later by the French physicist Charles Coulomb. (We have left out multiplicative constants since they aren't relevant in this context.) These formulas are identical, except that charges in the latter replace masses in the former. The analogy seems flawless on first sight, but Newton's formula gave rise to a serious problem concerning gravity. To illustrate it, we consider an extreme scenario.

Suppose our sun were to suddenly cease existing. It would take eight minutes for this disastrous piece of news to reach us on Earth (and for once the word "dis-aster" would be living up to its etymology); only after that period of time had elapsed would the sky go completely dark. The reason for the eight-minute delay is, of course, the finiteness of the speed of light, and this speed can be calculated directly from Maxwell's equations for electromagnetism. Unfortunately, the tight analogy between gravitational and electrical forces does not involve Maxwell's equations; it involves just Newton's law and Coulomb's law. Nothing comparable to Maxwell's equations had yet been found for gravitation. Newton's law — the lone equation that was then known to apply to gravity — did not predict that gravity could propagate across space. Thus, far from emerging as a consequence of known equations, the "speed of gravity" was an unheard-of notion; to suggest that gravity had a speed was to verge on spouting absurdities.

For this reason, a physicist of that era might well have declared, "It wouldn't be necessary to wait eight minutes for the bad news to reach us. Mother Earth would react *immediately* to the sudden dematerialization of the sun. After all, it would have no reason to continue to follow its quasi-circular orbit around a star that had ceased to exist and thus would no longer be exerting any tug on it. The earth would be like a dog whose leash had suddenly been cut: instant freedom!" On the other hand, another physicist of the era might well have argued the exact opposite — namely, that it would take time to detect the far-away sun's demise, a conclusion based on the intuitive belief that no event can have an instant effect on objects arbitrarily far away from it.

In any case, no experimental results or theoretical ideas were available to back up either side. And in the less disastrous (and more plausible) scenario of the sun's center of mass suddenly moving a little bit (perhaps on account of some kind of internal explosion), exactly the same sorts of questions ("How long would it take for the earth to 'find out' that the sun had moved?") could be asked, but to such questions no answer was offered by the physics of the day. In short, while gravity deeply resembles the electrical force in some ways, there are other ways in which the two forces seem to be profoundly different, and in those days no one had any idea how to write down a set of equations that would fully capture the phenomenon of gravitation.

In essence, the problem was to figure out how changes in gravity's intensity are propagated across space, and at what speed — finite or infinite? — such "news" travels from one point to another. This boils down to asking "What is the formula for the gravitational attraction between two *moving* objects?" This was a question that had already been of serious concern to Isaac Newton, the first person to offer a quantitative theory of gravitation, but in all the years since him, no one had yet solved the problem. Einstein, in an attempt to answer this riddle, turned to the analogy between Newton's formula for static gravitational attraction and Coulomb's formula for static electrical attraction (the two formulas we mentioned above), and threw in an extra term that seemed exceedingly natural, and which came from his theory of (special) relativity. This new term elegantly extended the analogy between gravity and electromagnetism so that it included objects moving relative to each other. This small but very tempting addition yielded a new Einsteinian theory of gravitation that had a Maxwellian flavor, in that gravity now had wavelike behavior, and among the new theory's consequences was that gravity propagated through space at a finite speed — in fact, at exactly the same speed as did light. According to Einstein's new theory, then, the earth would "learn", so to speak, that the sun had ceased to exist (or had suddenly moved a little bit) at exactly the same moment as our eyes would see it. Einstein's new theory of gravity meshed well with special relativity.

One Einsteinian Analogy Bites the Dust and Another One Replaces it

Although this analogy with electromagnetism was both natural and attractive, its discoverer, who was at the same time his own most severe critic, soon became aware that the formula that it gave for gravitational force had a fatal flaw: the force between two objects would no longer be proportional to the product of their masses (the numerator "$m \times M$" that we saw above). That property of gravity was so well established and seemed so central to the very nature of gravity that violating it was intuitively repugnant to Einstein. As a result, he abandoned the lines of work that came from this first analogy and began searching for a different analogy that would link gravitation and relativity.

In this new quest, he focused not only on gravitation, whose most stable and defining feature is its proportionality to the mass of each of the two objects pulling each other, but also on accelerated frames of reference, which are so fundamentally different from frames at rest (and ones that have a constant speed). Indeed, because the fruit of his first analogy had failed to respect gravity's proportionality to mass, this fact about gravity became one of the primary desiderata for a better theory.

This was the point at which some of the old ideas from courses at the ETH in Zurich started bubbling up — ideas involving accelerated frames of reference. In particular, the memory of fictitious forces came back to Einstein, for any fictitious force is likewise *proportional to the mass* of the entity it is acting on. Let us try to relive from a first-person perspective this Einsteinian mental process: "Hmm… Gravity *reminds me* of a fictitious force… Gravity *acts like* a fictitious force. Gravity *is analogous to* a fictitious force… Might, then, gravity actually *be* a fictitious force?" Here is a remarkably smooth mental glide that starts out with an innocent little case of reminding but that ends up being, once again, nothing less than a cosmic unification.

To see more clearly the implications of this idea, let's consider the canonical example of an accelerated frame of reference that Einstein himself used in order to explain his new analogy. Instead of imagining that we're in an accelerating bus or car, let's jump on board a cube-shaped interplanetary laboratory floating somewhere in empty space, far removed from any star, quite literally out in the middle of nowhere. And now let's throw into the mix a powerful rocket that starts to pull the lab by means of a cable attached to one of its six exterior walls. If the rocket pulls with a constant force, this will give rise to a constant acceleration of the lab (after all, as Newton told us, $F = ma$ — that is, a constant force induces a constant acceleration). Before the rocket started firing, people inside the lab were floating about between its six walls, and there was no reason to single out one particular wall and call it "the floor"; however, the moment the rocket started to pull, one of the six walls started approaching the lab-bound celestial travelers, and all of them banged into it and remained stuck against it, because the lab's constant acceleration broke the symmetry, putting an end to the possibility of floating in it. This particular wall thus became the lab's "floor".

Moreover, if one of the lab's denizens tossed a pencil into the air, the latter would "fall down" onto the "floor", just as Newton's famous (if apocryphal) apple fell down onto his head. Why would this be the case? Seen from *outside* the lab, it's clear as day: the floor is constantly moving towards the pencil (until they collide). By contrast, the people who are *inside* the lab and are ignorant of the rocket conceive of their lab as sitting absolutely still (or as moving at a constant speed) in the middle of outer space, and so for them the pencil is falling because a *gravitational pull* suddenly and inexplicably permeated their lab, affecting all people and objects in it, and of course that brand-new pull singled out a particular direction in space (which was then baptized "down"), and it made things *fall* in that direction.

This means, among other things, that if some Galileo copycat in the lab were to "stand up" on the new "floor" and were to "drop" two very different objects, such as a pencil and a cannonball, these items would start to "fall" side by side and would bang onto the "floor" at precisely the same instant. And why would this be? Once again, for external observers, it's obvious: the two objects aren't moving at all; rather, it's the floor that is moving "up" to greet them. So of course it will hit them at exactly the same instant. But from the viewpoint of the people inside the laboratory, the phenomenon arises because gravity has that key property that Galileo, creatively exploiting the leaning Tower of Pisa, was the first to demonstrate — namely, that all objects fall in the same way (*i.e.,* with the same acceleration), whatever their masses might be.

This "force of gravity" perceived by the denizens of the lab is a quintessential example of a fictitious force, but for them there is nothing fictional about it — for them, it is a *real* force; for them, there is a *real* floor and a *real* ceiling; for them, there is a *genuine* distinction between *up* and *down.* Unless they somehow manage to sneak a peek *outside* of their lab (which would violate the premises of Einstein's thought experiment), these voyagers have no way to tell apart the rocket-made gravitation from normal earthly gravitation, which they have known ever since their childhood. This means they have no way to tell whether their lab is constantly accelerating in empty space or is sitting still in the earth's gravitational field. The two situations are indistinguishable.

If you are picking up echoes of Galileo's principle of relativity, you're not mistaken. This is exactly what Einstein was up to. Like a top-drawer magician, he had started with the idea, obvious to everyone, that an accelerating frame of reference is *easily distinguishable* from a stationary frame, only to arrive at the diametrically opposite conclusion: that an accelerating frame of reference is *completely indistinguishable* from a stationary reference frame immersed in a gravitational field. What a fantastic trick! Galileo would surely have loved this ingenious combination of two of his greatest ideas.

Moreover, Einstein soon saw that there was a broad spectrum of indistinguishable labs; to get the flavor of them, you need merely imagine a lab on the surface of the moon (whose gravitational pull is much weaker than earth's), which at some point starts to be pulled upward by a slightly less powerful rocket than the previous one. Now the combination of the moon's feeble gravitational pull and this weaker rocket's pulling adds up to a result that, for people inside the lab, is exactly like the case we described earlier. They feel as if they are now in the *earth's* gravitational field. And it's obvious that we could twist the knobs that control (1) the strength of the "real" gravitational field in which the lab is sitting, and (2) the power of the rocket that is pulling the lab, in such a manner that in each case the resultant experience of gravity will be identical to the earth's gravitational field for the people inside the lab. All of these imaginary labs are indistinguishable from one another by any kind of mechanical experiment at all, as long as it is carried out entirely *within* the lab.

In summary, thanks to the memory of fictitious forces in classical mechanics that bubbled up in his brain, Einstein was able to breathe new life into his beloved principle of relativity, and he did so in a situation where any other physicist would have been sure that there was no hope of doing so.

The Principle of Equivalence (First Draft)

To the new and radically extended version of the principle of relativity that he had just found, Einstein gave the name "Equivalence Principle". By this, he meant that there was an equivalence or indistinguishability between gravity, on the one hand, and acceleration, on the other. In order to explore the consequences of his simple but counterintuitive hypothesis more deeply, he imagined another scenario involving his laboratory floating in space. This time, the lab is not out in the middle of nowhere, but is hovering 100 miles above the earth, where a magical angel is holding it up and perfectly still. The people inside have no qualms about using words like "floor", "ceiling", "up", and "down", because the earth's gravitational pull permeates their laboratory exactly as it would permeate a lab sitting on the surface of the earth, the only difference being that gravity is slightly weaker at an altitude of 100 miles than at sea level. If one ignores this detail, the people inside the lab could easily imagine themselves as being on the earth. But all at once the angel is stung by a randomly buzzing interplanetary bumblebee and lets go of the lab, which starts plummeting down towards the earth. What do the people inside it feel?

As soon as their cube starts its earthward descent, all the objects inside it also start to fall, and recall that, as Galileo first showed, they will all fall in the same way (recall his experiments at the Tower of Pisa). This means that any object that had previously been sitting on the floor no longer feels itself held down; it is suddenly free to float about in the lab. In fact, for the people inside the lab, there is no floor any longer, nor is there any ceiling; in a flash, the words "floor", "ceiling", "up", and "down" have been deprived of their meanings. The lab's inhabitants are experiencing weightlessness: that is, the sensation of *zero gravity*. Since everything in the lab is falling earthwards at exactly the same rate, its denizens have the curious impression that *nothing* is falling; for them, everything (themselves included) is now floating about inside their big room.

For Einstein, this realization was one of the most beautiful moments of his life, for as he started to glimpse it, a wonderfully fertile analogy suddenly sprang to his mind. He had been musing about electromagnetism for a long time, and he knew intimately that if one moves from one reference frame to another, the apparent values of the electric and magnetic fields change throughout all of space. For example, an observer who moves a magnet in a lab will be able to detect an electric field in the neighborhood of the magnet, and the more quickly the magnet is moved, the more intense will be the measured value of the electric field. (If the magnet is simply kept at rest, then it gives rise to no electric field, of course.) This important effect is called "electromagnetic induction", and was first observed in 1831 by the English physicist Michael Faraday. Now let us imagine a second observer sitting tightly attached to the magnet (which means the observer is in the magnet's reference frame). By definition, for this person, the magnet is perfectly at rest, and so Faraday's induction law says there is only a magnetic field. Put otherwise, from this person's viewpoint, there is no moving magnet to give rise to any electric field through electromagnetic induction. So we see that in electromagnetism, merely by changing viewpoint (*i.e.*, frame of reference), one can make an electric field completely disappear (or appear out of nowhere). (The same holds for a magnetic field, although we haven't described this case.) Even in his early adolescence Einstein had already been struck and fascinated by this mysterious effect.

Shortly after the memory of this effect in electromagnetism bubbled up, Einstein expressed his great joy thus: "At that moment the happiest thought of my life occurred to me — namely, the gravitational field, just like the electric field generated by a moving magnet, has an existence that is only relative." ("Relative" here meant that its existence depended on the frame of reference in which one was located, and in particular that in at least one frame, it didn't exist at all.)

And indeed, with his new scenario of the earthwards-plunging laboratory whose occupants feel that they now are experiencing no gravity, Einstein had found a situation where a perfectly real gravitational field in *one* reference frame can be made to totally vanish merely by jumping to *another* one. To an outside observer — say, someone on earth — the falling laboratory is still permeated by the earth's gravitational field (which is why everything in it is falling earthwards); and yet to the people inside it, there simply is no gravitational field at all, and nothing is falling.

This scenario is in some sense the flip side of the scenario featuring the laboratory in remote outer space being pulled by the powerful rocket, since in the latter scenario, the people inside feel, observe, and measure a gravitational field, while outside observers claim that there is *no* gravitational field — all they see is a rocket that is making the lab go faster and faster, with respect to the far-off stars.

Einstein Seeks and Finds a Deeper Analogy

We come now to a decisive moment in the story of general relativity. Above, we described Einstein's new principle as asserting that an accelerating reference frame is completely indistinguishable from a non-accelerating reference frame immersed in a gravitational field. However, in putting it this way we jumped the gun, because his original principle was significantly more limited than that — namely, it asserted that an accelerating reference frame should be indistinguishable, *by means of mechanical experiments*, from a non-accelerating reference frame immersed in a gravitational field. Einstein was keenly aware of the fact that the analogies that had led him to his new-found principle — the "happiest thought" of his life — involved only the *mechanical* behavior of imagined objects in various different imagined laboratories. That is, when he had imagined his various spacebound laboratories, he had considered only scenarios that involved concepts such as *speed*, *acceleration*, *rotation*, *gravity*, *friction*, *orbits*, *collisions*, *springs*, *pendula*, *vibration*, *tops*, *gyroscopes*, etc. — just the concepts of classical mechanics. He had not considered what might happen in the case of an *electromagnetic* experiment in an accelerating laboratory — say, an experiment that used light rays or electric or magnetic fields.

This was because he knew that he did not have the requisite knowledge, be it theoretical or experimental, that would allow him to predict what would happen in such a case. And that was why he knew he had arrived at a critical crossroads. His theoretical knowledge and his gift for imagining the consequences of various idealized physical circumstances (his famous "thought experiments"), even when aided by the cleverest reasoning, simply would not allow him to go any further. He had reached a crucial spot where he would have to take yet another daring step, once again a step that would rely solely on an esthetic motivation, a step grounded solely in his belief in the deep unity of the laws that govern the universe — that is to say, to his unshakable faith in the existence of very simple, general, and elegant principles.

As readers of this chapter are well aware, back in 1905 Einstein had already found himself, by chance, in an analogous situation — namely, when he had chosen to broaden the Galilean principle of relativity on esthetic grounds (his faith in the unity of the laws of physics), by replacing the phrase "any kind of *mechanical* experiment" with the phrase "any kind of *physical* experiment". In other words, he had already visited this region in the world of ideas, had already dared once earlier to make just this leap of analogical faith, and on that occasion his intuition had been richly rewarded — and so, why not do "exactly the same thing" in this analogous new situation?

Einstein thus extended his principle to run as follows: "An accelerating reference frame cannot be distinguished, *no matter what kind of physical experiment one might use*, from a non-accelerating reference frame immersed in a gravitational field." Once again we point out that from a certain point of view, replacing the word "mechanical" by the word "physical" (in other words, noting the analogy between mechanics and any other branch of physics) was the most trivial step Einstein could have taken, since that same analogical extension had already worked with flying colors one time earlier in his life (special relativity had already been confirmed by a good number of experiments) — and yet, from another point of view, it was an extremely audacious analogical leap into an utterly unknown world.

Let us listen once again to the words of Banesh Hoffmann on the subject of this jump that carried Einstein from the *restricted* principle of equivalence to the *extended* principle of equivalence:

> [The new principle] had artistic unity: for why should he needlessly assume *one* type of relativity for mechanical effects and a *different* one for the rest of physics?

Once again, we see an analogy-based conceptual leap that was apparently minuscule and elementary, and yet on the other hand turned out to be gigantic and brilliant. All of this came to him courtesy of his "instinct for cosmic unity", which was an almost inexhaustible font of rich analogies.

Consequences of the Extended Principle of Equivalence

We will give here an example of the unexpected consequences of this daring leap towards a truly general principle of relativity. Einstein imagined that there was, in his celestial laboratory being pulled through deepest outer space by the rocket, a perfectly horizontal flashlight (*i.e.*, parallel to the lab's "floor") that emitted a light ray.

Observers *outside* the lab will say that this ray is moving in a fixed direction with respect to the distant stars, while at the same time the lab surrounding it is "rising" at an ever greater velocity. From this constant "vertical" acceleration of the lab, it follows that observers *inside* the lab will perceive the light ray as *descending towards the floor* ever more rapidly as it crosses the lab at a fixed horizontal speed. In a word, for them, the light ray will follow a *curve* rather than a straight line. To be sure, the discrepancy from a horizontal trajectory will be tiny, because of the enormous ratio of the speed of light to the modest speed of the lab, but no matter; no sooner has the light ray emerged horizontally from the flashlight than its trajectory starts to bend downwards. At this juncture Einstein makes use of his newly conjectured equivalence principle generalized outwards so as to include electromagnetic scenarios. He reasons as follows.

If a nonmoving frame of reference in a gravitational field is indistinguishable *in every way* from an accelerating frame, then any effect that can be observed in an accelerated frame will also be observable in a lab on *terra firma* (since such a lab is obviously immersed in a gravitational field). The generalized equivalence principle thus told him that since a light ray in an accelerating lab in gravitation-free space follows a curved trajectory, then so must a light ray released by an observer standing still on the earth.

Einstein realized that this conclusion allowed him to predict certain celestial phenomena that had never been dreamt of, such as a tiny amount of bending of lightbeams coming from a distant star as they pass by our sun, whose gravitational field is, of course, very strong. However, for reasons too technical to go into here, this effect would be observable only during a total solar eclipse, and so, already in 1907, he urged that this effect be sought by astronomical observers. The German astronomer Erwin Finlay-Freundlich carefully examined many hundreds of photos of solar eclipses to find evidence of the minuscule effect, but found none. In fact it turned out to be necessary to wait twelve years longer, until 1919, for the confirmation of this prediction during a total eclipse observed by an English team led by the physicist Arthur Eddington from two islands in the south Atlantic Ocean.

The global effect of Eddington's team's confirmation was phenomenal. Not only did Einstein's prediction hit the bull's-eye, but the world, just emerging from under the dark pall cast by the "Great War", was thrilled that an English team had confirmed a fantastic prediction made by an "enemy" scientist (even if Einstein had renounced his German citizenship and become Swiss in order to distance himself from German militarism); indeed, many people saw Eddington's confirmation of Einstein's prediction as a moment of great glory for humanity as a whole. Soon Einstein would watch helplessly as he was transformed overnight into a world-famous celebrity.

The Noneuclidean Merry-go-round

To conclude this tour of some of the many analogies that undergird general relativity, we will give a capsule description of the key breakthrough that brought to light the appropriate branch of mathematics for Einstein's conception of gravitation. As we have just seen, all of Einstein's first thought experiments about how gravity and acceleration are related had to do with *linear* acceleration — scenarios in which a reference frame is moving in a fixed direction but with a speed that is changing. But another and equally important form of acceleration takes place whenever an object in motion undergoes a change in *direction* (though not necessarily in speed). The simplest and most canonical example is a disk spinning at a fixed number of revolutions per second. Each point on such a disk is accelerating because at every instant it is changing its direction of motion. Although applying the principle of equivalence to the rotating-disk scenario proved to be considerably harder than applying it to the linear scenario, this did not in any way discourage Einstein, who felt compelled to explore in depth the alluring case of the rotating disk.

The key turned out to be the phenomenon of *length contraction*, a subtle consequence of special relativity according to which the length of any object, when measured in a frame moving with respect to the object, is shorter than when measured in the object's *rest* frame. Roughly put, moving objects, when their dimensions are measured by observers who are *not* moving, undergo longitudinal but not transverse shrinking.

And turnabout is fair play: since an observer in either frame can validly consider *their* frame to be stationary and the *other* frame to be in motion, each one will see objects in the other frame — but not objects in their own — as being longitudinally contracted. This seems paradoxical at first, but Einstein showed why it is not. The following caricature analogy will help to suggest the subtle flavor of his resolution.

Chinese children might be tempted to say that Chilean children walk around upside down; symmetrically, Chilean children might be tempted to say that Chinese children walk around upside down. So who's right and who's wrong? Well, we who know that gravity pulls toward the earth's center understand that the direction called "down" is not the same direction everywhere on earth, despite what the naïve analogy would suggest; rather, it is relative — that is, *down* depends on where the observer is located. In analogous fashion, an object's *length* turns out not to be the same for all observers but dependent on the relative speed of object and observer; this is why two differently-moving observers of a given object will come up with disagreeing measurements of its length. But length, like down-ness, is relative, not absolute (that is, the naïve analogy is wrong in both cases), so the surface-level disagreement, though puzzling at first, does not constitute a true contradiction.

And with that, let's return to the rotating disk and general relativity. When a disk rotates, each point on it carries out motion that is circumferential but not radial; in other words, it moves round and round but not in and out. Special relativity then tells us that there is longitudinal but not transverse length contraction: this means there is contraction along the circumference but not perpendicularly (*i.e.*, along the radii). And so the disk's *circumference* that we measure, from our nonrotating vantage point, will be different in length from the one measured by observers in the frame of the spinning disk; on the other hand, our measurements of the disk's *diameter* will agree with theirs.

This idea was reached only very slowly and with great difficulty, for it profoundly violated intuition, but nonetheless Einstein stuck to his faith and proceeded to think it through with great care. And in so doing, he was led to a most peculiar finding: when a disk spins, non-spinning observers who measure its circumference-to-diameter ratio will discover that it is not equal to π. Moreover, the size of that ratio will depend on how fast the disk is spinning, with a disk that spins faster having a ratio that deviates further from π.

Even for Albert Einstein, who by this time was pretty used to coming up with intuition-defying ideas, this effect seemed very weird, and he and his colleagues referred to it as "Ehrenfest's paradox" (after the Dutch physicist who had first pointed it out), and they worked extremely hard to try to resolve it. It took several years, though. The key breakthrough came at last toward the end of the summer of 1912. Here is how Einstein described this special burst of clarity:

> I first had the decisive idea of the analogy between the mathematical problems connected with this theory and Gauss's theory of surfaces shortly after my return to Zürich.

(By "this theory", he meant his incipient theory of gravitation, which was equivalent — thanks to his well-named equivalence principle! — to the theory of accelerating frames of reference.)

To make sense of what he had just found about rotating frames, Einstein once again drew on long-dormant memories from his student days at ETH in Zürich, some twelve years earlier. At that time, he had taken a seminar on the geometry of two-dimensional curved surfaces, and in it he had learned that in this kind of non-Euclidean geometry, developed by Karl Friedrich Gauss and Bernhard Riemann in the nineteenth century, the ratio of the circumference of a circle to its diameter can be arbitrarily different from π. To get a feeling for how this can be, consider the earth's surface as a curved two-dimensional space. If we take the equator to be our sample circle, then a radius will be any line of longitude running due north from the equator all the way up to the North Pole. Such a line's length is one quarter of an equator long, and so a diameter equals *half* an equator, meaning that for observers who inhabit this curved space, the equator's circumference-to-diameter ratio will equal 2, not π.

At some point in late 1912, this theorem about non-Euclidean circles, which had long been buried in Einstein's memory, unexpectedly popped to mind — an old memory suddenly resuscitated by an event — much as had happened a few years earlier with the memory of fictitious forces. This very welcome new connection suggested to him that he could directly borrow all the formulas of Gauss's and Riemann's geometry of curved surfaces to characterize in mathematical terms the physics of an accelerating frame of reference, and even better — thanks to his equivalence principle — the physics of a world immersed in a gravitational field.

To be sure, events in our universe take place not in a two-dimensional plane but in three-dimensional space, and they take time to take place (so to speak). This amounts to four dimensions (three spatial ones, plus a temporal one). Although today the term "four-dimensional space-time" is very familiar — indeed, it has become a hackneyed cliché — the notion, when it was first proposed by the German mathematician

Hermann Minkowski, was a strange and momentous new revelation. Minkowski, who had been one of Einstein's professors at ETH, had noticed a striking analogy between the equations of Galilean relativity and those of Einsteinian special relativity, and this analogy (which, ironically, Einstein himself had somehow missed!) led him rapidly to the idea of four-dimensional space-time.

In developing *general* relativity, Einstein realized that what he needed to do was to take the two-dimensional theorems of Gaussian geometry that he'd studied years earlier at ETH (and which, back then, he'd considered purely mathematical and of no relevance to physics) and adapt them to Minkowski's more abstract space having three spatial dimensions and one temporal one. As he did so often, he was exploiting to the hilt an intuitively sniffed analogy — namely, he was "copying" in four space–time dimensions the ideas that his professors in Zürich many years earlier had taught him in two purely spatial dimensions — and what this analogy brought him was astonishingly rich and novel. It led him to the surreal notion of a *four-dimensional curved space* — and as if that weren't enough, the concept of *curvature* was now no longer limited to space but was being extended to the dimension of time. The idea of "curved time" was certainly pushing at the very limits of the human imagination.

The prolific set of analogies that we have just recounted — first, the analogy between a gravitational field and an accelerating frame of reference; second, the analogy (redeployed in a new context) between the laws of mechanics alone and the laws of physics as a whole; third, the analogy between rotating frames of reference and two-dimensional non-Euclidean geometries; and fourth, the analogy between two-dimensional and four-dimensional non-Euclidean geometries — provided Einstein with the crucial clue he had long sought, pointing the way to the tools with which to handle gravity mathematically. And so, over a period of several years, he was finally able to "tame" gravity, bringing it at last into the family of mathematical laws of physics. The resulting theory of general relativity was the most complex and daring accomplishment of Einstein's career, the crowning glory of all his astonishing discoveries.

Parallels that Meet

An entire book could easily be devoted to the fecundity of Albert Einstein's great analogies. Our goal in this chapter was more modest — it was merely to offer a sampling of them, in order to show that major advances in physics are not the result of virtuoso acts of stand-alone mathematical deduction and formal manipulation of equations, but that, quite to the contrary, they emerge as the fruit of analogies intuited by individuals who have the gift of seeing a unity where others see only diversity, individuals who have a keenly honed instinct for spotting the deep identity of phenomena that look extremely different from each other on the surface, individuals who trust their inchoate faith in such analogical links even more than they trust the imposing mathematical fortresses erected by prior generations, even if it means that extremely well-established, once rigorously worked-out ideas may possibly have to be uprooted and completely thought through anew.

Our characterization of Einstein's way of thinking portrays him as the polar opposite of the clichéd mathematical genius who launches an extraordinarily powerful calculating and deducing machine, which proceeds like a steamroller to mow down every obstacle that it encounters en route. Quite to the contrary, our discussion has shown that Einstein was not guided by phenomenal computational or reasoning skills. His brain did not house an enormous, lightning-fast supercomputer. Rather, he was driven by an unstoppable desire to seek out profound conceptual similarities, beautiful hidden analogies. Indeed, Einstein's primary psychological driving force was the quest for beauty, and he was spurred on by his certainty that the laws of nature are pervaded by the deepest, most divine beauty of all.

This kind of mindset would seem to belong to the Age of Enlightenment, to a thinker in the grand tradition of Leonardo da Vinci, Newton, Pascal, and Leibniz. Some might even suppose that Einstein must have been the last member of that special tribe of scientific geniuses that flourished in the era before today's hyper-specialization made earth-shaking insights just a relic of the past. But nothing could be further from the truth. We are dealing with phenomena that are much stabler than that. The same cognitive style is in fact the hallmark of the most outstanding scientists, whatever their discipline and whatever era they may live in. Today's most gifted researchers, even if their areas of research are so arcane that it would be impossible to explain them to a lay audience, are driven by the same inner fire as drove Albert Einstein, which is to say, by an unrelenting quest to uncover deep analogical links between phenomena.

For example, the French mathematician Cédric Villani, recent Fields medalist, in his book *The Living Theorem*, paints a portrait of his mathematical style that is nearly word-for-word identical with the portrait we have painted of Einstein's style. The excerpts below are in perfect resonance with our characterization of the great physicist:

> What made my reputation as a mathematician is the hidden links I've uncovered between different areas of mathematics. These links are so precious! They allow you to cast light on two different areas at once, in a game of ping-pong where every discovery made on one side flips back and gives rise to a corresponding discovery on the other side. ... Three years after I became a full professor, while working with my faithful collaborator Laurent Desvillettes, I found an unlikely connection... And right on the heels of that discovery, I came up with the theory of hypocoercivity, which was based on a new analogy...
>
> In the year 2007, I had a sixth sense that there was some hidden harmony, and I guessed that there was a deep relation... This connection seemed to have just come out of nowhere, and I proved it with Grégoire Loeper. ... Each time, it's a conversation that gives rise to some new insight. I really benefit from exchanges! And also I have faith in the existence of preexisting harmonies; after all, Newton, Kepler, and so many others showed us the way. The world is so chock-full of unsuspected connections.

Banesh Hoffmann, in his biography of Einstein, expressed a quite similar thought, although his passage exudes a flavor that is somewhat more mystical:

> Yet when we see how shaky were the ostensible foundations on which Einstein built his theory [of gravity], we can only marvel at the intuition that guided him to his masterpiece. Such intuition is the essence of genius... By a sort of divination genius knows from the start in a nebulous way the goal toward which it must strive. In the painful journey through uncharted country it bolsters its confidence by plausible arguments that serve a Freudian rather than a logical purpose. These arguments do not have to be sound, so long as they serve the irrational, clairvoyant, subconscious drive that is really in command.

This paragraph by Hoffmann portrays science as advancing solely through quasi-magical mental processes that are unlikely ever to be explained, and a reader of it might well be led to the conclusion that scientific genius, such as Einstein's, is an unfathomable mystery shrouded in the elusive depths of the Freudian unconscious. While we greatly respect Hoffmann's ideas, we do not have as mystical a viewpoint as the one above. We have, rather, tried to show that the sudden bolts from the blue that change the face of physics are invariably analogies, often even "tiny" analogies — that is, leaps that seem obvious once they have been pointed out.

Lifting the veil on Einstein's mental processes, as we have striven to do in this chapter, does not in any way diminish the depth of his genius or of his discoveries, because his lifelong obsession with "cosmic unity" pushed him to discover analogies at such a deep level that in the end he saw one single category of phenomena where even the other top scientists of his era had seen only unrelated things.

In this regard, it is interesting to read what Henri Poincaré, whom we cited earlier concerning the way mathematicians are guided by analogies, had to say about Einstein. In a letter of reference for the young man who had applied for a position as professor in 1911, he wrote:

> Mr. Einstein is one of the most original minds I have known. Despite his youth, he has already reached a very honorable rank among the top scientists of his era. What we must admire above all in him is the ease with which he welcomes new conceptions and finds ways to derive all possible consequences from them. He is not overly attached to classical principles and, in the presence of a physics problem, he very rapidly can imagine all the possibilities. This allows him to rapidly predict new phenomena that are likely to be confirmed by experiment as soon as there are ways to check them. [...] The future will show us ever more clearly what Mr. Einstein's value is, and any university that has the wisdom to hire this young master is assured of receiving great honor for having done so.

A contemporary of Johann Sebastian Bach once said that "as soon as Bach heard a theme, he was very quickly able to imagine all of its consequences". That sounds very much like the quality in Einstein that Poincaré praises so highly! It would thus seem that this ability to "imagine all the possibilities" in a very short time lies close to the core of human creativity at its highest levels.

If we examine the pathway that Einstein took to reach the extended principle of equivalence (which applies to all phenomena of physics, not just to those of mechanics), we see that he re-exploited, in a new context, an analogy that he had already exploited once. More specifically, he carried out the same conceptual extension in two different contexts, each time starting with the concept of *mechanical experiment* and ending up with the more abstract concept of *physical experiment.* Each of these extensions was of course due to an analogy — namely, the vertical analogical leap that amounts to the thought: "Physics is *analogous* to mechanics; it's just that it includes more." (Recall the similar vertical analogical leap of Doctor Gelenk, who thought, "Joints are analogous to knees, only they are a more general biological notion.") And Einstein used this analogical leap in two situations that were themselves analogous (first in extending the principle of Galilean relativity to yield special relativity, and later in extending the restricted principle of equivalence to yield the founding ideas of general relativity). Thus, the most advanced breakthrough of Einstein's life came out of an analogical leap that was analogous to another analogical leap — thus an analogy between analogies, or, if you will, a meta-analogy.

This recalls a remark once made by the Polish mathematician Stefan Banach, as recounted by his friend Stanislaw Ulam: "Good mathematicians see analogies between theorems or theories, but the very best ones see analogies between analogies." And along the same lines, Einstein's Scottish predecessor James Clerk Maxwell once observed that rather than being attracted by parallels between different principles of physics, he was attracted by parallels between parallels, which would certainly seem to be the quintessence of abstraction.

To sum things up, what Einstein's creative life illustrates so clearly is that the perception of profound and abstract analogies in the vast tree of science has the effect of shaking not just a twig or a branch, but the trunk itself. If ever anything made the earth tremble, it was the analogies discovered by Albert Einstein.

EPIDIALOGUE

Katy and Anna Debate the Core of Cognition

❧ ❧ ❧

Categorization versus Analogy-making

Our book ends with a dialogue in which two friends, Katy and Anna, argue about what lies at the core of cognition. Katy sees categorization as playing that role, and she is persuaded that it differs in many ways from analogy-making. Anna sees analogy-making as lying at cognition's core, though she agrees with Katy about categorization's importance; indeed, she does her best to show that analogy-making and categorization are one and the same thing by finding weaknesses, point by point, in Katy's arguments.

The following table summarizes the nine dimensions along which Katy claims that categorization and analogy-making differ significantly.

	Categorization	**Analogy-making**
Frequency	*Nonstop*	*Occasional*
Originality	*Routine*	*Creative*
Level of awareness	*Unconscious*	*Conscious*
Controllability	*Automatic*	*Voluntary*
Degree of similarity	*Disparities are undesirable*	*Disparities are desirable*
Focus of the activity	*Entities*	*Relations*
Levels of abstraction	*A jump between two levels*	*A bridge on just one level*
Degree of objectivity	*Objective*	*Subjective*
Trustworthiness	*Reliable*	*Suspect*

A careful examination of these potential distinctions leads, as will shortly be seen, into territory lying well beyond the question as to whether analogy-making and categorization constitute just one thing or different things. It raises issues concerning how we perceive the world, how we form concepts, how we understand, and how we communicate. In short, it opens up the entire question of the nature of thought.

But our two protagonists, in their spirited exchange, will surely make all this far clearer than we could, and so, without further ado, we'll give them the floor.

The telephone rings.

KATY: Hello! Who is it?

ANNA: Hello, Katy, it's Anna. I hope I'm not waking you, at this early hour.

KATY: Oh, no — not at all. As a matter of fact, just a few minutes ago I woke up from a strange dream in which you and I were having a lively telephone call. Actually, it wasn't an ordinary phone call — it was a heated argument! And it was taking place in Chinese, of all things. How *that* was possible, I don't have the foggiest idea (or the cloudiest, for that matter), since I don't speak a single word of Chinese!

ANNA: Are you serious? Exactly the same thing just happened to me!

KATY: How odd! But *exactly* the same?

ANNA: Well, sort of — you know what I mean. Something very similar — *almost* exactly the same. To be specific, I too just woke up from a strange dream, and in *my* dream, just as in yours, you and I were having a heated argument on the telephone! But in mine, our debate was all taking place in Russian — and as to how *that* could have happened, well, I don't have any concept (not even the tiniest one), since, as you well know, I couldn't utter a word in Russian to save my life!

KATY: What a weird coincidence! One dream, dreamt simultanously by two different people! It sounds like a fairy tale! What was the nature of our heated argument in your dream?

ANNA: To tell the truth, I don't recall at all. I guess that makes sense, since I literally didn't know what I was talking about, speaking as I was in Russian! And in *your* dream, what was our argument about, Katy?

KATY: Well, I have to admit that I don't remember a single word of it either, since everything I said was in Chinese, of which, as you well know, I am totally ignorant. Is truth not stranger than fiction?

ANNA: Well, I wouldn't know, but dreaming is certainly a strange phenomenon. The mind leaps about in such fantastic ways.

KATY: The human mind is profoundly mysterious, I agree. Thinking is the most elusive phenomenon under the sun, even though we do it all the time. Shouldn't its nature be crystal-clear to us, since it's the medium in which we swim? Or perhaps it is murky and miraculous precisely because it's so ubiquitous.

ANNA: I fully agree with you, Katy, and actually, this brings me right around to the reason that I phoned you so early in the morning. Lately, you see, I have been thinking a great deal about thinking, and I've come to the surprising conclusion that there is one special type of mental process that lies at the very core of it all. I wonder if you can guess what it is.

KATY: What a coincidence! I, too, have been thinking about thinking, Anna, and I, too, have identified a special mental process that I believe lies at thought's very core. Would it not be astonishing and wonderful if both of us had independently stumbled on the very same idea?

ANNA: Oh, yes — that would be a delightful surprise. So let me tell you straight off what my candidate mental process is. For me, the key mechanism underlying all of thinking is *the making of analogies* — the spotting of a link between something one is just experiencing now and something one has experienced before. An analogy can summon up any aspect of our past, and thus, when we face a new situation, we can bring to bear the closest experiences we have ever had. In a word, analogy-making is the mechanism at the basis of all thought.

KATY: Analogy-making, eh? Now there's an off-the-wall candidate! It's certainly not what I would have proposed. I see we have rather divergent views on the topic. So let me come straight out as well, and put my cards on the table: *categorization*, not analogy-making, is where I think the secret of our minds lies. Categories grow out of our experiences and they organize our mental libraries; categories are the building blocks of thought, and categorization is the magical key to all of thinking.

ANNA: Oh, really? So you would say that categorization is more important than analogy-making in the life of the mind? Tell me how you see things, Katy.

Categorization is a *constant necessity*; analogy is a *rare luxury*

KATY: Gladly! In the process of thinking, nothing is more pervasive or essential than the assignment of things and situations to known categories. To simplify the world, we constantly have to carve it up into standard, familiar pieces; otherwise, we would find ourselves overwhelmed by a tidal wave of constant novelty. After all, each moment and each situation that we face is different from all other ones we have already experienced, even if only slightly so. Each time you blink, shift your gaze, breathe in or breathe out, move your lips or nostrils or eyebrows, your face is a little different from how it was a hundredth of a second earlier. And each of these micro-movements you make changes, even if only infinitesimally, the perspective from which you see the table in front of us. And as our surroundings change from moment to moment, the only way for us to orient ourselves is to sort the incoming stimuli into familiar, reliable categories, such as *table*, even if we have never seen that specific table before. What is around us is always changing, and if we didn't continually categorize it all, thus simplifying it into stable regularities, our environment would seem to us like utter chaos. Everything would be novel and unknown, and our heads would forever be spinning. My ability to recognize a novel thing as being, say, a *chair*, a *table*, a *breeze*, a *look of surprise*, or a *thinly veiled threat*, is due to the fact that I have already developed these categories, and I have a huge repertoire of them. Without that storehouse, I would be unable to recognize any of the recurrent features of the world around me. So it's categorization that allows me to survive in this unpredictable world. That's my view, my dear Anna.

ANNA: You've made valid and insightful points, Katy, and you've clearly shown why *conception* and *perception* are such closely related phenomena. As a matter of fact, they are interchangeable notions, even if common usage tends to distinguish between them, with *perception* supposedly guided by the senses and *conception* supposedly guided by the mind.

KATY: All of this is well and good, but in my view, categorization must be ranked far above the making of analogies. I understand, Anna, that you hold analogies in very high esteem, and surely they merit your esteem, but nonetheless, just think for a moment: analogies are *rare* mental events, far rarer than the sorting of things into their proper categories. Analogies take place only when we are being inventive, when we are inspired to add sparkle and pizzazz to our mental life. When we make an analogy, we rock the boat, connecting two things in a way that we'd never dreamt of before. That lovely *frisson* when we spot a novel connection between ideas or objects or situations is a very gratifying feeling, but alas, it does not arise often. If we never made analogies, our mental lives would be a tad less zesty, no doubt, but our existence would not otherwise suffer in any way. Let me put it this way, Anna: if categorization is the meat and potatoes of cognition, then analogy-making is an exotic spice that one can easily do without. Coming up with an innovative analogy brings delight, that's for sure, but doing so is a cognitive luxury; one could easily live one's whole life without ever making a single analogy!

ANNA: I subscribe enthusiastically to how you have portrayed categorization, Katy. Categories are the heart and the breath, the motor and the fuel, the roads and the highways of thought, and it would take someone without an ounce of imagination to fail to recognize the utter pervasiveness and indispensability of categorization in the activity of cognition. Nonetheless, I think you delude yourself concerning analogy-making, for it is in fact ubiquitous, every bit as much as categorization.

KATY: That makes no sense to me. I make an analogy only once in a blue moon!

ANNA: You underestimate yourself, my friend. Do you recall how in Chapter 7 of this very book several quotations were presented, all taken from writings by various specialists, some of which characterized categorization and others of which characterized analogy-making, and yet the statements were virtually identical? Specifically, there it was stated that both of these mental processes cast new and strange things in a familiar light ("categories allow us to treat new things as if they were familiar" and analogy is what allows us "to make the novel seem familiar") — and being able to do that is an absolute necessity for survival, is it not? So you see, you make an analogy whenever you recognize the familiar in something unfamiliar.

KATY: Are you suggesting that analogies are as widespread as categorizations are? I don't see them everywhere — in fact, hardly anywhere. To be convinced, I would need some examples, because to me analogies seem as rare as hens' teeth.

ANNA: In that case, Katy, think of the most ordinary moments in our lives — if you look carefully, you will see that analogies abound in them, like flowers in a spring meadow. Given that analogies allow us to understand the new in terms of the old, they are every bit as common as the assignment of things to categories. Didn't you just point out, a few moments ago, that any new situation differs in all sorts of ways from all prior situations? Well, then, in order to make sense of a new situation, we have no choice but to relate it to things we have experienced before. This means that we have to make analogies left and right. Moreover, researchers who have studied the role of analogy in human thinking have often pointed out its ubiquity. For instance, the mathematician George Polya wrote: "Analogy pervades all our thinking, our everyday speech and our trivial conclusions as well as artistic ways of expression and the highest scientific achievements." Along similar lines, the theoretical physicist Robert Oppenheimer declared: "When faced with something new, we cannot help but relate to it except by comparing it with what is familiar and known to us." Psychologists who specialize in the study of analogy-making are also on the same page. Thus Dedre Gentner has written: "Analogies and metaphors are pervasive in language and thought." And Keith Holyoak and Paul Thagard have stated: "Analogy is ubiquitous in human thinking."

KATY: So far you've merely given me a bunch of generalities. They are good food for thought, but I told you I want some concrete examples.

ANNA: All right, let me try. On what basis do you walk up a staircase that you've never laid eyes on before? How do you know how to use a doorbell you've never encountered before? How do you know how to turn the knob of a door that you've never seen before? How do you know how to use a shower you've never used before? And what about sitting on a new chair for the first time, or picking up a magazine you've never touched before? How do you know how to flip its pages, effortlessly telling its ads from its articles? Life is filled to the brim with rudimentary analogies like these.

KATY: All these examples are not analogies; they're just *actions*, not thoughts.

ANNA: I challenge you to find the borderline between acting and thinking. Mental activity lies behind every action, and the term "mental activity" is simply a fancy synonym for "thinking". But if these examples don't satisfy you, then take a look at Chapters 1 and 2 of this book. They show how the choice of every single word or phrase comes out of analogy-making. What could be more ubiquitous than choosing what word to say next? Then Chapter 3 shows how, through analogy, we are constantly *reminded* by one situation of prior situations, and how, when we come out with such common phrases as "me too", "next time", "in general", "it won't happen again", "that's what always happens", and many others, we are relying on tacit analogies. For instance, if you say "me too", don't you mean "the analogous thing holds in my case"? And if you say "next time", what could you mean other than "as soon as an analogous event takes place"? Doesn't "like that" mean "analogous to that", just in more everyday language? And so, in short, analogies have no reason to feel outnumbered by categorizations; they are everywhere under foot, being neither rare luxuries nor exquisite delicacies. Analogies are not merely the icing on the cake of cognition — they are the full cake, including the icing!

Categorization is *routine*; analogy is *creative*

KATY *(still convinced that categorization and analogy-making are as different as day and night)*: I've followed your argument, Anna, and I'll withdraw my claim about the rarity of analogies. I'll concede that we make analogies all the time, even if most of them are trivialities. It would never have occurred to me before to call such mini-thoughts "analogies", but you've convinced me. Nonetheless, analogies and categorizations are very different beasts. You yourself implicitly admitted this point only moments ago, with your words "the icing on the cake of cognition". Let me explain. To categorize is to rely on preexisting mental categories. Let's say that a noisy, furry, four-footed entity is trotting down the street by your side and I assign it to the category *dog*. This connection to a previously acquired notion gives me instant access to many facts, such as that I'm dealing with a living entity, that it has many internal organs, that it could possibly bite and transmit rabies, that it tends to eat meat, that it enjoys being taken for walks, that it might smell bad, and so on. The moment I decide that something belongs to this category, I effortlessly retrieve all sorts of facts that I've accumulated over the course of my life about dogs in general or about certain breeds of dogs. But my use of all this knowledge has nothing creative to it. To the contrary, it's entirely humdrum.

ANNA: I have no objection to this; calling a dog "dog" is certainly very mundane.

KATY: Precisely. Whenever we categorize, we merely take advantage of facts acquired in our past. We rely on prior experiences, expecting them to repeat themselves, but we create nothing new. Although categorization pervades every moment of one's mental life, it is neither inspired nor inspirational. By contrast, your "icing on the cake of cognition" involves originality, creativity, inspiration, and flashes of insight, all of which go way beyond bland uses of prior knowledge. In this sense, analogy deserves praise, because it's what allows us to creatively connect things that one would tend to think are utterly unrelated. This is how analogy-making is so distinct from categorization. I admit that my observation saddens me a bit, because I would have liked categorization to play a more noble role in thought, but in any case this shows that categorizing and analogy-making are not cut from the same cloth. The former is *routine*, while the latter is *creative*.

ANNA *(all smiles)*: Well, Katy, I'm going to disagree with you once again, but this time I hope to boost your spirits, because you seem to hold categorization in very low esteem, and I'll show you that such a negative attitude is unjustified.

KATY: Oh, really? That would make me happy! Please tell me how.

ANNA: To begin with, I hope you'll agree with me that many analogies are anything but creative. In fact, I just gave you some examples — dealing with an unfamiliar staircase, doorbell, doorknob, shower, magazine, and so forth. We constantly make analogies on automatic pilot, without even noticing them, as effortlessly as we breathe. You'll have to agree that many analogies are dull as dishwater. As this shows, categorization is far from being the gold-medal winner for blandness; analogy-making will give it a good run for its money in that department. But if I understood you rightly, it's not in terms of dullness that you think the difference resides, but in terms of creativity.

KATY: Exactly. Your examples have shown me that analogies can be flat and uninspiring, but my point is different. It's that analogies *can* be very creative, at least now and then, whereas categorization, poor thing, is *never* creative.

ANNA: We agree that analogies can reach the heights of creativity; in fact, I would argue that every creative act is rooted in an analogy. One gets off the beaten track by seeing that two things are "the same thing" in a situation where their sameness has not yet been seen. Good examples are given at the end of Chapter 4, such as where Archimedes sees a crown as being "the same thing" as a human body, in that both occupy volume, or where a computer's mouse is seen as doing "the same thing" as a human hand, in that it can manipulate virtual objects that lie behind the screen. And in Chapter 8 we saw that great discoveries in mathematics and physics are rooted in analogies. The French mathematician Alain Connes declared, "The trips we take in the world of mathematics differ from trips taken in the physical world, and the main vehicle for such trips is analogy." And the philosopher Friedrich Nietzsche once defined intelligence as "a moving army of metaphors".

KATY: Yes, I've heard these ideas before, and in fact this is exactly what I'm trying to get you to see: analogies, being members of the cognitive aristocracy, have the potential to be creative, whereas humble categorization, bless its little heart, does not. And while we're quoting authorities, let me quote the French writer Georges Courteline: "Whoever first compared a woman to a rose was a poet; whoever next did so was a fool." That makes my point very well! The poet came up with a novel analogy between two entities, and this was a creative act; the other person merely took the category created by the poet and echoed the observation. It was a derivative act in which there wasn't one single shred of creativity.

ANNA: That's an interesting quote, but you interrupted me and didn't let me get to *my* point, which is that categorization, too, has great potential for creativity; no law says it's limited to the carrying-out of menial tasks. You quoted a mathematician talking about analogy, so now let me quote another mathematician talking about categorization. Henri Poincaré described mathematics as "the art of giving the same name to different things". That is, it's the art of discovering unusual and subtle new categories. And so, categorization should hold its head high; it should be proud of its ability to give rise to great novelty and high invention.

KATY: Are you joking?

ANNA: Not in the least! What makes a person, an idea, an opinion, or a remark original? Surely it's the fact that it involves a categorization that departs from the norm! For example, we think of certain architects or movie directors or novelists or fashion designers as being original because they've come up with an idiosyncratic universe, which means their way of categorizing things differs from those of their contemporaries. Don't you agree? Moreover, many opinions take the form of unconventional categorizations. Thus if I declare, "The current financial situation is a mess", I'm putting the financial situation in the category *mess.* If I say "Mathematics is a game", "Mathematics is a language", "Mathematics is a tool", "Mathematics is an art", "Mathematics is a cult", or "Mathematics is a passion", what am I doing if not assigning mathematics to the categories of *game*, *language*, *tool*, *art*, *cult*, and *passion*? If someone declares that my ideas are a *dead end*, or that their very voluble chatterbox friend is a *fire hose*, or that video games are a *scourge*, or that some up-and-coming young comedian is *another Woody Allen*, or that 60 is *the new 40*, or that the new tax law is *a disaster waiting to happen*, or that the president's proposal is just a *band-aid*, or that spell checkers are a *crutch*, or that a famous politician just caught in a lurid sex scandal is *history*, or that their other car is a *bicycle*, aren't all these just various cases of categorization, some more and some less original?

KATY: Well, I've always thought metaphorical language could be very creative.

ANNA: To be sure! And so, if you grant that the examples I've just given are creative, you are also granting that categorization can be creative. All of this shows, Katy, that, unlike what the naïve prejudice suggests, not only are there scads of boring analogies, but there are also scads of creative categorizations. To sum up, then, your dichotomy between *routine mental activity* and *creative mental activity* fails to provide a criterion for distinguishing analogy-making from categorization.

Categorization is *unconscious*; analogy is *conscious*

KATY: All right, I'll grant you your point: categorizing can involve discovery and originality, and isn't limited to bland acts of classification. Be that as it may, one should still beware of confusing analogy-making and categorization. Indeed, thanks to our lively interchange, my ideas about why and how these two cognitive activities are so distinct are becoming much clearer, and I thank you for this gift. What I'm now clearly realizing is that categorization is by its very nature an *unconscious* process, whereas analogies are inevitably produced via a *conscious* process.

ANNA: I don't know what would have led you to this hypothesis.

KATY: Well, let's look at an example. If I ask you how the structure of an atom resembles the structure of the solar system, you can conjure up images of the sun and an atomic nucleus, as well as images of planets and electrons, and then in your mind you can let all of these things execute their intricate orbital dances. What could be more conscious than that? Or let me take another example. If I propose that life is like a voyage, then in your mind's eye you'll see someone's birth and death as the origin and the destination of this trip, and you'll see the person's trials and tribulations as rough patches along the route. You might even see oases or green valleys as symbolizing the happier periods of life. All of this will take place consciously in your mind. But in stark contrast to that, whenever I categorize something, I am completely unaware of what's going on "down below". For instance, if I say "table" or "chair" in referring to the objects that I see around me in my bedroom, or if the noise in the neighboring apartment evokes the word "hullaballoo" in my mind, or if I can tell that the vowel in the word you just said is an "a", I do so without having any understanding whatsoever of the mental processes that have led me to do so; it's all opaque to me. Several times per second, categories just bubble up from nowhere, and I have no idea as to how or why they do so. You may be smiling at this thought, Anna, but how is it that I know that the word "smile" is the right one to apply to what I see on your face? This is a great mystery, in my view. I have no awareness of what is going on.

ANNA *(gently smiling)*: There's no doubt that one's intuition tends to agree with this distinction that you're proposing between analogy-making and categorization, but in my opinion, you've once again fallen into a trap that is simply due to the particular examples that you've chosen. All we need to do is choose some different examples, and I think you'll soon agree that there are plenty of unconscious analogies as well as conscious categorizations.

KATY: I am most dubious of this. I'm not able to find any counterexample at all.

ANNA: Well, then, let's see if I can help you. If you want some unconscious analogies, just look at Chapter 5, which is chock-full of examples. Also, that's the central idea of Chapter 7, devoted to naïve analogies. In both chapters, so many unconscious analogies are described that I wouldn't even know where to begin. For instance, take all the episodes described in which memories just sprang to someone's mind out of the clear blue sky, cases where the person had no understanding of how this reminding occurred to them. Here's the type of thing I mean. Suppose Margo slips on the sidewalk while rushing to her car in the rain and scrapes her elbow quite badly, and suppose that her telling me this one day brings to my mind a time many years earlier when I was hurrying to catch a plane and badly sprained my ankle as I recklessly dashed down a staircase. Well, all I can say about how such a reminding took place is that the two stories share a conceptual skeleton and that my hearing Margo's story woke up my story, which for years had been gathering dust on some remote shelf of my memory. Other than that, I can say nothing about what went on in my mind. What took place here, if not the totally unconscious making of an analogy?

KATY: Your example is provocative, but I don't agree with your conclusion, because the memory that bubbled up was a perfectly conscious one. You did indeed recall the old episode consciously, didn't you? And so the analogy you made wasn't all that unconscious.

ANNA: Well, yes, the retrieved *memory* was conscious, but the underlying *process* was unconscious, in just the same way as it is for categorizations. When I consciously recognize a table in a room or the vowel "a" inside a word, I am nonetheless unaware of the mental processes resulting in this categorization. Moreover, I was just about to give you some other examples of analogies that are so deeply unconscious that one doesn't even realize one has made them, and as a result they can only be detected by careful outside observers. Among these are the notorious "naïve analogies", described in Chapter 7. I'll mention just three cases: division seen as sharing, multiplication seen as repeated addition, and the equals sign seen as representing a process followed by a result. All these naïve analogies lurk hidden in the mind of nearly everyone, and they can insidiously lead us to wrong conclusions of which we are entirely unaware. Psychologists, however, have succeeded in coaxing these bashful analogies out of the woodwork by designing and carrying out careful experiments.

KATY: Your point is well taken as far as analogies go, but you're quite mistaken about categorizations. Whatever makes you think that such a basic cognitive process could be carried out in a *conscious* manner? The very idea makes me laugh!

ANNA: Well, you may be about to laugh on the other side of your face, because one can certainly choose to categorize various things, and at that point the act of categorization becomes completely conscious — every bit as conscious as the making of analogies that you referred to a few moments ago, such as one between an atom and the solar system, or one between a life and a trip.

KATY: Conscious categorization seems pretty unlikely to me, but I'll be glad to listen to you defend the implausible notion!

ANNA: It's not implausible at all! Here's an example. Is Pluto a planet? Do you remember that quite recently an elite international committee of astronomers pondered that question, deciding in the end that Pluto was no longer a member of that category? Their careful deliberation, which involved a drawn-out, intense, and even acrimonious debate, was very clearly a conscious act of categorization.

KATY: Now, now, Anna — Pluto's astronomical status is just a question of scientific fact, not one of mental processes. We're talking about the act of thinking, and let's not lose sight of that!

ANNA: But my dear Katy, scientists are thinkers *par excellence*! However, if this example strikes you as being too connected with science, then consider a legal trial in which the goal is to decide whether an accused person is guilty or innocent. Isn't the essence of such situations an extremely deliberate attempt to decide which label — "guilty" or "innocent" — to apply to the defendant? I hope you would agree that *guilty* and *innocent* are categories.

KATY: I admit that if you were to ask me to name a category, I would be more likely to think of a noun like "hammer" than an adjective like "guilty", but you're right: *guilty* and *innocent* are just as clearly members of the category *category* as is *hammer*.

ANNA: Your observation is spot-on, and I'll follow it up by giving a few more examples of conscious categorization. If a friend asks me whether a mutual acquaintance is *honest* or *reliable* or *generous*, I will start trying to recall various instances of their behavior and then I'll consciously use these carefully gathered recollections as the basis for making a category assignment. More generally, any time we have a doubt about what something really is (and this happens quite often), we start engaging our mind consciously in trying to figure out what category to assign it to. For example, if someone asks me if a certain painting is impressionistic, or if a certain film is a horror movie, or if a certain song is punk rock, or if a certain joke is Jewish, my thoughts on the topic are bound to be conscious attempts to figure out what category seems to best match the item in question. Do you see what I mean?

Categorization is *automatic*; analogy is *voluntary*

KATY: You are a skillful debating partner, Anna, and I will concede this point to you, but I am far from down and out! What you have made me realize just now is that the question is not so much one of conscious awareness, but one of *control.* It's just that I was a bit sloppy in formulating my ideas. The distinction I should have been focusing on was that of *automatic* versus *voluntary*, because categorization strikes me as being an automatic, involuntary process, while the crafting of an analogy is a process that one undertakes deliberately and intentionally. What I mean by this is that I have no way to prevent categorization from taking place, whereas I can certainly decide whether or not I wish to make an analogy. I'm the captain at the helm! I can *choose* to make an analogy, or I can choose *not* to pursue such a goal.

ANNA: I don't see what you mean.

KATY: Well, let me give you an example. If I tell you not to think about a table, you can't help disobeying me — the mere utterance of the word "table" makes you envision one! The process is automatic and irrepressible, and you can do nothing about it. Similarly, the categories that all the things around me belong to just jump to my mind unbidden, and luckily so, since without this automatic recognition process, I would be lost and disoriented, unable to make sense of my constantly changing environment. My categories are my guides to the world, and I can't prevent them from being triggered. On the other hand, I can easily reconsider the analogy between the solar system and an atom and may well conclude that it is misleading, just as I can choose to dismiss the "domino" analogy for the Vietnam war. In such cases I weigh the pros against the cons, and I thereby decide whether or not to accept a given analogy, whether or not to prefer a different one. So you see, whereas I'm at the mercy of my categorizations, which just bubble up from who knows where, I'm in full control of all my analogies. I'm their boss!

ANNA: Or so you claim! In truth, though, there are equal numbers of automatic, irrepressible analogies. And speaking of things that "just bubble up", think of how memories of analogous experiences from long ago just bubble up unbidden when strange things happen to us. Or think of all the analogies described in Chapters 5 and 7 of this book. I'm incapable of separating *division* from *sharing* in my mind, much as I'm incapable of separating *dishonesty* from *dirtiness.* And why do you think no one today would name their baby "Adolf"? It's because of the irrepressible analogy everyone would make with Hitler, of course. No matter how cute Baby Adolf might be, everyone's feelings about him would be contaminated by their feelings about the German dictator. This is because an analogy imposes a point of view on us, often without our being aware of it, let alone desiring it.

KATY: There you are certainly right. And this reminds me of the wedding that my fiancé and I recently attended together. As we watched the ceremony, I couldn't help but imagine him and me getting married. Each step of the way, I was seeing myself in the bride's shoes; the irrepressibility of such imagery made me chuckle inside. And I'm sure it was the same for my fiancé, even though neither of us breathed a word of it to the other. That thought was the elephant in the room!

ANNA: It seems to me that you have clearly understood the category *irrepressible analogy.* And there are a myriad other examples of this category — so many that one wonders if they aren't the rule rather than the exception. Once again, we come back to the fact that thought is all about using familiar notions to orient oneself in novel situations. Any story one hears will trigger memories, every conversation is built out of analogies following on the heels of other analogies, every perception involves analogies, and all these analogies come to us without our constructing them deliberately. If something very unusual happens to us (such as just barely missing an airplane), it will bring to mind some other unusual event that shares something deep with it. If something extremely commonplace happens to us (such as eating a pizza), it is interpreted by mapping it onto a mental structure that grew over years, as dozens, hundreds, or thousands of analogous events were superposed on each other. In either case, though, the mapping is an analogy. Without all these analogies churning unconsciously and irrepressibly in our heads and guiding us at all times, we would be completely unable to interact with our environment.

KATY: I concede that there are analogies that just bubble up and cannot be squelched, but analogy is at least *sometimes* under our control. This fact seals my argument, because categorization, in contrast, can *never* be controlled. Voilà! QED!

ANNA: But it certainly can, dear Katy! There are categorizations that are totally under our control. Only moments ago we were talking about the categories *innocent* and *guilty*, and also about the category *planet.* Such categories don't automatically jump to your mind all the time, and you can easily decide to ask, or to stop asking, whether a given entity belongs to a given category; this shows that the process can be controlled. For example, if you see a person walking down the street, you don't instantaneously try to figure out their blood type or their political affiliation. Depending on the context, you can decide either to do so or not to do so. And thus, Katy, once more, despite all your valiant efforts, you haven't shown me any clean line of demarcation separating analogy-making from categorization.

Categorization favors *similarities*; analogy favors *dissimilarities*

KATY: Enough said, Anna; your point is clear. But let me turn now to a criterion that seems to me to be absolutely central and undeniable — namely, *degree of resemblance.* The closer an item is to the core of a category, the easier categorizing it as such will be. That is, categorization is quickest when the resemblance is maximal, wouldn't you agree? Take my dining-room table. I effortlessly call it "table" because it has so many commonalities with my mental image of tables. If you show me typical members of various categories, such as a typical glass, chair, dog, or eagle, I'll have no trouble categorizing them. In a word, *resemblance* is crucial for categorization. But analogy-making is very different. In fact, it's the exact opposite! What makes the quality of an analogy is how *different* the two analogous situations are. To put it another way, what lends great strength to an analogy is if someone spots an essence shared by two situations that to most people appear wildly different. An analogy's power comes from the fact that it *sees beyond* the surface differences of two situations to reveal something deep and hidden they have in common. This is what gives an analogy power and interest. Do you see my point? At long last, I've put my finger on a rule of thumb fundamentally separating analogy-making from categorization!

ANNA: Eureka! Congratulations, Katy! And so, while we're speaking of fingers and thumbs, would you say that your right thumb and your left thumb are analogous?

KATY: Of course not! My thumbs look just alike. That doesn't make for an analogy!

ANNA: Really? Think back to Chapter 6's finger-wiggling puzzle. There it was self-evident that wiggling your left hand's thumb is a great analogue to wiggling your right hand's thumb. Doesn't that suggest that two thumbs are indeed analogous?

KATY: Surely you're joking! Any analogy worth its salt is a mental bridge between two situations that look very *different* from each other. Take the atom and the solar system, for instance, or the heart and a pump, or a brain and a computer, or the countries of South America and a big family. In each case, the two items don't look at all the same, and that's what makes them such great analogues. By contrast, a lack of similarity will weaken or even totally block a categorization. So in sum, my point is that the *more* differences there are between two situations, the stronger the analogy linking them will be, whereas in categorization it's the exact opposite: the *fewer* differences there are, the more apt the categorization will be.

ANNA: Whoa there, Katy! You seem to have forgotten that any situation has both superficial and deep aspects, and what matters most, whether we're talking about categorization or analogy, is that *deep* aspects be the same. In categorization no less than in analogy-making, there can be all sorts of superficial aspects that differ. My point is that in addition to obvious categorizations featuring resemblances at all levels, there are subtler categorizations where there are differences galore.

KATY: How could the existence of many salient differences ever be consistent with strong category membership?

ANNA: Well, tell me: what do a lion, a giraffe, a snake, a whale, an octopus, a bullfrog, a spider, a butterfly, and a millipede have in common?

KATY: They're all animals, obviously. But what are you driving at?

ANNA: What I'm driving at is that they all belong to a single category — that of *animal* — and yet they look very different. Superficially, they have differences galore. So you see, as soon as we start considering categories that are a bit general and abstract, then for their members to be highly variegated is par for the course! And this is not a rare phenomenon. Take the categories of *vegetable*, *mammal*, *clothing*, *furniture*, and *musical instrument.* Wouldn't you agree that each of these is a very familiar category whose members differ enormously from each other? None of them is a category whose members are like so many peas in a pod. And I've just scratched the surface of such cases. Think, for instance, of categories named by phrases, such as "slippery slope", or of categories named by proverbs, such as "beggars can't be choosers". In such cases, there are countless excellent members of the given category that differ from one another in myriads of superficial ways. Do you recall the long list in Chapter 2 of members of the abstract category *Once bitten twice shy*? There's no doubt that the proverb names a genuine category, but the list showed that there can be enormous differences between situations that bring the saying "Once bitten twice shy" to our lips.

KATY *(momentarily shaken but still confident)*: I have to say, Anna, that once again your points are well taken. What can I do but admit that *some* categories have members that are enormously different? And yet I am going to push onward because I am convinced that for an analogy to be strong, the existence of differences is *necessary*, whereas for a strong categorization, the existence of differences is merely *optional.*

ANNA: Come now, Katy! Do you seriously believe that your two thumbs are *not* analogous, whereas your thumb and big toe are *fairly* analogous, and your thumb and your little toe are even *more* analogous? That sounds backwards to me! Let me be more explicit. Some analogies are very simple and down-to-earth, in that they link objects or situations that look very much alike, and by making such analogies all the time we survive very well in our everyday world. Once again, think back to the novel staircase, novel doorbell, novel doorknob, and so on, which we spoke about not long ago. As we move about in the world, we make analogies left and right between novel situations we encounter and situations that we encountered in the past, and we don't place higher value on connections that are distant; indeed, the more closely the precedents that we retrieve from memory resemble the novel situations, the more confident we feel in our analogies. The rule of thumb "The fewer the differences, the better", which you proposed for categorization, holds just as much for these everyday, down-to-earth analogies. Unconscious, irrepressible, run-of-the-mill analogies are all around us, and they don't feature any major differences at all. Every time you say "me too", you've made one!

KATY: But Anna, who would ever be proud of having made analogies of that sort? They are flat and utterly devoid of interest.

ANNA: Agreed, but there's no law that says every useful analogy has to be a source of great pride. To be sure, these down-to-earth analogies are not ones that will make humanity swell with pride for having made them; they're not in the same league as highly subtle and abstract analogies that evoke feelings of astonishment. But even so, they still belong, beyond a shadow of a doubt, to the category *analogy*! Analogies linking Fido and Spot, linking my dining-room table and yours, linking my elbow to yours — these are all perfectly fine analogies, even if they fall short of being deep new insights. And while we're at it, let's not forget Chapter 3's analogy between my eyelid and yours, thanks to which I can tell you my technique for getting an annoying eyelash out of my eye (and thus out of yours, too). Analogous situations can have many superficial resemblances, but they can also have many superficial differences. Such considerations will determine how subtle or deep the analogy is. Just as huge differences are par for the course in abstract categorizations but are rarer in very concrete ones, so huge differences are standard in deep analogies but are rarer in rudimentary, everyday ones. The simplest bread-and-butter analogies, just like the simplest bread-and-butter acts of categorization, involve easily-seen resemblances and don't need dissimilarities to be thrown into the bargain.

KATY: Well, all right, then. I can see that categorizations, like analogies, can exist perfectly well despite all kinds of dissimilarities. So be it. But we come now to a key difference between analogy-making and categorization, and because of this key difference, no one should ever confuse the two.

ANNA: I'm eager to hear your new proposed distinction.

Categorization applies to *entities*; analogy to *relations*

KATY: I'll explain it with pleasure. The fact is that when one makes a categorization, one takes some *entity* from one's surroundings and associates it with a mental category, whereas when one makes an analogy, one takes the *relationships* in two situations and one finds that it's these *relations* that are shared. For instance, Anna, if I categorize the object I'm holding as a *hammer*, it's because I'm taking an element of my environment and linking it to a particular mental structure in my brain — the category *hammer*. So this object is matched with that category by a perceptual process that involves recognizing that it possesses many typical *hammer* traits. However, no relationships between parts of the hammer are involved.

ANNA: I can't help stating that this claim strikes me as extremely dubious.

KATY: Well, first let me finish, please, because by the time I'm done I bet you'll have changed your mind. No relationships are needed to categorize this object as a *hammer*. On the other hand, if I am told about the analogy between a heart and a pump, what goes on is that I see some *relationships* among the parts of a heart and also some relationships among the parts of a pump, and then I map those relationships onto each other. The relationships on either side of this specific analogy involve such concepts as *output*, *pressure*, *expansion*, *compression*, and so forth.

The analogy is convincing because the relationships on one side match up with the relationships on the other side. In the case of the atom and the solar system, the relationships among the parts on either side include such notions as *massive entity exerting an attractive force on lighter entities* and *rotation around a central entity*, among others, and once again, these relationships match up very nicely across the two sides. So you see, categorization pays no attention to relationships among parts of a situation, whereas analogy-making depends vitally upon them. This is surely a crucial distinction, wouldn't you agree?

ANNA: What I hear you saying is that if a process associates some entity to a mental structure, then it's a case of categorization, while if a process maps relationships in situation #1 to relationships in situation #2, then it's a case of analogy-making. Well, I'll simply say that this proposed distinction may seem appealing at first, but under scrutiny it falls apart, just as all the others have. In fact, your own examples will serve well to undermine your thesis! Let's take the heart–pump analogy. Is it really all that obvious that when one matches up the ideas of *heart* and *pump*, one is necessarily making an analogy and not a categorization? To be sure, a heart is analogous to a pump, but why not also assign a heart to the category *pump*?

KATY: Oh, I suppose you could do that, but you would just be making a metaphor.

ANNA: Really? Are you suggesting that hearts don't deserve membership in the category *pump*? I'm sorry, but that's quite an untenable viewpoint; all you need to do is open any random biology or anatomy textbook and there you'll see it printed in black and white that the heart pumps blood, that it's the pump of the circulatory system, and so on. The fact is, hearts *are* genuine pumps, pumps as canonical and exemplary as one could ever want. They just happen to be made out of muscle and nerve instead of metal and plastic — but then again, artificial hearts, which are made out of plastic and metal, are also pumps. That's quite obvious.

KATY: Well, all right, but I don't see what you're driving at.

ANNA: If a heart is a pump, the reason it is one is precisely because the *relationships* among parts of the heart and the blood passing through it are the same as those among parts of a metal pump and the liquid passing through it. So we see that categorization does indeed involve taking into account the relationships among the components of things, and that takes the wind out of your prior claim.

KATY: I see your point for the category *pump*, Anna, because a pump is a complex device having numerous moving parts, but I don't buy it for more basic categories, such as *hammer* or *plate*. Surely you aren't going to argue that there, too, category membership depends on relationships?

ANNA: Oh, yes, I surely am! For something to be a hammer, we expect it to have a handle that is grasped by a human hand, and a metal head that is raised (by the human hand) above the handle and then brought down onto a nail, allowing the latter to penetrate into various materials, such as wood. So here we are talking about a *handle*, a *head* (the hammer's), a *hand* (a person's), a *nail*, and a piece of *wood*, at the very least, and in order to describe the situation, one needs to refer to many relationships connecting these items. How does this differ from the case of the category *pump* (which, as we just saw, has internal parts and relationships among them)? It's quite the same.

KATY: Maybe that's true of *hammer*, but *plate* is different! All that counts for a thing to be a plate are its size, shape, and strength, and these are just *traits*, not relationships among parts. In fact, a plate has no parts to speak of. So my point is proven: at least *some* categories *don't* involve relationships, whereas *all* analogies *do.*

ANNA: Sorry, but you've jumped a little too fast, Katy. Relationships don't have to hold among the *pieces* of an object; they can connect an object to its *surroundings.* Thus for an object to be a plate, it has to be able to sit on a table without tipping over, and it has to be able to hold such things as gravy without the gravy leaking off. Moreover, it has to be washable, not poisonous, not prone to shedding small flakes, not damageable by the process of cutting, not prone to dissolve in water, and so forth. These key qualities that make something be a plate bring in many other notions, such as *tables*, *gravity*, *foods that flow*, *water*, *soap*, *human health*, *knives*, *cutting*, *dissolving*, and so forth. An object is a plate by virtue of its relationships to all these other things in the world. And so we see that any category, no matter how simple, is a mental structure that involves relationships among various entities in the world, and the act of categorization depends on taking these relationships into account. And thus, dear Katy, your hoped-for distinction goes down the drain.

KATY: Oh, my... You're quite right, Anna, that on more careful analysis, hammers and even plates are no less "relational" than are hearts or pumps, and so I have no choice but to withdraw my proposed dichotomy. This is getting frustrating and confusing, but luckily I have a new dichotomy up my sleeve, and with this one I'm pretty sure I'm going to put you into something of a pickle. Are you ready?

ANNA: Be my guest!

Categorization involves *two levels of abstraction*; analogy involves *just one*

KATY: In the examples you've just given, whenever an analogy is involved, such as between an atom and the solar system, or between a heart and a pump, the two entities involved are both situated at the same level of abstraction. But when you categorize something, that's not the case, because *by definition* any category is more abstract than all of its members. In other words, in contrast to what happens in analogies, there is a difference of abstraction levels between things that are being categorized and the categories into which they are getting placed.

ANNA: Shall we try to focus on a concrete example? That will help a lot, I think.

KATY: Sure! I assume you're sitting on a chair, right? That tangible object is a special case of the general category *chair*, which in turn is a special case of the more general category *furniture*, and on upwards it goes. And so there's a big difference between comparing two things of *equal* generality, such as a heart and a pump, and comparing two things that have *different* levels of generality, such as a specific chair and the abstract category *chair*, or the abstract category *chair* and the even more abstract category *furniture.* And by the way, there was a good reason that when we broached this topic, you carefully steered clear of the classic old analogy between the atom and the solar system: since both items are concrete things, thus at the same level of abstraction, there is no way that we could be dealing with a case of categorization here. To be specific, an atom is not a solar system, nor is the solar system an atom; they're at the same level. And so to summarize, in an analogy, source and target reside *at the same level of abstraction*, while in a categorization, the category is *more abstract* than the thing being categorized. All right; the ball's in your court now, Anna, and it's a foregone conclusion that I'm going to win this point!

ANNA: Don't be so sure, Katy. I believe I can handle your volley without too much trouble. Your claim that there is necessarily a level-distinction whenever one categorizes something is, I'm sorry to say, just a simplistic stereotype. In fact, this way of looking at things comes straight from the naïve analogy "categories are boxes", which suggests that one should look at categories as containers and entities in the environment as things to stick in them. Your image reflects the idea that categories in the mind are like Russian dolls, one nested inside another inside another, going many levels up or down. For example: Ollie is a *dog*; dogs are *canines*; canines are *mammals*; mammals are *animals*; animals are *living things*; and so forth. And indeed, if categories really were boxes containing their members, then your proposed distinction might be justified. But your idea collapses as soon as it is examined carefully. I suggest you go back and take a look at the end of Chapter 3, where the mental process of categorization or analogy-making — call it whichever of these two labels you prefer — is portrayed as the construction of a mental bridge between two structures that themselves are also mental entities. This happens not only when we make a link between a heart and a pump (or more precisely, between mental structures representing them) but also when we assign an object on a shelf to the category *plate.* In either case, we take a fresh new mental representation of something and we link it to some preexisting mental structure. And as I just stressed, this process of building a link between two mental structures can be called either "analogy-making" or "categorization" — whichever label one prefers.

KATY: Good grief, Anna — are you implying that categorization and analogy-making are *exactly the same thing*, and that there is no difference *at all* between them?

ANNA: I'm glad to see that you're starting to catch my drift, Katy. Hopefully this will make you more receptive to my ideas. Do you recall the category of faces that looked alike, back in Chapter 3? Seeing Mark Twain's face gave rise to a first mental structure, and subsequently, seeing Edvard Grieg's face enriched this initial structure, making it more complex. Each time a new face is seen that resembles the previous ones, the evolving mental structure gets further enriched. There is thus no box here, but simply a mental structure growing in complexity, and as the number of faces involved in this process increases, the mental structure starts to feel more abstract. If we step back and look at this type of process, we see that there is no box representing a category, with members neatly placed inside it, and that there is no upwards leap of abstraction involved in an act of categorization that makes it differ from the process of making an analogy.

KATY: I'm not convinced. After all, human faces are very concrete, and mentally superimposing several faces has very little to do with the way categories are normally built up. More *typical* categories are much more abstract than that.

ANNA: All right; your healthy resistance pushes me to seek a more abstract example to get my point across. Here's an attempt. When I was little, our family went to an amusement park where we boarded a boat that sailed down a river through a jungle filled with scary beasts. I was thrilled by this great adventure, but when we got off the boat, my father pointed out that our vessel wasn't really floating down the river but was actually rolling on wheels on some underwater tracks, which forced it down a predetermined pathway. I realized that our boat was in truth more a *trolleycar* than a *boat*, which made me sad. The disillusionment struck me, but I didn't dwell on it for long. Several years later, when I was in high school, some friends convinced me to take part in a talent show in which they were doing a tango number. They were all good tango dancers but I knew nothing, and I had to work like the devil in order to get the routine down. Finally I learned it, and on the big day our number came off pretty well. Everyone congratulated me for my tango-dancing talent. But I knew I didn't have any such talent! When I explained that the only bit of tango I knew was this tiny two-minute routine that I could perform only in the most inflexible way, I suddenly recalled that boat in the park that looked as if it was freely sailing down the river but was actually rigidly constrained by tracks. My apparent tango skill was just another boat held rigidly on its course by invisible tracks and capable of taking in naïve viewers!

KATY: That's an amusing analogy, but what's the point?

ANNA: Well, you've called it an analogy, which is fine with me, but the way I just described it, my deceptive tango skill was a new member of my childhood category *fake-boat-on-tracks*.

KATY: Oh, now I see your point! *My* interpretation was that you made an *analogy* between two deceptive events in your life, whereas *yours* was that your deceptive tango skill seemed to you to belong to an old *category* that you knew — that of *boats that look free but that are actually constrained by hidden tracks*.

ANNA: To be more precise, *my* interpretation is that the making of the analogy was tantamount to a categorization. The two acts are one and the same thing.

KATY: Aha — one can look at things either way. I frankly have to admit, that's most provocative… I can even see how this is parallel to the example of the faces. The boat rolling on tracks established a new category during your childhood; that event in your life is parallel to seeing a photo of Mark Twain. Then your tango-dancing fakery a few years later was the second member of the category, which is parallel to seeing a photo of Edvard Grieg, which makes the original category broader and richer. Now that you've transmitted to me the *constrained-boats* category, I can even think of a pretty good member of this very category in my *own* life!

ANNA: I'd like to hear it.

KATY: Well, one time I was on a plane sitting next to an elderly gentleman from Chile who hardly spoke a word of English. Just for the fun of it, I — who had never taken even a week of Spanish — recited a couple of lines from a poem by Pablo Neruda that a high-school friend of mine who loved South American literature had once taught me by rote because I was so taken with those lines' beautiful sonority when she recited them. The Chilean gentleman jumped to the conclusion that I was a fluent speaker of his native language, and I think he was quite disappointed when he realized that my Spanish was limited to following that very short stretch of hidden tracks. Don't you think that my Spanish on tracks is a nice example of your *fake-boat* (and also *fake-tango*) category?

ANNA: Indeed! Let me be the first to welcome this new member into my childhood category, Katy. I think you now can see that any category, abstract or concrete, is launched by a first experience and then builds up gradually as, over one's life, one runs into various analogous entities. And it's crucial to see that there is no critical moment when the first memory suddenly switches status and turns into a category. It's more like the process whereby a hamlet turns into a village, which may grow into a town, and possibly into a city. There isn't a sharp baptismal moment at which one must henceforth say "town" or "city", because the metamorphosis is gradual. And likewise, there is no hierarchical difference, no sudden jump in abstraction, between an initial memory and a category. The initial memory founds the category, just as a hamlet founds a potential city. We may be tempted to think that there is a qualitative difference between a category and its members, but that is an illusion coming from the naïve analogy "categories are boxes".

Categorization is *objective*; analogy is *subjective*

KATY: I see I'm going to have to give up on this dichotomy, because your argument is persuasive. But I'm not done with you yet, Anna! I'm holding a pen in my hand. If I tell you it's a ballpoint pen, that's an objective fact. No one could say the opposite. Much the same holds for that sheet of paper and the paper clip that I see sitting over there on my desk. *Ballpoint pen*, *sheet of paper*, and *paper clip* are all categories that have the quality of total objectivity. An elephant is an elephant, an apple is an apple, and Paris will always be Paris. There are no two ways about it. When we categorize, we do something that is objective.

ANNA: I don't agree with you, but I think you might have Plato on your side, because I believe he argued, in *Phædrus*, that every human being is given the opportunity to look at situations ranging from the most general to the most specific, and to carve up the world in an objective fashion while making very fine distinctions. Am I not more or less correct in my memory of Plato?

KATY: You're absolutely right, and I can even cite the exact passage. It's where Socrates tells Phædrus that humans have "the ability to separate things according to their natural divisions, without breaking any of the parts the way a clumsy butcher does." This is how biologists proceed when they make taxonomies, for example, and it's also how cultures and civilizations evolve, gradually moving towards the capability of sorting all the situations in the world into the categories to which they objectively belong.

ANNA: Well, I'm sorry, but I have to contradict you (and needle your friend Plato as well): categorization is not objective, as you would have it, but is profoundly subjective. For instance, if I assert "Donald Trump is a troublemaker", is it not the case that in thus categorizing Donald Trump I am making a highly subjective judgment?

KATY: I understand your example, but *troublemaker* is an extremely blurry category. You deliberately chose a category that is as blurry as possible! I even think you did it just to be a troublemaker!

ANNA: Me, a troublemaker? Never! And the truth is that my example is hardly unusual. Blurriness is par for the course with categories. With the greatest of ease we can find categorical blurriness all about us. Edgar Allan Poe's poem "The Raven" might be seen by one literary critic as a *masterpiece* and by another as *a piece of junk*; I might be judged *reliable* by one person and *flaky* by another; chicken liver might be considered *mouth-watering* by one person and *disgusting* by another. Doesn't this show that categorization is enormously subjective? And there are so many other cases. Is our friend Virginia *good-looking*? Is she *cordial*? Is our friend Stanley an *artist*? Is he *fluent* in German? Was George's retort to Virginia's jab *appropriate*? Are Virginia's clothes *matching*? Is George's brother a *sleazeball*? Is it *sprinkling* outside, or is it *raining*? Is that a *hill* or a *mountain* over there? Was what Jane just said an *insult* or a *joke*? Is Jim *impulsive*? Is he *straightforward*? Is he a *patriot* or is he a *hypocrite*? Is he *ambitious*, *pushy*, or *driven*? And think about bright people vehemently arguing over the nature of *progress* or about whether something is a piece of *kitsch* or not. Don't you see how deeply blurriness and subjectivity pervade categorization?

KATY: Well, perhaps I misspoke myself, because you've pointed out very effectively that *some* categories are subjective. But that doesn't affect the crux of my point, which is that *analogies* are *always* subjective. For example, wouldn't you agree that when one event reminds you of another, it's an analogy, and that such remindings are totally subjective because they depend completely on the idiosyncratic memories that you've built up over the course of your life?

ANNA: I'll certainly grant you that *some* analogies are subjective.

KATY: Excellent! I see that we're on the same wavelength. So let me continue. We can come up with analogies between anything and anything else, depending on what's recently been passing through our minds. And if one is focused on one thing in particular, then analogies by the bucketful will come to mind. Back in Chapter 5, we saw how an obsession of any sort — golf, dogs, physics, a video game, or who knows what — can trigger a raft of analogies with just about anything that one encounters. One can unwittingly wind up in a kind of analogy-mania! And it doesn't even take an obsession for this kind of thing to happen. Just yesterday, in fact, something of the sort took place. A friend told me about a science-fiction story he'd read and enjoyed. The gist of it was that some guy was listening to the news on the radio and he heard that a woman in a nearby town had been killed when she'd been hit by a car driven by a wild driver. For some reason he heard this piece of news at 7:30 PM but it said that the accident had taken place at 8 PM. So maybe you can anticipate what happened in the story?

ANNA: Mmm, not really…

KATY: Well, the guy dashes to his garage, jumps in his car, and makes a beeline for the nearby town in order to warn the victim to get far from all roads. But as he's approaching the main square, he loses control of his car and careens right into the woman, killing her at exactly 7:59 PM.

ANNA: That's a striking and original story. But where is it leading us?

KATY: Well, if you think about it, your first thought is probably going to be that it's paradoxical, and if you think a little further about what lies at its essence, you will come to the conclusion that it's the idea of *an unfortunate incident brought about by the very act of trying to avoid it.* Such a thing might seem to be totally unique — a clever fantasy dreamt up by an inventive author. And yet, if you start to look around you with this idea in mind, you'll find a rich harvest of analogous events.

ANNA: Oh, so you came up with a bunch? That sounds interesting.

KATY: Yes, and it wasn't even very hard. They just spontaneously flooded my mind, and I couldn't rightly say why. For example, the first memory that came to my mind was from way back when I was a little girl. One day I saw a tall vase sitting on a table that I knew was rickety, and so, trying to be helpful, I reached out to try to grab hold of the vase to make it stabler. What do you know, I accidentally banged the table with my hand, and immediately the vase toppled and shattered!

ANNA: That anecdote about little Katy clearly has something central in common with the science-fiction story — they share a conceptual skeleton.

KATY: To be sure. That's exactly why it came to mind. But that wasn't the only memory that came to me. I next remembered another time when I offered a friend a gift to make up for an argument we'd had, but for some reason she found my gift offensive, and this eventually led to a total break between us. And then I suddenly remembered that time a while back that you, Anna, baked a delicious cake for a party you were throwing, and how at the last minute you decided to brown it to make it even tastier, but you got distracted by a few guests arriving early and you overcooked it, thus ruining your cake. You'll never forget that incident, will you?

ANNA: How could I? It was such a disappointment to me! And I agree with you that it is analogous to the other events that you were reminded of. I think anyone would. So doesn't that show the objectivity of this analogy?

KATY: No, no! All these analogies are totally *subjective.* They came to my mind only because I was thinking about the story of the woman killed by the driver who was doing his utmost to prevent her death, and if no one had told me that story, none of these memories would have been triggered in my mind. None of them would have bubbled up, although any of them could have bubbled up in some completely different context. So you see how subjective analogy-making is!

ANNA: Well, not so fast. I greatly appreciate the variety of the episodes in your life that this story reminded you of, but as we both just agreed, they all share a conceptual skeleton, which you formulated as *an unfortunate incident brought about by the very act of trying to avoid it.* These episodes are all seen by everyone as analogous because it's clear that they all share this central essence. This is why I insist that analogy is not always subjective — no more than categorization is.

KATY: All right, all right... I see that the common core is clear here, but maybe this analogy, or this family of analogies, is a special case. Maybe this wasn't the best possible example.

ANNA: No, that's not the problem. The same would hold for analogies all across the board. For instance, think of the analogy in Chapter 6 between a thin wooden stick offered for stirring coffee and some javelins offered to row a boat in a lake. Who could ever claim that these situations have nothing in common? No one, since their shared essence is clear as a bell! Or take the analogy between sound waves in air and ripples in water — where's the subjectivity in that? Or the analogy between a point (x, y) in a plane and a complex number $x + yi$, or the analogy between lungs and gills, or that between a bullet and an arrow, or between a table in your house and a table in my house. What could be more objective than these analogies?

KATY: But I still insist that there is something highly subjective about analogy-making, because the memories that flashed to my mind when I heard that science-fiction story were completely idiosyncratic, and depended entirely on the chance events of my life and on how they happened to be stored in my memory. Yes, they all share the same conceptual skeleton with the science-fiction story, but they wouldn't have occurred to anyone but me! They are products of my brain, and my brain alone! And so this definitively proves the subjectivity of analogy-making!

ANNA: No one could deny, Katy, that what the science-fiction story brought to your mind — that specific set of curious and paradoxical episodes — is completely unique to you. As you say, this is a set of memories that is yours and yours alone, and it's a result of the random vicissitudes of your life, the events that chanced to happen to you or to your friends, or perhaps events that you had read about or seen in films. And whenever you hear any story, analogous situations will come floating up to your consciousness, and they will certainly be a function of your personal experiences. Your unique experiences and how you encode them will determine when they will come to mind on later occasions. What this clearly shows is that categorization (or analogy-making, whatever you want to call it) comes from the perspective one adopts on a situation. And for this reason, categorization is profoundly subjective.

KATY: Your words are confusing me, Anna. I carefully chose an example to show you that *analogy* is subjective, and yet you're turning my own example against me, to argue that *categorization* is subjective! What kind of sleight of hand is this, anyway?

ANNA: There's nothing underhanded or tricky going on here, Katy. It's just an inevitable outcome of clear thinking. Your example was all about a set of stories that we agreed are all *analogous*, stories that all share a conceptual skeleton — and it was also all about a set of stories that all belong to a single *category*, a category whose core and whose fringes were very nicely fleshed out by your highly diverse set of examples. The verbal label "unfortunate incidents brought about by the very act of trying to avoid them" attempts to pinpoint the subtle shared essence that makes them all analogous.

KATY: I can easily envision a little category centered on the story of the woman whose death was caused by the man whose goal was to save her.

ANNA: Why call it "little"? It wouldn't be hard to add members galore to this category. For instance, one might think of a person strangled by their seatbelt, or an attempt to stave off a war between two hostile countries that goes wrong and winds up triggering the feared war. One could also think of a face deformed by plastic surgery that came out unexpectedly, of a singer who loses her voice on the very morning of her recital as a result of too much practicing, of the big stain in a tablecloth left by a stain remover, of a résumé that is so jam-packed with lists of achievements and honors that all potential employers are immediately put on their guard, and who knows what else. This shows how diverse are the potential members of such a category, and all the situations that you and I described could be looked at in plenty of other ways and thus could be seen as members of plenty of other categories. For instance, the singer whose voice went hoarse as a result of too much practicing could be assigned to the category *too much of a good thing*, and the grotesque face caused by plastic surgery could be seen as a member of the category *should have left well enough alone*. This kind of shift in point of view gives rise to a shift in categorization. In short, I believe that I've just demonstrated the profoundly subjective nature of categorization.

KATY: So you think that my example demonstrates the subjectivity of categorization just as much as the subjectivity of analogy-making?

ANNA: Absolutely! And it's precisely its subjective nature that lends categorization so much interest. Our categorizations are influenced by many factors: the place we happen to be in; our current goals; our knowledge; our culture; our emotional state; our obsessions; and who knows what else!

Categorization is *reliable*; analogy is *suspect*

KATY: Touché! I can see now that I was a bit off base in focusing on the distinction between *objective* and *subjective*, although I was getting close to the target. Now though, thanks to your help, I have at last located the bull's-eye, the true distinction between analogy-making and categorization. Indeed, it's obvious, after the fact! It all comes down to the question of *certainty* versus *riskiness.* Comparing a categorization's reliability to that of an analogy is like comparing day to night. When I make a categorization, there is no chance of error, because I am just connecting something in my environment with a category that it matches; in such a cognitive act, there is no risk-taking. When I recognize a table, a chair, a piano, a melody, or what-have-you, I'm not blue-skying it; I'm just *perceiving* something as it is, end of story. I call a spade a spade; that's categorization for you! Doubt is not on my radar screen when I categorize. However, making an analogy is always a gamble. You're taking two situations and *hoping* that they will match, but there is no guarantee whatsoever that your guess will be right. The activity of analogical guesswork is a minefield of uncertainty. To make an analogy is to make a risky bet, knowing clearly that one may well lose one's shirt. Take, for instance, the political analogies discussed in Chapter 6 — the ones that guided the course of the Vietnam war, at least on the American side. While some were on target, others were way off course, and nobody at the time could tell the wheat from the chaff. That's the nature of analogy; to make an analogy is to put one's good money on one's very unreliable intuitions. What say you, my friend?

ANNA: Good try, but once again no soap. To start with, you're fooling yourself if you think that categorization is always reliable and certain. Didn't we just agree a moment ago that categorization is not objective? Well, for similar reasons, categorization often leads one into error, which means it cannot be relied on.

KATY: Can you give me some examples?

ANNA: Very gladly! You might put salt into your coffee, having taken one white powder for another. You're thus the victim of a mistaken categorization! Then again, having been instructed to "take the second left", you might turn down a driveway instead of a road. Wrong category once again! Or you might rub shaving cream in your hair, having taken it for hair lotion. Or you might drink rubbing alcohol, if it has been poured into a bottle labeled "cream soda". Or you might happily chomp into a hot pepper, thinking it is merely a red pepper. Or you might mistake a flower bulb for an onion and chop it up in your salad. Or you might look up into the sky and think you see a bird when actually it's a plane. Or you might take Mars to be a star instead of a planet. All of these events are mistaken categorizations. And think of stereotypes based on such things as sex or race or nationality or age or profession or religion — they too are a kind of categorization, made overhastily or unreflectively. Most people buy wholly into their stereotypes without realizing that such coarse-grained judgments of other people are often way off the mark. In short, stereotypes are a frequent source of deeply erroneous categorizations.

KATY: Your examples are quite convincing, Anna, I agree. All of the categorizations that you've just listed do amount to errors, and I accordingly acknowledge my mistake. A cognitive process that often leads to errors can certainly not be claimed to be always reliable! One point more for you, Anna, but I nonetheless must slightly take the edge off your glee by pointing out that in all the cases that you cited, the deluded individual is persuaded of their categorization's correctness until some event in the world reveals that it was wrong. Thus we are always persuaded that our current categorizations are correct, even if later we come to realize that we were mistaken. For this reason, I stand behind my claim that a *feeling* of certainty — even if it's just a subjective, fallible feeling — constitutes the dividing line between categorization and analogy-making, because one always believes lock, stock, and sinker — er, lock, stock, and *barrel*, that is! — in one's categorizations, whereas one is always distrustful, as well one should be, of one's own analogies.

ANNA: You've put it very well, Katy, but now I must cast cold water on your idea. All categories have their zones of uncertainty, and exactly the same thing can be said for analogies. If you take a category — any category — and you try to trace its borders, then gray zones will immediately start appearing. Take the category of *clothing*, for instance. You wouldn't hesitate for a split second in saying that a coat, a jacket, a pair of pants, a skirt, and a sweater belong to this category, while a blanket, a pistol, and a cell phone do not.

KATY: I don't see the slightest uncertainty about membership in the category of *items of clothing*.

ANNA: If you'll let me finish, perhaps you will... Just focus for a moment on the fringes of this category and you'll see that your sense of security starts to wobble. Is a hat an article of clothing? What about a belt or a scarf? What about gloves or socks or a headband? What about ski goggles or flippers for swimming? Or consider the category of *furniture*. Is a piano a member of the category? What about a toy piano? A beanbag chair? How about a laundry basket? A chandelier? A toy chest? A coatrack?

KATY: Well, you've raised extreme examples. It would take me a while to make up mind about some of these cases.

ANNA: Of course! Things are far from black and white! There's no doubt that you will go back and forth in some of these cases, and even after making up your mind about them all, if you ask your friends, it's guaranteed that you'll get back all sorts of diverging opinions, everyone having their arguments ready at hand. At some point your head will start spinning and you won't be sure any longer what you think about these marginal cases. And we can also recall some cases we mentioned earlier, such as whether Pluto is a planet or not (on this topic, it would seem wise to keep an open mind), or whether an accused person is guilty or innocent (indeed, it's precisely because of marginal cases of category membership like these that the professions of lawyer and judge exist).

KATY: I must admit that I'm coming to feel that I'm less and less sure about what I thought I was sure about, alas. I now concede that uncertainties exist in categorization, but they always lie out at the fringes of categories, while most members of categories are close to the core and far from the margins, and thus are safely removed from the battle zones of doubt. So in categorization, uncertainty, although possible, is rare. But in analogy-making, uncertainty is pretty much the rule, not the exception. Are you going to disagree with me once more?

ANNA: Well, I'm afraid that once again you've fallen for a stereotype. Just go back and think of all the automatic, unconscious, run-of-the-mill analogies that I brought up earlier, such as the doorknob that easily turns, rather like hundreds of others that you've already turned, or Joanie, who, always true to herself, is perennially late to appointments, or this French fry, crunchy and warm, just like its plate-mates, or this elevator, which can be relied on to work, just like so many similar ones in similar buildings. These are certainly all analogies, but you'll grant also that they are very certain and that one does not tremble in one's boots in relying on them.

KATY *(laughing)*: What can I say? Once again I find I'm in agreement with you, Anna. You've knocked down my pins. You've rolled a strike! At last we're in harmony.

The telephone rings.

KATYANNA *(groggily)*: Hello... Ah, uh... who is it?

A male voice replies: Good morning, Katyanna! It's Dounuel. I'm so relieved to have reached you! I just woke up from a very strange and upsetting dream, and if you have a moment, I'd appreciate it if you were willing to talk about it a little with me.

KATYANNA: You woke me up, you know? But it's all right — I'm coming to my senses, and that's good. So go ahead and tell me about your troubling dream.

DOUNUEL: Well, in my dream I was divided into two people who were both extremely stubborn, and they were having a huge argument. Being split in two in that way was a very weird feeling, I tell you.

KATYANNA: What a peculiar dream! And what were these opponents arguing about?

DOUNUEL: It was most unusual. Although intellectually they were bitterly opposed to each other, they were actually very good friends, and they had decided that they would write a book together, but they hadn't yet started it. One of them was vehemently insisting that the book would have to be written in French, while the other one was equally ardently demanding that it be written in English. It was a ferocious battle, and yet they were both using the most polite and friendly language with each other.

KATYANNA: My goodness! What a peculiar nightmare! But this imaginary book that they were hoping to write together — what was it going to be about?

DOUNUEL: Oh, you know, my standard old hobbyhorse — the unity of analogy-making and categorization. No surprise there!

KATYANNA: Now there's a theme that I know like the back of my hand! At least you were pretty much yourself in your dream.

DOUNUEL: Yes, luckily. But it was awful to feel myself split into two pieces that were fighting so intensely over the choice of language. It was such a relief when I finally woke up and realized I was just one ordinary person, not plagued by schizophrenia. So thank you, Katyanna, for having let me let off some steam about my disturbing dream. And now, thanks to you, I'm feeling fresh as a daisy. It's time for me to get back to work on my book, which, I'm glad to report, is nearly done (I'm just putting the finishing touches on a dialogue that comes at the very end, like icing on a cake), and since today is an odd-numbered day, I'll write in French, as is my custom — and then tomorrow I'll write in English, and so forth and so on.

KATYANNA: What a coincidence, my dear brother! Can you believe that exactly the same thing just happened to me?

DOUNUEL: No! Tell me about it! I'm all ears!

KATYANNA: Yes, indeed, dear brother, I had a parallel dream, a similar dream, a comparable dream, an analogous dream…

DOUNUEL: You wouldn't mean a dream belonging to the very same category?

KATYANNA: Ah, yes — just the phrase I was looking for! You couldn't have known it, but your phone call woke me up from my own very upsetting nightmare. In it, I too was split into two people who were having at each other like two angry little gremlins, although always using extremely polite words. And something of your hobbyhorse must have rubbed off on me after all these years, because one of the angry gremlins was insisting that analogy-making is the core of cognition, while the other one maintained with equal vehemence that categorization played that role. It was all nonsense, of course, as I now realize clearly, but at the time it really seemed as if it made perfect sense for these two strange gremlins to be arguing that these are two mental processes that differ from each other.

DOUNUEL: How droll it is to see ourselves taken over at night by wild fantasies, making us believe in notions that by light of day are clearly sheer nonsense! But luckily you're wide awake now, just as I am, and as they say, all's well that ends well! And so before hanging up, let me just wish you a lovely day, my dear sister. Good-bye!

KATYANNA: And likewise to you, my dear brother! Good-bye!

And thus, just as the sun was rising, after this troubled night during which she had imagined herself split into two subselves plunged in a bitter struggle, Katyanna arose with a feeling of serenity, happiness, and inner peace, thanks to her recovered unity.

NOTES

Prologue

Page 8 *There is a famous Russian poem…* Selvinsky (1920).
Page 16 *One day… the idea came to me…* Poincaré (1908), p. 52.
Page 21 *likeness is a most slippery tribe…* Plato (1977), p. 231a.
Page 21 *a mobile army of metaphors…* Nietzsche (1873), p. 46.
Page 22 *The light of human minds, is perspicuous words…* Hobbes (1651), Chapter V, 36.
Page 22 *Expressing oneself with metaphors has the quality…* Alberic (1973), pp. 146–147.
Page 22 *A science that accepts images is…* Bachelard (1934), p. 47.

Chapter 1

Page 38 *Censorship is the mother of metaphor…* Jorge Luis Borges, quoted in Manea (1992), p. 30.
Page 38 *Leisure is the mother of philosophy…* Hobbes (1651), IV, 46.
Page 38 *Death is the mother of beauty…* From the poem "Sunday Morning", in Stevens (1923).
Page 60 *I'll tell you something…* Robert Pond, "Fun in metals", *John Hopkins Magazine*, April 1987, pp. 60–68, quoted in Murphy (2002), p. 18.
Page 61 *A language is a dialect with an army…* Max Weinreich (1945). "The YIVO and the problems of our time". Lecture delivered at the Annual YIVO (Yiddish Scientific Institute) Conference, New York, 5 January 1945.

Chapter 2

Page 97 *Our goose is cooked…* Chiflet (1985) and Whistle (2000).
Page 112 *A famished fox observed some grapes…* Morvan de Bellegarde (1802).
Page 112 *Driven by hunger, a fox was lusting…* Phædrus (1864), Book 4, Fable 3.
Page 112 *We can't have all we seek, alas…* Benserade (1678), p. 108.
Page 112 *A certain fox from Normandy…* La Fontaine (1668), Book III, p. 11.
Page 132 *When a dog eats the flesh of a goose…* Henri Poincaré, as quoted by Roger Apéri in Dieudonné, Loi, and Thom (1982), pp. 58–72.
Page 133 *There Is No Word…* Tony Hoagland, *Poetry Magazine*, July–August, 2012.

Chapter 3

Page 138 *Items to save when one's house is burning down…* Barsalou (1991).
Page 154 *Phelps is pretty much my double…* Mark Spitz, quoted in an article by Drew Van Esselstyn in the *New Jersey Star-Ledger*, 15 August, 2008.

Page 155 *Back in 1972, they didn't have a 50-meter race…* *Ibid.*
Page 160 *Karnak Caps…* From *Egypt Sweet* by Kellie O. Gutman (privately issued, 2005), p. 9.

Chapter 4

Page 188 *Ireneo Funes, the main character in Jorge Luis Borges' short story…* Borges (1962), p. 114.
Page 208 *As both émigré and physicist, Dr. Teller was aware of the Nazis' lengthening shadow…* Mark Feeney, in the obituary "Bomb pioneer Edward Teller dies", in *The Boston Globe* ("Nation" section), 10 September, 2003.
Page 221 *Of course, the historical figure of mathematical fame…* Swetz and Kao (1977), p. 7.

Chapter 5

Page 274 *Genius is 1 percent inspiration…* Thomas Edison in 1902, as reported in the September 1932 edition of *Harper's Monthly Magazine.*
Page 301 *If the only tool you have is a hammer…* Maslow (1966), p. 5.
Page 313 *That memory is knowledge, that knowledge is going to interfere…* Jiddu Krishnamurti, "First conversation with Dr. Allen W. Anderson" in San Diego, California, 18 February, 1974.
Page 313 *"Freedom from the Known", the title of one of his most famous works…* Krishnamurti (1969).
Page 315 *There are magic links and chains…* James Falen, "Odelet in Praise of Constraints", in Hofstadter (1997), p. 272.

Chapter 6

Page 330 *In their book Mental Leaps…* Holyoak and Thagard (1995), p. 139.
Page 334 *The 1930s is a composite analogy…* Khong (1992), p. 59.
Page 334 *The analogy of Munich raised the stakes…* Khong (1992), p. 184.
Page 337 *One of the most interesting findings of researchers…* Khong (1992), p. 217.
Page 338 *These findings may leave us feeling…* Gentner, Rattermann, and Forbus (1993), p. 567.
Page 368 *When I look at an article in Russian…* Personal letter from Warren Weaver to Norbert Wiener, quoted in Weaver (1955).
Page 363 *Parfois, le succès ne fut pas au rendez-vous…* Bertrand Poirot-Delpech, in the obituary "Sagan, l'art d'être soi", in *Le Monde*, 26 September, 2004.

Chapter 7

Page 388 *All summer long, without a care…* La Fontaine (1668), Book I, p. 1.
Page 400 *The real problem with the interface is…* Norman (1990), p. 210.
Page 408 *I will sette as I doe often in woorke use, a pair of paralleles…* Recorde (1557).
Page 413 *Multiplying one number by another is…* Bezout (1833), p. 12.
Page 415 *As long as one considers numbers as abstract entities…* Bezout (1833), p. 13.
Page 420 *To divide one number by another means…* Bezout (1833), p. 21.
Page 420 *The number to be divided is the dividend…* Bezout (1833), p. 21.
Page 420 *One's goal in doing a division is not always to find out…* Bezout (1833), p. 21.
Page 436 *Categories let people treat new things as if…* Spalding and Murphy (1996), p. 525.
Page 436 *Analogy is what allows us to see the novel as familiar…* Gick and Holyoak (1983), p. 1.
Page 436 *In an analogy, a familiar domain is used…* Clement and Gentner (1991), p. 89.
Page 436 *If one establishes that a given object belongs to a certain category…* Anderson (1991), p. 411.

Chapter 8

Page 438 *Could anyone think… that they have always marched forward…* Poincaré (1911), p. 31.
Page 440 *"8 – 3" is easily understood; 3 can be taken from 8…* De Morgan (1831), pp. 103–104.
Page 443 *an elegant and marvelous trick found in the miracle of Analysis…* Leibniz (1702), p. 357.
Page 454 *I saw that mathematics was split up…* Einstein, cited by Banesh Hoffmann (1972), p. 8.
Page 461 *That he may sometimes have missed the target…* Planck, quoted in Stehle (1994), p. 152.
Page 467 *That a principle of such broad generality…* Einstein (1920), p. 17.
Page 473 *With his instinctive sense of cosmic unity he now tosses off…* Hoffmann (1972), p. 81.
Page 477 *In his paper of 1905 Einstein said that all energy…* Hoffmann (1972), p. 81.
Page 495 *[The new principle] had artistic unity…* Hoffmann (1972), p. 113.
Page 498 *I first had the decisive idea of the analogy…* Einstein, quoted in Stachel (2001), p. 255.
Page 500 *What made my reputation as a mathematician is…* Villani (2012), p. 146–147.
Page 501 *Yet when we see how shaky were the ostensible foundations…* Hoffmann (1972), pp. 127–128.
Page 501 *Mr. Einstein is one of the most original minds I have known…* Hoffmann (1972), p. 99.
Page 501 *A contemporary of Johann Sebastian Bach once said…* David and Mendel (1966), p. 222.
Page 502 *Good mathematicians see analogies between theorems or theories…* Ulam (1976), p. 203.

Epidialogue

Page 506 *categories allow us to treat new things as if…* Spalding and Murphy (1996), p. 525
Page 506 *analogy is what allows us…* Gick and Holyoak (1983), p. 1.
Page 507 *Analogy pervades all our thinking…* Polya (1957), p. 37.
Page 507 *When faced with something new, we cannot help…* Oppenheimer (1956), p. 129.
Page 507 *Analogies and metaphors are pervasive…* Gentner and Clement (1988), p. 307.
Page 507 *Analogy is ubiquitous in human thinking…* Thagard, Holyoak, Nelson and Gochfeld (1990), p. 259.
Page 509 *The trips we take in the world of mathematics…* Alain Connes, in the short film *Mathématiques, un dépaysement soudain*, produced in November of 2011 by Raymond Depardon and Claudine Nougaret at the Fondation Cartier in Paris.
Page 509 *a mobile army of metaphors…* Nietzsche (1873), p. 46.
Page 509 *Whoever first compared a woman to a rose was a poet…* Georges Courteline.
Page 509 *Henri Poincaré described mathematics as…* Poincaré (1908), p. 29.
Page 523 *the ability to separate things according to their natural divisions…* Plato (1950), 265d–265e.

BIBLIOGRAPHY

❧ ❧ ❧

We have divided our bibliography into eleven sections, one for each of the ten main parts of our book, with one extra section at the beginning providing a list of references that have global relevance to the ideas that we are exploring in our book. Since some of the works cited below are relevant to more than just one chapter of our book, certain entries appear in more than one section of the bibliography. We have preceded each section with a few very general comments about the books and articles listed in it.

General

We open our list of references with a set of books that are relevant to every aspect of our own work. In particular, the book by Fauconnier and Turner, like ours, places conceptual mapping at the center of cognition and also uses a rich and highly variegated array of examples to flesh out the key themes. The volume by Holyoak and Thagard and the anthology edited by Gentner, Holyoak, and Kokinov have become standard references for the field of analogy, seen from a cognitive-science perspective. Both Helman's compilation and that by Vosniadou and Ortony tackle analogy from a number of angles, the former placing it in an interdisciplinary perspective and the latter focusing on links to similarity. Although these two works are not recent, many of their chapters are still highly relevant. Murphy's book is an excellent resource on categorization, while Lakoff and Johnson's study has greatly enhanced the recognition of the systematic role played by metaphor in human thought. Lastly, Sander's book and that by Hofstadter and the Fluid Analogies Research Group paved the way for the present volume, although they are somewhat more traditionally academic in style.

Fauconnier, Gilles and Mark Turner (2002). *The Way We Think: Conceptual Blending and the Mind's Hidden Complexities.* New York: Basic Books.

Gentner, Dedre, Keith J. Holyoak, and Boicho N. Kokinov, editors (2001). *The Analogical Mind: Perspectives from Cognitive Science.* Cambridge, Mass.: MIT Press (Bradford Books).

Helman, David H., editor (1988). *Analogical Reasoning: Perspectives of Artificial Intelligence, Cognitive Science, and Philosophy.* Boston: Kluwer Academic Publishers.

Hofstadter, Douglas and the Fluid Analogies Research Group (1995). *Fluid Concepts and Creative Analogies: Computer Models of the Fundamental Mechanisms of Thought.* New York: Basic Books.

Holyoak, Keith J. and Paul Thagard (1995). *Mental Leaps.* Cambridge, Mass.: MIT Press (Bradford Books).

Lakoff, George and Mark Johnson (1980). *Metaphors We Live by.* Chicago: University of Chicago Press.

Murphy, Gregory L. (2002). *The Big Book of Concepts.* Cambridge, Mass.: MIT Press.

Sander, Emmanuel (2000). *L'Analogie, du naïf au créatif: Analogie et catégorisation.* Paris: L'Harmattan.

Vosniadou, Stella and Andrew Ortony, editors (1989). *Similarity and Analogical Reasoning.* New York: Cambridge University Press.

Prologue

The selections by Alberic of Monte Cassino, Bachelard, Bartha, Bouveresse, Hobbes, Lloyd, Nietzsche and Plato deal with the question of analogy's reliability as a mode of thinking. The far-ranging studies of chairs by Danto and Lévy, by Fiell and Fiell, and by Samaras, as well as the large collections of typefaces by Weinberger and by Jaspert, Berry, and Johnson, are aimed at revealing the astonishing variety of the categories in question. The article by Anderson stresses the psychological value of categorization, while that by Tversky and those by Medin, Goldstone, Gentner and colleagues explore the psychological process of observing similarity.

Alberic of Monte Cassino (1973). "The Flowers of Rhetoric". In Joseph M. Miller, Michael H. Prosser, and Thomas W. Benson (eds.), *Readings in Medieval Rhetoric.* Bloomington: Indiana University Press.

Anderson, John R. (1991). "The adaptive nature of human categorization". *Psychological Review*, 98, pp. 409–429.

Bachelard, Gaston (1934). *The Formation of the Scientific Mind: A Contribution to a Psychoanalysis of Objective Knowledge*, translated by Mary McAllester Jones. Manchester: Clinamen Press.

Bartha, Paul (2010). *By Parallel Reasoning: The Construction and Evaluation of Analogical Arguments.* New York: Oxford University Press.

Bouveresse, Jacques (1999). *Prodiges et vertiges de l'analogie.* Paris: Raisons d'agir.

Danto, Arthur C. and Jennifer Lévy (1988). *397 Chairs.* New York: Harry N. Abrams.

Fiell, Charlotte and Peter Fiell (1997). *1000 Chairs.* Cologne: Benedikt Taschen Verlag.

Goldstone, Robert L. (1994). "Similarity, interactive activation, and mapping". *Journal of Experimental Psychology: Learning, Memory, and Cognition*, 20, pp. 3–28.

Goldstone, Robert L. and Son, Ji Yun (2005). "Similarity". In Keith J. Holyoak and Robert G. Morrison (eds.), *Cambridge Handbook of Thinking and Reasoning.* New York: Cambridge University Press, pp. 13–36.

Hobbes, Thomas (1651). *Leviathan.* Reissued by Cambridge University Press (New York), 1996.

Hofstadter, Douglas (1985). "Analogies and roles in human and machine thinking". In Douglas Hofstadter, *Metamagical Themas: Questing for the Essence of Mind and Pattern.* New York: Basic Books, pp. 547–603.

Jaspert, W. Pincus, W. Turner Berry, and A. F. Johnson (1983). *The Encyclopædia of Type Faces.* Poole, Dorset: Blandford Press.

Lloyd, G. E. R. (1966). *Polarity and Analogy: Two Types of Argumentation in Early Greek Thought.* New York: Cambridge University Press.

Medin, Douglas L., Robert L. Goldstone, and Dedre Gentner (1993). "Respects for similarity". *Psychological Review*, 100, pp. 254–278.

Nietzsche, Friedrich (1873). "On Truth and Lies in a Nonmoral Sense". In *The Portable Nietzsche*, translated by Walter Kaufmann, 1976 edition. New York: Viking Press.

Plato (1977). *The Sophist.* In Jacob Klein (ed.), *Plato's Trilogy.* Chicago: University of Chicago Press.

——— (2008). *The Republic*, translated by R. E. Allen. New Haven: Yale University Press.

Poincaré, Henri (1908). *Science et méthode.* Paris: Flammarion.

Samaras, Lucas (1970). *Chair Transformation.* New York: Pace.

Selvinsky, Il'ya L. (1920). "К вопросу о русской речи". In Il'ya L. Selvinsky, *Selected Works.* Leningrad: Sovetskii Pisatel' (Biblioteka Poeta, Bol'šaya seriya), 1972.

Sternberg, Robert J. (2005). *Barron's Miller Analogies Test.* Hauppauge, New York: Barron's Educational Series.

Tversky, Amos (1977). "Features of similarity". *Psychological Review*, 84, pp. 327–352.

Weinberger, Norman (1971). *Encyclopedia of Comparative Letterforms for Artists & Designers.* New York: Art Directions.

Chapter 1

The selections by Carey, Gelman, Gentner, Goldin-Meadows, Keil, Malt, Mandler, Oakes, Pinker, Prinz, Rakison, and Wolff are important works on conceptual development and language, and are relevant throughout the chapter. The pieces by Bruner, Collins, Hull, and Smoke explain the classical approach to categorization, while those by Barsalou, Glucksberg, Goldstone, Hampton, Lamberts, McCloskey, Medin, Nosofsky, Osherson, Pothos, Richard, and Smith, as well as their colleagues, represent a more contemporary approach, which was launched largely by the article by Wittgenstein. The works by Boroditsky, Gibbs, Johnson, and Lakoff and Turner are relevant to this chapter's sections on the metaphorical usage of words. Duvignau's article concerns semantic approximations by children, and that by Atran and Medin concerns the influence of culture on categorization. The article by Ma is about syntactic analogies, while that by Huth and colleagues investigates the neural substrate of semantic spaces. Finally, Kalużа's monograph is an in-depth study of the categories denoted by the definite and indefinite articles in English.

Aitchison, Jean (1994). *Words in the Mind: An Introduction to the Mental Lexicon* (second edition). Oxford: Blackwell Publishers.

Atran, Scott and Douglas L. Medin (2008). *The Native Mind and the Cultural Construction of Nature.* Cambridge, Mass.: MIT Press.

Barsalou, Lawrence W. (1985). "Ideals, central tendency, and frequency of instantiation as determinants of graded structures in categories". *Journal of Experimental Psychology: Learning, Memory and Cognition*, 11, pp. 629–654.

Barsalou, Lawrence W. and Douglas L. Medin (1986). "Concepts: Static definitions or context-dependent representations?" *Cahiers de psychologie cognitive*, 6, pp. 187–202.

Boroditsky, Lera (2000). "Metaphoric structuring: Understanding time through spatial metaphors". *Cognition*, 75 (1), pp. 1–28.

Bruner, Jerome, Goodnow, Jacqueline J. and Austin, George A. (1956). *A Study of Thinking.* New York: John Wiley and Sons.

Carey, Susan (2009). *The Origin of Concepts.* New York: Oxford University Press.

Collins, Allen M. and M. Ross Quillian (1969). "Retrieval time from semantic memory". *Journal of Verbal Learning and Verbal Behavior*, 8, pp. 240–247.

Duvignau, Karine (2003). "Métaphore verbale et approximation". *Revue d'intelligence artificielle*, special issue 5–6, pp. 869–881.

Duvignau, Karine, Marion Fossard, Bruno Gaume, Marie-Alice Pimenta, and Élie Juliette (2007). "Semantic approximations and flexibility in the dynamic construction and 'deconstruction' of meaning". *Linguagem em Discurso*, 7 (3), pp. 371–389.

Duvignau, Karine and Bruno Gaume (2005). "Linguistic, psycholinguistic and computational approaches to the lexicon: For early verb-learning". *Cognitive Systems*, 6 (1), pp. 255–269.

Garrod, Simon and Anthony Sanford (1977). "Interpreting anaphoric relations: The integration of semantic information while reading". *Journal of Verbal Learning and Verbal Behavior,* 16, pp. 77–79.

Gelman, Susan A. (2005). *The Essential Child: Origins of Essentialism in Everyday Thought.* Oxford: Oxford University Press.

Gentner, Dedre (2003). "Why we're so smart". In Dedre Gentner and Susan Goldin-Meadow (eds.), *Language in Mind: Advances in the Study of Language and Thought.* Cambridge, Mass.: MIT Press, pp. 195–235.

Gentner, Dedre and Susan Goldin-Meadow, editors (2003). *Language in Mind: Advances in the Study of Language and Thought.* Cambridge, Mass.: MIT Press.

Gibbs, Raymond W. (1994). *The Poetics of Mind.* New York: Cambridge University Press.

Hampton, James A. (1979). "Polymorphous concepts in semantic memory". *Journal of Verbal Learning and Verbal Behavior,* 18, pp. 441–461.

Hobbes, Thomas (1651). *Leviathan,* revised student edition. Reissued in 1996 by Cambridge University Press (New York).

Hofstadter, Douglas R. (2001). "Analogy as the Core of Cognition". In Dedre Gentner, Keith J. Holyoak, and Boicho N. Kokinov (eds.), *The Analogical Mind: Perspectives from Cognitive Science.* Cambridge, Mass.: MIT Press (Bradford Books).

Hull, Clark L. (1920). "Quantitative aspects of the evolution of concepts". *Psychological Monographs,* XXVIII (1.123), pp. 1–86.

Huth, Alexander G., Shinji Nishimoto, An T. Vu, and Jack L. Gallant (2012). "A continuous semantic space describes the representation of thousands of object and action categories across the human brain". *Neuron,* 76 (6), pp. 1210–1224.

Kaluża, Henryk (1976). *The Articles in English.* Warsaw: Panstwowe Wydawnictwo Naukowe.

Keil, Frank C. (1979). *Semantic and Conceptual Development: An Ontological Perspective.* Cambridge. Mass.: Harvard University Press.

——— (1989). *Concepts, Kinds, and Cognitive Development.* Cambridge, Mass.: Harvard University Press.

Lakoff, George (1987). *Women, Fire, and Dangerous Things: What Categories Reveal about the Mind.* Chicago: University of Chicago Press.

Lamberts, Koen and David Shanks (1997). *Knowledge, Concepts, and Categories.* Cambridge, Mass.: MIT Press.

Ma, Yulei (2011). *Analogy and Its Use in Grammatical Constructions: A Cognitive–Functional Linguistic Perspective.* Saarbrücken: Verlag Dr. Müller.

Manea, Norman (1992). *Felix Culpa.* New York: Grove Press.

Malt, Barbara and Phillip Wolff (2010). *Words and the Mind: How Words Capture Human Experience.* Oxford: Oxford University Press.

Mandler, Jean M. (2004). *The Foundations of Mind: Origins of Conceptual Thought.* Oxford: Oxford University Press.

Markman, Ellen M. (1991). *Categorization and Naming in Children: Problems of Induction.* Cambridge, Mass.: MIT Press (Bradford Books).

McCloskey, Michael E. and Sam Glucksberg (1978). "Natural categories: Well defined or fuzzy sets?" *Memory and Cognition,* 6, pp. 462–472.

Medin, Douglas L. and M. M. Schaffer (1978). "A context theory of classification learning". *Psychological Review,* 85, pp. 207–238.

Mervis, Carolyn B., Jack Catlin, and Eleanor Rosch (1976). "Relationships among goodness of example, category norms, and word frequency". *Bull. of the Psychonomic Soc.,* 7, pp. 283–294.

Nosofsky, Robert M. (1986). "Attention, similarity, and the identification–categorization relationship". *Journal of Experimental Psychology: General*, 115, pp. 39–57.

Osherson, Daniel and Edward E. Smith (1981). "On the adequacy of prototype theory as a theory of concepts". *Cognition*, 9, pp. 35–58.

Pinker, Steven (1997). *How the Mind Works.* New York: Norton.

——— (1999). *Words and Rules: The Ingredients of Language.* New York: Basic Books.

——— (2007). *The Stuff of Thought.* New York: Viking.

Poitrenaud, Sébastien, Jean-François Richard, and Tijus, Charles A. (2005). "Properties, categories, and categorisation". *Thinking and Reasoning*, 11, pp. 151–208.

Pothos, Emmanuel M. and Andy J. Wills (2011). *Formal Approaches in Categorisation.* New York: Cambridge University Press.

Prinz, Jesse J. (2002). *Furnishing the Mind: Concepts and their Perceptual Basis*, Cambridge, Mass.: MIT Press.

Rakison, David H. and Lisa M. Oakes (2008). *Early Category and Concept Development.* New York: Oxford University Press.

Rosch, Eleanor (1975). "Cognitive representations of semantic categories". *Journal of Experimental Psychology: General*, 104, pp. 192–233.

——— (1976). "Classifications d'objets du monde réel : Origines et représentations dans la cognition". *Bulletin de psychologie*, special issue edited by Stéphane Ehrlich and Endel Tulving, pp. 242–250.

——— (1978). "Principles of categorization". In Eleanor Rosch and Barbara Lloyd (eds.), *Cognition and Categorization.* Hillsdale, New Jersey: Lawrence Erlbaum Associates, pp. 27–48.

Rosch, Eleanor and Carolyn B. Mervis (1975). "Family resemblances: Studies in the internal structure of categories". *Cognitive Psychology*, 7, pp. 573–605.

Rosch, Eleanor, Carolyn B. Mervis, Wayne D. Gray, David M. Johnson, and Penny Boyes Braem (1976). "Basic objects in natural categories". *Cognitive Psychology*, 8, pp. 382–439.

Ross, James F. (1981). *Portraying Analogy.* New York: Cambridge University Press.

Smith, Edward E. and Douglas L. Medin (1981). *Categories and Concepts.* Cambridge Mass.: Harvard University Press.

Smoke, Kenneth Ludwig (1932). "An objective study of concept formation". *Psychological Monographs*, XLII (191), pp. 1–46.

Stevens, Wallace (1923). *Harmonium.* New York: Alfred E. Knopf.

Turner, Mark (1987). *Death is the Mother of Beauty: Mind, Metaphor, Criticism.* Chicago: University of Chicago Press.

Wittgenstein, Ludwig (1953). *Philosophical Investigations.* Oxford: Basil Blackwell.

Woo-Kyoung, Ahn, Robert L. Goldstone, Bradley C. Love, Arthur B. Markman and Phillip Wolff (2005). *Categorization Inside and Outside the Laboratory: Essays in Honor of Douglas L. Medin.* Washington, D.C.: American Psychological Association.

Chapter 2

The books by Aitchison, Braitenberg, Itkonen, Malt and Wolff, and Pinker, as well as the article by Gentner and that by Hofstadter, are relevant to the chapter as a whole. The humorous volumes by Chiflet and Whistle (actually just one person), as well as the books by Glucksberg and by Langlotz, deal with idiomatic expressions; Brézin-Rossignol, Schank, and Visetti consider the categories of proverbs and fables, while Benserade, La Fontaine, Morvan de Bellegarde and Phædrus are relevant to our section on Æsop's fable "The Fox and the Grapes".

Festinger's book is a classic on cognitive dissonance. The books by Carroll and by Sapir deal with the Sapir–Whorf hypothesis, and the volume by Atran and Medin covers the way that culture channels human language and thought. Finally, the books by Flynn and by Sternberg tackle the topic of intelligence.

Aitchison, Jean (1994). *Words in the Mind: An Introduction to the Mental Lexicon* (second edition). Oxford: U.K.: Blackwell Publishers.

Anderson, Poul (1989). "Uncleftish Beholding". *Analog Science Fiction*, 109 (13), pp. 132–135.

Atran, Scott and Douglas L. Medin (2008). *The Native Mind and the Cultural Construction of Nature.* Cambridge, Mass.: MIT Press.

Benserade, Isaac de (1678). *Fables d'Ésope en quatrains dont il y en a une partie au labyrinte de Versailles.* Paris: Sébastien Mabre-Cramoisy.

Braitenberg, Valentino (1996). *Il gusto della lingua: Meccanismi cerebrali e strutture grammaticali.* Merano, Italy: Alpha&Beta.

Brézin-Rossignol, Monique (2008). *Dictionnaire de proverbes* (second edition). Paris: La Maison du dictionnaire.

Carroll, John B., editor (1956). *Language, Thought, and Reality: Selected Writings of Benjamin Lee Whorf.* Cambridge, Mass.: MIT Press.

Chiflet, Jean-Loup (1994). *Sky! My Husband.* Paris: Éditions du Seuil.

Dieudonné, Jean, Maurice Loi, and René Thom (1982). *Penser les mathématiques. Séminaire de philosophie et mathématiques de l'École normale supérieure.* Paris: Éditions du Seuil.

Festinger, Leon (1957). *A Theory of Cognitive Dissonance.* Stanford: Stanford University Press.

Flynn, James R. (1987). "Massive IQ gains in 14 nations: What IQ tests really measure". *Psychological Bulletin*, 101, pp. 171–191.

——— (2009). *What Is Intelligence? Beyond the Flynn Effect.* New York: Cambridge University Press.

Gentner, Dedre (2003). "Why we're so smart". In Dedre Gentner and Susan Goldin-Meadow (eds.), *Language in Mind: Advances in the Study of Language and Thought*, pp. 195–235, Cambridge, Mass.: MIT Press.

Glucksberg, Sam (2001). *Understanding Figurative Language: From Metaphors to Idioms.* New York: Oxford University Press.

Hofstadter, Douglas R. (1995). "Speechstuff and Thoughtstuff: Musings on the Resonances Created by Words and Phrases via the Subliminal Perception of their Buried Parts". In Sture Allén (ed.), *Of Thoughts and Words: The Relation between Language and Mind* (Proceedings of Nobel Symposium 92). London: Imperial College Press.

——— (1997). *Le Ton beau de Marot : In Praise of the Music of Language.* New York: Basic Books.

Itkonen, Esa (2005). *Analogy as Structure and Process.* Amsterdam: John Benjamins Publishing Company.

La Fontaine, Jean de (1668). *Fables choisies mises en vers.* Paris: Barbin et Thierry.

Langlotz, Andreas (2006). *Idiomatic Creativity: A Cognitive-Linguistic Model of Idiom-Representation and Idiom-Variation in English.* Amsterdam: John Benjamins.

Malt, Barbara and Phillip Wolff (2010). *Words and the Mind: How Words Capture Human Experience.* New York: Oxford University Press.

Morvan de Bellegarde, Jean-Baptiste (1802). *Les Cinq Fabulistes ou les Trois Cents Fables d'Ésope, de Lockmann, de Philelphe, de Gabrias et d'Avienus.* Paris: Poncelin.

Phædrus (1864). *Fables de Phèdre*, translated by M. E. Panckoucke. Paris: Garnier Frères.

Pinker, Steven (2007). *The Stuff of Thought: Langage as a Window into Human Nature.* New York: Viking.
Sapir, Edward (1921). *Language: An Introduction to the Study of Speech.* New York: Harcourt, Brace.
Schank, Roger C. (1982). *Dynamic Memory: A Theory of Reminding and Learning in Computers and People.* New York: Cambridge University Press.
———— (1999). *Dynamic Memory Revisited.* New York: Cambridge University Press.
Sternberg, Robert J. (1994). *Encyclopedia of Human Intelligence.* New York: Macmillan.
Visetti, Yves-Marie and Pierre Cadiot (2006). *Motifs et proverbes. Essai de sémantique proverbiale.* Paris: Presses universitaires de France.
Whistle, John Wolf (2000). *Sky ! Mortimer !* Paris: Mots et compagnie.

Chapter 3

The studies by Barsalou explain and explore the notion of ad hoc categories. Schank's books deal with reminding and the mechanisms responsible for it. The article by Bower deals with the centrality of emotions in reminding, while Kanerva's book and the articles by Foundalis, by Gentner and her colleagues, by Kahneman and Miller, and by Thagard, Holyoak, Nelson, and Koh concern the mechanisms underlying memory retrieval. The books by Csányi and Horowitz describe the mental life of dogs and the nature of canine categories. The monographs by French and by Mitchell, along with the chapter by Hofstadter and Mitchell, are relevant to our sections on "me too" analogies.

Barsalou, Lawrence W. (1983). "Ad hoc categories". *Memory and Cognition*, 11, pp. 211–227.
———— (1991). "Deriving categories to achieve goals". In Gordon H. Bower (ed.), *The Psychology of Learning and Motivation.* New York: Academic Press, 27, pp. 1–64.
Bower, Gordon H. (1981). "Mood and memory". *American Psychologist*, 36 (2), pp. 129–148.
Csányi, Vilmos (2005). *If Dogs Could Talk: Exploring the Canine Mind.* San Francisco: North Point.
Foundalis, Harry (2013). "Unification of clustering, concept formation, categorization, and analogy-making". Technical Report, Center for Research on Concepts and Cognition, Indiana University, Bloomington.
French, Robert M. (1995). *The Subtlety of Sameness: A Theory and Computer Model of Analogy-Making.* Cambridge. Mass.: MIT Press (Bradford Books).
Gentner, Dedre, Jeffrey Loewenstein, Leigh Thompson, and Kenneth D. Forbus (2009). "Reviving inert knowledge: Analogical abstraction supports relational retrieval of past events". *Cognitive Science*, 33 (8), pp. 1343–1382.
Hofstadter, Douglas and Melanie Mitchell (1995). "The Copycat project: A model of mental fluidity and analogy-making". In Douglas Hofstadter and the Fluid Analogies Research Group, *Fluid Concepts and Creative Analogies.* New York: Basic Books, pp. 205–267.
Horowitz, Alexandra (2010). *Inside of a Dog.* New York: Scribners.
Kahneman, Daniel and Dale T. Miller (1986). "Norm theory: Comparing reality to its alternatives". *Psychological Review*, 93 (2), pp. 136–153.
Kanerva, Pentti. *Sparse Distributed Memory.* Cambridge, Mass.: MIT Press.
Mitchell, Melanie (1993). *Analogy-Making as Perception.* Cambridge, Mass.: MIT Press.
Schank, Roger C. (1982). *Dynamic Memory: A Theory of Reminding and Learning in Computers and People.* New York: Cambridge University Press.
———— (1999). *Dynamic Memory Revisited.* New York: Cambridge University Press.
Thagard, Paul, Keith J. Holyoak, Greg Nelson, and David Gochfeld (1990). "Analog retrieval by constraint satisfaction". *Artificial Intelligence*, 46, pp. 259–310.

Chapter 4

The contributions by Bowdle and Gentner, Geary, Gibbs, Glucksberg, Indurkhya, Jones and Estes, Lakoff and Turner, Ortony, and Pinker are relevant mainly to our sections on metaphor. The works by Chi, Ericsson, Feltovich, Hoffman, Johnson, Mervis, Ross, Tanaka, and Taylor and their colleagues are relevant to our discussion of expertise in a broad sense. The works by Greenberg, Sander (with Dupuch), and Politzer discuss the phenomenon of marking. The article by Collins and Quillian shows the classical approach to abstraction, while Poitrenaud's monograph and the articles by Laurence and Margolis and by Richard and Sander offer more recent views of the phenomenon. Chrysikou, Duncker, Nersessian, Richard, and Ward treat abstraction and creativity and their role in problem-solving. The book edited by Laurence and Margolis concerns artefacts, while Casati's fascinating study is devoted to shadows in a very wide sense of the term.

Blessing, Stephen B. and Brian H. Ross (1996). "Content effects in problem categorization and problem solving". *J. of Experimental Psychology: Learning, Memory, and Cognition*, 22, pp. 792–810.

Borges, Jorge Luis (1962). "Funes the Memorious", translated by Anthony Kerrigan. In Jorge Luis Borges, *Ficciones.* New York: Grove Press, p. 114.

Bowdle, Brian F. and Dedre Gentner (2005). "The career of metaphor". *Psychological Review*, 112 (1), pp. 193–216.

Casati, Roberto (2004). *Shadows: Unlocking Their Secrets, from Plato to Our Time.* London: Vintage.

Chi, Michelene T. H., Paul J. Feltovich, and Robert Glaser (1981). "Categorization and representation of physics problems by experts and novices". *Cognitive Science*, 5, pp. 121–152.

Chi, Michelene T. H., Robert Glaser, and Marshall J. Farr (1988). *The Nature of Expertise.* Hillsdale, New Jersey: Lawrence Erlbaum.

Chrysikou, Evangelia G. (2006). "When shoes become hammers: Goal-derived categorization training enhances problem solving performance". *Journal of Experimental Psychology: Learning, Memory, and Cognition*, 32, pp. 935–942.

Collins, Allen M. and M. Ross Quillian (1969). "Retrieval time from semantic memory". *Journal of Verbal Learning and Verbal Behaviour*, 8, pp. 240–248.

Duncker, Karl (1945). "On problem solving". *Psychological Monographs*, 58, pp. 1–110.

Dupuch, Laurence and Emmanuel Sander (2007). "Apport pour les apprentissages de l'explicitation des relations d'inclusion de classes". *L'Année psychologique*, 107 (4), pp. 565–596.

Ericsson, K. Anders, Neil Charness, Paul J. Feltovich, and Robert R. Hoffman, editors (2006). *Cambridge Handbook of Expertise and Expert Performance.* New York: Cambridge University Press.

Geary, James (2012). *I is an Other: The Secret Life of Metaphor and How It Shapes the Way We See the World.* New York: Harper Perennial.

Gibbs, Raymond W., editor (2008). *Cambridge Handbook of Metaphor and Thought.* New York: Cambridge University Press.

Glucksberg, Sam and Boaz Keysar (1990). "Understanding metaphorical comparisons: Beyond similarity". *Psychological Review*, 97, pp. 3–18.

Glucksberg, Sam, Matthew S. McGlone, and Deanna Manfredi (1997). "Property attribution in metaphor comprehension". *Journal of Memory and Language*, 36, pp. 50–67.

Greenberg, Joseph (1966). *Language Universals, with Special Reference to Feature Hierarchies.* The Hague: Mouton.

Indurkhya, Bipin (1992). *Metaphor and Cognition: An Interactionist Approach.* New York: Springer.

Johnson, Kathy E. and Carolyn B. Mervis (1997). "Effects of varying levels of expertise on the basic level of categorization". *Journal of Experimental Psychology: General*, 126, pp. 248–277.

Jones, Lara and Zachary Estes (2005). "Metaphor comprehension as attributive categorization". *Journal of Memory and Language*, 53, pp. 110–124.

Lakoff, George (1987). *Women, Fire, and Dangerous Things: What Categories Reveal about the Mind.* Chicago: University of Chicago Press.

Lakoff, George and Mark Turner (1989). *More than Cool Reason: A Field Guide to Poetic Metaphor.* Chicago: University of Chicago Press.

Laurence, Stephen and Eric Margolis (2012). "Abstraction and the origin of general ideas". *Philosophers' Imprint*, 12 (19), pp. 1–22.

Margolis, Eric and Stephen Laurence (2007). *Creations of the Mind: Theories of Artifacts and Their Representation.* New York: Oxford University Press.

Nersessian, Nancy J. (2008). *Creating Scientific Concepts.* Cambridge, Mass.: MIT Press.

Ortony, Andrew. (1993). *Metaphor and Thought.* New York: Cambridge University Press, revised edition.

Pinker, Steven (2007). *The Stuff of Thought: Language as a Window into Human Nature.* New York: Viking.

Poitrenaud, Sébastien. (1995). "The PROCOPE semantic network: An alternative to action grammars". *International Journal of Human–Computer Studies*, 42, pp. 31–69.

Poitrenaud, Sébastien. (2001). *Complexité cognitive des interactions homme-machine. Modélisation par la méthode ProCope.* Paris: L'Harmattan.

Politzer, Guy (1991). "L'informativité des énoncés : Contraintes sur le jugement et le raisonnement". *Intellectica*, 11, pp. 111–147.

Richard, Jean-François (2004). *Les Activités mentales. De l'interprétation de l'information à l'action.* Paris: Armand Colin.

Sander, Emmanuel (2006). "Raisonnement et résolution de problèmes". In Serban Ionescu and Alain Blanchet (eds.), *Nouveau cours de psychologie. Psychologie cognitive et bases neurophysiologiques du fonctionnement cognitif* (coordinated by Daniel Gaonac'h). Paris: Presses universitaires de France, pp. 159–190.

——— (2008). "En quoi Internet a-t-il changé notre façon de penser ?" In Philippe Cabin and Jean-François Dortier (eds.), *La Communication. État des savoirs.* Auxerre: Éditions Sciences humaines, pp. 363–369.

Sander, Emmanuel and Jean-François Richard (1997). "Analogical transfer as guided by an abstraction process: The case of learning by doing in text editing". *Journal of Experimental Psychology: Learning, Memory, and Cognition*, 23, pp. 1459–1483.

——— (1998). "Analogy-making as a categorization and an abstraction process". In Keith J. Holyoak, Dedre Gentner, and Boicho Kokinov (eds.), *Advances in Analogy Research: Integration of Theory and Data from the Cognitive, Computational, and Neural Sciences.* Sofia: New Bulgarian University Series in Cognitive Science, pp. 381–389.

——— (2005). "Analogy and transfer: Encoding the problem at the right level of abstraction". *Proceedings of the 27th Annual Conference of the Cognitive Science Society*, Stresa (Italy), pp. 1925–1930.

Swetz, Frank J. and T. I. Kao (1977). *Was Pythagoras Chinese? An Examination of Right Triangle Theory in Ancient China.* University Park: Pennsylvania State University Press.

Tanaka, James W. and Marjorie Taylor (1991). "Object categories and expertise: Is the basic level in the eye of the beholder?" *Cognitive Psychology*, 23, pp. 457–482.

Ward, Thomas B. and Yulia Kolomyts (2010). "Cognition and creativity". In James C. Kaufman and Robert J. Sternberg (eds.), *The Cambridge Handbook of Creativity.* New York: Cambridge University Press, pp. 93–112.

Chapter 5

The contributions by Barsalou, Clark, Damasio, Gibbs, Glenberg, Harnad, Johnson, Lakoff, Overton, Pecher and Zwaan, Sweetser, Varela, Zhong, and their colleagues are concerned with the role of embodiment in cognition. On the topics of generalization and induction, the books by Bartha, Feeney and Heit, George, Kahneman, and Sloman, as well as that by Holland, Holyoak, Nisbett, and Thagard, are relevant, as are the articles by Holyoak and colleagues, by Osherson and Smith, and by Rips. The works by Arnaud, Baars, Cutler, Erard, Fromkin, Hofstadter and Moser, Rossi and Peter-Defare, and Rumelhart and Norman deal with speech errors and action errors and the psychological mechanisms underlying those phenomena. Bassok, Clément, Novick, and Richard and Zamani explore the role of unconscious presumptions in problem-solving. Finally, the books by Chu, Krishnamurti, and Serafini explore, each in its own highly personal fashion, the limits of human imagination.

Arnaud, Pierre J. L. (1997). "Les ratés de la dénomination individuelle : Typologie des lapsus par substitution de mots". In Claude Boisson and Philippe Thoiron (eds.), *Autour de la dénomination.* Lyon: Presses universitaires de Lyon, pp. 307–331.

Baars, Bernard J., editor (1992). *Experimental Slips and Human Error: Exploring the Architecture of Volition.* New York: Plenum.

Barsalou, Lawrence W. (1999). "Perceptual symbol systems". *Behavioral and Brain Sciences*, 22, pp. 577–609.

Bartha, Paul (2010). *By Parallel Reasoning: The Construction and Evaluation of Analogical Arguments.* New York: Oxford University Press.

Chu, Seo-Young (2010). *Do Metaphors Dream of Literal Sleep? A Science-Fictional Theory of Representation.* Cambridge, Mass.: Harvard University Press.

Clark, Andy (2011). *Supersizing the Mind: Embodiment, Action, and Cognitive Extension.* New York: Oxford University Press.

Clément, Évelyne and Richard, Jean-François (1997). "Knowledge of domain effects in problem representation: The case of tower of Hanoi isomorphs". *Thinking and Reasoning*, 3 (2), pp. 133–157.

Cutler, Anne, editor (1982). *Slips of the Tongue and Language Production.* New York: Mouton.

Damasio, Antonio R. (1994). *Descartes' Error: Emotion, Reason, and the Human Brain.* New York: G. P. Putnam's Sons.

Erard, Michael (2007). *Um… Slips, Stumbles, and Verbal Blunders, and What They Mean.* New York: Pantheon Books.

Feeney, Aidan and Evan Heit (2007). *Inductive Reasoning: Experimental, Developmental, and Computational Approaches.* New York: Cambridge University Press.

Fromkin, Victoria A., editor (1980). *Errors in Linguistic Performance: Slips of the Tongue, Ear, Pen, and Hand.* New York: Academic Press.

George, Christian (1997). *Polymorphisme du raisonnement humain.* Paris: PUF.

Gibbs, Raymond W. (2006). *Embodiment and Cognitive Science.* New York: Cambridge University Press.

Glenberg, Arthur M. (1997). "What memory is for: Creating meaning in the service of action". *Behavioral and Brain Sciences*, 20, pp. 1–55.

Goldstone, Robert L., Sam Day, and Ji Yun Son (2010). "Comparison". In Britt Glatzeder, Vinod Goel, and Albrecht von Müller (eds.), *On Thinking*, Vol. II: *Towards a Theory of Thinking.* Berlin: Springer, pp. 103–122.

Goodman, Nelson (1972). *Problems and Projects.* Indianapolis: Bobbs-Merrill.

Harnad, Stevan (1990). "The symbol grounding problem". *Physica D*, 42, pp. 335–346.

——— (1990). *Categorical Perception: The Groundwork of Cognition.* New York: Cambridge University Press.

Hofstadter, Douglas (1997). *Le Ton beau de Marot: In Praise of the Music of Language.* New York: Basic Books.

Hofstadter, Douglas R. and David J. Moser (1989). "To err is human; To study error-making is cognitive science". *Michigan Quarterly Review*, 28 (2), pp. 185–215.

Holland, John H., Keith J. Holyoak, Richard E. Nisbett, and Paul R. Thagard (1986). *Induction: Processes of Inference, Learning, and Discovery.* Cambridge, Mass.: MIT Press.

Holyoak, Keith J., Hee Seung Lee, and Hongjing Lu (2010). "Analogical and category-based inference: A theoretical integration with Bayesian causal models". *Journal of Experimental Psychology: General*, 139 (4), pp. 702–727.

Johnson, Mark (1987). *The Body in the Mind: The Bodily Basis of Meaning, Reason, and Imagination.* Chicago: University of Chicago Press.

——— (2007). *The Meaning of the Body: Æsthetics of Human Understanding.* Chicago: University of Chicago Press.

——— (2008). "The meaning of the body". In Willis F. Overton, Ulrich Mueller, and Judith Newman (eds.), *Developmental Perspectives on Embodiment and Consciousness.* Hillsdale, New Jersey: Lawrence Erlbaum, pp. 19–43.

Kahneman, Daniel (2011). *Thinking, Fast and Slow.* New York: Farrar, Straus and Giroux.

Krishnamurti, Jiddu (1969). *Freedom from the Known.* New York: Harper & Row.

Lakoff, George and Mark Johnson (1999). *Philosophy in the Flesh: The Embodied Mind and Its Challenge to Western Thought.* New York: Basic Books.

Maslow, Abraham (1966). *The Psychology of Science.* Chapel Hill: Maurice Bassett.

Norman, Donald A. (1981). "Categorization of action slips". *Psychological Review*, 88, pp. 1–15.

Novick, Laura R. and Miriam Bassok. (2005). "Problem solving". In Keith J. Holyoak and Robert G. Morrison (eds.), *Cambridge Handbook of Thinking and Reasoning.* New York: Cambridge University Press, pp. 321–349.

Osherson, Daniel N., Edward E. Smith, Ormond Wilkie, Alejandro López, and Eldar B. Shafir (1990). "Category-based induction". *Psychological Review*, 97, pp. 185–200.

Overton, Willis F., Ulrich Mueller, and Judith Newman, editors (2008). *Developmental Perspectives on Embodiment and Consciousness.* Hillsdale, New Jersey: Lawrence Erlbaum.

Pecher, Diane and Rolf A. Zwaan (2005). *Grounding Cognition: The Role of Perception and Action in Memory, Language, and Thinking.* New York: Cambridge University Press.

Raban, Jonathan (1991). *Hunting Mister Heartbreak.* New York: Harper Collins.

Richard, Jean-François (2004). *Les Activités mentales. De l'interprétation de l'information à l'action.* Paris: Armand Colin.

Richard, Jean-François, Poitrenaud, Sébastien, and Tijus, Charles A. (1993). "Problem solving restructuration: Elimination of implicit constraints". *Cognitive Science*, 17, pp. 497–529.

Richard, Jean-François and Mojdeh Zamani (2003). "A problem-solving model as a tool for analyzing adaptive behavior". In Robert J. Sternberg, Jacques Lautrey, and Todd I. Lubart (eds.), *Models of Intelligence: International Perspective.* Washington, D. C.: American Psychological Association, pp. 213–226.

Rips, Lance J. (1975). "Induction about natural categories". *Journal of Verbal Learning and Verbal Behavior*, 14, pp. 665–681.

Rossi, Mario and Peter-Defare, Evelyne. (1998). *Les Lapsus, ou, Comment notre fourche a langué.* Paris: Presses universitaires de France.

Rumelhart, David E. and Donald Norman (1982). "Simulating a Skilled Typist: A Study of Skilled Cognitive–Motor Perforamnce". *Cognitive Science*, 6 (1), pp. 1–36.

Serafini, Luigi (1981). *Codex Seraphinianus*. Milan: Franco Maria Ricci.

Shiffrin, Richard M. and Walter E. Schneider (1977). "Controlled and automatic human information processing: II. Perceptual learning, automatic attending, and a general theory". *Psychological Review*, 84, pp. 127–190.

Sloman, Steven (2009). *Causal Models: How People Think About the World and Its Alternatives*. New York: Oxford University Press.

Sweetser, Eve (1990). *From Etymology to Pragmatics: The Mind-as-Body Metaphor in Semantic Structure and Semantic Change*. New York: Cambridge University Press.

Varela, Francisco, Evan T. Thompson, and Eleanor Rosch (1991). *The Embodied Mind*. Cambridge, Mass.: MIT Press.

Zhong, Chen-Bo and Katie Liljenquist (2006). "Washing away your sins: Threatened morality and physical cleansing". *Science*, 313, pp. 1451–1452.

Zwaan, Rolf A. and Carol J. Madden (2005). "Embodied sentence comprehension". In Diane Pecher and Rolf A. Zwaan (eds.), *The Grounding of Cognition: The Role of Perception and Action in Memory, Language, and Thinking*. New York: Cambridge University Press.

Chapter 6

The works by Bassok, Dunbar, Forbus and Gentner (and their colleagues), Gick, Holyoak, Keane, Novick, Ross, Sander, and Thagard concern the classical source–target paradigm and its limitations, as well as the relationships between external surface and internal structure. This question is also dealt with in the contribution by Chalmers, French, and Hofstadter. The books by French and by Mitchell, as well as the article by Hofstadter and Mitchell, describe the Copycat microdomain and the related Tabletop microdomain, and the computational modeling of the creation of analogies. The latter topic is also dealt with by Falkenhainer, Forbus, and Gentner. The set of studies by Coulson, by Fauconnier, by Fauconnier and Sweetser, and by Fauconnier and Turner collectively present a rich vision of frame-blending and conceptual integration and demonstrate the pervasiveness of these phenomena in human thought. Khong, Record, and Suganami deal with the use of analogies in politics, while the books by Kahneman, by Bonneforn, and by Sloman, as well as that by Holland, Holyoak, Nisbett, and Thagard, deal with the importance of non-deductive reasoning in cognition. The books by Locke and Booth and by Weaver, as well as those by Hofstadter, are dedicated to the topic of translation as carried out by humans and by machines. Finally, Grothe contains a wide sampling of analogies, including a fair number of caricature analogies.

Bassok, Miriam and Keith J. Holyoak (1993). "Pragmatic knowledge and conceptual structure: Determinants of transfer between quantitative domains". In Douglas K. Detterman and Robert J. Sternberg (eds.), *Transfer on Trial: Intelligence, Cognition, and Instruction*. Norwood, New Jersey: Ablex, pp. 68–98.

Bassok, Miriam and Karen L. Olseth (1995). "Object-based representations: Transfer between cases of continuous and discrete models of change". *Journal of Experimental Psychology: Learning, Memory, and Cognition*, 21, pp. 1522–1538.

Bassok, Miriam, Ling-ling Wu, and Karen L. Olseth (1995). "Judging a book by its cover: Interpretative effects of content on problem-solving transfer". *Memory and Cognition*, 23, pp. 354–367.

Bonnefon, Jean-François (2011). *Le Raisonneur et ses modèles : Un changement de paradigme dans la psychologie du raisonnement*. Grenoble: Presses universitaires de Grenoble.

Chalmers, David J., Robert M. French, and Douglas R. Hofstadter (1992). "High-level perception, representation, and analogy: A critique of artificial intelligence methodology". In Douglas Hofstadter and the Fluid Analogies Research Group, *Fluid Concepts and Creative Analogies.* New York: Basic Books, pp. 169–193.

Coulson, Seana (2001). *Semantic Leaps: Frame-Shifting and Conceptual Blending in Meaning Construction.* New York: Cambridge University Press.

Day, Sam and Robert L. Goldstone (2011). "Analogical transfer from a simulated physical system". *Journal of Experimental Psychology: Learning, Memory, and Cognition*, 37, pp. 551–567.

Dunbar, Kevin (2001). "The analogical paradox: Why analogy is so easy in naturalistic settings, yet so difficult in the psychology laboratory". In Dedre Gentner, Keith J. Holyoak, and Boicho N. Kokinov (eds.), *The Analogical Mind.* Cambridge, Mass.: MIT Press, pp. 313–334.

Falkenhainer, Brian, Kenneth D. Forbus, and Dedre Gentner. (1989). "The structure-mapping engine: Algorithm and examples". *Artificial Intelligence*, 41, pp. 1–63.

Fauconnier, Gilles (1984). *Les Espaces mentaux.* Paris: Éditions de Minuit.

——— (1985). *Mental Spaces: Aspects of Meaning Construction in Natural Language.* Cambridge, Mass.: MIT Press (Bradford Books).

——— (1997). *Mappings in Thought and Language.* New York: Cambridge University Press.

Fauconnier, Gilles and Eve Sweetser, editors (1996). *Spaces, Worlds, and Grammar.* Chicago: University of Chicago Press.

Fauconnier, Gilles and Mark Turner (2002). *The Way We Think: Conceptual Blending and the Mind's Hidden Complexities.* New York: Basic Books.

Forbus, Kenneth D., Dedre Gentner, and Keith Law (1995). "MAC/FAC: A model of similarity-based retrieval". *Cognitive Science*, 19, pp. 141–205.

French, Robert M. (1995). *The Subtlety of Sameness: A Theory and Computer Model of Analogy-Making.* Cambridge, Mass.: MIT Press (Bradford Books).

Gentner, Dedre (1989). "The mechanisms of analogical learning". In Stella Vosniadou and Andrew Ortony (eds.), *Similarity and Analogical Reasoning.* New York: Cambridge University Press, pp. 199–241.

Gentner, Dedre, Mary Jo Rattermann, and K. Forbus (1993). "The role of similarity in transfer: Separating retrievability from inferential source". *Cognitive Psychology*, 25, pp. 524–575.

Gentner, Dedre and Cecile Toupin (1986). "Systematicity and surface similarity in the development of analogy". *Cognitive Science*, 10, pp. 277–300.

Gick, Mary L. and Keith J. Holyoak (1980). "Analogical problem solving". *Cognitive Psychology*, 12, pp. 306–355.

——— (1983). "Schema induction and analogical transfer". *Cognitive Psychology,* 15, pp. 1–38.

Grothe, Mardy (2008). *I Never Metaphor I Didn't Like: A Comprehensive Compilation of History's Greatest Analogies, Metaphors, and Similes.* New York: Harper.

Hofstadter, Douglas (1997). *Le Ton beau de Marot : In Praise of the Music of Language.* New York: Basic Books.

——— (2009). *Translator, Trader: An Essay on the Pleasantly Pervasive Paradoxes of Translation.* New York: Basic Books.

Hofstadter, Douglas and Melanie Mitchell (1993). "The Copycat project: A model of mental fluidity and analogy-making". In Douglas Hofstadter and the Fluid Analogies Research Group, *Fluid Concepts and Creative Analogies.* New York: Basic Books, pp. 205–267.

Holland, John H., Keith J. Holyoak, Richard E. Nisbett, and Paul R. Thagard (1986). *Induction: Processes of Inference, Learning, and Discovery.* Cambridge, Mass.: MIT Press.

Keith J. Holyoak (2012). "Analogy and relational reasoning". In Keith J. Holyoak and Robert G. Morrison (eds.), *The Oxford Handbook of Thinking and Reasoning.* New York: Oxford University Press, pp. 234–259.

Keith J. Holyoak and Kyunghee Koh (1987). "Surface and structural similarity in analogical transfer". *Memory and Cognition*, 15, pp. 332–340.

Kahneman, Daniel (2011). *Thinking, Fast and Slow.* New York: Farrar, Straus and Giroux.

Keane, Mark (1997). "What makes an analogy difficult? The effects of order and causal structure on analogical mapping". *Journal of Experimental Psychology: Learning, Memory, and Cognition*, 23, pp. 946–967.

Khong, Yuen Foong (1992). *Analogies at War.* Princeton: Princeton University Press.

Locke, W. N. and A. D. Booth (eds.) (1955). *Machine Translation of Languages.* Cambridge, Mass.: MIT Press.

Macrae, C. Neil, Charles Stangor, and Miles Hewstone (1996). *Stereotypes and Stereotyping.* New York: The Guilford Press.

Martin, Shirley A. and Miriam Bassok (2005). "Effects of semantic cues on mathematical modeling: Evidence from word-problem solving and equation construction tasks". *Memory and Cognition*, 33 (3), pp. 471–478.

Mitchell, Melanie (1993). *Analogy-Making as Perception.* Cambridge, Mass.: MIT Press.

Novick, Laura R. (1988). "Analogical transfer, problem similarity, and expertise". *Journal of Experimental Psychology: Learning, Memory, and Cognition*, 14, pp. 510–520.

——— (1995). "Some determinants of successful analogical transfer in the solution of algebra word problems". *Thinking and Reasoning*, 1, pp. 5–30.

Novick, Laura R. and Keith J. Holyoak (1991). "Mathematical problem solving by analogy". *Journal of Experimental Psychology: Learning, Memory, and Cognition*, 17, pp. 398–415.

Record, Jeffrey (1998). *Perils of Reasoning by Historical Analogy: Munich, Vietnam, and American Use of Force since 1945.* Technical Report 4, Center for Strategy and Technology, Air War College, Maxwell Air Force Base.

Ross, Brian H. (1987). "This is like that: The use of earlier problems and the separation of similarity effects". *Journal of Experimental Psychology: Learning, Memory, and Cognition*, 13, pp. 629–639.

——— (1989). "Distinguishing types of superficial similarities: Different effects on the access and use of earlier problems". *Journal of Experimental Psychology: Learning, Memory, and Cognition*, 15, pp. 456–468.

Sander, Emmanuel (2008). *De l'analogie aux inférences portées par les connaissances.* Habilitation report, University of Paris VIII.

Sloman, Steven (2009). *Causal Models: How People Think About the World and Its Alternatives.* New York: Oxford University Press.

Suganami, Hidemi (2008). *The Domestic Analogy and World Order Proposals.* New York: Cambridge University Press.

Thagard, Paul, Keith J. Holyoak, Greg Nelson, and David Gochfeld (1990). "Analog retrieval by constraint satisfaction". *Artificial Intelligence*, 46, pp. 259–310.

Weaver, Warren (1955). "Translation". In W. N. Locke and A. D. Booth (eds.), *Machine Translation of Languages.* Cambridge, Mass.: MIT Press, pp. 15–23.

——— (1964). *Alice in Many Tongues.* Madison, Wisconsin: University of Wisconsin Press.

Wharton, Charles M., Keith J. Holyoak, Paul E. Downing, Trent E. Lange, Thomas D. Wickens, and Eric R. Melz (1994). "Below the surface: Analogical similarity and retrieval competition in reminding". *Cognitive Psychology*, 26, pp. 64–101.

Chapter 7

The books and articles by Carey, Chi, Draaisma, Hatano and Inagaki, Keil, Lautrey, Leary, Viennot, and Tiberghien deal with naïve analogies in scientific domains, while those by Bryant, Clement, English, Fayol, Fischbein, Ginsburg, Hudson, Kieran, Kintsch and Greeno, Lakoff and Núñez, Linchevski and Vinner, Nesher, Nunes, Resnick, Schliemann, Schoenfeld, Silver, Thevenot, Tirosh, Vergnaud, and Verschaffel concern naïve analogies in arithmetic. The studies by Bassok, Brissiaud, Gamo, Richard, Sander, and Taabane are relevant to our sections on mental simulations and situations that "do the thinking for us". Carr, Fauconnier and Turner, Norman, Sander (2008), Serres, and Tricot are relevant to our sections on computers and associated technologies. Aubusson *et al*, Ausubel, Bastien, Bruner, Mahajan, and Richard focus on educational perspectives, while Anderson, Gentner, Spalding and Murphy, and Holyoak and Thagard (and colleagues) are relevant to the section concluding this chapter.

Anderson, John R. (1991). "The adaptive nature of human categorization". *Psychological Review*, 98, pp. 409–429.

Aubusson, Peter J., Allan G. Harrison, and Stephen M. Ritchie (2005). *Metaphor and Analogy in Science Education.* New York: Springer.

Ausubel, David P. (1968). *Educational Psychology: A Cognitive View.* New York: Holt, Rinehart, and Winston.

Bastien, Claude (1997). *Les Connaissances de l'enfant à l'adulte. Organisation et mise en œuvre.* Paris: Armand Colin.

Bastien, Claude and Mireille Bastien-Toniazzo (2004). *Apprendre à l'école.* Paris: Armand Colin.

Bassok, Miriam (1996). "Using content to interpret structure: Effects on analogical transfer". *Current Directions in Psychological Science*, 5 (2), pp. 54–58.

———— (2001). "Semantic alignments in mathematical word problems". In Dedre Gentner, Keith J. Holyoak, and Boicho N. Kokinov (eds.), *The Analogical Mind: Perspectives from Cognitive Science.* Cambridge, Mass.: MIT Press, pp. 401–433.

Bassok, Miriam, Valerie M. Chase, and Shirley A. Martin (1998). "Adding apples and oranges: Alignment of semantic and formal knowledge". *Cognitive Psychology*, 35, pp. 99–134.

Bezout, Étienne (1833). *L'Arithmétique de Bezout, à l'usage de la marine et de l'artillerie.* Paris: L. Tenré.

Bosc-Miné, Christelle and Emmanuel Sander (2007). "Effets du contenu sur la mise en œuvre de l'inférence de complément". *L'Année psychologique*, 107 (3), pp. 61–89.

Brissiaud, Rémi (2002). "Psychologie et didactique : Choisir des problèmes qui favorisent la conceptualisation des opérations arithmétiques". In Jacqueline Bideaud and Henri Lehalle (eds.), *Traité des sciences cognitives. Le développement des activités numériques chez l'enfant.* Paris: Hermès.

———— (2003). *Comment les enfants apprennent à calculer.* Paris: Retz.

Brissiaud, Rémi and Emmanuel Sander (2010). "Arithmetic word problem solving: A situation strategy first framework". *Developmental Science*, 13 (1), pp. 92–107.

Bruner, Jerome (1960). *The Process of Education.* Cambridge, Mass.: Harvard University Press (reissued in 1977).

Carey, Susan (1985). *Conceptual Change in Childhood.* Cambridge, Mass.: MIT Press.

Carr, Nicholas. (2011). *The Shallows: What the Internet Is Doing to Our Brains.* New York: Norton.

Clement, Catherine A. and Dedre Gentner (1991). "Systematicity as a selection constraint in analogical mapping". *Cognitive Science*, 15, pp. 89–132.

Clement, John J. (1982). "Algebra word problem solutions: Thought processes underlying a common misconception". *Journal for Research in Mathematics Education*, 13 (1), pp. 16–30.

Draaisma, Douwe (2001). *Metaphors of Memory: A History of Ideas about the Mind.* New York: Cambridge University Press.

English, Lyn D., editor (1997). *Mathematical Reasoning: Analogies, Metaphors, and Images.* Hillsdale, New Jersey: Lawrence Erlbaum.

Fabre, Jean-Henri (1897). *Souvenirs entomologiques*, Series V. Paris: Delagrave.

Fauconnier, Gilles and Mark Turner (2002). *The Way We Think: Conceptual Blending and the Mind's Hidden Complexities.* New York: Basic Books.

Fayol, Michel (1990). *L'Enfant et le nombre.* Neuchâtel: Delachaux et Niestlé.

Fischbein, Efraim (1987). *Intuition in Science and Mathematics: An Educational Approach.* Dordrecht: D. Reidel.

——— (1989). "Tacit models and mathematical reasoning". *For the Learning of Mathematics*, 9, pp. 9–14.

——— (1994). "Tacit models". In Dina Tirosh (ed.), *Implicit and Explicit Knowledge: An Educational Approach.* Norwood, New Jersey: Ablex Publishing, pp. 97–109.

Fischbein, Efraim, Maria Deri, Maria Nello, and Maria Marino (1985). "The role of implicit models in solving verbal problems in multiplication and division". *Journal for Research in Mathematics Education*, 16, pp. 3–17.

Gamo, Sylvie, Emmanuel Sander, and Jean-François Richard (2010). "Transfer of strategies by semantic recoding in arithmetic problem solving". *Learning and Instruction*, 20, pp. 400–410.

Gamo, Sylvie, Lynda Taabane, and Emmanuel Sander (2011). "Rôle de la nature des variables dans la résolution de problèmes additifs complexes". *L'Année psychologique*, 111, pp. 613–640.

Gick, Mary L. and Keith J. Holyoak (1983). "Schema induction and analogical transfer". *Cognitive Psychology*, 15, pp. 1–38.

Ginsburg, Herbert P. (1977). *Children's Arithmetic.* New York: Van Nostrand.

Hatano, Giyoo and Inagaki, Kayoko (1994). "Young children's naive theory of biology". *Cognition*, 50, pp. 171–188.

Hudson, Tom (1983). "Correspondences and numerical differences between disjoint sets". *Child Development*, 54, pp. 84–90.

Keil, Frank C. (1989). *Concepts, Kinds, and Cognitive Development.* Cambridge, Mass.: MIT Press.

Kieran, Carolyn (1981). "Concepts associated with the equality symbol". *Educational Studies in Mathematics*, 12, pp. 317–326.

Kintsch, Walter and James G. Greeno (1985). "Understanding and solving word arithmetic problems". *Psychological Review*, 92, pp. 109–129.

Lakoff, George and Rafael Núñez (2000). *Where Mathematics Comes From: How the Embodied Mind Brings Mathematics into Being.* New York: Basic Books.

Lautrey, Jacques, Sylvianne Rémi-Giraud, Emmanuel Sander, and Andrée Tiberghien (2008). *Les Connaissances naïves.* Paris: Armand Colin.

Leary, David E. (1990). *Metaphors in the History of Psychology.* New York: Cambridge University Press.

Linchevski, Liora and Shlomo Vinner. (1988). "The naive concept of sets in elementary teachers". *Proceedings of the Twelfth International Conference, Psychology of Mathematics Education* (Veszprem, Hungary), vol. II, pp. 471–478.

Mahajan, Sanjoy (2010). *Street-Fighting Mathematics: The Art of Educated Guessing and Opportunistic Problem Solving.* Cambridge, Mass.: MIT Press.

Nesher, Pearla (1982). "Levels of description in the analysis of addition and subtraction word problems". In Thomas P. Carpenter, James M. Moser and Thomas A. Romberg (eds.), *Addition and Subtraction: A Cognitive Perspective.* Hillsdale, NJ: Lawrence Erlbaum, pp. 25–38.

Norman, Donald A. (1990). "Why the interface doesn't work". In Brenda Laurel (ed.), *The Art of Human-Computer Interface Design.* Reading, Mass.: Addison-Wesley, pp. 209–219.

——— (1993). *Things that Make Us Smart.* Reading, Mass.: Addison-Wesley.

Novick, Laura R. and Miriam Bassok (2005). "Problem solving". In Keith J. Holyoak and Robert G. Morrison (eds.), *Cambridge Handbook of Thinking and Reasoning.* New York: Cambridge University Press, pp. 321–349.

Nunes, Terezinha and Peter Bryant (1996). *Children Doing Mathematics.* Oxford: Blackwell.

Reiner, Miriam, James D. Slotta, Michelene T. H. Chi, and Lauren B. Resnick (2000). "Naive physics reasoning: A commitment to substance-based conceptions". *Cognition and Instruction*, 18, pp. 1–34.

Resnick, Lauren B. (1982). "Syntax and semantics in learning to subtract". In Thomas P. Carpenter, James M. Moser and Thomas A. Romberg (eds.), *Addition and Subtraction: A Cognitive Perspective.* Hillsdale, New Jersey: Lawrence Erlbaum, pp. 136–155.

Richard, Jean-François and Emmanuel Sander (2000). "Activités d'interprétation et de recherche de solution dans la résolution de problèmes". In Jean-Noël Foulin and Corinne Ponce (eds.), *Les Apprentissages scolaires fondamentaux.* Bordeaux: Éditions CRDP, pp. 91–102.

Riley, Mary S., James G. Greeno, and Joan I. Heller (1983). "Development of children's problem solving ability in arithmetic". In Herbert P. Ginsburg (ed.), *The Development of Mathematical Thinking.* New York: Academic Press.

Sander, Emmanuel (2001). "Solving arithmetic operations: A semantic approach". In *Proceedings of the 23rd Annual Conference of the Cognitive Science Society* (Edinburgh), pp. 915–920.

——— (2002). "L'analogie, source de nos apprentissages". *La Recherche*, 353, pp. 40–43.

——— (2007a). "Manipuler l'habillage d'un problème pour évaluer les apprentissages". *Bulletin de psychologie*, 60, pp. 119–124.

——— (2007b). "Processus cognitifs, analogie et conception/évaluation de sites". In Serban Ionescu and Alain Blanchet (eds.), *Nouveau cours de psychologie. Psychologie sociale et ressources humaines* (coordinated by Marcel Bromberg and Alain Trognon). Paris: Presses universitaires de France, pp. 479–488.

——— (2008). "En quoi Internet a-t-il changé notre façon de penser ?" In Philippe Cabin and Jean-François Dortier (eds.), *La Communication. État des savoirs.* Auxerre: Éditions Sciences humaines, pp. 363–369.

——— (2011). "Les mécanismes de la pensée dans les apprentissages". In Nicolas Balacheff and Michel Fayol (eds.), *Apprendre et transmettre. Des idées, des savoir-faire, des valeurs.* Paris: Autrement.

Schliemann, Analucia, Claudia Araujo, Maria Angela Cassunde, Suzana Macedo, and Lenice Niceas (1998). "Use of multiplicative commutativity by school children and street sellers". *Journal for Research in Mathematics Education*, 29, pp. 422–435.

Schoenfeld, Alan H. and Douglas J. Herrmann (1982). "Problem perception and knowledge structure in expert and novice mathematical problem solvers". *Journal of Experimental Psychology: Learning, Memory, and Cognition*, 8, pp. 484–494.

Serres, Michel (2012). *Petite Poucette.* Paris: Le Pommier.

Silver, Edward A. (1981). "Recall of mathematical problem information: Solving related problems". *Journal of Research in Mathematical Education*, 12, pp. 54–64.

Spalding, Thomas and Gregory L. Murphy (1996). "Effects of background knowledge on category construction". *Journal of Experimental Psychology: Learning, Memory, and Cognition*, 22, pp. 525–538.

Thagard, Paul, Keith J. Holyoak, Greg Nelson, and David Gochfeld (1990). "Analog retrieval by constraint satisfaction". *Artificial Intelligence*, 46, pp. 259–310.

Thevenot, Catherine, Michel Devidal, Pierre Barrouillet, and Michel Fayol (2007). "Why does placing the question before an arithmetic word problem improve performance? A situation model account". *Quarterly Journal of Experimental Psychology*, 60 (1), pp. 43–56.

Thevenot, Catherine and Jane Oakhill (2005). "The strategic use of alternative representations in arithmetic word problem solving". *Quarterly Journal of Experimental Psychology*, A, 58 (7), pp. 1311–1323.

Tiberghien, Andrée (2003). "Des connaissances naïves au savoir scientifique". In Michèle Kail and Michel Fayol (eds.), *Les Sciences cognitives et l'école. La question des apprentissages.* Paris: PUF.

Tirosh, Dina and Anna O. Graeber (1991). "The influence of problem type and common misconceptions on preservice elementary teachers' thinking about division". *School Science and Mathematics*, 91, pp. 157–163.

Tricot, André (2007). *Apprentissages et documents numériques.* Paris: Belin.

Vergnaud, Gérard (1982). "A classification of cognitive tasks and operations of thought involved in addition and subtraction problems". In Thomas P. Carpenter, James M. Moser and Thomas A. Romberg (eds.), *Addition and Subtraction: A Cognitive Perspective.* Hillsdale, New Jersey: Lawrence Erlbaum, pp. 39–59.

Verschaffel, Lieven, Brian Greer, Wim van Dooren, and Swapna Mukhopadhyay, editors (2009). *Words and Worlds: Modelling Verbal Descriptions of Situations.* Rotterdam: Sense Publications.

Viennot, Laurence (1979). *Le Raisonnement spontané en mécanique élémentaire.* Paris: Herman.

Vosniadou, Stella and William F. Brewer (1992). "Mental models of the earth: A study of conceptual change in childhood". *Cognitive Psychology*, 24, pp. 535–585.

Chapter 8

The books by Bartha, Changeux and Connes, Hesse, Fischbein, Lakoff and Núñez, Nersessian, Oppenheimer, Poincaré, and Polya, which cover the process of scientific discovery from epistemological, philosophical, or psychological points of view, are relevant to the chapter as a whole. Those by De Morgan, Dunham, Kasner and Newman, Leibniz, Sawyer, Stewart, Stillwell, Timmermans, and Ulam are rich resources concerning the evolution of ideas in mathematics. The books by Born, Holton (and Brush), Miller, Pais, Pullman, Segrè, Stehle, and Tomonaga are marvelous gems documenting the history of ideas in physics in general, while those by Einstein, Hoffmann, Holton (2000), Miller, Pais (1982), Rigden, and Stachel focus on the more specific story of Albert Einstein's ideas. McAllister, Stewart, and Wechsler explore the role of esthetics in scientific discoveries, while Weiner recounts the ever-present role of analogies in the story of his own life as a physicist. The books by David and Mendel, Ulam, and Villani are all quoted in the closing section of the chapter.

Bartha, Paul (2010). *By Parallel Reasoning: The Construction and Evaluation of Analogical Arguments.* New York: Oxford University Press.

Bernstein, Jeremy (2006). *Secrets of the Old One: Einstein, 1905.* New York: Copernicus Books.

Born, Max (1936). *The Restless Universe.* New York: Harper and Brothers.

Changeux, Jean-Pierre and Alain Connes (1998). *Conversations on Mind, Mathematics, and Matter.* Princeton: Princeton University Press.

David, Hans T. and Arthur Mendel (1966). *The Bach Reader.* New York: W. W. Norton & Company.

De Morgan, Augustus (1831). *On the Study and Difficulties of Mathematics.* Reprinted in 2004 by Kessinger Publishing, Whitefish, Montana.

Dunham, William (1991). *Journey through Genius: The Great Theorems of Mathematics.* New York: Penguin.

Einstein, Albert (1920). *Relativity: The Special and the General Theory.* Reprinted in 1961 by Crown Publishers (New York).

Everitt, C. W. F. (1975). *James Clerk Maxwell, Physicist and Natural Philosopher.* New York : Charles Scribner's Sons.

Fischbein, Efraim. (1987). *Intuition in Science and Mathematics: An Educational Approach.* Dordrecht: D. Reidel.

Hesse, Mary (1966). *Models and Analogies in Science.* South Bend: Notre Dame University Press.

Hoffmann, Banesh (1972). *Albert Einstein: Creator and Rebel.* New York: Viking.

——— (1983). *Relativity and its Roots.* New York: Scientific American Books.

Holton, Gerald (1988). *Thematic Origins of Scientific Thought: Kepler to Einstein.* Cambridge, Mass.: Harvard University Press.

——— (1998). *The Scientific Imagination.* Cambridge, Mass.: Harvard University Press.

——— (2000). *Einstein, History, and Other Passions: The Rebellion against Science at the End of the Twentieth Century.* Cambridge, Mass.: Harvard University Press.

Holton, Gerald and Stephen G. Brush (2001). *Physics, The Human Adventure.* New Brunswick: Rutgers University Press.

Kao, T. I. and Frank J. Swetz (1977). *Was Pythagoras Chinese? An Examination of Right Triangle Theory in Ancient China.* University Park: Pennsylvania State University Press.

Kasner, Edward, and James Newman (1940). *Mathematics and the Imagination.* New York: Simon and Schuster.

Lakoff, George and Rafael Núñez (2000). *Where Mathematics Comes From: How the Embodied Mind Brings Mathematics into Being.* New York: Basic Books.

Leibniz, Gottfried Wilhelm von (1702). "Specimen novum analyseos pro scientia infiniti, circa summas & quadraturas". Reprinted in 1858 by C. I. Gerhardt as *Leibnizens Mathematische Schriften*, Sec. 2, I, No. XXIV. Halle: Verlag H. W. Schmidt.

McAllister, James W. (1996). *Beauty and Revolution in Science.* Ithaca: Cornell University Press.

Miller, Arthur I. (1985). *Frontiers of Physics: 1900–1911.* Boston: Birkhäuser.

——— (1986). *Imagery in Scientific Thought: Creating 20th-Century Physics.* Cambridge, Mass.: MIT Press.

——— (1997). *Albert Einstein's Special Theory of Relativity: Emergence (1905) and Early Interpretation (1905-1911).* New York: Springer.

——— (2000). *Insights of Genius: Imagery and Creativity in Science and Art.* Cambridge, Mass.: MIT Press.

Nersessian, Nancy J. (2008). *Creating Scientific Concepts.* Cambridge, Mass.: MIT Press (Bradford Books).

Oppenheimer, J. Robert (1956). "Analogy in science". *American Psychologist*, 11, pp. 127–135.

Pais, Abraham (1982). *Subtle Is the Lord: The Science and Life of Albert Einstein.* Oxford: Clarendon Press.

——— (1986). *Inward Bound: Of Matter and Forces in the Physical World.* Oxford: Oxford University Press.

——— (1991). *Niels Bohr's Times, in Physics, Philosophy, and Polity.* Oxford: Clarendon Press.

Poincaré, Henri (1908). *Science et méthode.* Paris, Flammarion.

——— (1911). *La Valeur de la science.* Paris, Flammarion.

Polya, George (1954). *Induction and Analogy in Mathematics.* Princeton: Princeton University Press.
——— (1954). *How to Solve It.* Princeton: Princeton University Press.
Pullman, Bernard (1998). *The Atom in the History of Human Thought.* Oxford: Oxford University Press.
Recorde, Robert (1557). *Whetstone of Witte.* New York: Da Capo Press (reprinted 1969).
Rigden, John S. (2005). *Einstein 1905: The Standard of Greatness.* Cambridge, Mass.: Harvard University Press.
Sawyer, Walter W. (1955). *Prelude to Mathematics.* Baltimore: Penguin Books.
Segrè, Emilio (1980). *From X-Rays to Quarks: Modern Physicists and Their Discoveries.* New York: W. H. Freeman.
——— (1984). *From Falling Bodies to Radio Waves: Classical Physicists and Their Discoveries.* New York: W. H. Freeman.
Stachel, John (2001). *Einstein from 'B' to 'Z'.* Boston: Birkhäuser.
Stachel, John, editor (2005). *Einstein's Miraculous Year: Five Papers that Changed the Face of Physics.* Princeton: Princeton University Press.
Stehle, Philip (1994). *Order, Chaos, Order: The Transition from Classical to Quantum Physics.* New York: Oxford University Press.
Stewart, Ian (2007). *Why Beauty is Truth: A History of Symmetry.* New York: Basic Books.
Stillwell, John (1989). *Mathematics and its History.* New York: Springer.
——— (2001). *The Four Pillars of Geometry.* New York: Springer.
Timmermans, Benoît (2012). *Histoire philosophique de l'algèbre moderne : Les origines de la pensée abstraite.* Paris: Classiques Garnier.
Tomonaga, Sin-itiro (1997). *The Story of Spin.* Chicago: University of Chicago Press.
Ulam, Stanislaw M. (1976). *Adventures of a Mathematician.* New York: Charles Scribner's Sons.
Villani, Cédric (2012). *Théorème vivant.* Paris: Grasset.
Wechsler, Judith, editor (1978). *On Aesthetics in Science.* Cambridge, Mass.: MIT Press.
Weiner, Richard M. (2008). *Analogies in Physics and Life: A Scientific Autobiography.* Singapore: World Scientific.

Epidialogue

The contributions by Dietrich, French, Hofstadter, Mitchell, Sander, and Turner deal with the question of how analogy-making and categorizaton are related, and the possibility of finding differences between the two. Fischbein, Oppenheimer, and Polya demonstrate the ubiquity of analogy-making in science. Murphy and Ross's article is concerned with how and when inferences can be made as a result of categorization, while Anderson and Thomson's article looks at abstraction hierarchies, and that by Barsalou examines the possible influences of context on categorization.

Anderson, John R. and Ross Thomson (1989). "Use of analogy in a production system architecture". In Stella Vosniadou and Andrew Ortony (eds.), *Similarity and Analogical Reasoning.* New York: Cambridge University Press, pp. 267–297.
Barsalou, Lawrence W. (1982). "Context-independent and context-dependent information in concepts". *Memory and Cognition*, 10, pp. 82–93.
Dietrich, Eric (2010). "Analogical insight: Toward unifying categorization and analogy". *Cognitive Processing*, 11 (4), pp. 331–345.
Fischbein, Efraim (1987). *Intuition in Science and Mathematics: An Educational Approach.* Dordrecht: D. Reidel.

French, Robert M. (1995). *The Subtlety of Sameness: A Theory and Computer Model of Analogy-Making.* Cambridge, Mass.: MIT Press.

Gentner, Dedre and Catherine Clement (1988). "Evidence for relational selectivity in the interpretation of analogy and metaphor". In G. H. Bower (ed.), *The Psychology of Learning and Motivation: Advances in Research and Theory*. New York: Academic Press, vol. 22, pp. 307–358.

Gick, Mary L. and Keith J. Holyoak (1983). "Schema induction and analogical transfer". *Cognitive Psychology*, 15, pp. 1–38.

Hofstadter, Douglas (2001). "Analogy as the core of cognition". In Dedre Gentner, Keith J. Holyoak, and Boicho N. Kokinov (eds.), *The Analogical Mind: Perspectives from Cognitive Science.* Cambridge, Mass.: MIT Press, pp. 499–538.

Holyoak, Keith J. and Paul Thagard (1995). *Mental Leaps: Analogy in Creative Thought.* Cambridge, Mass.: MIT Press.

Mitchell, Melanie (1993). *Analogy-Making as Perception: A Computer Model.* Cambridge, Mass.: MIT Press.

Murphy, Gregory L. and Brian H. Ross (1994). "Prediction from uncertain categorizations". *Cognitive Psychology*, 27, pp. 148–193.

Oppenheimer, J. Robert (1956). "Analogy in science". *American Psychologist*, 11, pp. 127–135.

Plato (1950). *Plato's Phædrus.* Millis, Mass.: Agora Publications (reissued in 2009).

Polya, George (1957). *How to Solve It.* Princeton: Princeton University Press.

Sander, Emmanuel (2003a). "Les analogies spontanées : analogies ou catégorisations ?" In Charles A. Tijus (ed.), *Métaphores et analogies.* Paris: Hermès, pp. 83–114.

——— (2003b). "Analogie et catégorisation". *Revue d'intelligence artificielle*, 17 (5–6), pp. 719–732.

Spalding, Thomas and Gregory L. Murphy (1996). "Effects of background knowledge on category construction". *Journal of Experimental Psychology: Learning, Memory, and Cognition*, 22, pp. 525–538.

Thagard, Paul, Keith J. Holyoak, Greg Nelson, and David Gochfeld (1990). "Analog retrieval by constraint satisfaction". *Artificial Intelligence*, 46, pp. 259–310.

Turner, Mark (1988). "Categories and analogies". In David H. Helman (ed.), *Analogical Reasoning.* Amsterdam: Kluwer Academic Publishers, pp. 3–24.

Wisniewski, Edward J. (1995). "Prior knowledge and functionally relevant features in concept learning". *Journal of Experimental Psychology: Learning, Memory, and Cognition*, 21, pp. 449–468.

INDEX

❧ ❧ ❧

— B —

— C —

— G —

—H—

—I—

—N—

— O —

— P —

— S —

—T—

—U—

—V—

—W—

—X—

—Y—

—Z—